RSA Reliability and Maintenance
Newsletter Vault Collection

**Supplementary Series on World Class
Maintenance Management - The 12 Disciplines**
(Volume 8)

By Rolly Angeles

1

<u>**RSA Reliability and Maintenance Newsletter Vault Collection**</u>
Supplementary Series on World Class Maintenance Management – The 12 Disciplines (Volume 8)

10 9 8 7 6 5 4 3 2 1

First Edition: Printed in the Philippines by **Central Books**
Head Office: Phoenix Bldg. 927 Quezon Avenue, Quezon City, Philippines 1101

The National Library of Philippine Catalogue

Kindle Asin Amazon   B0998FBJX6
Paperback Amazon   979-8535935348
Hardcover Amazon   979-8535938233
Paperback Ingramspark   979-8885259965
Hardcover Ingramspark   979-8885260008

Published by:

Published by Amazon KDP for Kindle and Paperback 2021
Published by Ingram Spark for Hardcover Format 2021

<u>This book was designed and produced by:</u>

RSA Reliability and Maintenance Consultancy Firm, 2021
Sta. Rosa, Laguna, Philippines 4026
Website: http://www.rsareliability.com
Email: rollyangeles@rsareliability.com

First Printing: July 2021

Table of Contents

Table of Contents

About the Author

Rolly is a seasoned international maintenance and reliability consultant with over 30 years of solid experience in the field. He has been invited to different countries and has conducted reliability and maintenance training in United Arab Emirates, Qatar, India, Malaysia, Indonesia, Brunei, Thailand, Nigeria, Bangladesh, South Africa, China, and Botswana. His maintenance training portfolio includes maintenance and reliability courses on TPM, Lubrication, Tribology, Condition-Based Maintenance, RCM, RCFA, Planned Maintenance, World Class Maintenance Management, The 12 Disciplines, Oil Contamination Control, Maintenance Indices and KPI's, Maintenance and Reliability Management Strategies and much more. Rolly previously worked with Amkor Technology Philippines as a TPM Senior Engineer, an industry engaged in the manufacture of Integrated Circuit products, and spearheaded their Planned Maintenance organization, composed of maintenance managers and engineers. He was also responsible for the dramatic reduction of unplanned breakdowns in their TPM Journey and RCM implementation on their Facilities AHU units and their substation equipment. Rolly is currently working as an independent reliability and maintenance consultant.

Rolly is a graduate of Mechanical Engineering from Mapua Institute of Technology in the Philippines, batch 1985, and passed the licensure board examination the following year in 1986. With 30 years of solid experience, he had worked in various industries from shipping, woodworking, foundry, cast-iron machining, assembly lines, semiconductor manufacturing, and the mining industry. Here, he gained hands-on experience and understanding of TPM and RCM, respectively, a strategy from both the west and the east. His last corporate employment was in 2002, where he worked as a technical training specialist at Lepanto Consolidated Mining Industry. In 2005, Rolly retired early from the industry and decided to establish his own consulting business, RSA Reliability and Maintenance Consultancy Firm, where he dedicates his time and passion for working as an independent reliability and maintenance consultant. He provides in-house training, consultation, and facilitation to different maintenance and reliability best practices. Rolly Angeles can be reached through his email at rollyangeles@rsareliability.com, or you can visit his website at http://www.rsareliability.com.

Preface

This is my eighth on a series of books I wrote and am still writing, based on the original concept of my first book on World Class Maintenance Management, The 12 Disciplines, as shown on page 3 of this book. In as much as I can, I have written all my books in the simplest way to understand for the benefit of the readers. Here is feedback to one of my readers, which I truly value and inspire me to write more books about our common passion, reliability, and maintenance.

Perhaps the reader would want to know what RSA is. The first and the last stand for my name, which is Rolly Angeles. The letter "S" is my middle name, or better yet, it also stands for Stones, which is my favorite band that I always featured and discussed in my previous books. This book is a collection of all my reliability and maintenance newsletters. I wrote, which I started from May 2007 until December 2020. It contains around 164 newsletters on different topics of our common interest, all about reliability and maintenance. The majority of these newsletters are included in my books based on their particular maintenance discipline.

This is a supplementary book to my sequel on World Class Maintenance Management – The 12 Disciplines, unlike my other books specifically dedicated to a particular discipline on World Class Maintenance Management, which is why it is in series. This book covers all the disciplines into one reading. I have used most of these newsletters in my other books depending on the particular maintenance discipline it fits in. The chapters of this book are chronologically arranged according to the year the newsletter was written, starting from May 2007 to December 2020.

Unlike my other books in which you will start to read from Chapters 1, 2, 3, and so on, in this book, I would recommend the reader start reading a particular newsletter that interests them or is directly related to their current work or activities. I have created a summary of all the titles of this newsletter below and just mark them with either done, ok, checkmark, or just indicate the date you have completed reading in the column on "**Completed**" for the reader to have their own record on all the reliability and maintenance newsletters you have read in this book after some time. On the other end, you can treat this like any other book and start at the very beginning. The choice will actually be up to the reader. This book also summarizes all the reliability and maintenance newsletters that you have read and are about to read for your own reference.

This book contains many topics, yet they are connected to the original concept of the 12 disciplines on maintenance. Just scroll down on the table of contents below, and you will find interesting topics on Root Cause Failure Analysis, Preventive, Predictive Maintenance, Total Productive Maintenance, Reliability-Centered Maintenance, Maintenance Measurements, and KPIs, Lubrication, Oil Analysis, MRO Spare Parts, Autonomous Maintenance, Planned maintenance, and many more interesting topics on maintenance. I hope you enjoy reading them. I would greatly appreciate it if you could provide some feedback about your experience reading this book.

Preface

No.	Month/Year	2007 Title of Newsletter	Completed
1	May 2007	Learning's From The Principles of Equipment Reliability	
2	June 2007	Can Equipment Failures Really Be Eliminated?	
3	July 2007	How To Make Training Work For Your Industry	
4	August 2007	Why Are Most Industries Reactive?	
5	September 2007	A Different Root Cause Failure Analysis Experience	
6	October 2007	Where Does RCFA Fit In The RCM Strategy?	
7	November 2007	Why TPM Is Hard To Implement?	
8	December 2007	Does Oil Wear Out and Needs To Be Changed?	

No.	Month/Year	2008 Title of Newsletter	Completed
9	January 2008	Where Do We End Our Probe On Root Cause Analysis?	
10	February 2008	An Inconvenient Truth About Preventive Maintenance	
11	March 2008	The Lifeblood of Root Cause Analysis	
12	April 2008	My TPM Experience, A Successful Failure	
13	May 2008	Top 10 Problems Experience on Preventive Maintenance	
14	June 2008	Comparing RCM and TPM, Which Is The Better Strategy?	
15	July 2008	Why do Most Root Cause Initiatives fail?	
16	August 2008	The Case of Infant Mortality and Random Failures	
17	September 2008	How To Fit RCM Into The TPM Strategy	
18	October 2009	Maintenance - The Last Man Standing	
19	November 2008	Is There Still Value In Training?	
20	December 2008	A Typical Maintenance Culture	

No.	Month/Year	2009 Title of Newsletter	Completed
21	January 2009	Is Reliability Just a Numbers Game?	
22	February 2009	TPM Planned Maintenance Pillar	
23	March 2009	TPM and RCM Do They Contradict or Complement Each Other?	
24	April 2009	How Much of Industry's Problems are Man-made?	
25	May 2009	13 Grave Mistakes in TPM Implementation	
26	June 2009	Why Operators are Important in the Reliability Strategy	
27	July 2009	Should There be an Eight Question? Part 1	
28	August 2009	Should There be an Eight Question? Part 2	
29	September 2009	Is Root Cause Failure Analysis Reactive or Proactive?	
30	October 2009	Operations and Maintenance - Will the Feud Ever Stop?	
31	November 2009	Cost Reduction: The Wrong Way to Save Maintenance Costs	
32	December 2009	Why More PM Will Just End Up to More Problems	

No.	Month/Year	2010 Title of Newsletter	Completed
33	January 2010	Why Most RCM Initiatives Fail?	
34	February 2020	What Does it Take to Change the Maintenance Culture?	
35	March 2010	Why RCFA and MTTR Do Not Blend?	
36	April 2010	What Does it Take to Initiate a Plant-wide Improvement?	
37	May 2010	The Golden Rule on Root Cause Failure Analysis	
38	June 2010	The Importance of Life Cycle Costing in the Reliability Strategy	
39	July 2010	Understanding Human Errors	
40	August 2010	What Maintenance is All About	

No.	Month/Year	Title of Newsletter	Completed
41	September 2010	Most Common Mean Time Indicators Part 1	
42	October 2010	Most Common Mean Time Indicators Part 2	
43	November 2010	Most Common Mean Time Indicators Part 3	
44	December 2010	Autonomous Maintenance for Industries Part 1	
No.	Month/Year	2011 Title of Newsletter	Completed
45	January 2011	Autonomous Maintenance for Industries Part 2	
46	February 2011	Are Zero Breakdowns Really Possible?	
47	March 2011	Top 10 Problems Experienced on Preventive Maintenance - Revisited	
48	April 2011	Why Operations Can Never Be Maintenance Customers	
49	May 2011	Different Maintenance Strategies Explained	
50	June 2011	The Two Sides of Failure	
51	July 2011	Tips on Performing Reliability-Centered Maintenance (RCM)	
52	August 2011	Tips on Performing Total Productive Maintenance (TPM)	
53	September 2011	Survey on Top Problems on MRO Spare Parts	
54	October 2011	Decision on Whether to Stock or Not To Stock Spare Parts	
55	November 2011	Do Industries Know What Maintenance Is All About?	
56	December 2011	Most Common Types of Mechanical Wear Explained	
No.	Month/Year	2012 Title of Newsletter	Completed
57	January 2012	Condition-Based Maintenance Explained	
58	February 2012	Is OEE A Perfect Measurement?	
59	March 2012	Who Should Manage Your MRO Spare Parts Storeroom?	
60	April 2012	The Importance of Conducting Oil Analysis	
61	May 2012	The 12 Wisdom on Maintenance - Part 1	
62	June 2012	June 2012: The Problem with Predictive Maintenance	
63	July 2012	Why Maintenance is A Complete Unknown?	
64	August 2012	Why Operations and Maintenance Went Their Separate Ways?	
65	September 2012	Why Establishing Basic Equipment Condition Matters Most	
66	October 2012	Frequently Asked Questions On Oil	
67	November 2012	Why Planned Maintenance Should Be the Strongest Pillar in TPM	
68	December 2012	Planned Maintenance Phase 0 Preparatory Stage	
No.	Month/Year	2013 Title of Newsletter	Completed
69	January 2013	Definition and function of Lubricating Grease	
70	February 2013	Benefits of Condition-Based Maintenance for Industries	
71	March 2013	Step by Step Details on Implementing RCM	
72	April 2013	Planned Maintenance Phase 1 Restoration Stage Part 1	
73	May 2013	Planned Maintenance Phase 1 Restoration Stage Part 2	
74	June 2013	Honesty is the Best Policy" does not apply to MRO Spare Parts	
75	July 2013	Equipment 6 Big Losses Part 1	
76	August 2013	Equipment 6 Big Losses Part 2	
77	September 2013	Lessons on Total Productive Maintenance (TPM)	
78	October 2013	Total Productive Maintenance - Part 1	
79	November 2013	Total Productive Maintenance - Part 2	
80	December 2013	Total Productive Maintenance - Part 3	

No.	Month/Year	2014 Title of Newsletter	Completed
81	January 2014	Total Productive Maintenance - Part 4	
82	February 2014	Total Productive Maintenance - Part 5	
83	March 2014	A Passion for Reliability and Maintenance	
84	April 2014	The Right Reasons To Measure Performance	
85	May 2014	What Is Machine Breakdown In The First Place?	
86	June 2014	Step by Step Activities in Improving your MRO Storeroom	
87	July 2014	Top Reasons Why We Need to Perform RCFA	
88	August 2014	Analytical Problem Solving Techniques and RCFA	
89	September 2014	Step by Step Approach on How to Perform RCFA	
90	October 2014	Planned Maintenance Phase 2 Flow of Activities	
91	November 2014	Why do Many CBM Programs fail In Other Industries?	
92	December 2014	Difference Between RCFA, RCA, and Failure Analysis	

No.	Month/Year	2015 Title of Newsletter	Completed
93	January 2015	Maintenance - A Scapegoat for Operations Problems	
94	February 2015	Industry's Procedure for FMEA/FMECA Part 1	
95	March 2015	Industry's Procedure for FMEA/FMECA Part 2	
96	April 2015	Industry's Procedure for FMEA/FMECA Part 3	
97	May 2015	Why Safety Cannot Be First	
98	June 2015	Is It Possible to Eliminate Human Errors in Maintenance?	
99	July 2015	Maintenance Induced and Non-Maintenance Induced Errors	
100	August 2015	Tips in Conducting Autonomous Maintenance for Industries	
101	September 2015	Reducing Human Errors in Maintenance (Part 1)	
102	October 2015	Reducing Human Errors in Maintenance (Part 2)	
103	November 2015	Guidelines in Conducting Focused Improvement	
104	December 2015	Why Japan Industries are Successful Compared to Other Countries	

No.	Month/Year	2016 Title of Newsletter	Completed
105	January 2016	Strengthening Operator and Maintenance Partnership	
106	February 2016	Tips in Conducting Root Cause Failure Analysis	
107	March 2016	The Need for a Structured Asset Management System ISO 55001	
108	April 2016	Guidelines in Setting-up a CBM Organization in your Industry	
109	May 2016	Reasons Why We Need to Measure KPI	
110	June 2016	Where MTTR Should Be Used	
111	July 2016	Why is the Study of Tribology Important to Industries	
112	August 2016	Small Problems Matters Most	
113	September 2016	Guidelines for the Autonomous Maintenance Audit	
114	October 2016	Common Causes of Bearing Failure	
115	November 2016	Understanding the Difference Between Poka-Yoke and RCFA	
116	December 2016	The Common Thing RCM and TPM Believes	

No.	Month/Year	2017 Title of Newsletter	Completed
117	January 2017	Wisdom on Maintenance - Part 2	
118	February 2017	Most Common Operator's Indices	
119	March 2017	Tips in Setting Up Operators and Maintenance Indices	
120	April 2017	Wisdom on Maintenance - Part 3	

No.	Month/Year	Title of Newsletter	Completed
121	May 2017	Wisdom on Maintenance - Part 4	
122	June 2017	Wisdom on Maintenance - Part 5	
123	July 2017	Recommended MRO Storeroom Measurements and KPI's	
124	August 2017	Consolidate All Improvement Initiatives in the Plant	
125	September 2017	The Physical, Human, System, and Latent Cause of the Problem	
126	October 2017	What to Do with Obsolete Parts in your Storeroom	
127	November 2017	Skills We Need for Autonomous Maintenance Operators	
128	December 2017	The Roots of TPM	
No.	**Month/Year**	**2018 Title of Newsletter**	**Completed**
129	January 2018	How to Initially Implement CMMS in Maintenance	
130	February 2018	Two Types of Improvements	
131	March 2018	World Class Maintenance Management, the 12 Disciplines Review	
132	April 2018	Maintenance - Roadmap to Reliability Review	
133	May 2018	Reliability- A Shared Responsibility for Operators and Maintenance	
134	June 2018	Why is Cheap Actually Expensive?	
135	July 2018	The Importance of having a Strategic Planning for Maintenance	
136	August 2018	Preventive Maintenance Explained	
137	September 2018	Selecting the Correct Interval for Preventive Maintenance	
138	October 2018	Steps in Adapting an In-house Oil Analysis Laboratory in Industries	
139	November 2018	Benefits of Oil Analysis	
140	December 2018	The sinking of RMS Titanic - A Classic Case of Human Error Part 1	
No.	**Month/Year**	**2019 Title of Newsletter**	**Completed**
141	January 2019	The sinking of RMS Titanic - A Classic Case of Human Error Part 2	
142	February 2019	Switching Redundancy and Standby Tasks	
143	March 2019	Aftermath Additives as Seen on TV	
144	April 2019	The Evolution of Maintenance	
145	May 2019	The Roots of RCM	
146	June 2019	Different Awards on TPM Given by JIPM	
147	July 2019	What Empowerment can do to Operators?	
148	August 2019	Precision Maintenance Explained	
149	September 2019	Requirements for Precision Maintenance	
150	October 2019	Different Types of Lubricating Film	
151	November 2019	The Use of Barcoding for MRO Spare Parts and Storeroom	
152	December 2019	New ISO 55010 Alignment of Financial and Non-Financial Functions	
No.	**Month/Year**	**2020 Title of Newsletter**	**Completed**
153	January 2020	What Maintenance Need is Knowledge and Not Experience	
154	February 2020	Industrial Internet of Things (IIoT) and Maintenance 4.0	
155	March 2020	Reliability is Everyone Responsibility in the Plant	
156	April 2020	Is Industries Ready for Maintenance 4.0?	
157	May 2020	Risks, Criticality, and Consequences (Part 1)	
158	June 2020	Risks, Criticality, and Consequences (Part 2)	
159	July 2020	An Advice to Storekeepers	
160	August 2020	Why the Term Usage is Different from the Parts Withdrawn	

161	September 2020	What to Do with Non-Moving Parts	
162	October 2020	Six Myths About Lubrication	
163	November 2020	Effects of Excessive Moisture on Lubricating Oil	
164	December 2020	The TPM and RCM Crossroads	

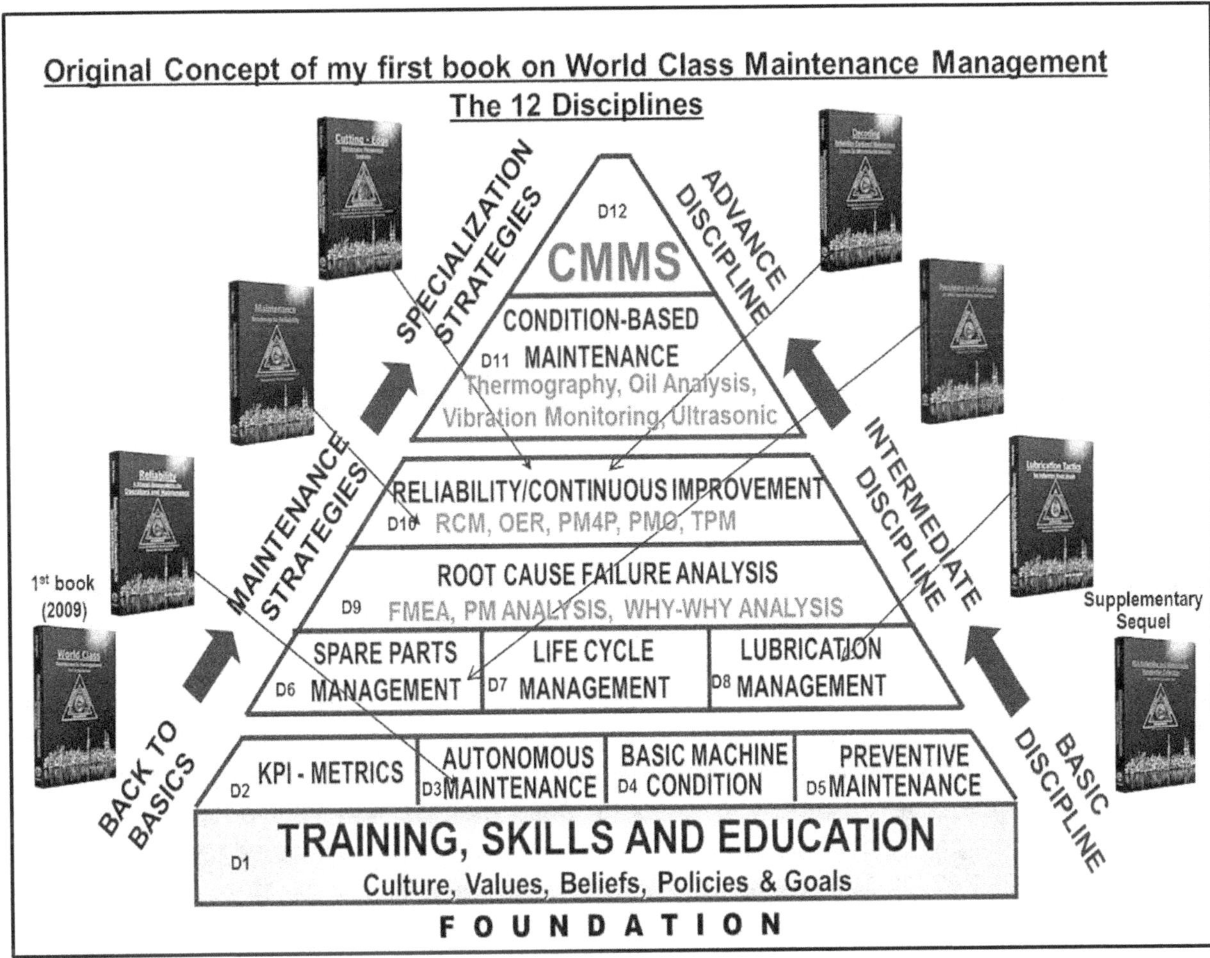

Figure B: RSA Books and the Original World Class Maintenance Management, The 12 Disciplines

Year **1**

2007 RSA Reliability Newsletter Vault Archive

> *There is a thin line between the term maintenance and repair. Maintenance are tasks and activities done before the failure happens, on the contrary, repair are activities done after a failure occurs. Remember that it is more expensive to react than to act before the failure is about to occur.*

1.1: May 2007: Learnings from the Principles of Reliability

I have devoted much of my time to developing these maintenance training courses because of my passion for preaching maintenance and reliability to industries. Most of the time, equipment failures struck us outright in our faces, and in most cases, we are unprepared. We need to understand that there are no shortcuts or silver bullet solutions for improving equipment reliability. We need to understand the principles behind them before we can even start to apply them. Let me share these reliability maxims with you and try to reflect on each one of them for a while. Are we applying them, or are we doing the opposite, which eventually leads us to be reactive most of the time? I urge you to share these maxims with your people. Improving reliability is not going to be done overnight. It's a long journey, just like starting to play the guitar for the first time, but as the saying goes, every journey begins with a single step.

Maxim No. 1

The focus must be on reliability and not cutting costs because if reliability starts to improve, the cost will definitely go down. It cannot be the other way around. Remember, there will be times that are focusing on reducing cost will hurt reliability, a lesson we all should reflect upon.

Maxim No. 2

Never accept failures in your plant. Troubleshooting and repair is no longer an effective strategy. In today's competitive world for industries, the analyst finds real solutions to their equipment problems. Remember that when our people become really good at repairing failures, then something is definitely wrong with our organization since they are doing it much too often, but when we expect a different result from the same things that we are doing, it just ain't possible; the Chinese called this insanity.

Maxim No. 3

The best time to address a problem is when it is small; it is very hard to advance to any specialized maintenance activities and improvement efforts if Basic Equipment Condition had not been established; always remember our equipment remains a shared responsibility for operators and maintenance people, a lesson we must all learn from the Japanese.

Maxim No. 4

In a reactive environment, we always complain that we lack manpower resources to address equipment failures, but when equipment starts to improve, our people are now visible, and we always wonder where they had been in the first place.

Maxim No. 5

There is no silver bullet solution, program, or strategy that can transform a plant's reliability overnight. All will start with its basic foundation, and that is through "education" and changing our people's mindset on maintenance.

Maxim No. 6

The real challenge in any equipment reliability initiative is improving in a reactive world with the same amount of resources and time. Remember that all best-in-class and world-class industries started from being reactive themselves.

Maxim No. 7

The best maintenance strategy to adopt is to learn when to use the different maintenance strategies simultaneously with the aid of a decision diagram or an algorithm and that the degree of maintenance requirements should always be based upon the consequences of failure itself.

Maxim No. 8

There is only one secret for equipment that fails due to lubrication, just keep the oil clean. If the oil is kept clean, there is no reason for the oil to oxidize and no need to change it. Oil should be changed not based on the number of hours it has run but by the number of contaminants and impurities.

Maxim No. 9

Reliability is not a program with an end but a culture without an end. It is the same as any continuous improvement philosophy.

Maxim No. 10

The distinction between a true-blooded maintenance and a mechanic is that maintenance uses a balance of his hands and brain, while a mechanic uses his hand much of the time. Let us start to treat our people as maintenance and not as mere mechanics.

Maxim No. 11

The best way to change a culture is to focus on results. Remember, what gets measured gets done. If we don't measure our performance, we are just another person with an opinion, and opinions don't last, but measurements will impact.

Maxim No. 12

Always remember that in any reliability improvement initiative, the focus must be on the people.

Provide them with the education and skills they need, and these skills will be used to improve their equipment. People will improve their machines, and it is not the other way around.

1.2: June 2007: Can Equipment Failures Really be Eliminated

I have asked some of my friends from reliability forums and my email lists if all failures can really be eliminated; some say yes, and others simply say that we can only eliminate a fraction of them. Hence, I tried to rephrase my question by asking which of these following statements would you think would be more appropriate, relevant, true, and meaningful:

• **1st:** We can eliminate failures by analyzing them through RCA and RCFA methodologies.
• **2nd:** We cannot eliminate the likelihood of failure, but we can only prevent or predict the failure from occurring on its own.
• **3rd:** Failures cannot be eliminated. The best that maintenance people can do is to reduce the frequency of failures.
• **4th:** Failures cannot be eliminated; the best thing we can do is delay or prolong the process of failure from happening.

First and foremost, all failures are not created equal since each failure will have its own unique consequence. The word failure itself is very broad and diversified and can have different meanings or interpretations. Sometimes confusion happens as when do we call it a failure. Before answering my question, I would like to explain failure in the way I understand it.

Patterns of Failure – How the Failure Occurs

Infant Mortality Failures - These are failures that can occur at the beginning of its life. Others call them start-up failures, commissioning failures that are likely to occur after a major overhaul or Preventive Maintenance had been initiated.

Random Failures - These are failures that can occur at any given period, and this is where our routine Preventive Maintenance will be at its weakest point. The recommended task that can be used for random failures will be Predictive Maintenance if the failure provides a warning or potential failure and that it is on the verge or process of occurring. If Predictive Maintenance is not feasible, then we resort to modification or simply run it until it fails if the consequences of failure will be minimal.

Wear-Out Failures also known as age-related failures where parts will eventually survive to the age that it is supposed to fail consistently.

John Moubray, the author of RCMII book, further indicates in his book that these three failure patterns mostly occur in combination by studying parts behavior and led to what we called the 6 failure patterns; but eventually, those 3 patterns of failure (bathtub curve) can be seen as existing in statistical books and Weibull Distribution.

Classifications of Failures

Hidden Failures are failures that will not become evident to the operator or maintenance when it occurs on its own such as failure of a standby pump or failure of protective devices. The only time hidden failures will be known if a secondary failure takes its toll.

Evident Failures are failures that will become evident to the operator or maintenance when it occurs independently.

Different Types of Failures

Function Loss Failure or failure of the primary function is when a failure occurs, the equipment will totally stop and definitely halt operations. RCM termed this as a failure of the primary function.

Function Loss Reduction or failure of secondary functions is simply when the failure occurs, the equipment can still be capable of running. Every piece of equipment has its own primary and secondary functions that we must be aware of. One thing to note is that there are cases that the failure consequences for secondary functions are far much worse than the failure of the primary function.

From TPM Books, references and resource breakdowns and failures can either be planned or unplanned and what maintenance can eliminate or zero out is the unplanned breakdowns and not to mention cases of chronic and sporadic failures and breakdowns.

Planning for an activity to be performed on our equipment, such as routine Preventive Maintenance, parts replacement, or lubrication, will constitute a planned downtime. An example of this is when the CBM or Predictive Maintenance group noted that a potential failure is now in progress, and the next step is scheduling the equipment for some intervention, which will also constitute a Planned Downtime and Breakdown. An unplanned breakdown simply means that maintenance is unaware and caught by surprise when a failure occurs. Hence, going back and answering these statements:

1st: We can definitely eliminate failures by analyzing them through a thorough RCFA: Disagree; failures cannot be totally eliminated by analyzing them through RCFA or RCA. We must consider various causes to understand, and every single failure has its own unique causes. When we treat a single cause, then there is a likelihood of that same part failing again due to a different cause. Remember, when we speak about Root Cause Failure Analysis and Root Cause Analysis, we deal with evidence of what really has caused the part to fail and not all the probable causes that might have caused the part to fail. As my good and dear friend, Bob Nelms from Failsafe Network, once told me that you cannot move an ocean with a fork (not even a spoon), simply concurs that it is possible to remove or eliminate the causes, but not every single one of them.

A bearing can fail for multiple varieties of reasons, and taking care only of a single cause in which perhaps the analysis shows that bearing failure was due to a lubrication failure will not eliminate its recurrence since the bearing can fail in the future due to some other reasons such as false brinelling, pitting fatigue, spalling, misalignment, careless handling, over or under lubrication, and many more. RCM was founded on the belief that its purpose was to eliminated or reduce accidents per million take-offs by studying every single part and its behavior. Today, the plane still crashes for some other reasons, such as terrorism or human error. First and foremost, all failures are not created equal since each failure will have its own unique set of consequences. The word failure itself is very broad and diversified. Different people can have different meanings or interpretations. Sometimes confusion happens as when do we declare a failure. Before answering my question, I would like to explain failure in the way I have to understand it.

2nd: We cannot eliminate the likelihood of a failure, but we can only prevent or predict the failure from occurring: Agree for some failures, it is possible to prevent them, and this is true for age-related failures or parts and components which inhibits some sort of wear out mode and which constitute around 20 to 30% of overall equipment failures, and likewise, we can predict failures for those who shows signs of potential failures. But this is not the goal of maintenance, if a bearing fails prematurely, yes, we can predict that it is on the verge of failing, but the life of the bearing had not been maximized at all; perhaps we were able to predict it from failing, and we have advised operations about it, but we should not stop from here. Data from Predictive Maintenance can be most useful in analyzing why components fail. Predictive maintenance is just one step towards achieving Proactive Maintenance. Not to mention that not all failures can be predicted nor prevented.

3rd: Failures cannot be eliminated, and the best that maintenance can do is reduce them: Disagree, some think that when we have experienced zero unplanned breakdowns or have reduced failure tremendously as per seen on our breakdown indices for the past couple of years, we think that we have eliminated the failure completely, but come to think of it technically, have we really eliminated the likelihood of a failure, or rather we have just delayed the failure process to occur? Just try to think of buying a second-hand car and driving it while giving it the best maintenance ever. Eventually, there will still be many parts that will be subject to wear out, and when parts wear out, then they have actually and eventually failed in the first place?

4rth: Failures cannot be eliminated. The best that maintenance can do is prolong or delay the failure itself from occurring: Agree, we must not be misled that maintenance can eliminate failures. The truth of the matter is, failures cannot be eliminated. They will happen, and they will occur. The best that maintenance can do is delay its process, control the timing of failure or eventually prolong the occurrence of failure, yet in the end, failure and breakdowns will occur in our equipment, and we must be ready for it.

To conclude, failures cannot be eliminated, and failures vary in consequences. The best maintenance and reliability people can do is delay or prolong the process and control the failure's timing but eventually expect failure to occur in the future. Suppose you are experiencing a no-failure situation in your equipment today; it is just temporary because of your good maintenance system. The truth is you are just delaying the process of failures. I've been in TPM for so many years, and its goal is to zero out breakdowns is idealistic and, in its technical sense, next to impossible if we really understand how part really behaves. TPM is a dreamer. It dreams of an ideal factory with no breakdowns, accidents, and defects and wants industries to reach that stage, which never exists in the real world, but they will come close to it. When parts wear out, in its technical sense, it had failed. Hence, we are not really eliminating failure itself but just doing our best to prolong the part by maximizing its life cycle.

1.3: July 2007: Why Most Industries are Reactive

Have you ever figured out why our equipment keeps failing despite your best efforts on Preventive Maintenance? I have a list of questions below, and try to answer them honestly on your own.

• Are you always pressed for time, and you do not have the luxury of time to attend training because your boss will not allow you to attend to them? Am I right on this one?
• Do you need to turn on your cell phone at night to received late calls from other shifts wanting you to go to the plant in the middle of the night or early morning for a problem they cannot seem to fix?
• Do you often hear from maintenance people that they are always outnumbered by the failures, and maintenance always complains that they lack the manpower resources to fix all failures?
• Do you have to sacrifice some family time to keep working overtime, and when you return home, everyone is asleep, including your wife with your food waiting on the table with a sign from your wife to wash the plates?
• When equipment is newly overhauled, or a major Preventive Maintenance overhaul was recently done, do you ever wonder why operators always complain about why it isn't running and make a grin on their face and whisper something like say, no wonder this machine ain't running; they PM it?
• Do you need to cannibalize some parts from other stationary equipment since there is no stock or spare parts in the storeroom? Does this happen all the time?
• Is your time consumed by attending too many meetings almost every day when you look at your watch, it is almost 5:00 pm, and everyone starts packing to go home? In short, you have done nothing for the day but attend meetings
• Have you experienced working straight for 24 hours or even more without changing clothes or changing your underwear?
• The worst part is, do you have nightmares in the middle of the night of your work or your boss yelling at you, asking for the root cause that wakes you up in the middle of the night?

If you answer yes to almost any or all of these questions, don't be surprised; you are not alone in industries. Let me explain to you when I was still studying all this stuff about reliability and maintenance; I use to have a friend and mentor who have worked for a very long time in the industry; they retire him because he was old. I called him Mang Tibo. In the Philippines, when you address someone as "**Mang,**" followed by his first name, you address an elderly person. I guess that would be the trend for industries when they reached their age limit. But anyway, I remember asking him what the best maintenance strategy that we can adopt for industries. He told me that if I wanted to know the answer to my question, he said to come with him. And so I followed him, we walk and sit down in a park near his place. The park had a basketball court, and he asks me to sit down and watch the basketball game since there was an ongoing league at that time. We sat and watched the whole game, and not a single word from him was ever spoken. We just watch the game. The basketball game ended, and it was getting a little dark, and he told me to go home since my place was far from here. I told him why we watched this game and asked him to answer my question about the best maintenance strategy to adopt.

Calmly, he looked at me, smiled, and said, Rolly, if you watch the game carefully, then you have already answered your question. I was shocked since I really have no idea what he meant when he said that.

I usually give this story to people I teach and wonder if they can figure it out themselves. I guess, like me, they find a difficult time what the old man was talking about or if he was making some sense into it. Now, the clue lies in the June 2007 Edition, and let me dished them out for you. Below are the patterns of failure; this is how your components or parts behave. John Moubray, the author of RCMII, further indicates in his book that these three failures mostly occur in combination by studying parts behavior and led to what we called the six failure patterns, but eventually, those 3 patterns of failure (bathtub curve) can be seen as existing in TPM books, Statistical books, and Weibull Distribution.

Mang Tibo said that maintenance is a game of balance, just like a basketball game. He said that a basketball team comprises two guards, two forwards, and one center. Now let's speculate and imagine if your first five were all center as tall as Yao Ming. In this case, your team is not balanced. Like in maintenance, it is like assuming that all parts will eventually wear out, which is unlikely to happen since only around 15 to 30% of most parts and components will have these patterns. Imagine putting a basketball team of five people, all assuming the position of guard. Again, it will not be balanced. Your team may have the speed but not the defensive or rebound capability. Again in maintenance, it is like assuming that all parts will fail randomly; again, there are still infant mortality failures and wear-out patterns of failures. Therefore, the best maintenance strategy to adopt is not about being proactive or applying the best Predictive or Preventive maintenance strategy solely. It is about creating a balance and utilizing every possible maintenance task available, but the key is understanding when to use each of these maintenance tasks respectively since every failure has its own unique pattern. Every failure pattern should have its own specific maintenance task to adopt, and not to mention that we must also look upon the consequences of failure itself.

Understanding these patterns is the key to uncovering why most industries are reactive, or is it really a mystery or simply just a lack of solid understanding? But first, let us define the following maintenance strategies so we can have a common understanding.

Reactive Maintenance: A strategy that tells us that when a machine fails, maintenance simply fixes the failure. Maintenance is done at a point when there is an actual breakdown. It occurs when repair action is taken in the event of a failure. Other terms used include breakdown maintenance, unplanned breakdowns, run to fail, run to destruction, band-aid maintenance, no scheduled maintenance, which simply means that maintenance will fix it when something fails.

Preventive Maintenance: These are maintenance activities performed on a fixed operating schedule to extend the equipment's life. It assumes that the machine's condition is correlated with time, which means that the part or component can be expected to operate reliably until a period and is expected to wear out. Other terms used are Calendar-Based, Time-Based, Schedule Discard, Schedule Overhauls, Scheduled Outages, and the time can be referred to as hours, days, strokes, or running time depending on the industry.

Predictive Maintenance: These are maintenance activities geared toward indicating where a piece of equipment is on its critical failure curve and predicting its remaining useful life. This is done with the aid of non-destructive instruments. It also checks the equipment for potential signs of failure. Other terms used are Condition-Based Maintenance, On-condition tasks, or Reliability-Based Maintenance. In Predictive Maintenance, performing maintenance will be based upon the equipment's condition and not based on time or schedule, differentiating it from Preventive Maintenance. This can be done both inline and offline.

Proactive Maintenance: Just the opposite of reactive maintenance. Proactive maintenance is about analyzing why failures keep on recurring through techniques such as Root Cause Failure Analysis, FMEA, or other analytical problem-solving tools. In most cases, parts are modified or redesigned to lengthen the part's lifespan. I would say that modifications, redesign, analyzing failures all belong to this category.

Ok, now we know the different maintenance strategies and how each failure occurs at different patterns, so let's just tie it up with basketball. Most industries, I know, rely too heavily on their Preventive Maintenance efforts, yet even with the best structure and software, many failures still occur. Why?

Preventive maintenance will only apply to age-related failures or those whose parts directly relate to the operating age itself. When the failure is random in nature, there is no Preventive Maintenance that can solve this problem. In fact, this is where Preventive Maintenance will be at its weakest point. Second, too many activities on Preventive Maintenance, most especially replacements and overhauls, will increase the chances of infant mortality failures. A quote from the original manuscript of Stanley Nowlan and Howard Heap on Chapter 1, page 3 states that a maintenance policy based exclusively on operating age (he was referring to Preventive Maintenance) would, no matter what the age limit, has little or no effect on the failure rate.

Putting too much effort into your current Preventive Maintenance activities will only lead you to more failures since PM can only accommodate failures with a wear-out pattern. This is like putting five centers in one basketball team. Your team won't work; they won't blend. This is why most industries are reactive since they place all their center players in one team. They assume that their current PM tasks will capture all kinds of failures, which in the real world cannot. Eventually, they are wrong in every sense of the way. Hence by studying how a part behaves and fails, we can adopt a more appropriate and suitable maintenance task that will be much more effective. So try to observe the parts and how they fail. Here are some of my recommendations:

- Suppose the part or component has worn out and consistently survives that particular period, it means that there is a direct relationship between its operating age and the rate of wear. In that case, the best maintenance task to adopt is to undergo a Preventive Maintenance replacement or overhaul.

- When the part fails randomly, which means that the part can fail at any given period, you have three options. The first is to allow the failure to happen or have a run-to-fail mode for failures with minor consequences or parts with redundancies. Second, the use of Predictive

Maintenance for random failures that have potential failures or signs that it is on the verge of failing. Lastly, suppose run to fail, or Predictive Maintenance is not feasible; in that case, our last option is to Modify or Redesign that particular part or item.

• Remember that around 15 to 30% of parts will fail due to wear-out mode and that the majority will fail randomly; hence, let us not rely too much on our center players. A good basketball coach always knows when to place the right players at the right time, and a good maintenance structure is knowing when to adopt the different maintenance strategies at hand.

1.4: August 2007: How to Make Training More Effective in your Industry

We all know that training plays a vital and very important role in any organization and industry. Yet, most of the time, it is not taken very seriously. Worst of all, it had been the subject of cost-cutting measures and initiatives by top management. They say that people are a company's greatest asset. Still, I disagree with this statement since I believe that not all people are assets and that only the right people are the company's greatest asset. In contrast, to acquire the right people, they must be trained and educated to build their skills and perform their work better; as technologies change, our skills need to be upgraded to cope with technology. The correct statement should be that the right people are the company's greatest asset, and the wrong people are liabilities.

Most industries are looking for a quick-fix solution for every problem they have, and they want it done fast and quickly. Hey, let's just copy the best practices from this plant and let us not re-invent the wheel, and so a group of people was sent to a benchmarking journey to copy the plant's success and apply them in their plant, and sad to say it did not work. Did you know how many years they tried to achieve that stage? Did you know how many failures that plant has undergone just to reach that level? Is their culture, values, beliefs goals the same as your industry? Did you know what changes their people made to make that adjustment? Is the operating condition or operating context their equipment is having the same as yours?

Ok, it did not work, so let's try another strategy, so you just hired someone who can do the job. I can't tell what word you use in your country, but here we termed this **pirating**, which means one industry getting key people from other industries and offering them much better compensation, benefits, and salary. I used to know someone with a tremendous amount of knowledge on TPM (Total Productive Maintenance and it's not me), which was pirated by one industry so that the strategy can be implemented in their industry; after a year, this guy left his work since no one seems to be following what he was recommending because everyone was busy doing their own day-to-day things fast and quickly.

Again let me state this for the record, there is no silver bullet solution or rocket science strategy that can transform a plant's reliability overnight. Everything will start from its very basic foundation, and that is through training and education.

Ok, let's say that I have convinced you at this very moment and that you are willing to send your people finally for training, and if you do, then this is just the start of a bigger problem. A group of people was sent to training, and they have absorbed the learnings, and when they return to their work, nothing happens. Management then noted that nothing happens and concludes that training is not the answer and considered it just a waste of time and money. Again they resulted once more in their usual fire-fighting and reactive mode of doing things in the plant.

Let's take some time and try to absorb the message that we need to slow things down to make things fast. Let's just simply take things one step at a time. Training is an investment, whether you have your own training department or the training will be performed by an outside consultant like me. There will be cost involved such as materials, food, snacks, cost of consultant, handouts, overhead, facilities, electricity, lodging if the consultant needs a place to stay, space if you will be renting or if the training will be done inside your plant, etc. Try to plan ahead before spending your money on training.

- **First,** we need a specific reason why we need our people to have this training.
- **Second,** management must understand what the training is all about. I have often received feedback from people that I trained if their management had been trained in the things I discussed. I think that they should be the first to attend the training before sending their people to it.
- **Third,** management must provide time for their people to practice what they have learned. Support is not enough. There must be management commitment. Support and commitment are two different things. Your management can support yet never commit to your initiative.
- **Fourth,** management must provide time to review the logistics required to complete the home run. Is the consultant still needed in the initial stage of implementation?
- **Fifth,** people want to be recognized for their efforts; it gives them a sense of pride, motivation, and enthusiasm to work better. Although I am not speaking about giving them money in return, acknowledge them for the results of their efforts.

I have been involved in training people for the past 8 years, in which 5 years were spent when I was still employed and the last 3 years conducting training and seminars as an independent reliability and maintenance consultant. I realized a big difference between training people as an employee and training industries as an independent consultant. Since I started this training and consultation venture of my own, up to this point in writing, I have trained around 2,635 people, 165 batches (I started to record only last 2002 up to the present, so I know it's more), 14 subjects all on reliability and maintenance courses. Let me share with your some highlights of my career in training.

At Amkor Technology

- A dramatic reduction of breakdowns in their facilities nine substation equipment from 49 times in 2000 to only three breakdowns for 2001 by implementing Reliability-Centered Maintenance Strategy.
- A dramatic reduction from 888 failures from 436 equipment from January 2001 to only 14 breakdowns as of September 2001 by implementing TPM Planned Maintenance 4 Phases through Zero Unplanned Breakdown.

At Lepanto Consolidated Mining Industry

From their 2003 Annual Report, reduced maintenance cost from an average of 57 million pesos (1 dollar is around 45 pesos) during the 1998 to 2000 period to an average of 37 million pesos during the 2001 to 2003 period.   The reduction in maintenance cost was due to our Preventive and lately, with Predictive Maintenance and Condition-based Maintenance Strategy.

Figure 1.1: Early Training Days as a Plant Trainer and as an Independent Consultant

I say again that there was quite a big difference when I was still employed and training maintenance people as an independent reliability consultant, although the courses I train are entirely the same. When I was still employed, I trained the maintenance people, but I also gave a short presentation to management. I try to check the people I train from time to time, provide them advice on how they will implement it, and provide some inexpensive yet memorable recognition schemes. I provided certificates to teams that successfully implemented what I taught. Being a maintenance consultant is different; an industry will hire me to train, and after I deliver the training, my hands and feet are tied, and that's the end of my services with them. Unless otherwise, they will ask me for some additional guidance and facilitation after the training. Unlike before, I have all time and opportunity to check the people I trained as often as I want and hold meetings if necessary to complete their projects. I can do these things during my employment days but not independently since I am only paid to deliver the training. I think of this as my greatest challenge that I have no sure answer to until this point in time.

Let me put it in another perspective, let's say that I am providing a class lecture on "How to Drive" in which I teach everything from the function of every single part of a car from the most important ones such as breaks, clutch, the engine, fuel system, the features of the car, the signs on the road, everything you need to know about the car and how to drive it. After the class lecture, can I tell you to drive me to the airport or elsewhere? In the initial stages, I need to be with you, right? This is until the time that you can be on your own completely with confidence. Absorbing the learnings from a lecture is a different thing from actually implementing it. If you're new, you need to be guided accordingly.

As much as I can, I tried to put in my best effort to teach as an independent reliability and maintenance consultant. I know that after my last closing words have been spoken, I'm done. It's up for that industry to implement the learnings unless otherwise we consultants will be included to guide them in their initial stages until we feel confident enough to make it on their own. I hope that industries can reflect my perspective and that all we want would be best for both of us. I hope other consultants like me share the same sentiments as well.

I always say to myself that the best marketing strategy any maintenance teacher like me can have is not based upon the testimonies or feedback we have received from our past participants indicating how good we deliver the subject matter, but rather the benefits that the industry had derived from as a result from the training that we have provided them and good to say I have a few of them. Again, most industries make the mistake of just sending people to training without expecting something from them. What is important is that before taking a journey and embarking your people to go to training, we also need to plan what to do after completing the course. What is expected of them? Why have they been chosen in the first place to attend this specific training? And so on.

Training and education can be the most powerful weapon in your maintenance arsenal if you plan it correctly, or it could mean nothing. Both your time and money will only be wasted if you have no plans on what to do after the training had been completed. Simultaneously, the people who attended will just update their cvs' and resumes to start looking for a greener pasture. Remember that whatever investment you have made on your training, the management likewise expects a return from it, but the return will only come when we plan things carefully before sending our people to training.

1.5: September 2007: A Different Root Cause Failure Analysis Experience

We often hear the word Root Cause Analysis and Root Cause Failure Analysis, yet I wonder if we really know what it truly means. Almost every industry has its own unique Root Cause Analysis techniques to follow.

I have been in the business of teaching industries about reliability and maintenance courses, and one of the courses I truly love to teach Root Cause Failure Analysis. Recently, last July of this year 2007 I attended one of my good old friend course on this subject and after the training ended I just can't help to think for a while that for the first time it seems clear to me what Root Cause Analysis truly means which I thought I thoroughly understand. It allows me to reflect on my perspective and the basic lessons of what life is all about. During my employment days, we solve problems, thinking that we are pinpointing the root cause of the problem. After hearing from Bob, I came to realized that I knew nothing about root cause analysis

Some of the basic questions being raised in the first place that needs an answer are; how far should we go on with our analysis? Although there is a wide range of analytical tools on the market that had been provided, such as Ishikawa or Fishbone Diagram, FMEA, Pareto Analysis, Kepner Trego, Eight Disciplines, TOPS, P-M Analysis, FMECA, 5 Whys, Fault Tree, etc., the real question raised is that are these tools really meant to address the root cause of the problem or only the most likely cause? Each of these tools will claim yes, but I really doubt if they do.

First, let us define what Root Cause Analysis is all about. A free encyclopedia states that Root Cause Analysis (RCA) is a problem-solving method to identify the root causes of problems or events. RCA's practice is predicated because problems are best solved by correcting or eliminating the root causes instead of merely addressing the immediately obvious symptoms. By directing corrective measures at root causes, it is hoped that the likelihood of the problem recurrence can be minimized. ***However, it is also recognized that complete prevention of recurrence by a single intervention is not always possible.***

Let me explain the last sentence discussed in my recent newsletter; if failures can really be eliminated? Complete elimination of failure in our equipment is impossible; the best maintenance can do is delay, prolong, or merely anticipate its recurrence. In its technical sense, failures are inevitable and are meant to happen. If a part wears out, then it had actually failed.

Root Cause is not a silver bullet or rocket science strategy that will eliminate all known problems. This tool can only be useful if we truly understand what its intent and purpose are all about. Yet, most people from industries abuse the word Root Cause, thinking that it is meant to end their problems. All I can say is that in every way, they are wrong. We need to understand that Root Cause is being performed to learn from the things that go wrong.

The problem lies in how we understand the problem and how we analyze it. Pareto will say that 80% of the effects come from 20% of the causes. My question is the 20% really the root cause or just some probable causes? The answer is it may or it may not. Why-why states that by performing why-why five times, and the team finds it hard to answer, the bottom line is that the root cause had been defined. But worse than that, when we find the guilty person, who performed the error, the root cause was finally identified. Given all of these, let me shed some light on what a true and meaningful Root Cause Analysis is all about.

Lessons about Root Cause Failure Analysis

First: Root Cause is not about failure modes or probable causes. It is always based on facts, and the facts are always based on the evidence gathered. Once the evidence is in place, we can only perform a sequence of events that ultimately lead to the failure. Root Cause will always have to be based on pure evidence. Every failure had some sort of clue to leave as to why it occurred. Do we talk to people who were involved in the problem? Did we examine the part that failed? Did we find anything unusual about the event that took place? Are our RCA efforts based purely on evidence or not?

Second: Root Cause is about learning from the things that go wrong. This statement had to change the way I think about failures. Let me put it this way. Have you ever heard the word benchmarking other industries' success, or is it most worthwhile to benchmark other industries' failures? Can we learn from other people's success, or can we learn from our own failures and adversities? Let me get a little soft and philosophical about this matter so we can understand it better. Rethinking failure is not bad after all. Whether from the lessons of life or our assets, failure can be our greatest teacher if we can learn from them. The same principle applies in Root Cause Failure Analysis. Let me share with you some quotes about people who learned from their own failures.

- There is a tendency to walk away from failure and leave it buried. An enormous amount of institutional learning gets buried because failures don't get analyzed. So the real learning is what's learned from failure, *by Andrew Grove, Chairman, and Co-founder of Intel Corporation*
- You don't learn about yourself through your success. You only learn through your failures and your mistakes, *by Wynonna (The Judds)*
- Most people make a common mistake by thinking of failure as the enemy of success. You've got to put failure to work for you. Go ahead and make a mistake. Make all you can. Remember, that's where you'll find success on the far side of failure *by Thomas Watson of IBM*
- What people see of my success is only one percent but what they don't see is the 99 percent, my failures. *By Soichiro Honda, President of Honda Corporation*

Third: People commit mistakes and errors; almost if not all failures or problems will lead to an error or mistake done by a human being. Even with the best system in place, people err, and the analyst must realize that not all mistakes people make are within their control. In this regard, I would strongly emphasize that Root Cause is not a tool to blame and punish someone. This will only make people more defensive. Some industries which truly understand what Root Cause is provided some sort of amnesty program and emphasized it clearly at the beginning of any RCA investigation process that we only would perform a thorough Root Cause Analysis because we want to learn from the things that go wrong and not to blame or punish someone. We all contribute to problems, yet we are like Pontius Pilate, washing hands most of the time, thinking that we ain't part of the problem.

Fourth: A true and meaningful Root Cause Analysis is done on three levels. First, determine the Physical Cause, then analyze the Human Cause and determine the problem's latent cause. There will also be cases that between the Human and Latent causes are system causes. All Physical Failures are triggered by humans, but humans are negatively influenced by Latent Causes. Therefore Root Cause analysis ends when the Latent Cause of the problems had been exposed. Latent Causes are not only about flawed systems, flawed procedures, no training, incorrect policies, decision making but rather simply humbling us and asking the following;

- What is it about how we contribute to our problems?
- What is it about the way I am that contributes to our problems?

Are we really sure that if we are in the shoes of the person who committed the mistake, will we do otherwise or also commit the same mistake? What was in the person's mind that eventually lead him to commit that mistake? A classic example here would be the Challenger Explosion, where 74 seconds after lift-off, the Challenger Shuttle exploded. The flight on January 28, 1986, ended the life of the six astronauts and one civilian. What amount of pressure had management been under that led them into this tragedy since it is being covered worldwide? Are we sure that we had done it differently or otherwise the same if we are in their shoes on that particular time they made that decision? Think about it. People make systems, management make decisions, and we are all humans. Latency understands why people did what they did, which eventually leads to this problem. We can only learn from the things that go wrong if we accept that we are also part of the problem itself. This RCA philosophy tells us that RCA has no room for pinpointing mistakes and blaming others. It simply has no room in any

RCA or RCFA.

Fifth: The greatest lesson so far, one can benefit from truly understanding Root Cause, is that it is always better to analyze failures than fix them every time. If we become good at fixing failures, then something is definitely wrong with our organization. Why? Simple, because we are doing it much too often. Troubleshooting is no longer an effective strategy. In the real world, mostly in manufacturing, or whatever type of industry, we need people who can analyze problems and not just fix them. This eventually makes us understand the difference between a maintenance person from a mere mechanic. A mechanic mostly uses his hands as his tools to fix problems, while maintenance mostly uses his brain and mind to understand and analyze the problem instead of using his hands too often. Sometimes I find it funny how most industries think. Most industry often yells that they have no time to perform an RCA or RCFA Analysis, yet they have all the time in the world to fix the problem repeatedly, which makes their people really good at fixing failures.

Here are simple guidelines that can help us to determine your basic RCA requirements.

- Does everyone in the organization understand the objective of performing a Root Cause Analysis Investigation? Are they united in their purpose, or they have their own agenda?
- Is management willing to be trained in Root Cause Analysis? Does management truly understand what RCA can and cannot do?
- Will, the people, investigating the problem allowed to complete the RCA analysis, and will their recommendation be implemented or fall on deaf ears of management?
- Is 3rd party consultation being required in the initial process of implementation?
- Are we willing to learn from the things that go wrong and be part of the learning process?

To conclude, what makes Failsafe Latent Cause Analysis different from the other analytical approaches? Well, LCA takes your analysis a step further from the rest.

1.6: October 2007: Where Does RCFA Fit in the RCM Process

First, let me note how RCM (Reliability-Centered Maintenance) is being performed to start this topic. RCM is defined as a process used to determine any physical asset's maintenance requirements in its present operating context. To those new to this concept, this is simply being performed by answering the 7 basic questions of RCM.

- What are the functions and associated standards of performance of the asset in its present operating context?
- In what ways does it fail to fulfill its function?
- What causes each functional failure?
- What happens when each failure occurs?
- In what ways does each failure matter?
- What can be done to prevent or predict the failure?
- What should be done if a suitable proactive task cannot be found at all?

RCM consists mainly of two parts; the first part derives the FMEA of the asset being analyzed through what it termed as the RCM Information Worksheet. The second part includes addressing each failure mode, effects, and the consequences of each failure so the RCM team can define the most appropriate and the most feasible maintenance task to use.

Difference between a Failure Mode and Root Cause

When equipment fails, there is a wide variety of causes as to why it fails, and when we speak of this vast amount of causes that prompt the equipment to fail, we speak about its failure modes. As the late John Moubray said, a failure mode refers to an event that could cause a functional failure. These refer to the probable causes of failures that might have occurred before or could occur. On the other hand, Root Cause is about understanding why things go wrong so the team can learn from the failure themselves. The depth of the analysis in Root Cause will always tell us that for every failure, there is always a corresponding Human Cause as well as a hidden cause or Latent Cause behind the problem. Failure Modes and Root Cause are two different terms. Its difference lies in the depth and breadth of the analysis.

Both RCFA and RCM are two powerful strategies, yet they can only complement each other when we know how and when to use them, respectively, and this is where we can derive the most benefit from its application. Again both RCA and RCM should not contradict each other but rather complement one another. Failure Modes are like having a machine gun, where Al Capone is shooting in all directions in the dark, while a root cause is much like a rifle with a scope equipped with a laser and aiming for one target at a time with pinpoint accuracy.

RCFA termed failure modes as probable causes or hypotheses and is usually used in the initial analysis. Each failure mode is verified to proceed to the next level. The root cause will always be based on the evidence found during the investigation and analysis. Any analysis which tends to end up at the part or component level is still not considered as the Root Cause, but rather it is simply termed as the physical cause of the failure or most commonly known as its Failure Analysis.

Let me provide an example to clarify this, a pump failed to fulfill its function since it is not discharging any fluid. This failure to discharge is known as a functional failure. Let's think of the different reasons or probable causes of why this pump is not discharging any fluid. We are speaking about failure modes and not the Root Cause. Therefore the failure modes of the pump on why it is not discharging fluid at all will be as follows:

Failure Modes Include:
- Valve is totally closed
- Motor totally burnt out
- Bearing seizure (stuck-up)
- Strainer totally clogged
- Broken pump impeller
- Supply tank is empty
- Clogged impeller
- Driver imbalance

• Insufficient suction pressure

I could continue with the list of failure modes or probable causes that could warrant a pump not to discharge fluid at all, but let's assume all the failure modes have been listed so far. Failure modes are the probable causes that will affect the pump from not discharging any fluid at all. You may even include lighting to struck the pump or an earthquake that severely damage the pump. On the other hand, to determine the root cause, we need shreds of evidence to conclude the reason for its failure. This means that the pump must actually fail so we can determine the failure. Second, the failure should be fresh, which means that it just happened recently. We cannot get the root cause of this pump whose failure happened two years ago, but we can guess the probable reasons. This is what makes Root Cause Failure Analysis reactive in the first place since a failure must occur before performing it.

On the other hand, Root Cause Failure Analysis is, at the same time, Proactive because by knowing the cause of the failure, we can learn from it and address similar situations in the future when it arises. Therefore, by analyzing the pump's failure, we need to verify what failure mode actually happened during the time the pump had failed, and when the analyst had verified each failure mode and found out, for example, that the actual cause of the pump's failure to fulfill its function is a bearing seizure, then we need to analyze what caused the bearing to seized, the other failure modes will be disregarded. After we analyze the cause of the bearing seizure, a failure analysis will be written.

Failure Analysis will be: Bearing Seizure due to lack of lubricant in the raceway. Failure analysis will stop at the component level. However, a Root Cause Failure Analysis will still proceed with the analysis. It will only conclude when the Latent Cause of the problem had been identified. Hence, a complete Root Cause Failure Analysis will be written as follows:

Root Cause Failure Analysis will be:

• **Level 1: Physical Cause of the failure:** Bearing Seizure due to lack of lubricant in the raceway
• **Level 2: Human Cause of the failure:** Maintenance had used the wrong lubricant for the bearing
• **Level 3: Latent Cause of the failure:** It is not clearly specified in the PM procedure as to what type of lubricant to be used for this type of bearing, which had to cause the maintenance to use the wrong type of lubricant.

We have differentiated failure modes from failure analysis and Root Cause Failure Analysis; the question raised is: where does RCFA fit in the RCM Strategy? Some say that RCFA can fit into some failure modes that are already the subject of some Root Cause Failure Analysis investigation. I disagree with this since if the failure mode is subject to investigation, then the correct term is Failure Analysis and not Root Cause Failure Analysis. RCFA will go much deeper and would expose the hidden causes of failure. Second, suppose a Root Cause Analysis had already been performed in any of the failure modes. In that case, it should not be written in the RCM Analysis lists of failure modes since a recurrence is unlikely to happen if a successful RCFA had been concluded so far. Most RCFA will default to redesign or modification.

John Moubray, the RCMII author, asks that which comes first, redesign or maintenance? He simply stated that maintenance should come first before the redesign. I agree with his statement since there is always a temptation to modify or redesign the system without reviewing the current maintenance tasks at hand. On the other hand, most people are tempted to redesign or modify the equipment without even performing a thorough RCFA investigation, leading to new problems. RCM is being performed on an asset so that the people involved, which are the operators and maintenance, can better understand the different failure modes that can affect their asset and the consequences of each failure itself when it actually occurs. Not all failure modes can be addressed by maintenance, and when a maintenance task is not feasible to use, then the team will default to modification or redesign. A redesign will include changing a component's specification, adding a new item, changing a component's size, shape, material, or even changing a process or procedure.

Figure 1: 2: Myself with my friend Bob Nelms from Failsafe Network at Jakarta Indonesia

There is a limitation to maintenance itself; this is seen in the last question on the RCM's seven basic questions. What should be done if suitable proactive tasks cannot be found? One of the options to use here is to redesign. A redesign is warranted and feasible to use if the consequences of failure are simply not acceptable to the user. RCFA is used after performing an RCM Analysis and not before starting an RCM Analysis. I believe this is where RCFA will fit in the RCM process. Hence, after performing an RCM Analysis, and there are still failure modes that keep repeating themselves in which the consequences are not acceptable, an RCFA should be conducted. The second point is that equipment failures attributed to human errors should be included in the Root Cause Failure Analysis Investigation.

I have friends who consult RCM and those that consult RCFA and RCA, and I sometimes see them in a conflicting mode as to which strategy is best to adopt. I believe that both strategies are not perfect yet. When combined, it will create a much more powerful strategy for the maintenance function, but this can only be possible when used from the right perspective. Again, they should not contradict each other, but rather they should complement one another.

1.7: November 2007: Why is TPM Hard to Implement?

After spending at least seven years as a TPM Senior Engineer in one of the biggest semiconductor industries here in the Philippines, it gave me some insight and an indebted understanding of why TPM (Total Productive Maintenance) is not easy as you think to implement. Many books had been published on TPM with industries reaping its rewards and having a dramatic increase in their productivity; decrease in defects, breakdowns, and not to mention the reduction in their operating and maintenance cost, but on the other side, many industries initiating their TPM journey in their plant had miserably failed and abandoned the process completely. I truly believe in the TPM process without a doubt, but TPM can only be successful if you know how to implement it correctly. Hence, before embarking or initiating a TPM strategy, note the following because if these factors are not taken seriously, all TPM efforts are doomed to fail in your industry. Therefore, for those plants initiating a TPM strategy in your plant, consider the following:

1st: TPM is a Top-Down Approach: TPM can never be a down-up approach. TPM should be initiated by the company's highest member, your CEO, General Manager, or President. As noted from TPM Twelve Developmental Steps, a formal announcement from the Top should help get your TPM started. My advice is that do not implement TPM if this will not be endorsed from the top.

2nd: TPM requires an Office and not a Facilitator: TPM requires full-time staff and not a part-time facilitator. Assigning one person to handle TPM would surely lead you to its doomsday and downfall. How many people will be staffed in the TPM Office will depend upon the size of the plant. Just to give you an idea, the plant I've worked with has around 12,000 employees, and 8 of us belong to the TPM Office, and each of us directly reports to the TPM Manager. I was assigned to carry on the pillars of Planned Maintenance and Early Equipment Management. TPM Office will provide the details and legwork on how each of the pillars is being implemented, initial audits and certification per step or phase, training on their respective pillars, team recognition, and, most importantly, consolidating everything from the teams implementing TPM. Remember, TPM requires everybody's involvement, and all employees of the plant will have a specific pillar to be part of.

3rd: TPM is a long-term approach, never a short-term: Operations and maintenance want everything fast; therefore, they result in quick-fix solutions that will eventually resurface again shortly. Management must understand that TPM is a long-term and not a short-term approach. Benefits will be realized first on a small scale, which will eventually be felt in other plant areas. There is no shortcut to this process. TPM is being implemented on a step-by-step approach.

4rth: TPM will require a Budget: The most difficult part of the TPM implementation would be the Preparatory Stage or start-up. Most management is sensitive to the issue of cost. TPM is costly in the initial stages of implementation, and believe me when I tell you. Some of the expenses in the initial TPM stages will include training and educating all employees on TPM and their respective pillars. Initial cleaning for Autonomous Maintenance in which the teams try to correct abnormalities in their equipment. Restoration activities for Planned Maintenance: TPM aims to bring back the equipment as close as possible to its original basic condition.

5th: TPM and company goals must be aligned: This is one of the most important aspects of any TPM implementation that is mostly overlooked. Both TPM and company goals must be aligned; they cannot be different. Otherwise, TPM will be seen as a separate activity and not as a plant directive. This will include their indices and KPIs being tracked so the company's progress can be known. OEE or Overall Equipment Effectiveness should be part of the company's primary measures.

6th: TPM is heavy on documentation: What is being done should be documented in the TPM process. Documents should be placed on activity boards and not on binders and folders so that everyone not yet involved in the TPM journey can see the benefits and results of their TPM implementation. The activity boards should be updated regularly by the team involved in the TPM activity.

7th: Managers Model Machine for Managers: Managers must be involved in this shared responsibility in the TPM process. Most especially, they should be part of a team whose role is to act as a coach or mentor, not part of the actual team. This will be an initial requirement for any industry aiming for TPM Certification and, therefore, essential to any TPM implementation success. Part of their role is to see that their pillar is moving and advancing to the next step. Remember that each TPM pillar is performed on a step-by-step approach, and there are no shortcuts to it.

8th: AM must be done in parallel with PM: Although Autonomous Maintenance (AM) would be the largest pillar in population on the number of people involved, Planned Maintenance (PM) should be the strongest pillar in any TPM implementation. The equipment is always a shared responsibility for both operations and maintenance together. One of Planned Maintenance's utmost responsibilities will be to educate the operators on their equipment as operators learn to accept some minor responsibilities in their equipment, such as doing the basics, maintaining the equipment clean, proper lubrication, and tightening bolts. These things can only be appreciated by operators with the aid of Planned Maintenance as their mentors. If the Planned Maintenance structure is weak, so will be its Autonomous Maintenance. Planned Maintenance should serve as the Autonomous Maintenance coaches and mentors in their equipment. On the other side, maintenance can only advance to other maintenance activities and get out of the reactive mode of firefighting if Autonomous Maintenance will take part in their shared responsibility on their equipment.

9th: Management Commitment and not Support: Management can support but never commit to any TPM activities. In TPM, what we need are managers who can provide commitment and be part of the process. This is a very important lesson in TPM. These two words are entirely different, "support and commitment." Remember that TPM must be done not as a separate activity, but rather management must realize that their day-to-day activities must be part of the TPM process. TPM is not a once-a-week meeting or activity but rather a continuous process.

10th: TPM is all about people and not machines: Most industries that implement TPM think that it is all about improving equipment, yet little regard is being provided in educating its workforce. Many say that people are the companies' greatest asset; I hate to disagree with this statement. And how can we have the right people? Only if they are equipped with the right knowledge to perform their jobs correctly. People will improve their equipment, and it's not the other way around. Hence, people must continuously be educated, and their knowledge upgraded from time to time.

11th: OEE is the impact of all TPM Pillars combined: Having a high OEE is not a product of only one pillar, but rather it is the consolidated effort of all pillars involved. Improving OEE means improving the six major equipment losses, and all pillars will have their corresponding role in improving them. Below are the six major equipment losses and the pillar that will impact them most.

• Breakdown losses by Planned Maintenance
• Set-up, adjustment, and conversion by Kobetsu-kaizen or Focused Improvement
• Start-Up Losses by Planned Maintenance, Early Equipment Management
• Idling and Minor Stoppages (chokotei) by Autonomous Maintenance, Focused Improvement and Planned Maintenance
• Design Speed Loss by Focused Improvement Pillar
• Defects and Rework Losses by Focused Improvement and Quality Maintenance Pillar.

12th: TPM is not culture-bound: Although TPM originated from the east, Japan, it will work in any industry since it is not bounded by culture and beliefs. Japanese have a certain way of doing things, and what is important is not to imitate the Japanese but rather to be flexible and transparent in adopting TPM to the culture that suits your industry.

I would like to dedicate this newsletter to my former TPM team in which some of which I am still in contact with and some I don't. I have been asked several times by industries to work with them and initiate a TPM journey, which I decline and tell them that hiring me full-time to do TPM would only be possible if I worked with the former TPM team. Wherever you are right now, it was an honor and privileged working with you.

1.8: December 2007: Does Oil Wear Out and Needs to be Changed?

Contrary to the notion that most industries change their oil on time-dominated frequency, they replace oil-based from running hours. The question raised is, should oil really be changed on a time-dominated pattern or based upon its condition? Our typical belief in changing oil is that oil should be replaced regularly since more parts will fail if we do not change it.

Let us go back with the basics. Oil comprises of two ingredients, the base, which can either be petroleum-based (extracted from nature) or synthetic (man-made oil), and then we have the additives. The base oil is then treated with chemicals and additives to form the different lubrication grades to suit a wide variety of applications. Once the additives deplete, there is no more reason to use the oil, and, therefore, it should be changed since it had already failed in its technical sense.

But what causes the oil to break down?

First and foremost, we have a depletion of additives: Additives deplete due to contamination present in the oil. Contamination does not only come in solid particles, but water and the air also form contaminants themselves.

The second is heat: Contaminants induce heat. One indicator of heat tolerance will be the oil's Flash Point. The lower the Flash Point, the greater the chance of vaporization loss. Once this happens, viscosity increases and oil oxidization is imminent to happen. Oil, therefore, becomes acidic. This means that TAN (Total Acid Number) increases while TBN (Total Base Number) decreases.

Our belief about contamination: Contamination is the main reason for oil degradation. Hence, the need to change the oil on a time-dominated frequency is essential. More contamination means more failures and the need to change the oil. But what is important is to analyze the oil based on the amount of contaminants present and not by the frequency of changing the oil itself. By knowing what elements and contaminants are present in the oil, maintenance can strategize ways to improve its fluid cleanliness and lengthen its drain interval. Remember that clean oil does not necessarily mean new oil. New oil contains a lot of contamination, even if the containers are sealed. The bottom line is that there is a direct relationship between the life of the oil and the number of contaminants present. The more contaminants present in the oil, the more chances of failure. Therefore, if oil can be maintained clean, then there is no reason to change the oil and follow the OEMs requirement to change the oil based on specific running hours. Studies show a direct relationship between the amount of contaminants present and the lifespan of parts. Here are what some industries experience in maintaining clean oil in their equipment and assets.

- **General Motors Corporation** shows that 82% of internal wear comes from abrasive particles less than 40 microns in size.
- **SKF Bearing** states that improper lubrication account for 54% of bearing failure.
- **Canadian National Research Council** studies show that contamination was found out to be the leading cause of wear in a variety of industries investigated.
- **British Hydromechanics Research Association**, which covered a 3-year study on 117 hydraulic machines composed mainly of injection molding, machine tools, mobile, construction, marine, metals, concluded a dramatic relationship between fluid contamination and service life.
- **Cummins Technical Center** indicates that wear can be reduced by 91% using a by-pass filter combined with a full-flow filter.
- **SAE** states that contamination in the lubricant of engines, transmissions, and hydraulic systems causes 70% of equipment failures.
- **Nippon Steel** reduced bearing failures by 50% through aggressive contamination control.
- **International Paper's Pine Bluff Mill** reported a 90% reduction in bearing failures through an aggressive contamination control program.
- **Caterpilla**r states that dirt and contamination are the number one cause of hydraulic system failures. Their case study states that they must be kept clean concerning hydraulic systems at all times.

Can contamination be removed from the oil? First, solid contaminants are measured in microns. One micron is about 39 millionths of an inch. Perhaps a grain of iodized salt is around 60 microns. A hair follicle may be in the range of around 100 microns. A red blood cell is around 7 microns. An average human eye with a perfect vision of 20/20 can only see particles in the range of 40 microns and above. Hydraulic equipment and engines' primary defense against these contaminants will be their filtration system or oil filter. The typical function of the oil filter is to remove abrasive particles from the oil to achieve the desired fluid cleanliness level.

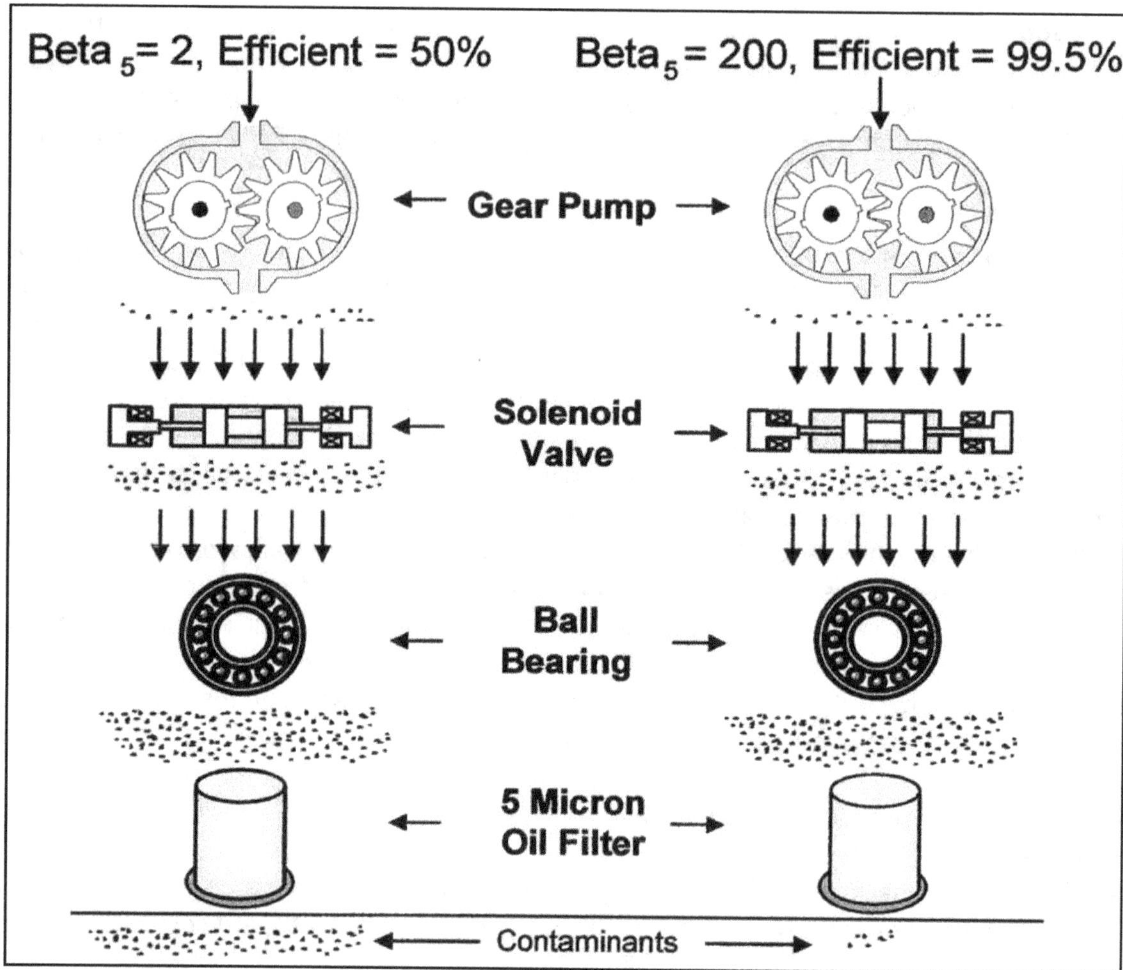

Figure 1.3: Comparing Absolute and Nominal Filtration

The typical function of an oil filter is to clean the oil, but the real purpose is to reduce maintenance operating costs by reducing abrasive, and fatigue wear on its components. But we need to understand that oil filters can be classified as nominal or absolute in ratings, and there is a big difference between the two. The absolute rating refers to the smallest size particle that will be removed during filtration, while a nominal filtration rating will refer to the average particle size that will remain in the fluid after filtration. Tests have shown that a 200-micron size filter can pass through a nominal 10-micron rating filter. To understand what this means is that every filter has its corresponding Beta Rating and efficiency. An absolute filter rating will have its lowest efficiency of 98.5% or a Beta Rating of around 75, while a Nominal Filter Rating will have a Beta Rating of 2 or efficiency of around 50%.

This means that if we have around 100,000 particles entering the filter (upstream) and our filter rating is 50% efficiency or beta of 2, only 50,000 will be trapped by the filter. The remaining 50,000 will pass through the filters upstream and downstream and go back again to the system. There is a high probability that these particles can be trapped between clearances that can cause abrasion to occur in our system. Although there are many absolute filters today in the market, this newsletter intends not to promote them but to let our readers understand how they function. Perhaps, this illustration will further prove my point.

Clearly, from this comparison, we can see where frequent failures can be experienced. We should also note that fluid pressure will play a critical role in selecting the correct micron size filters. These high-efficiency filters should not replace the primary filter as a drop in pressure and flow would definitely create havoc in your system. These high-efficiency filters' role is not to replace the original filters but rather to be part of a secondary filtration process, most commonly known as by-pass filters.

Let's discuss a typical by-pass filter; every single vehicle has its own primary full-flow filter. These filters can only remove particles around 30 to 40 microns in size and higher; hence, they cannot remove sooth, combustion by-products, and contaminants 30 microns and lower. A secondary by-pass filter should be installed separately from the original full-flow filter to remove these abrasive particles. Although I have seen how this type of filter works, one must fully understand its usefulness before installing them in your vehicle, especially from an economic perspective. These filters cost 5 to 20x the cost of the original full-flow oil filter. These by-pass filters would be feasible for corporations with many vehicles rather than installing them in a single car. The oil filter cartridge, once clogged, is replaced instead of the oil filter, plus a regular oil analysis check will be done on the oil. This is the secret of extended oil drain, in which logically, if the oil is maintained clean, then there is no reason for the additives to deplete and, moreover, no need to change the oil itself. If you understand how this works, then extending oil change is possible.

Does oil really wear out? The base oil really does not wear out. The additives only become depleted. We change the oil in our cars, generators, engines, and hydraulic systems because our assumption is that the oil is not fit for use anymore. Maybe we are right, but maybe we are wrong technically. Only an oil analysis check can dictate if our oil needs to be changed or not. Even if the oil is dirty, we have so many ways to clean it and check it again if it is still fit for use. Such cleaning oil methods include electrostatic cleaning, centrifuge, bypass, and offline filtration with fine micron rating filters. Suppose all vehicles are equipped with secondary by-pass filters, and almost all gas stations had an independent oil analysis laboratory where ordinary motorists like you and me can check their oil; in that case, I believe that the 3000 miles oil change can be disregarded for good. Hence, to conclude, if contaminants can be removed from the oil, there is no reason to change the oil itself. Here are some extracts I have researched that have the same conclusion;

- **The US Air Force says**: Despite the popular notion that oil breaks down or wears out and is unfit for further use. Its permanent value has been known to the U.S.A.F since World War 1.
- **Lubrication Engineering, Volume 17:** Oil does not wear out; Standard Oil of New York states that many times, the question has been asked, does lubricating oil wear out? The question should be answered negatively.
- **Mobile Oil Technical Bulletin # 863:** Oil does not wear out, break down, or otherwise deteriorates to such an extent that it needs to be changed. The oil becomes contaminated with water, acids, carbon particles, and sludge. The average oil filter can remove solid particles above a certain size. It cannot

remove water, acids, carbon particles, all of which pass through the oil filter just as readily as the oil.

- **The U.S. Standard Bulletin # 86:** Oil does not wear out but only gets dirty.
- **Theory and Practice of Lubrication for Engineers, 2nd edition P590-591:** Oil like any mineral and cannot wear out. The oil becomes dirty and contaminated, but like copper, iron, or silver, they are new when reprocessed.
- **Proactive Maintenance Solution Newsletter says:** Oil does not wear out; it simply becomes contaminated due to inadequate filtration, that to the point that it needs to be drained. The actual base stock never wears out; only the oil additives become depleted due to contamination.
- **Australian Government – Department of Environment and Heritage**: Oil doesn't wear out; it just gets dirty. It can be cleaned, re-refined, and used again and again.

Figure 1.4: Does the Oil Really Wear Out?

If I explain lubrication in the simplest way I can, it's just a battle between the good and the bad. Simple as we may think but read again, here's the catch. Unknowingly, maintenance people become collaborators with these terrorists, I mean contaminants. It will start with how we stored our lubricants. New oil already contains a terrorist cell, but they are in a small group and cannot do anything. Still, since maintenance collaborates with the terrorist, we store them anywhere, mix them, over-lubricate, under-lubricate, and then oxidation happens. The terrorist cell becomes bigger and stronger, killing these agents one by one; that is why 95% of the time, the terrorist wins, and the agents lose and are killed. Then we change the oil, and the same thing repeatedly happens for the rest of our lives, but the truth of the matter is the base never wears out.

These are just a few of the many businesses globally that state that the oil can be re-refined and used again. Publication number (Patent Number)

- US4101414A: Re-refining of used motor oils
- US4512878A: Used oil re-refining
- EP1025189B1: Method of re-refining waste oil by distillation
- US5514272A: Process for re-refining used oil
- US3305478A: Process for re-refining used lubricating oils
- EP0618959A1: Process to re-refine used oils
- WO2002018523A1: Method for reclaiming used motor oil for further use
- US3819508A: Method of purifying lubricating oils
- FR2368534A1: Method for recovery of lubricating oils Holding
- US5922189A: Process to refine petroleum residues to other petroleum products
- US4502948A: Phillips Petroleum Company on Reclaiming used lubricating oil

Year **2**

<u>**2008 RSA Reliability Newsletter Vault Archive**</u>

> *People are not the company's greatest asset. The right people are the company's greatest asset, and the wrong people are called liabilities. We can only have the right people if they are equipped with the right knowledge and skills to do their jobs right the first time around.*

<u>**2.1: January 2008: Where Do We End Our Probe on Root Cause Analysis?**</u>

One of the biggest confusion in performing a thorough Root Cause Analysis is to understand how deep we should pursue our analysis or simply state where do we stop our investigation in performing a Root Cause Analysis? Going too deep will lead us to the bible, Timothy 6:10, for the love of money is a root of all evil, and going too shallow will allow the problem to recur again and again.

Although a lot of analytical tools are currently being used by industries in the market today. Are they really meant to uncover the root cause or simply what we termed as the physical cause of the problem only? One industry found out that John Doe was responsible for closing the valve that wrecked one of the turbines. John Doe was sentenced to be suspended from work for one month without pay. The lawyers, insurance people, and managers were happy that the culprit was finally condemned. A year later, the same problem occurred, and this time by a man named **Johnny Thorr**, and he quote, I just use to sweep dirt around here, but the supervisor instructed me to close the valve, and I don't know what valve so I close them all. The question being raised here is John Doe and Johnny Thorr, the Root Cause of the problem? Are we certain if John Doe is punished, the problem will be gone for good? Before we begin any further with our analysis, we need to ask ourselves, in the first place, why do we need to perform a Root Cause Analysis?

• Do we simply do it to comply with customer's requirements, such as manufacturing plants?
• Do we perform a Root Cause Analysis because our management wants to know what really happened?
• Do we need to perform a Root Cause Analysis because we want to know what simply caused the problem?

The reason for asking your industry is to perform a Root Cause Analysis for the right reasons, *and there is no better reason I can think of than learning from the failure itself. We need to learn from the things that go wrong.* Again, do we learn from failure when we discipline someone, or we just expand the gap between our people and us? I recall a couple of years back when I was teaching Root Cause Failure Analysis for two days in one plant when after the end of the first day, one of the participants approached me and said, Sir, I'm afraid I won't be able to attend the 2nd part of your training tomorrow and when I asked why? The person said that he will be serving his suspension starting tomorrow for a week's duration. My curiosity aroused me, so I asked the person who was a maintenance technician what happen, and he told me that the operator committed an error, and both of them will have to serve their suspension tomorrow. I asked why he was included if it was the operator who committed the mistake, and he said that it was the procedure management wants. And the worst part is that their names, mistakes, and even photos had been published on a bulletin board detailing their mistake committed for everyone to see. **VERY HUMILIATING INDEED!** This really irritates and pissed me off. I have met and experience many managers like this. They are a complete, you know what I mean. By the way, this maintenance technician had already been in that industry serving for 12 years, and it was his first taste of discipline in the number of years he worked for in that industry.

Boss: John, close the valve
John: OK boss
Boss: What have you done?
John: You didn't say which valve, so I
close them all.

And so John Doe closed the wrong valve and was punished for wrecking the turbine. Most organizations change the human condition but what they should be changing is the condition under which the people are working.

Figure 2.1: Blaming John Doe for Closing the Wrong Valve

Human Cause in Root Cause Analysis

Root Cause Analysis believes that all failures are caused by humans, but all humans are prone to commit mistakes or errors. Either someone did the wrong job or simply did the job wrong. People commit slips and lapses. Believe me when I tell you that people are likely to commit errors even with the industry's best procedures, and the smartest person can actually perform the worst error.

- December 12, 2002: A small plane belonging to the Philippine Air Force crashed into a manufacturing plant run by IBIDEN Philippines, killing one person and injuring eight people.
- In one of the hospitals here in the Philippines, one patient switches beds with another patient. The patient who was sleeping on the opposite bed was supposed to be given a blood transfusion. Since the nurse did not check the record of the person who was sleeping, the wrong blood was given, causing the death of the person who switched beds. Investigation indicates that the person who died switched to another person since he wanted to be near a window. The nurse thought that this was the person that needs to have the blood transfusion.

People commit errors. Human errors can be classified as slips and lapses. Slip is when somebody does something incorrectly, such as an electrician rewinds a motor incorrectly that it runs backward. A lapse is trying to miss out on a step in a key sequence of events, such as leaving a tool behind after an extensive and exhaustive Preventive Maintenance in equipment. Both are human errors, and both are unintentional in the first place. Many factors can be attributed to causing these kinds of errors, including fatigue, pressure, environment, inattention to details, just to name a few, or the person was simply daydreaming.

We must understand that most human error is not necessarily the fault of the person who actually committed the mistake or error. We need to understand that either the error was caused by external circumstances far beyond our control or by flawed rules and systems that need to be changed. We also need to understand why the error was committed in the first place to learn from them. In fact, are we convinced that if we were in the shoes of the person who committed the error, will we do things differently? But the most important lesson to ask is, do we really learn from the failure itself by blaming and punishing people?

<u>Latent Cause in Root Cause Analysis</u>

Underneath every Human Cause lies a deeper cause, which is called the Latent Cause. These are concealed and hidden causes that eventually cause the human error to be committed. The only way to address these Latent Causes is to expose them, and we can only expose them if we truly understand what it is all about. Latent Cause Analysis is not just about system flaws and procedures that eventually led a person to commit the mistake. It is not just about organizational management weaknesses. It is not all about flawed management decisions, but rather Latent Causes mean understanding that we are also part of the problem. People create the system, people make decisions, and if the decision or system is flawed, then a human error is most likely to occur without a doubt. If we just go after the person, then sad to say, we never did learn from the things that go wrong.

Learning from the things that go wrong is not as easy as we think it is. The only way to learn from the things that go wrong is to see ourselves in the mirror and admit that we are all part of the problem. Collectively, we all have our share that eventually caused a person to commit the mistake. The sad thing about this is when this happens, we isolate the person who commits the mistake and put all fingers and blame it on him because we always wanted a fall guy.

John 8:2—11 And the scribes and Pharisees brought unto Him a woman taken in adultery; and when they had set her in the midst, they say unto him, Master, this woman was taken in adultery, in the very act. Now Moses in the law commanded us, that such should be stoned: but what sayest Thou?" "This they said, tempting him that they might have to accuse him." "But Jesus stooped down and with his finger wrote on the ground, as though he heard them not. So when they continued asking him, he lifted up himself and said unto them. ___He that is without sin among you, let him first cast a stone at her.___

Before we can understand Latency, we need to ask ourselves the following:
• What is it about the way we are that contributed to the problems?
• What is it about the way I am that contributed to the problems?

If we are part of the problem, we must be responsible for being part of the solution. Engineers, technical people, maintenance can easily arrive at the physical cause of the problem. Suppose some mechanical component such as a bearing failed; these people could dig up evidence that will eventually lead to their physical cause. Still, digging deeper into the latent cause, I strongly recommend a third party, an independent person, or someone who is unbiased, unless you can carry out the probe yourself.

Example of Physical Cause: Evidence on the bearing's raceway shows fatigue and spalling. After an investigation, the team found out that the motor and pump misalignment seems to be the problem's physical cause.

Example of Human Cause: After studying the shreds of evidence that lead to this misalignment problem, the team found out that the person performing the alignment does not possess the skills and does not have the tools to perform such practices. He was only using his eyesight.

Probe on Latency: Investigation shows no instrument (Laser Alignment) and no training on why this equipment needs to be aligned. We asked why the management refused to allocate the budget for training due to some ridiculous cost-cutting scheme. And when we probe on with the Latent Causes, we understand that each of us is part of the problem.

• **Training Department:** After this incident occurred, we at the training department realized that a course on misalignment should be provided for our maintenance people to understand.
• **Maintenance Department:** It is normal for management to slash the budget and reduce operating costs, but this instrument is what we need to justified improving our equipment uptime. We requisitioned this instrument a year ago, but it was denied by Management. I believe we can justify this instrument if we really want to by performing a cost study in the first place and presenting an ROI to management; I'm pretty sure that it won't be denied.
• **Purchasing Department:** We always adhere to management to cut costs. To the extent that we change vendors on cheaper parts, I think part of our fault is not to consult this with our technical people for evaluation purposes.
• **Top Management:** We should have provided the budget for training and instrument. After this severe downtime, we learned to realize that this instrument maintenance is asking is not a nice to have, but a must-have after all.

And again, tell me if we are all part of the problem, or is the person who performs the misalignment be condemned "**alone**" in the first place? Let us all be in the shoes of this person who made a mistake and ask ourselves, without any instruments and training, are we certain that we have done it differently?

Latency is not about system causes but rather understanding how we contribute to the problem since we are taught that we serve in the company's best interest. I consider latency to be a higher form of human cause. Again, exposing these Latent Causes is the only way to understand a true and meaningful Root Cause or Root Cause Failure Analysis. Hence, to answer the question, where do we end our probe on Root Cause Analysis? The answer is only when we reach the Latent Cause of the problem.

2.2: February 2008: An Inconvenient Truth about Preventive Maintenance

Most industries today have a form of Preventive Maintenance being performed in a routine and scheduled fashion, yet despite the very best and noble efforts performed by our maintenance craftspeople, reactive maintenance is still what most industries are experiencing today. This leads us to the question, Can Preventive Maintenance really capture failures, or does Preventive Maintenance cause these failures?

First, let us define what Preventive Maintenance is. Preventive Maintenance is basic maintenance performed on the equipment and facilities. The main goal of performing the task on a scheduled basis is to extend the equipment's life and assure its capacity to support the plant's goals and targets. It is also a series of tasks performed at a defined frequency dictated by the passage of time, running hours, or volume, which extends the asset's life or detects that an asset is on its critical wear curve and is in the process of failing. Preventive Maintenance activities include regular routine checking, inspection, routine cleaning, lubrication, planned component replacement, minor and extensive overhauls.

Problems with Preventive Maintenance: Maintenance people believe that all part's after consistently use will reach a point of wear and tear, hence, overhauling or replacing the part before it fails on a specific fixed schedule will ensure the reliability of the equipment; therefore, the concept of Preventive Maintenance will help solve the problem of an unexpected failure. If you agree with this statement, then this is where you got it all wrong. Let me explain.

• Here is a quote from John Moubrey in his book on RCMII on page 143: It is also borne out by the machine operator who said that every time maintenance works on it over the weekend, it takes up to Wednesday to get it going again.

• Manager: Hey, what's the problem with your equipment?
• Operator: I don't know, boss, it's been acting crazy since I started it this morning
• Manager: Did maintenance perform some PM on it?
• Operator: I think so, boss, there's a stamp on the equipment
• Boss: No wonder it's running crazy; they touch it!

Isn't it if we performed Preventive Maintenance in our equipment, then it should be the other way around? Are we missing something here? These things simply are called Infant Mortality Failures, and many factors contribute to them. Infant Mortality Failures are most likely caused by human errors committed during performing scheduled overhauling or PM on a piece of equipment.

Most maintenance people believe that the more often an item is overhauled, the less likely it is to fail. In reality, this is just the opposite. The truth is that scheduled overhauls or Preventive Maintenance may increase failures by introducing what we call Infant Mortality Failures into otherwise stable systems. This means that if we are **"not"** equipped with the correct knowledge and skill to disassemble and assemble systems in our equipment or lack some training or tools to perform the Preventive Maintenance tasks, the best thing to do is to just leave it alone. Don't disturb stable systems, **period**, unless we are really equipped with the tools, skills, and knowledge to do it.

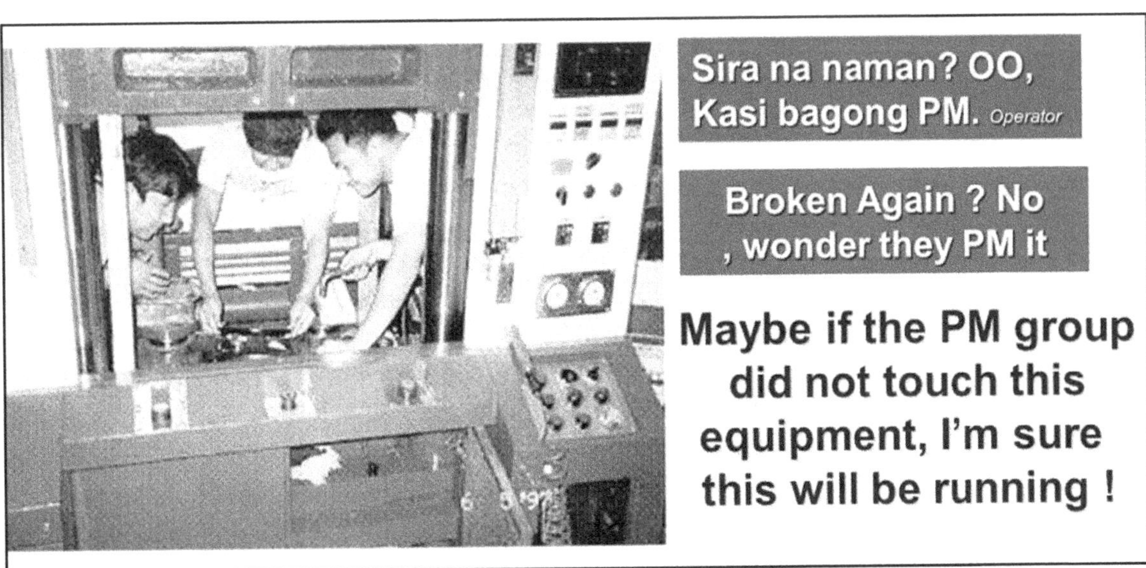

Figure 2.2: Are You a Victim of Infant Mortality Failure?

Another problem with Preventive Maintenance is that most maintenance assumes that similar assets will require similar maintenance activities without considering that each equipment has its own unique operating context. Hence, Preventive Maintenance is used for all similar assets without considering that different consequences may apply in different operating contexts. This results in many Preventive Maintenance schedules that are wasted not because they are wrong in the technical sense, but in reality, they achieve **nothing** but an increase in maintenance costs.

I used to provide this sample to explain the last paragraph so an ordinary layman can grasp it. I used to ask if they have a bulb in their home they use in the evening, perhaps around 30 watts, and most of them would say yes. When I ask them if they perform some inspection on the bulb daily, if it is working or not, then my audience simply laughs and says, **"of course not."** They will say that when it gets busted, then they just replaced it with a new one. When I told them that I also have the same bulb in my home, and we used to inspect this bulb daily, not once, not twice but 6x a day. When I asked them if I was stupid or something for doing this, their reply was a resounding **yes**! Then I told the delegates of my class that if I can prove beyond a reasonable doubt that inspecting the bulb 6x a day is right, then you need to pay me an extra 100 USD, and the delegates will disagree. Of course, I would not do that, and I was just kidding, but not for the bulb.

I tasted my first work in 1986 as an oiler or helper of a shipping company, and I used to be a crew for a cargo ship named MV Sea Raider. It was my home where I considered the crew as my family and the ship captain as our father. My duty was from 4 am to 8 am and again from 4 pm to 8 pm in the afternoon. There were six shifts where we require working two shifts a day, totaling 8 hours. I belong to the engine room and inside the engine room were 2 generators and two 750 hp Diesel Engines, which means that this cargo ship has 2 propellers. On the wall were 6 bulbs, 3 bulbs located on the port (left-side), 3 bulbs at the starboard side (right-side) of the wall, and when our shift came to go to the engine room. The first thing we do is get the logbook and check if these bulbs were working or not; the bulbs are color-coded green, yellow, and red. If the green bulb is on, then the ship simply moved forward at a designated speed, but if the green bulb blinks continuously, then the captain or officers on deck wants us to increase the rpm and speed of the ship, and we comply. If the yellow light blinks, we simply need to stay alert as either the green or red bulb will be lit afterward. When the red bulb lights up, we need to stop the engine; if the two red bulbs from both port and starboard side light up, both engines should be stopped. But if the red bulb blinks continuously, then the command was to have the engines in full reverse. You need to perform this without wasting any precious second; otherwise, the ship might hit something that can cause all the crew's lives on board. Remember that ships have no breaks, and you need to reverse the engines to stop a ship. After giving this story, I asked the delegates again if I am stupid or something for inspecting the bulb 6x a day, and they answer no. Why? Because the consequence of failure on the red bulb greatly differs from the failure of the bulb you have at home. But these are the same parts. Imagine failing to inspect the red bulb, and it got busted; what do you think will happen to us. Well, I'm still here today writing down this newsletter because we did inspect the bulb religiously 6x a day. This same bulb where we just color-coded it is the same bulb you have in your home. This simply explains what operating context is all about.

Let's take a sample of these three identical pumps. One is a stand-alone pump, while the other pump has a standby pump connected in parallel with the duty pump. Suppose we are to derive the most suitable tasks for each pump. In that case, I say that for Pump A, the most feasible task is to Predict or use Preventive Maintenance since if this pump fails, it will affect operations since this is a stand-alone pump, while we can run or better create a switching pattern Pump B and C since if it fails operations won't be affected since we have a back-up pump. Pump C, a standby pump, can be run once in a while to check if it is still running. This simply explains that the conditions in which equipment is being operated should be considered in determining the most feasible maintenance tasks and that similar equipment might require different tasks depending on its operating context.

Another problem encountered in Preventive Maintenance is replacing perfectly good parts with new ones since it is assumed that the part will eventually fail. And most of the time, the assumption is wrong. This is what makes Preventive Maintenance costly. Preventive Maintenance does not guarantee that the parts to be replaced really need to be replaced. If the equipment is in good condition, it will be a good thing not to touch it and just perform the basics such as keeping the equipment clean, checking lubrication, and routine inspection.

Lessons from Nowlan, Heap, and Matteson about Reliability: The message of Stanley Nowlan and Howard Heap dates back to the late '60s, which started in the airline industry which states that first, scheduled maintenance has little or no effect on the reliability of a complex item unless the item has a dominant failure mode, and second, there are many parts, items, components for which there is no effective form of scheduled or Preventive Maintenance. Their discovery led to the existence of the six failure patterns, which I believe every single maintenance should understand.

These 6 failure patterns simply tell us how a part will eventually fail after all. A study by Nowlan and Heap from the aviation industry reveals that 11% of an aircraft's components conform to patterns A to C, in which the rate of wear is directly proportional to its age. But around 89% of components will fail randomly with some presence of infant mortality failure. This means that if the part's failure pattern conforms to either pattern A, B, or C, then this is where Preventive Maintenance or Scheduled Maintenance is feasible to use. But when the part or item's failure conforms to either pattern D, E, or F, then this is where Preventive Maintenance will not be feasible to use. It is not recommended to use Preventive Maintenance on these failure patterns, but rather other feasible options that can be used for this failure pattern includes a run to fail when consequences are low, Predictive Maintenance when there is evidence of a potential failure, or simply redesign the system if both runs to fail and Predictive Maintenance is not feasible. The 6 failure patterns simply tell us that not all parts will fail or eventually wear out.

Hence, to answer the question, why don't Preventive Maintenance reduce Reactive Maintenance being performed in your plant? It is because we simply abuse the use of our Preventive Maintenance system. My experience tells me that our PM checklists' activities continue to grow by the day due to deviations from audits, customers, quality control, or simply important people in your plant who visited your operations. We must understand that the lesser the PM is performed on your equipment, the better your equipment will run. Intrusive or forced maintenance will only create havoc in our operations. What is important is that our people must understand and realize the lessons from Nowlan and Heap that in the real world of doing maintenance, there are not one, two but 6 types of failure patterns, and each part, spare or item will fail according in any of the six failure pattern. This means that not all parts will eventually wear out; since most parts will fail randomly. Therefore, if we are encountering random failures, PM is not the best option to use. Remember that Preventive Maintenance was designed around the theory that equipment failures are directly related to the equipment's age. In reality, only around 20% of equipment failures fit this pattern, and that 80% of equipment failures are not being effectively managed by doing time-based or scheduled Preventive Maintenance Activities.

Conclusion about Preventive Maintenance

It is important to understand that Preventive Maintenance has its own unique limitations, and understanding when to use and not to use it will derive the most benefit from its application. When applied correctly, PM will increase equipment's uptime; however, if we abuse, misuse or overuse will cause us more harm by being reactive instead of being proactive in our assets. Preventive Maintenance is feasible when the part or component will wear out directly concerning the age of the part, and its usage will definitely survive this defined age.

We need to resound the learning's from Nowlan and Heap, which they discovered in the 1960s since it will create a lasting impact if we really absorb and apply the learnings. The lessons they provided us are of great value for every person responsible for taking care of their equipment and assets. There is no substitute for training since training is where we acquire this knowledge and that whatever learning we gain should be shared among our people.

48 years had passed, yet most industries I've been through have really not absorbed the learning. What Stanley Nowlan and Howard Heap had discovered is not about rocket science or some silver bullet solution on maintenance but rather understanding that Preventive Maintenance has its own limitations and must not take this for granted.

I hope that this newsletter edition will serve as an eye-opener to reflect on it for a while and think that the first step in improving reliability is to acquire the correct knowledge on how to maintain our equipment correctly. Preventive Maintenance will always have a place in the reliability world if we understand how and when to use them correctly.

2.3: March 2008: The Lifeblood of Root Cause Failure Analysis

There are two mistakes industries make in performing a Root Cause Analysis. First, people performing Root Cause Analysis think that the most important factor in doing a Root Cause Analysis is to have sufficient data; without data, it is seemingly impossible to carry out such an analysis. Second, most industries, especially in manufacturing, think they can finish their Root Cause Analysis in 24 or 48 hours. If you seem trapped in these predicaments, then hear me out first and **read on**.

Those who rely on data miss out on the most important aspect of any Root Cause Failure Analysis Investigation. Hence, if data is not the most important factor in performing a Root Cause Analysis, what is it? The answer is **"evidence."** Uncovering the evidence is the most important factor in any Root Cause Analysis investigation. In fact, it can be said that evidence is the **lifeblood** of any Root Cause Analysis; without evidence, it is entirely impossible to carry on such a probe.

First, let us define what evidence is all about. Evidence is a piece of information that supports a conclusion. Wikipedia defines evidence in its broadest sense to include anything used to determine or demonstrate the truth of an assertion. Philosophically, evidence can include propositions presumed to be truly used to support other propositions that are presumed to be falsifiable. The term has specialized meanings concerning specific fields, such as policy, scientific research, criminal investigations, and legal discourse. Without evidence present in our RCA efforts, all we have will be inconclusive and biased opinions, hypotheses, gossip, and arguments.

Failures do not just happen without leaving a trace or clue. Failures will always leave some sort of clue for the investigator or analyst to uncover. Root Cause Analysis separates the facts from fiction. RCA is not about trial and error and seeing which one works and not. Although there are many techniques in analyzing a problem that provides a quick answer. It does not mean the answer is correct every time. A true and meaningful Root Cause Analysis will definitely take the time to prove that what we have uncovered will be based on facts and that every single fact we find will be based on evidence. When the facts are finally backed up by evidence, then we now have a crystal clear understanding and appreciation of what Root Cause is all about.

Case Study: Evidence of Missing Money: A Car wash manufacturer sold one of his turn key car wash systems to a client in Maryland. This includes a change machine for people who wish to wash their cars. After quite some time, the owner realized that he was losing a significant amount of money from this machine each week and insinuates that the manufacturer's employees have a spare key and are stealing the money. The problem started when the new franchise owner complained to Bill, the car wash owner, that he lost a significant amount of money from this coin machine every time he opened it. Bill cannot believe that his employees are stealing the money since they have worked with him for a long time. So the owner decided to install a hidden camera and found out who the real culprit was, and definitely, it was not the employees who were stealing the money. The owner followed where the bird went and found around USD 4,000.00 in the car wash roof and more under a nearby tree. The new owner apologized to the people he used to blame for stealing the money. Evidence from the surveillance camera clearly shows that not only one bird but also several of them are getting money from this car wash system's coin machine. The only problem that was not solved here is why the birds are stealing the money and what the birds want to buy with the stolen money.

• **But the RCA group had not given up and try to monitor their surveillance camera and this is what they found out . . .**

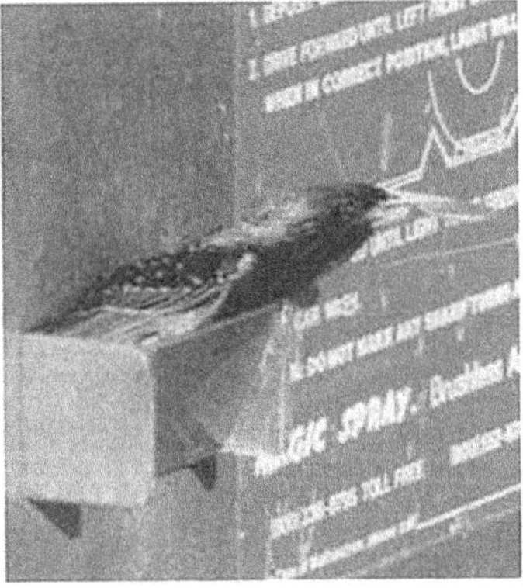

That's a bird sitting on the change slot of the machine and it had to go down into the machine but why ?

That's 3 quarters he has in his beak, another amazing thing is that it was not just one bird but several of them

There goes another bird this time taking only 1 quarter

• *Once they identify the thieves, they found over $ 4,000.00 in the roof of the car wash and more under a nearby tree, therefore, the case of the stolen money was solved thanks to Root Cause Analysis . . .*

Figure 2.3: RCFA Case Study on the Missing Money

Types of Evidence

A friend of mine, Charles Robert Nelms, founder of Failsafe Network, one of the Root Cause Analysis leading providers in the US, classifies evidence into three categories. He named it the 3P's, which are People, Physical, and Paper evidence. This evidence is arranged according to its importance.

People Evidence: The most important evidence to consider is people. People who witness the failure should be interviewed immediately after the failure. People have senses; they see things, hear them, and feel them. They all have opinions as to what went wrong in the first place. People also tend to forget things easily; that's why interviewing people should be prioritized after a failure had occurred. Remember that people's evidence is the easiest to evaporate. What is important is to interview and ask them the right questions, such as:

• What did you hear, or how did it sound like?
• What happens first, and then what?
• What were you doing at the time when the failure occurred?
• Were there others with you that witness the failure?
• Was there any signs before the failure occurred?
• What do you think went wrong with it?
• How do you feel about your job?
• What do you think people working here feels about their jobs?

Interview as many people as you can, a minimum of 5 to 10 people, and spent an hour with each person. Have a set of questions on hand and write down their response. Tape recorders can only be used with the person you will interview with the consent and approval of the person you are interviewing. The interview should be done as close as possible to the interviewee's event during the failure if needed only. After interviewing several people, you will be quite surprised by the information you have gathered since a sequence of events will emerge from the interview. Be compassionate and friendly when interviewing people; the more relaxed the person is during the interview, the more information can be gathered. Tell them that no one will be punished as a result of the interview, and the reason for conducting this interview is to know the truth as to why the failure occurred in the first place. Another thing is to remain focused and unbiased during the interview. Remember that there are always two sides to a coin.

<u>Physical Evidence:</u> Physical evidence is the second most important evidence of all. It can conceivably include all or part of any object. This includes pictures and photographs taken, actual part or component that failed, sample or debris that scattered, video or surveillance camera during the time of failure, laboratory results of the parts being analyzed. For example, the actual bearing that failed will be part of the physical evidence. One of the important things to consider in physical evidence will be to preserve them. When equipment fails, maintenance try to restore them at once, for operations to continue and downtime to be on a minimum

On the other hand, it is also important for the team to perform the analysis and investigation to freeze up the evidence as quickly as possible to uncover the cause of the problem while still fresh. The restoration team must understand that parts that failed should not be thrown away and should preserve every time equipment fails; otherwise, the parts' physical evidence will be impossible for the investigative team. There will always be traces of evidence left on the part that failed.

Positional evidence, likewise, is included in the Physical Evidence. This includes mapping out the position during the time of failure. What are the gauges' readings during the time of failure, temperature, what valves are open or close, where the leak seems to be coming out, what buttons are pushed, etc.? Map everything, if possible. As the restoration team tries to bring back the equipment up, most of the positional evidence is lost. It is important for the investigative team to extract the evidence on the equipment before restoring them back to normal again. As much as possible, try to draw and record every piece of information you can when uncovering positional evidence. Try to analyze this figure with Detective Varnike and solve the missing puzzle.

Detective Varnike was playing golf in a hotel garden when a shot was suddenly heard in one of the rooms. The detective appeared near the first floor of the hotel he was booked. He was a few steps from where he heard the shot and slightly pushed and opened the door and squeezed himself into the room. Help, someone who wanted to kill me! Cried addressing Varnike, a well-known movie star who was in the hotel's room. There was a man in a mask here. I was attacked by him. I protected myself as I could, then he shot. Probably, during the struggle, I had snatched the gun out from his hands. After that, he rushed to the door and disappeared into a corridor. Please, call the police! After Detective Varnike observed the room, he said, maybe, I will invite the correspondents sitting in the hallway and call the media people right now but not the police. Detective Varnike said, there was no one here except you. I believe you did all this yourself. Probably, they are only what you wish, perhaps to gain some attention from the public. The police hardly should be disturbed because of such worthless performance!

Figure 2.4: Why did Detective Varnike refuse to pursue the criminal?

Question: Why did Detective Varnike refuse to pursue the criminal? It is because Detective Varnike saw something on the floor that proved beyond reasonable doubt that no one entered the room. What evidence in the above convinced the detective that no one attacked the actress and that the mess was all her doing? (Answer will be provided in the next edition of this newsletter.)

Paper Evidence: When we speak about paper evidence, all recorded documents that we can gather, such as history records, recent PM records, strip charts, manuals, operators' logbooks, training records, specs and procedures used, and condition-monitoring records, equipment drawings, and schematics. Paper evidence should be summarized by the team or person doing the probe or investigation. The investigator should collect all paper evidence as much as possible for a thorough review of the process. In summary, the paper evidence we should look for will include operations, maintenance, and personnel-related data. What is important for the investigator is to determine which data are relevant and important at the time of failure.

I know that we have many chronic failures in nature, and up to now, continue to exist in our plant. We must understand that the most important factor in performing a Root Cause Analysis is to gather evidence to truly understand, once and for all, why the failure occurs in the first place. We need to become one with failure and understand them. Most people think that failure is bad for business; failure is telling us something in the first place. Failure is the best teacher we can ever have, as we can all learn from them.

Before they become successful, they failed miserably

Thomas Alva Edison

Thomas Edison boyhood teacher told him that he was too stupid to learn anything. He was known today as one of the worlds greatest inventors

Elvis Presley

King of Rock and Roll was fired after just one show at the Grand Ole Opry and was told "that you ain't going nowhere son"

Michael Jordan

Before joining the NBA Jordan was just an ordinary person that he was cut from high school basketball team because of his lack of skills

Albert Einstein

His parents thought he was Mentally retarded. His grades at school were poor that his teacher ask him to quit and said that he will never amount to anything

Soichiro Honda

Honda was turned down by Toyota during a job interview as an engineer during WWII and again rejected by Toyota as a supplier of piston rings

The Wright Brothers

They say only birds can fly, heavier than air flying machines are impossible to fly according to Lord Kelvin 1895

Bill Gates

Before starting Microsoft he was a dropout at Harvard University and started his software by purchasing it from someone at $ 50.00

Jackie Chan

When Bruce Lee died, he along others were picked up to fill the vacuum and failed miserably, so instead of being him (Bruce Lee), he decided to be himself

Figure 2.5: People that Failed Miserably before they become Successful

- *People see my success as only one percent, but what they don't see the 99 percent, which is my failure. From Soichiro Honda, President of Honda*
- Failure is only the opportunity to begin again more intelligently. *From Henry Ford*
- I have not failed; I've just found 10,000 ways that don't work. *From Thomas Edison*

Conclusion: 80% of the time spent on Root Cause Analysis must be based on evidence gathering, including the 3P's of evidence. Root Cause is not about quick solutions and countermeasures but rather understanding why the problem occurred to learn from them and understand how and why it occurred. We need to dig up evidence. I really find it very difficult to conduct a Root Cause Analysis in just 24 hours or a couple of days and arrived with the problem's Root Cause. Having worked in various industries such as shipping, assembly lines, manufacturing, and mining industries, and with all the training I've been through in all these years on problem-solving tools, it allows me to think that Root Cause is all about solutions and countermeasures. I am definitely dead wrong. We all have a human intuition that when we encounter a failure, we want a solution fast to get rid of the failure, and most of us have the temptation to modify without understanding the Root Cause of the problem.

2.4: April 2008: My TPM Experience, A Successful Failure

Answer to March 2008 Puzzle: The **brush.** If the actress really claims that an assailant was in her room and went out of the door, then the brush will not be in its original place since it will also be moved as the door was swayed. It was impossible for the brush not to move if someone goes out of the door. In Root Cause, this is known as positional evidence.

• *As Tom Hanks said in his role in the movie "Apollo 13", our mission was a successful failure, successful that we return safely to the earth but failure since we never landed on the moon.*

I think and feel the same way; spending seven years in TPM was quite frustrating yet rewarding. This was where I truly learn what TPM is all about. It is all about people and not machines. The real challenge is making it a way of life for all people and a part of their culture. Because of the sensitivity of this newsletter, I'll just call this company Bravo. In 1995, I worked in Company Bravo as a production supervisor for 3 months; until I was transferred to one department, we called the TPM Office. It was a newly established department in the organization. Initially, there were six of us, including the TPM Manager, her secretary, and four engineers. The three engineers were assigned on the following pillars, one engineer handling Autonomous Maintenance, another handling Office and Administrative TPM, the other engineer handling Planned Maintenance. Since I was new at this plant, I was assigned to be an assistant to all of them, Rolly, can you Xerox these 10 copies, please? Can you call all of these people for a meeting and all sorts of this and that?

Since we were ignorant of what TPM is all about and how to go about each step, my TPM Boss decided to purchase several books on TPM, divided them among the engineers, and spent our days reading. In the afternoon, there was always a brainstorming session that lasted for hours on how to go about each of the TPM Pillars. I was always listening to them and sometimes was called upon to ask my perspective; I think that is where I was allowed to speak up.

Six months had passed, and a sad thing happened. The Planned Maintenance engineer died of cancer, and so a replacement was to be in place, until one morning, my TPM Boss called me to his office and told me, Rolly, I want you to handle Planned Maintenance; there is no other person I can think off and see fit to handle this than you. You will handle Planned Maintenance, and these are the people in the maintenance department you need to see. My expression was overwhelming; wow, simply, wow. That was my reaction. That was the only motivation I got from my boss, yet I can remember his words vividly in my mind until now. As we grow old, there are things worth remembering, even if we forget many things in our lifetime.

And so this is how I started in TPM. Since changes in this organization were abrupt, I recall serving with at least 5 TPM bosses. They were replaced all the time, especially when they are not performing well. My last boss was formerly the training department manager and later was transferred to the TPM Office. Later on, our TPM efforts were focused on two plants; the other plant was 3 to 4 times larger regarding the number of employees. The TPM office was now serving at least around 8,000 people from the two plants. Two people were added to the office to take care of it. Our TPM Office staffs were now 8, including the secretary. This time, I was also asked to handle the EEM or Early Equipment Management pillar of TPM and partly Focused Improvement.

TPM days were tough; you need to push people and remind them "**always**" of the journey to excellence. We worked along with the other pillars, such as Autonomous Maintenance and Focused Improvement, and, later on, Early Equipment Management. Our mission was how to get our pillars moving and get the people on doing TPM. Our CEO was committed to TPM and appreciated the reports and progress the different TPM pillars were providing.

Through the years in our Planned Maintenance journey, we held yearly strategic planning. We have recognitions and rewards for teams that performed well. We even have in-house symposiums exclusively for the maintenance people for sharing their improvement, which was in the form of competition. It was really a lot of work for the TPM staff, but the rewards and fruits are beyond expectation; it was such a success.

Morale was very high, especially with the Planned Maintenance teams. In fact, all members meet weekly to discuss each and everyone's progress, which they report to me, which I summarized and consolidated. We have the highest attendance record compared to the rest of the TPM Pillars at 92%. Each member knows what he will be missing if he was not present. Each of the Planned Maintenance Committee members knows that it's a mortal sin to be absent from the Planned Maintenance meeting. No one was late during our meetings; because of one simple rule we imposed, the last person to enter the room will have to pay for all the bills on the snack. Each of these people represents their department. Most of them were senior engineers; others are managers. The Planned Maintenance meetings opened doors and communication lines for each other. During the meeting, there were several times where they share their people's resources since they are not fully loaded in their operations and even spares. We learn several spare parts that are the same in specs through these meetings, used by the same equipment but have a different part number since we have much common equipment in every operations department. The bottom line is that we learn so much by keeping our communication lines open; I guess this was one reason why they attend the meetings plus the food, of course.

By the year 2000, we decided to challenge the JIPM Awards for TPM Excellence Awards, 2nd category, and our dream to be the 2nd industry from the Philippines to achieve it. We hire a JIPM consultant and an interpreter since the consultant can hardly speak English. The JIPM Consultant comes in every quarter, spending around two days of online tours and details on what we need to accomplish, which he wanted to see on his next visit.

There was so much pressure on the TPM Office since everyone is doing TPM; I thought previously that the pressure is when the people are stubborn in doing TPM. Each of us (8,000) people in the plant from the three shifts knows that we were aiming for the TPM Awards. At the TPM Office, we were so busy, especially with departments that were not up to speed in their roadmap. Our CEO was excited and even called us a couple of times at the TPM lobby where all the awards, trophies, and recognition of this plant are displayed and show us a spot where he wanted the JIPM TPM Award trophy to be placed and ask us when he can see the award? I think being in my shoes, you can feel the excitement and pressure. Everyone was doing their part on TPM. It was a dream come true; we have change people. Scores of improvements were evident not only on my pillar but on each TPM Pillar; well, of course, our Planned Maintenance pillar garnered the most number of improvements in four years, with a total of 9,272 improvements and a cost savings of Php 5,787,369.72, not to mention the dramatic reduction in our repairs and maintenance costs which still excludes the savings generated by the Predictive Maintenance group had initiated. MTBF improve, and the number of breakdowns reduced dramatically. This was our actual graph where BDO stands for Breakdown Occurrences or Number of breakdowns on their equipment.)

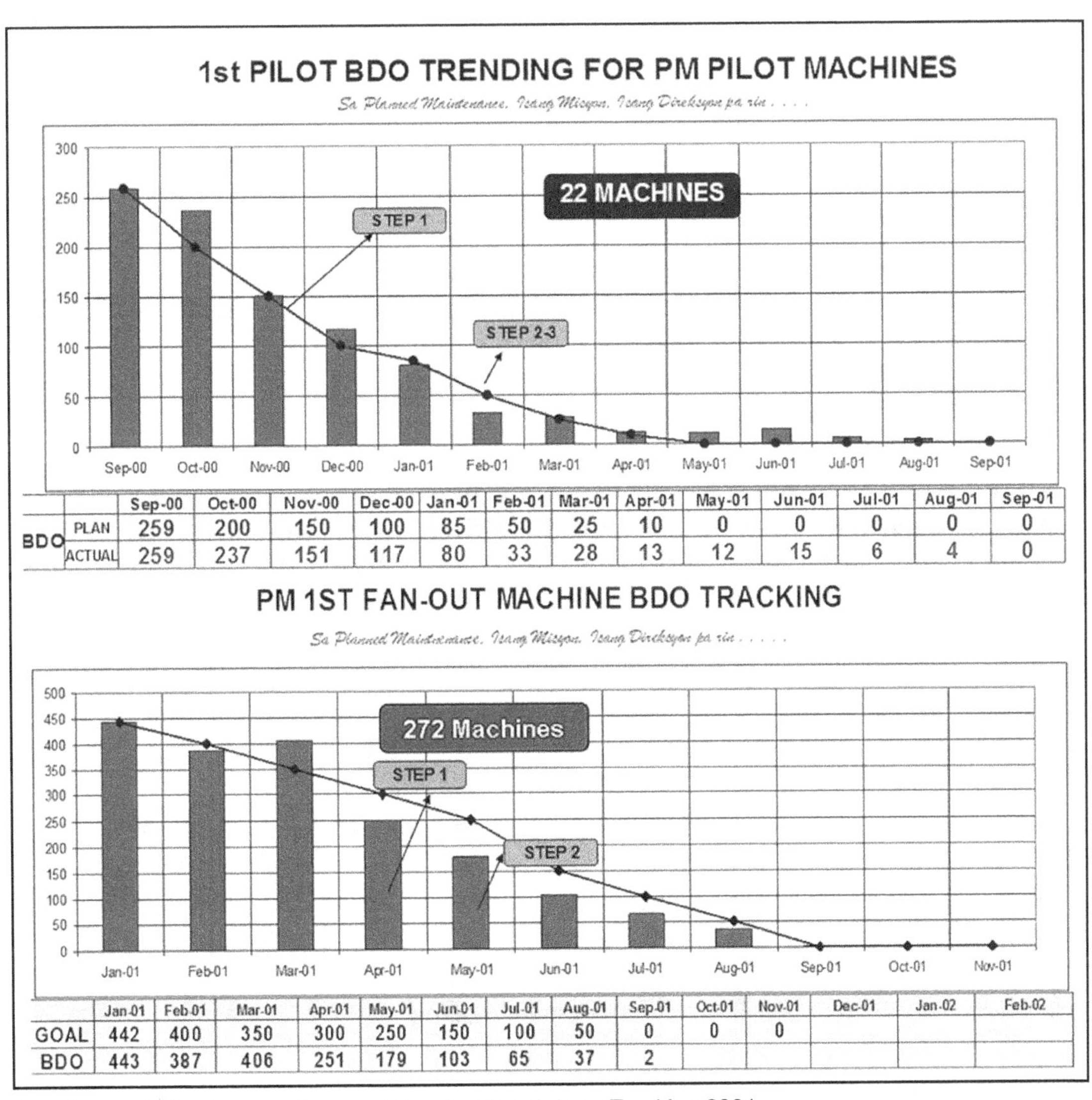

BDO	Sep-00	Oct-00	Nov-00	Dec-00	Jan-01	Feb-01	Mar-01	Apr-01	May-01	Jun-01	Jul-01	Aug-01	Sep-01
PLAN	259	200	150	100	85	50	25	10	0	0	0	0	0
ACTUAL	259	237	151	117	80	33	28	13	12	15	6	4	0

	Jan-01	Feb-01	Mar-01	Apr-01	May-01	Jun-01	Jul-01	Aug-01	Sep-01	Oct-01	Nov-01	Dec-01	Jan-02	Feb-02
GOAL	442	400	350	300	250	150	100	50	0	0	0			
BDO	443	387	406	251	179	103	65	37	2					

Figure 2.6: Planned Maintenance Pilot and Fan-Out Breakdown Tracking 2001

Our consultant's last visit was in 2001, as far as I can recall. It was the 2nd quarter of 2001. At the end of every visit, he would meet with all TPM Pillars and members in a big room to assess his results. He would talk about his findings on every pillar, and when it was time to discuss the Planned Maintenance pillar, he simply put his two hands and made a sign of two thumbs up. He said Planned Maintenance, very good. I believe that was his message. His last words, as translated by the interpreter Company Bravo now ready for JIPM Awards; I go to Japan now and tell JIPM, you are ready. I will report this to my office, and it is time for you to make the JIPM Book. There was a round of applause from all of us. It was a relief for us all at the TPM Office; after all the preparations, meetings, training, and hustle, we were ready for the awards. Our JIPM Consultant was confident that we will pass the initial assessment since two assessments are needed to achieve the most coveted TPM Excellence Awards. And that was the last time I saw the consultant. I told the TPM consultant that someday, I want to be like him; he said, you are good, Rollysan.

One day, my TPM colleagues said, have you heard the news? I said, what news? Our CEO is being replaced by a new one. I thought it was not such a big deal and work as usual, but I guess I was dead wrong. One day, our boss told us that our senior VP, which was the second in highest command, wanted to talk to us and asked us all in the TPM office to come to his conference room, and so we did. He was frank and straight to the point, and I honestly do not want to accept what I was hearing. He said that TPM Office will be abolished. Our new president does not believe in TPM. You are given two options, resign or transfer to another department. I simply stood up and said, Sir, we are just a few miles from the finish line in getting the awards; let's go for it. He looked at me and said, I know, but I am just following orders. After the meeting, six transferred to other departments, two resign, myself included. There were tears in my eyes when I left that room, yet I said if company bravo does not want TPM, then I will look for a plant that wanted TPM, simply stated, I can't let go of the feeling since I've seen people change because of TPM.

News easily spread that the plant will no longer be pursuing TPM and that the TPM Office was to be abolished. My phone kept ringing, verifying if the news was true and that my members are asking me if we were still going to do Planned Maintenance; I simply said, the big guys wanted us to drop it, so let's just drop it. It was like the World Trade Center collapsing in the blink of an eye, and then suddenly, everything was gone. I feel like Nicolas Cage trapped beneath the rubbles of the World Trade Center. TPM was my life, and it collapses right before my very eyes.   I believe that the new president believes more in Lean Organization and not TPM.

To cut the story short, I officially resign from this company in the 4rth quarter of 2001. We were given three months free where we don't have to report to work anymore and still received our benefits and salary. In 2002, I ended up in a mining firm as a training specialist, where I develop and study further reliability best practices, and certainly, Lean Manufacturing is not one of them. I resigned from this Mining industry in 2005 and decided to carry on my own consulting business, which, up to now, is still what I am currently doing.

Last February of 2005, I was acquired by company bravo and provided a contract to teach reliability courses and improvement initiatives, and guess what? One of the courses they want me to teach is TPM and the pillars. I had learned that they finally realized their mistake and looking forward to accepting TPM once again. But they have to start from the beginning; this was their setback. I also learn that only 1 person from the original TPM Office was still at this plant, while the rest had resigned and are now working in other industries. I was asked a couple of times if I wanted to have my old job back at TPM, and I said much willingly "yes," but only on one condition, I will only work with my old team, if you can bring them back, then I will come back.

My experience on doing TPM is worth sharing. We were successful. As proven in our records, we have achieved dramatic improvements in our breakdown reduction, increased MTBF, and reduced maintenance costs. I see people change because of the results they achieve. Those who were pessimistic about TPM were the ones in the frontline and even preaching TPM with their people. TPM changes attitudes; it changes lives. Our Planned Maintenance team moved mountains.

I failed because we did not achieve the JIPM Awards for reasons way beyond the scope of my responsibility during those times. Yet, in my heart and mind, there will always be a place for TPM. My experience with TPM is that it was simply a case of a successful failure, and I think that's all I have to say about that.

2.5: May 2008: Top Ten Problems Experienced on Preventive Maintenance

A survey was conducted among the people who attended my World Class Maintenance Management training to survey the top problems and issues maintenance people are currently experiencing in their Preventive Maintenance. To be exact, 132 people took the survey in different batches. The instructions were to select at least 3 problems and issues they experienced with their current Preventive Maintenance based on the lists provided below. The results of the survey follow;

• Add on PM checklists syndrome - where PM checklists and activities seem to grow
• Infant mortality failures - problems arise after a PM replacement and overhaul
• Replacement of good parts to conform to PM specs and procedures
• Random failures - where random failures are included in the PM checklists
• Aging workforce - nearing retirement
• Lack of training on the maintenance function
• Still reactive and a lot of corrective maintenance even with a sound PM program
• Frequent reorganization in the plant - where the new boss makes a new system
• Lack of or poor documentation in Preventive Maintenance
• PM is waived. Operations won't give equipment for the PM to cope with production

(43 Votes): Lack of or poor documentation in PM. This tops the list. What is important in PM documentation is recording what is relevant and not. PM makes decisions from these documents, and if these documents are flawed, incomplete, or half-baked, then our decision on PM will be affected likewise. History records, an interview from operators, measurement readings, and troubleshooting guides are just some of the relevant information that must be documented. Some industries have CMMS or Computerized Maintenance Management Software to ease up with the tasks of documentation. What is important is that before acquiring such software, maintenance must understand what information most relevant to them is needed by their people.

(41 Votes)Maintenance is still reactive with a lot of corrective maintenance even if a sound Preventive Maintenance program is in place. So, despite the very best and noble efforts and activities industries spend on Preventive Maintenance, why is there still many firefighting and breakdown maintenance performed? The starting point in developing an effective Preventive Maintenance strategy is to first understand the different failure patterns. Failures are diversified in nature. Preventive Maintenance overhauls and replacement are only applicable if the parts will wear out directly concerning the age or period used, and the part will definitely survive this period. Most Preventive Maintenance people are misguided by this principle. When the failure is not age-related or random, stop using PM because Preventive Maintenance will not address random failures.

(39 Votes): Introduction of Infant Mortality Failures. This simply explains that scheduled overhauls can actually increase the chances of overall failures by introducing infant mortality failures into otherwise stable systems. Infant Mortality Failures are usually experienced right after a maintenance intervention or during the commissioning of the equipment. Infant Mortality Failure exists, and being aware of its existence is the first step in reducing its likelihood. If our PM team is not equipped with the proper knowledge, tools, and skills to perform a replacement or major overhaul, better to think twice before trying to dismantle your equipment just to comply with your PM activities.

(37 Votes): Lack of training on the maintenance function. Are maintenance deprived of training, or do they have no time to attend to it? Has Training Needs Analysis been conducted on the maintenance function to determine their needs regularly or not? Training plays a crucial role in any reliability improvement initiative. This is where maintenance acquires knowledge to understand their equipment better. Technology is moving at a fast rate and rapidly changing. It is traveling at the speed of light, but have we changed the way we do maintenance on our equipment or not? After so many decades of frustration from a firefighting mode, it is sad to note that many are still on this type of maintenance mode. Thanks to the caveman who invented fire. Maintenance people need to be trained, and management must not deprive their people of this privilege.

(35 Votes): Add on PM Checklists Syndrome - PM checklist's activities seem to grow. Remember when you were just new with your industry or when the equipment has just been newly commissioned in your plant? The vendors left you a list of what to do with your equipment, from the simple task of inspecting and replacing some parts in the equipment likely to wear out. However, through the years, new activities have been added to the growing list of your PM checklists, and today you have a long list of activities to perform, and the list continues to grow minute-by-minute, day-by-day. Is this a familiar issue with your Preventive Maintenance? This list simply comes from daily audits conducted on your plant. When maintenance fails to catch a failure, maintenance is highlighted for this issue. The corrective actions will simply add something to the never-ending and growing list of PM activities. When the boss said, "I don't want this thing to happen again," the default will simply add things to their PM checklists. The Golden Rule on Preventive Maintenance is basic and quite simple. Our belief that the more Preventive Maintenance we perform on our equipment, the less likely it is to fail no longer holds valid. The airline industry was the first to discover this truth more than 50 years ago. However, sad to note that their study has not reached many of us in maintenance for land industries because we are too busy putting out the fire in the first place.

(33 Votes): Random Failures are included in the PM checklists. Again, random failures are failures, which can occur at any given period. There is little or no relationship between how long the component or equipment has been in the service and the likelihood of failure. A suitable replacement period is not feasible at this stage simply because the failure is indeed random in nature. Electronic, electrical parts fail randomly. Hence, if the failure is random, there is no scheduled Preventive Maintenance to prevent this type of failure. Maintenance tasks for random failures can either be Predictive Maintenance if the component provides a potential failure or provides a sign or signal that it is on the verge of failing. Other options are to allow it to fail or a run-to-fail situation if the failure has minimal consequences. The only consequences will be the direct cost of repair or using Root Cause Failure Analysis to address why these parts and components keep on failing.

(28 Votes): Replacement of good parts to conform to PM specs. One of the most controversial questions raised with Preventive Maintenance is to determine whether the part we are going to replace is actually on the verge of failing or not. The assumption is yes, but most of the time, the assumption is wrong. The part's lifespan has not been reached because we replace it at an early stage. When we replace good parts that are still capable of running, then we waste money. When the part or component has no problem, then we don't need to replace them. Do we? We do not want to take away Preventive Maintenance credibility here. That is not my intention. The maintenance and management team must understand the diversity and the different patterns of failure to confidently decide on their activities. There is always a thin line in doing PM. Remember, it is not their fault that they are deprived of training, as seen in this survey. All I am saying is that allowing maintenance people to understand these things so they can transition once and for all and improve the way we perform maintenance on their equipment and assets.

(22 Votes): Ageing maintenance workforce, nearing retirement. When a maintenance person retires, his experience goes with him to the grave. What is important is that these people should be grouped together with newbies in the plant. Transfer of knowledge is essential for the PM to work effectively. This aging workforce probably knows better and has experienced what works and what does not work for them. Try to document what they know before they leave the plant and retire. Examples of this are repair and troubleshooting guides based on how they perform them on their equipment since they are the most experienced people in the plant. Most mistakes of industries are not documenting in the process while these good people are still around in their plants.

(20 Votes): PM is waived; operations will not give the equipment for PM. There are several cases when production waived PM to cope with the production demand. Feud exists between these two groups. From a production point of view, PM is waived because operations lagged behind on productivity. They need to catch up on this, while, from the maintenance point of view, production should expect to have more failures in their equipment when PM is waived. When failures occur, these are when confrontation and pressure seem to build upon both of them. Remember that your equipment is always a shared responsibility for both operations and maintenance working together towards a common goal. If the demand for production is high and PM will likely take around 6 hours or more on the equipment, have your Predictive Maintenance group intervened. Let them check what is likely to fail so that the PM can focus on those critical parts of the equipment. Secondly, if your industry does not have a Predictive Maintenance group, schedule the PM into intervals of one or two hours per day and prioritize the work that needs to be performed.

(17 Votes): Frequent reorganization in the plant where systems always change. There are plants and industries where reorganization is very dynamic and rampant. Whenever a new PM boss takes place, a new systems eventually unfold. The duration of the new system lasts until the new boss is around. When a PM system that has been developed is effective, then don't change it. Explain this to the new boss on why this system is being used in the first place and what does it address. Focus on changing systems that are less effective in your current Preventive Maintenance program.

2.6: June 2008: Comparing RCM and TPM, which is the Best lb. for lb. Strategy?

To compare these two World Class Improvement Strategies, I just can't help thinking how these two well-known improvements will fair in a boxing match if they would fairly square it out evenly. Also, not to mention that I'm still a bit overwhelmed by the victory of the Philippines' very own Manny "Pacman" Pacquiao over his latest opponent David Diaz.

TPM has its origin from the east (Japan), while RCM originated from the west (United Airlines), where its origin can be traced to the works of Stanley Nowlan and Howard Heap based on a study they had generated from the airline industry and later on adopted by the late John Moubray for use in the Nuclear Facilities which later on spread to industries. I hope the reader doesn't mind me writing this newsletter in this fashion. So let me hand you over to Michael Buffer for the introduction.

Ladies and gentlemen in attendance and to the millions of fans watching. Let's Get Ready To Rumble! Fighting for 12 rounds for the Undisputed Heavyweight Improvement Strategy of the World. In the red corner from my left, hailing from the eastern part of the world, Seiichii Nakajimaaaaaaaaaaaa (audience applauding), and his opponent from the blue corner hailing from the western part in the UK the great legend John Moubrayyyyyyyyyyyyyyyy (audience applauds).

Round 1: How it's being approach and which is easier to implement: One of the main differences in approach regarding these two improvement initiatives is that TPM is always a Top-Down Approach and a plant-wide improvement initiative, while RCM can be done from the ranks or Bottom-Up Approach. Without management commitment, it is very difficult for TPM to succeed. When we say top-down, it means that it should start from the highest-ranking person in the plant to announce that TPM will be implemented effective this date. On the other hand, RCM is much easier to implement than TPM. I give this round to RCM. (RCM 10 points, TPM 9 points)

Round 2: Primary measure of performance: Although many indices will be improved if any of these two initiatives will be successfully implemented. The main indices used to measure TPM performance is OEE or Overall Equipment Effectiveness. OEE is computed by multiplying the equipment's Availability by its Performance Rate and Quality Rate. OEE addresses the six major equipment losses encountered on the equipment. This will mainly include breakdowns, set-up, conversion, start-up losses, idling and minor stoppages, design speed loss, defect and rework losses. For RCM, the main measure of performance used is MTBF, mostly related to equipment failures. Although Moubray has some issues, he wrote about OEE, which states that:

- *There is often a tendency to focus too heavily on primary functions when assessing maintenance effectiveness. This is a mistake because trivial secondary functions embody bigger threats to the organization if they fail than secondary functions. As a result, every function must be considered when setting up maintenance effectiveness and measures. The OEE, as defined above, only relates to the primary functions of the asset.*

What Moubray meant is that even if your OEE is high (85% for World Class Companies), and as long as availability, performance rate, and quality rate are good, it does not guarantee you that the equipment is indeed reliable most especially if there are secondary functions that are in a failed state such as protective devices. But comparing these two primary measurements, OEE is still wider in scale than MTBF since failures and breakdowns are just a subset of OEE. I hope I am not confusing the readers as there are so many indices that can impact a plant when these two methodologies are successfully implemented; I am just referring to the two's main indices. Given this, I give this round to TPM. (RCM 9 points, TPM 10 points)

Round 3: What RCM and TPM believe: TPM believes that basic equipment conditions must be carried out first before advancing to any improvement initiative. This basic equipment condition includes cleaning, proper lubrication, bolting, and equipment free of leaks. Why don't you complete all the screws and bolts of your equipment before performing any vibration analysis on it? On the other hand, RCM states that the first step is to change the way people think and apply this change thought to their assets. I like what John Moubray says because even before basic equipment conditions should be established, both operators and maintenance should change how they think about their assets if they want their equipment to improve. I give RCM more credit to this round. (RCM 10 point, TPM 9 points)

Round 4: Goal on maintenance: TPM's goal is to zero out unplanned breakdowns. RCM's goal is not about eliminating breakdown but understanding that each failure has its own unique set of consequences. If the consequences of the failure are not acceptable, all efforts must be exhausted to reduce the likelihood of failure. Maintenance is not about zeroing out or eliminating failures but simply trying to understand that it is more important to know the consequences of failure than entirely eliminate them. In reality, it is impossible to eliminate failures. What maintenance can do is anticipate, prevent, predict, control, manage, or prolong the failure's duration. RCM's explanation here is much deeper than TPM; hence, I give this round to RCM. (RCM 10 points, TPM 9 points)

Round 5: Operators' involvement in maintenance: One of the similarities between these two improvement initiatives is that RCM and TPM value operators' importance in maintenance. In any RCM analysis, operators must be part of the team doing the analysis since it is the operators and not maintenance who will directly experience the failure. TPM is much broader in scope as it composes of 8 pillars and one of these pillars is Autonomous Maintenance, where operators learn to understand that their equipment is both a shared responsibility for operators and maintenance. Autonomous Maintenance is performed in 7 steps excluding Step 0. It has a detailed structure. Maintenance teaches operators about their equipment, and operators learn to operate and take good care of their equipment. TPM's role is to change the mindset that operators are not just here to operate while the maintaining function performs repairs on their equipment. In Autonomous Maintenance, the operators understand that they have a role in maintaining their equipment by performing the basics such as cleaning the equipment, monitoring its lubrication, and checking for any loose and incomplete bolts. In short, the role of operators is much more detailed and structured in TPM. Hence, I give this round to TPM. (RCM 9 points, TPM 10 points)

Round 6: Its belief about continuous improvement: I believe that this is where both TPM and RCM have some form of contradiction. TPM is heavy on improvements. It believes that the equipment should be continuously improved, while, on the other hand, RCM focused more on understanding the consequences of each failure that can impact the equipment with the aid of a decision diagram or algorithm. The team selects the most appropriate tasks for each failure mode based on its consequences. In fact, John Moubray, in his book RCMII on page 188, considers that maintenance should be first before redesign or modification. He stated that most organizations are faced with many more apparently desirable design improvement opportunities than are physically or economically feasible. Most improvements take time to accomplish. Some may even take several months to complete. Hence, a person on duty today must maintain the equipment as it exists today and not what it should be there in the future. Before considering improving and redesigning the equipment, have we asked ourselves if the asset is here to stay there for a long time, or will it be decommissioned soon? I have seen many improvements being wasted on their equipment because the people who performed the improvement did not know that their equipment will be decommissioned. Hence, before improving or redesigning the equipment, we must have a concrete answer to these questions: If you answer yes to all these questions, then redesign is feasible.

• Does the failure involve any major operational consequences?
• Is the cost of scheduled breakdown maintenance high?
• Are there specific costs that can be eliminated by the design change?
• Does the improvement have no harmful effects after the design change?
• Is the asset to stay for a long time and will not be decommissioned out soon?

I believe this credit belongs to RCM. I give RCM this round. (RCM 10 points, TPM 9 points)

Round 7: RCM and TPM initial approach: RCM starts by determining the asset's functions and failure modes. On the other end, TPM believes that the best way to start is to first address the equipment's basic condition. Many failures start from small things and these small things that are often left neglected cause the bigger problems. TPM believes that big failures can be prevented if we address the basics and small problems. Some failure modes can be reduced if the equipment's basic equipment condition is well established. I give TPM credit for this round.

(RCM 9 points, TPM 10 points)

Round 8: Plant involvement: As its journey continues, TPM aims to involve everyone from the organization, from the lowest to even the CEO. They will each have their TPM pillar to be busy with. People from the offices such as HRD, Finance, Accounting, and Administration will be part of Administrative or Office TPM, aiming to improve their systems and conduct focused administrative improvement. During our TPM days, when we audit our HRD department, where we asked for a 201 document file of a particular employee, we time them using a stopwatch on how fast they can retrieve the dossier. Have you gone to your admin department lately to ask for some files and tell me to come back in a couple of days?

On the contrary, RCM involvement would be limited to operators, maintenance, engineers, and from time to time, vendors mostly that have a direct impact on the equipment being analyzed. I doubt if HRD can be involved in the RCM analysis. Again, as for plant involvement, I give this round to TPM. At this point in the round, both RCM and TPM are dead, even with 76 points. Staggering left hook by TPM, RCM is stunned (RCM 9 points, TPM 10 points).

Round 9: Focus on maintenance: Although I like to limit this discussion to the equipment side, TPM will address the 6 big equipment losses. By reducing these equipment losses, OEE will improve. On the other hand, RCM focus on both the primary and secondary functions of the equipment. RCM highlights the importance of the equipment's secondary functions and allows us to understand cases where secondary functions' failure poses a bigger threat than the failure of a primary function. Imagine multiple failures occurring because the protective device, a secondary function, is in a failed state. I think RCM and TPM have good points on this, and I consider this round to be a draw. (RCM 10 points, TPM 10 points)

Round 10: Applicability on the industry: Speaking about its applicability to industries, I believe that RCM has an edge on this as its process can be applied and can suit almost any type of industry from the Nuclear, Oil, and Gas, Mining, Metals, Airline Industry where it originated, Shipping, Power plants, Automotive, or manufacturing as long as there are equipment and assets to maintain. TPM application is mostly suited to manufacturing industries; I just won't recommend TPM and doing initial cleaning for a sub-station. Hence, for industrial applications, I think RCM is more diversified than TPM. I'll give this round to RCM (RCM 10 points, TPM 9 points)

Round 11: Flexibility of combining TPM and RCM: If we buy a pizza, TPM itself is the whole pizza, while RCM is just a mere slice. I see RCM fit perfectly into the higher phases of Planned Maintenance, which is one of the main pillars of TPM. In fact, several TPM case studies include doing RCM on their equipment. According to the author of RCM, John Moubray, no mention of alignment with other improvement initiatives was mentioned in his book. He also provides a standard SAE JA1011 to differentiate his classical RCM from those of streamlined RCM versions. Because of this, I think TPM is more flexible than RCM. Therefore, I give this round to TPM (RCM 9 point, TPM 10 points)

Round 12: What RCM and TPM want to achieve in maintenance? TPM aims for Maintenance Prevention, while RCM aims for Proactive Maintenance. TPM's aim is more idealistic since, in reality, we really cannot eliminate maintenance. Our equipment is not a plug-

and-play asset where we expect it to operate smoothly at all times by doing nothing. TPM wants us to think about an ideal plant with zero breakdowns, accidents, defects, and what it would take to bring our company closer and closer to it. An ideal plant had been chosen by TPM so that its people will continuously improve. TPM does not want us to remain in the status quo, but to improve gradually step by step, drip by drip. People believe that TPM is a continuous journey, and if they have one word that is out of their vocabulary, it is the word "BEST" because they believe that the best can still be improved.

On the contrary, RCM aims to be Proactive, finding the most suitable maintenance tasks for each failure mode uncovered. Doing RCM increases the asset's integrity by considering both environmental and safety implications before considering the effects on operations. RCM believes that maintenance is about understanding the consequences of failure rather than eliminating the failure itself. RCM is much more realistic because it allows us to understand what maintenance can and cannot do.

I think that both RCM and TPM clearly explain what they want to achieve in maintenance. Both initiatives understand that the key to all of this is their people. If people change, their assets will be improved. Hence, I consider this round to be a draw. What a match. Honestly, I believe in these two methodologies. We really cannot conclude which reliability strategy has an edge until we detail it through the rounds. Both have their strengths and weaknesses as well. The results of this comparison are purely based upon my experience and knowledge with these two powerhouse improvement strategies and might differ from other opinions. I hope that you enjoy reading this reliability newsletter. Therefore, to finally tally up the score, back to Michael Buffer, judge Rolly Angeles, score the bout TPM 115 and RCM 115. The match is a draw, and the people wild on their feet as they applaud the two great warriors.

ROUNDS	1	2	3	4	5	6	7	8	9	10	11	12	TOTAL
RCM	10	9	10	10	9	10	9	9	10	10	9	10	115
TPM	9	10	9	9	10	9	10	10	10	9	10	10	115

Figure 2.7: Comparing RCM and TPM, Which is the Best Pound for Pound Improvement Strategy

2.7: July 2008: Why Most Root Cause Initiatives Fail

Many industries have some form of Root Cause Analysis or problem-solving tools to analyze problems, yet many seem misguided about what root cause analysis really is all about. I admit that I had only known its true meaning when my good old friend Charles Robert Nelms invited me to attend his Latent Cause Analysis training in March 2007. In this issue of my reliability newsletter, I would like to provide some shed of light on what Root Cause Analysis is all about and why most industries fail in their Root Cause initiatives.

Wrong Belief: To perform Root Cause, we must have DATA: Data, data, data; without data, we can't perform a thorough Root Cause Analysis. I heard this message a lot, and I just cannot emphasize how wrong they are. The lifeblood of any Root Cause Analysis is "**EVIDENCE**" and not data. Data is just a part of the evidence. When something goes wrong, there is a cause and effect behind every bit of the problem. To determine the cause, bits of puzzles are gathered to determine what the big picture really is. Data is just part of the evidence, but we can understand what caused the problem and what Root Cause is all about when you tie things together. To understand the cause of failure is to envision an **ICEBERG**. Failures are evident, but underneath, the failure remains hidden. Our only link on what is underneath the failure will be the evidence.

Figure 2.8: The Fast and the Slow Team in RCFA

Most of the Evidence is being washed up by the fast team: When an equipment failure occurs, two teams are going to the equipment, the fast team and the slow team. The fast team refers to the restoration team, which will restore the equipment so that operations can be up and running again. They must be fast, and spares must be at hand. The clock is ticking. They will be timed about how fast they can repair the equipment; otherwise, the longer they take to repair, the more shadows (people watching them) will be at their back dictating you what to do until you really get pissed off. There is always pressure on the fast team. Since they have been doing this often, their MTTF (Mean Time to Repair) had improved by 30% for the last 5 years, and they are keen on reducing their repair time always. The longer the downtime, the bigger the revenue that will be lost. They are considered heroes all the time because without them, then operations are delayed. Does this sound familiar?

On the other hand, we have a slow team. This will be the team that will analyze the failures. The problem is that when the slow team arrives at the scene of the failure, the equipment is already running. Most of the evidence had already been washed up and evaporated since the machine had already been repaired. So how in the world can we perform a Root Cause Analysis in this situation? Tell me then.

Many will disagree with me, but I live by my principles and what I believe is right. MTTF or

Mean Time to Repair can never be a measure of Reliability; when people become too good at repairing, it only means one thing that the failure keeps on repeating itself. Rather than repair the same failure repeatedly, I would rather take my time, understand, and learn from this failure. Before fixing the failure, my recommendation is simple, have the restoration team take photographs of the equipment, inside and out, and a close-up photograph of the affected part or component of the equipment. Secure the part or component that failed. Place them on a plastic bag and any foreign parts or debris that they can locate near the part that failed. Endorse them to the slow team, which is the team that will investigate and analyze the problem. Make this a habit before repairing the equipment!

How deep should we probe Root Cause Analysis: A true and meaningful Root Cause Analysis always believes that all failures have a physical cause. Still, all physical causes are triggered by humans. This means that there is always a human error involved. Still, humans are negatively influenced by latent forces. Therefore the goal of any root cause analysis is to expose and identify these latent causes of failures. The depth of probe on any Root Cause Analysis efforts should end once the Latent Causes is reached.

During one of my Lubrication and Oil Contamination Control Class training, a manager hands me an oil analysis report on one of their compressors. When I scanned through the report, I was totally shocked that I almost slipped the report out of my hands. I told the person if their compressor failed frequently. He said, not this time. I said that I do not believe you unless otherwise, you are giving me the wrong report. He said that their air compressor fails a lot previously when they were using mineral-based oil. They talked to the vendor about it. The vendor recommended to them to use synthetic oil. After using synthetic oil, the failure seemed to disappear like magic. My question was the root cause of the problem determined? I believe no.

You see, the oil analysis report indicates a moisture content of around 7000 ppm. The standard sample use for any oil analysis is a 100 ml bottle of the sample. The allowable moisture content for oil is around 100 ppm. When the amount of moisture is around 700 ppm, the oil needs to be changed. Still, moisture reading is 10 times the warning limit for moisture, and the oil seems capable of handling it. Why? Because the oil is synthetic, and this is what it is designed for. Synthetic oil can withstand the heat, but you must also be capable of withstanding the cost of synthetic oil. When I ask how much the synthetic oil cost, he told me it is around Php 120,000.00 (around 2,667 USD) per barrel, but the mineral or petroleum oil they use before cause around 15,000 Php (333 USD). Again we are comparing 1 barrel against one drum of mineral oil. There are 55 gallons in a drum and 42 gallons in a barrel.

Surely, when we probe our Root Cause analysis, it will have something to do with the oil that causes the compressor to fail. As we said, all failures are triggered by humans. When we visited the oil storage room, the lids were either missing or open on this oil and the other drums. Perhaps the maintenance did not close the lid when he went to get some oil in the barrel. Again my second question is, do we blame the maintenance for not closing the container's lid that led to the compressor failure? NO!

But when we probe deeper into the Latent Cause of the problem, the drums and barrels were stocked out in the open, exposed to sun, dust, rain, and moisture. I bet that even if you have those lids covered, there will still be water present due to the condensation effects of moisture, most especially during a cold temperature. The drums are piled up on the wall, and every employee who enters the plant for work and leaves may have noticed this small problem that the lids of the oil container are missing. No one seems to care, or no one seems to understand its effect that causes the failure of the air compressor, which had paralyzed 1/2 of their plant a year back. Can we blame the employees or maintenance people? I don't think so since they were never trained on lubrication and contamination control. If you asked why the maintenance was not trained on lubrication, the company has no budget.

Nevertheless, they have the budget to repair the failure again and again. Latency lies in each one of us, yet sad to note that when the who or person in the root cause is known, most organizations end their probe, and the guilty one is punished. I cannot emphasize it more.

Root Cause takes much more than 24 hours to complete: Root Cause is not about some fancy software that provides you a list of failure modes to select where we can just click here and there. To undergo a Root Cause is a slow and painful process. This will require time to complete and for the people investigating to absorb the learning. We can only learn from failure if we acknowledge that we are also part of the problem. As part of my work, I teach industries on selecting courses on reliability and maintenance I offered. In one of my classes, a participant from Top Level Management approached me at the end of the day and said, Rolly, come with me. I will show you something, so I followed him to his office, and he handed me a report on Failure Analysis done by one of their clients, so I glimpse through the report. As usual, the analysis ends on the physical cause of the failure. After I placed the report on his desk, he handed me another report, the same failure. Although the difference was that this report was 3 years ago. It was an exact replica word for word. The report was just pasted so that the engineers have something to show; the only difference between the current and recent reports was the date of the report and the person who did the report. The dates of these two reports were three years apart.

You see, Root Cause takes more than 24 hours to undertake. In fact, depending on the magnitude and gravity of the problem, it will take several days or others, perhaps a week or more. This will take time, most especially in verifying the causes of the problem based on the evidence unfolded. If the evidence doesn't show up, then the people investigating the problem need to dig more evidence until they verify each cause. I came from manufacturing and spent quite several years working. My experience tells me that most industries must complete their analysis in 24 hours or even less. This is most especially true if their customer is the one that complained to them about a defective product. We want to impress them on how fast we can resolve and identify the cause of the problem, but even if you ask the Root Cause experts, it simply cannot be completed in just 24 hours. Again, Root Cause is a slow and painful process. We must bear with it to get to the truth that causes the problem to occur.

Root Cause is not about probabilities of failure and failure modes: There is somewhat confusion regarding what methods and tools would determine a Root Cause Analysis, and some are led to the belief that these tools will eventually lead them to the root cause of the problem.

Fault Tree Analysis was developed in the 1950s by BOEING Aerospace Engineer for use in the design process's development stages. It is a mathematical tool and yields probabilities. Its primary intent is to predict the probability of a specific failure. This is a design tool. *Source: What You Can Learn From the Things That Go Wrong by Bob Nelms*

Failure Mode and Effects Analysis discipline was developed in the United States Military. Procedure MIL-P-1629, titled Procedures for Performing a Failure Mode, Effects and Criticality Analysis, dated back November 9, 1949. It was used as a reliability evaluation technique to determine the effect of system and equipment failures. Failures were classified according to their impact on mission success, personnel, and equipment safety.

Pareto's 80/20 Rule: Quality Management pioneer, Dr. Joseph Juran, working in the US from 1930 to 1940, recognized a universal principle called the vital few and trivial many. He reduced it to writing. In an early work, a lack of precision on Juran's part made it appeared that he was applying Pareto's observations about economics to a broader work body. The name Pareto's Principle stuck, probably because it sounded better than Juran's Principle. As a result, Dr. Juran's observation of the "vital few and trivial many" is based on the principle that 20 percent of something is responsible for 80 percent of the results. Later on, it became popularly known as the Pareto's Principle or the 80/20 Rule.

Ishikawa Diagram or Fishbone Diagram, developed by Kauro Ishikawa in 1969. A fishbone is constructed and assigned 4M's and 1E. The m's stand for man, machine, method, and materials, while the e stands for the environment. Most of the probable causes related to the problem are listed accordingly. The team will brainstorm and focused on the most likely or probable cause of failure.

There is no disrespect to the people who develop these tools and those who use them; in fact, these are powerful problem-solving tools, but they are simply not designed to determine the Root Cause but rather the probable or the most likely cause of the failure. These two are entirely different, a Root Cause and a Probable cause. A Root Cause Analysis will always depend on the evidence found to determine the problem's precise cause. It is not about a selection of the trivial few, it is not about brainstorming, it is not about the probability of failure, it is not about experience, but rather it is all about **evidence** that we must unfold that will eventually lead us to the truth behind the cause of failure.

<u>Root Cause's focus should be on small problems and not big problems:</u> Big problems are caused by small problems. These small problems accumulate over time and cause us chaos in our operation. Remember my example about the air compressor causing a certain portion of the plant to shut down. Lose its revenue because the lid on the oil container was missing allowing moisture and contamination to penetrate. A power plant was shut down for several days. Its triple redundancy protective device failed over time because the plant never inspects them and is confident that this would not cause them any problem because the protective device had two more backups. In a manufacturing plant, I previously worked, one of the main customers of this plant pulled out all its work because the products they shipped were off specs. It was traced later on to a tiny proximity sensor that was clogged due to dirt. This sensor's function is to determine whether the product is off specs or not. I can go on and on, but these things have one thing in common. These small things caused big problems to come up. Root Cause is reactive because we can only perform a root cause analysis if a failure occurred, but if people can learn from Root Cause, especially the little things that are frequently neglected, we can only understand the causes of these big problems. When people in the plant start to care, then this is what makes Root Cause Analysis proactive.

Root Cause is not about "WHO" caused the problem: Root Cause Analysis is not designed to determine who caused the problem but rather understand why and how it manifested itself. While it is true that there is always a human cause for every failure, there is always a hidden cause that caused this person to commit the mistake. In fact, according to Bob Nelms, the Golden Rule of any Root Cause Analysis is, we will try to understand to such an extent that we are convinced that we would have done the same thing if we were in that person's shoes. It must be clear to all people in the organization that no one should be punished or reprimanded for any root cause analysis investigation unless this is a clear case of sabotage. Our goal is to understand what led that person to commit that mistake to learn from it. This means that before we punish someone, let us ask ourselves, are we sure that we have done it differently if we were in their shoes? Blaming and finger-pointing have no room for any Root Cause Analysis Investigation. Suppose our intent for any Root Cause Analysis is to simply punish people; in that case, if this is the end goal, then our efforts will be entirely useless, and the industry will never learn from the things that go wrong in their industry.

2.8: August 2008: The Case of Infant Mortality and Random Failures Explained

Infant Mortality Failures can best be explained from page 143 on the book of John Moubray on Reliability-Centered Maintenance; he quotes:

• *It is also borne out by the machine operators who say that every time the maintenance works on it over the weekend, it takes up to Wednesday to get it going again.*

Have you been a victim of this phenomenon? Operators often say sarcastically that if you have not performed your regular overhauling and replacement on this piece of equipment, I bet it will be running without any problems. But isn't the main purpose of Scheduled or Preventive Maintenance is to ensure that the equipment performs as intended? But the opposite seems to happen whenever maintenance does something on the equipment. Is it profound? No! You are just a victim of what you called Infant Mortality Failures. In almost all cases, Infant Mortality Failures are caused by human errors and intrusive or forced maintenance. In fact, many things can go wrong when we try to dismantle equipment for overhauling purposes. Human errors such as slips and lapses can occur on the part of the maintenance performing the overhauls.

Infant Mortality Failures are failures that occur at the beginning of life; others refer to them as commissioning failures, start-up failures, or debugging failures. Many factors affect and caused Infant Mortality Failures, including poor equipment design, poor quality manufactured, incorrect installation, incorrect commissioning, incorrect operation, unnecessary maintenance, slip and lapses, human errors, or simply bad workmanship. The case of Infant Mortality Failure, Pattern F of the Six Failure Pattern, starts off with a high incidence of early failures, which eventually drops to a constant or very slow, increasing conditional probability of failure and ending in a no wear-out zone.

Studies done by Nowlan and Heap on civil aircraft showed that 68% of items that failed conformed to pattern F, which is the case of Infant Mortality Failures. An example of some of the benefits of this learning on infant mortality failures done by the civil aircraft industry was a dramatic reduction in their scheduled overhauls in their DC-8 aircraft from 339 items for an overhaul to only 7 items DC-10. One of the items no longer subject to overhaul was their turbine engines. Likewise, on the initial program developed for Boeing 747, it took United Airlines 66,000 man-hours on major structural inspections to reach an inspection interval of 20,000 hours. Compared to their traditional 4 million man-hours inspection on smaller and less complex aircraft such as DC-8. Truly, it had been a remarkable achievement and feat which saved them millions of dollars on maintenance without compromising its safety. But perhaps the greatest benefit of learning from Infant Mortality Failures is that it saves lives. Below is the actual record for the worldwide commercial fleet from 1959 to 1995, indicating the rate of accidents per million take-offs.

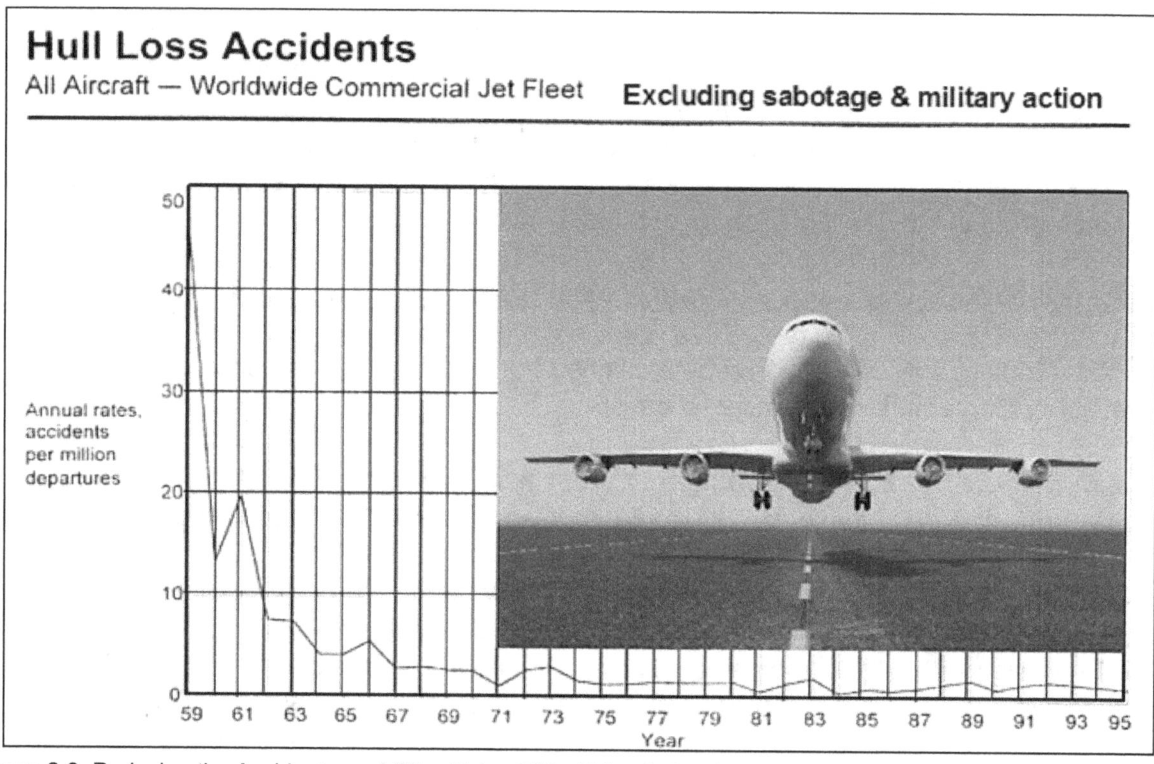

Figure 2.9: Reducing the Accidents per Million Take-Off in Airline Industries

Performing too much overhaul on our equipment to comply with PM specs and activities guarantees that equipment will be reliable in the future. It is similar to the logic that a fat child is not exactly a healthy child; healthy and fat are not synonymous. In fact, there are cases when a child is more prone to sickness when they are fat. Or eating plenty of food each meal is not an indication of a healthy person. Eating the right balance of food can make us healthy. A similar thing holds through with maintenance. The more activities we perform on our equipment, the more likely it will fail. I have seen many cases when equipment is subject to overhaul and returned to operations. The operators have a hard time running the equipment. There are also many cases where maintenance tends to forget some items that should be in the equipment, especially if the part is just a small item. Most people are shocked when I say that the More PM you perform on your equipment, the more problems you will encounter; the less PM, the fewer problems you will experience. This is where I end. I never said, No PM, No Problem. There will always be an amount of scheduled maintenance needed on the equipment. The key is to provide the correct amount of maintenance which is not more and not less.

If Infant Mortality Failure exists, the question raised, is can we eliminate them? Many factors contribute to infant mortality failure. It is difficult to eliminate them, but still possible, the best we can do is reduce their likelihood. Let us look at the following perspective:

Design Stage: No equipment is perfect by design; there is always a design weakness. Some parts fail prematurely during commissioning and even in actual operation. We need to identify these parts and conduct a physical investigation and Root Cause Analysis investigation as to why these parts failed prematurely. Only when we fully understand the cause of the failure can we recommend a redesign, such as changing the materials' shape, dimension, or strength.

Commissioning Stage: Problems occur during the commissioning of equipment in our plant. Either some human error was involved during set-up or start-up. Infant Mortality Failures are why vendors and manufacturers provide some form of warranty on their equipment, especially during commissioning and start-up activities.

Scheduled Overhauls and Replacement: When equipment is operational, infant mortality failures occur more frequently when maintenance intervenes to perform their time-based scheduled overhaul and replacement. There is always the assumption that when maintenance dismantles the equipment, they can put it back confidently together in the same condition. It is not the intent of this newsletter to discredit Preventive Maintenance, but rather to educate each one that there are many cases when performing overhauls on equipment can introduce infant mortality failures into otherwise stable systems. There are many ways to check the condition of the equipment without dismantling it.   These techniques are known as Condition-Based Maintenance. For example, instead of overhauling an engine, why not check its oil. Normally when the equipment is not in good shape and needs reconditioning or overhauling, it will be indicated in the oil in particulates, mostly metals. Metals present in the oil can indicate which part is actually on the verge of wearing out so that maintenance can focus greatly on which parts to replace.

With this in mind, we should rethink our strategy again when performing PM overhauls and replacement on a scheduled frequency because this is when the equipment is more vulnerable to infant mortality failures. Perhaps, we might not get a high score on our PM completion. For manufacturing plants, there might be possibilities that Quality Control will provide a non-conformance to the maintenance department because some activities on the PM lists were ignored. Suppose we intentionally missed PM due to the high incidence of Infant Mortality Failures in our plant; in that case, I believe our reasons are valid and that sometimes it is best for our equipment not to be disturbed.

Random failures are failures that occur at any given period. This means that the probability that an item will fail in any one period is the same as in any other given period. This means that the conditional probability of failure is constant. To further illustrate my point on random failures, imagine that you are driving your car, and a small piece of rock hit your windshield, which causes a very small fracture. My question is, when do you think it will happen again in the future? That is a difficult question in which the answer will always be a form of guess or probability. To know the occurrence of random failures is like catching a bolt of lightning with a Polaroid camera on that precise second it shows up. The example above shows a list of failure distributions of the same bearing that failed during operations. Every time this same bearing failed, maintenance will record the date when it failed. In this case, we are speaking of the same bearing. Our history showed that 5 bearings failed in the first year, 15 in the second year, etc. No bearings achieved a life of more than 9 years. With this distribution, my question is, when is the best period and time to replace this bearing? If you answer in the first year, our maintenance effort would be very costly since most bearings will last for more than 1 year. This means that we are replacing good parts or parts that are not yet on the verge of failing, then we are wasting money on maintenance. Suppose you answer that the best time to replace the bearing will be in the 8th year; in that case, we have reactive maintenance or run-to-fail situation in our operations. Perhaps this would be a good idea if the consequences of failure can be limited to the cost of repair or redundancy is in place. But if this is not the case, then a run-to-fail situation would not possibly be the best strategy. My point is no amount of scheduled maintenance can fully address a random failure. So if this is the case you're experiencing today in your equipment, the following are some of the options to address random failures:

Use of Predictive Maintenance Strategy: Predictive Maintenance is valid if the part whose failure is random poses a potential failure. A potential failure is actually a symptom or indication that a part is on the verge of failing, although it is not yet in its failed state or has not yet reached its functional failure. Sound, noise, excessive heat, vibration, increase in temperature are signs of Potential Failure that can be detected with an ailing bearing. Predictive Maintenance instruments can help us determine the potential failure of certain parts in our equipment, such as the bearing on this case. Once a potential failure is spotted, we need to determine the P-F Curve and perform some maintenance before reaching its functional failure.

Analyze Using Root Cause Failure Analysis: Understanding the causes of failures through Root Cause Failure Analysis. Failures happen for a reason, and there is always a cause and effect for each failure. Most people mistakenly jump to redesign or modification without understanding the cause behind the failure itself and realize that a new problem emerges caused by the redesign and modification. Performing a thorough investigation will provide us some shed of light in understanding why the failure occurred based on the evidence. Hence, when performing a Root Cause Failure Analysis, the probe's depth should always end up on the Latent Cause of Failure.

Run to Fail Strategy: This is feasible and applicable to use if the consequence of failure will be limited to the direct cause of failure and would not compromise safety, environment, and operations. This is also applicable if the system has some form of backup or redundancy, making the failure tolerable and safe, as in the Airline Industry's case. If something fails when the plane is airborne, I cannot provide you a rope to check the engine's condition or perform some sort of Predictive Maintenance. They allow failure to happen simply because they have redundancy, making failure tolerable and safe.

Hence, to conclude, for random failures, never use Preventive Maintenance or time-based strategy overhauls and replacements. This is where this strategy will not be feasible. Applicability of using Preventive Maintenance overhauls and replacement only applies to parts that conform to a wear-out pattern. Just remember, the more automated and complicated your equipment is, the more cases of Infant and random failures we will encounter. What is important is understanding which maintenance tasks are most feasible to use.

2.9: September 2008: How to Fit RCM into the TPM Strategy

First of all, I would like to dedicate this newsletter to all my maintenance brothers in India and International Business Conference to conduct this Reliability-Centered Maintenance training. TPM or Total Productive Maintenance had already made its mark in India, where many industries implement its philosophy and principles. There is quite a list of certified industries by JIPM (Japan Institute of Plant Maintenance), where a TPM Award (5 different types of awards to be exact) is achieved by industries who have matured, reaped the rewards and benefits of their TPM implementation in their plant. I also noted that they have their own TPM Club India, which I believe is a joint venture from JIPM, which promotes TPM and provides references, consultation, seminars, conferences, and research among TPM Industries to strengthen its ties with industries implementing TPM. In fact, TPM has a very strong foundation in India. It has helped many industries achieve a World Class level on how they do and operate their business.

Total Productive Maintenance (TPM) is strongly recommended for manufacturing industries, yet some oil and gas industries in India likewise implement TPM. I am curious about how they compute their OEE or Overall Equipment Effectiveness when considering the Quality Rate component since this can only be calculated if the industry produces products.

My first travel to India was in the year 2008. Reliability-Centered Maintenance (RCM) is a new concept in India. Some may have heard of it, yet others hardly knew what it's all about. Its concept and origin started from the airline industry and have a strong foundation in the western part. It is a strategy that highlights achieving equipment reliability by focusing on how we do maintenance on our equipment and assets. It is based on its original standard SAE JA1011. This standard answers the 7 Basic questions of RCM as follows:

• What are the functions and associated performance standards of the asset in its present operating context?
• In what ways does it fail to fulfill its functions?
• What causes each functional failure?
• What happens when each failure occurs?
• In what ways does each failure matters?
• What can be done to predict or prevent each failure?
• What should be done if a suitable proactive task cannot be found?

I have spent around eight years of my life implementing TPM as a senior TPM Engineer, so I guess I have to say that TPM will always have a place in my heart and mind. So the question to raise is, can they fit together, or will they blend together or not? To answer this question is to understand how both are taken up into a focal point of strategy. If a plant already has a strong TPM foundation, then RCM can fit into the system, which I will explain, but if a plant without TPM would want to initiate RCM, then for it to provide good results, both operations and maintenance should address the basic equipment condition which includes, correct lubrication, keeping the equipment clean, and equipment with complete bolts.

TPM is for everyone in the organization, while RCM is not design for everyone. In fact, only those from operations and maintenance with extensive knowledge of the asset they are operating should be part of the RCM team. If we speak about the military, these are the elite forces like Delta Force or Special Ops. TPM will always be the bigger program, and RCM will be part of the Planned Maintenance pillar of TPM, yet if RCM is implemented with TPM, other TPM pillars would benefit from it such as Environmental, Health, and Safety, Autonomous Maintenance, Focused Improvement for RCM decisions that will default to redesign or modifications, Initial Flow Control Activities or Early Equipment Management, but the biggest impact of dong RCM will be the TPM Pillar of Planned Maintenance.

How TPM Planned Maintenance Pillar is Implemented:

There is a wide range of how to effectively apply the Planned Maintenance system, just to give you a glimpse of how the TPM Planned Maintenance pillar is applied in industries worldwide. Planned Maintenance implementation based on my experience depends on the JIPM consultant industry hires. Somehow it will always follow a general pattern which includes, preparation, restoration, modification, and sustaining the restoration process. Here are some of the variations on how TPM Planned Maintenance is implemented by industries;

Reference: TPM Industries by Tokutaro Suzuki
- Step 1: Evaluate equipment and understand the current conditions
- Step 2: Restore deterioration and correct design weaknesses
- Step 3: Build an information management system
- Step 4: Build a periodic maintenance system
- Step 5: Build a predictive maintenance system
- Step 6: Evaluate the planned maintenance system

Reference: Nissan Motor Co., Ltd, Yokohama Plant by Yuichi Suzuki taken from TPM World Congress C4-2-8
- Phase 0: Establishment of the aims of planned maintenance
- Phase 1: Understanding the actual conditions of equipment and work
- Phase 2: Restoration of deterioration
- Phase 3: Recurrence prevention and countermeasure for weak points in the design
- Phase 4: Establishment of periodic maintenance
- Phase 5: Improvement of maintenance efficiency
- Phase 6: Horizontal replication
- Phase 7: Condition management of facilities

Reference: Idemetsu Kosan Co, Ltd. Hokkaido Refinery by: Akira Kitayamna
- Phase 1: Deterioration, repair, and elimination of forced deterioration
- Phase 2: Extension of characteristic life span
- Phase 3: Identify and repairing internal deterioration
- Phase 4: Pursue predictive maintenance and forecasting techniques

Reference: JIPM TPM Instructor's Course Second Revision March 1996

- Phase 1: The dispersion of reduced or extended MTBF (Mean Time Between Failure)
  - Restoration of unattended deterioration
  - Removal of forced deterioration and establishing key operating conditions
- Phase 2: Life span should be extended
  - Extend life expectancy and reduce sporadic failures
- Phase 3: Time-Based restoration of deteriorated portion to its original state
  - Assume lifespan and time-based restoration extend life expectancy
  - Understand irregularity for internal deterioration through the five senses
- Phase 4: Prediction of failure by the equipment diagnostic techniques
  - Failure mode for diagnosis and extension of lifespan by technical analysis of catastrophic Failures and breakdowns

Reference: As per my previous employment and JIPM Consultant

- Step 0: Preparatory stage and understanding the need for Planned Maintenance System
- Step 1: Perform initial cleaning
- Step 2: Restore deteriorations uncovered
- Step 3: Preparation of standard documents
- Step 4: Countermeasure for weak points in the design
- Step 5: Periodic – Preventive Maintenance
- Step 6: Overall audit and diagnosis
- Step 7: Machine ultimate utilization

Many versions of how TPM's Planned Maintenance Pillar is implemented may vary from one industry to another. Whatever Phases or Steps are being adopted to comply with the JIPM consultants, they will end up doing the 4 basic activities, which are as follows: ***RCM will fit perfectly in the higher Phases of Planned Maintenance, which is Phase 3 and Phase 4.***

- Phase 1: Establishing basic equipment conditions through restoration
- Phase 2: Address design weaknesses through improvement
- Phase 3: Periodically restore deterioration with a Maintenance Algorithm or Decision Diagram
- Phase 4: Achieve a Predictive Maintenance stage

Phase 1: Establishing Basic Equipment Condition through Restoration: Both Autonomous Maintenance and Planned Maintenance begin to accept the concept that their equipment is always a shared responsibility for both of them. These TPM pillars address both abnormalities and deterioration of their equipment. As Autonomous Maintenance addresses the exterior part of the equipment, the Planned Maintenance pillar addresses its interior parts. The goal of establishing Basic Equipment Conditions in their equipment is to transition from a stage where accelerated deterioration is rampant to natural deterioration for age-related parts in nature. This Phase on Planned Maintenance aims to reduce the unplanned downtime caused by machine-related problems, mostly equipment breakdowns. ***Hence, when we perform Phase 1, we are dramatically reducing the occurrences of unplanned breakdowns.***

Phase 2: Address design weaknesses through improvement: Once accelerated deterioration had been reduced, equipment will suffer from natural deterioration. There are spares and parts of equipment that will deteriorate naturally. The team exposed themselves to study several parts of the equipment with an inherent short natural lifespan and challenge themselves to correct design weaknesses by improving the part's dimension, the strength of materials, construction, dimensions, etc. An MP (Maintenance Prevention) is usually used for this activity and later on looped back to IFCA (Initial Flow Control Activity Group) or Early Equipment Management (EEM) so that when the company decides to purchase future equipment, these improvements are discussed with the designers to be included in the new equipment purchase so that existing problems would no longer exist. A cycle must be established where MP Design improvements must be looped back to the TPM pillar of IFCA. Correcting design weaknesses can prevent major breakdowns from recurring unexpectedly. Teams are trained on advanced analytical tools such as P-M Analysis for a more detailed approach to addressed chronic Breakdowns. Breakdowns are caused by human errors; hence, both operators and maintenance must upgrade their skills to reduce or mitigate the chances of human errors. Application of Poka-Yoke solution may solve human errors but not necessarily improve the mistake level caused by the person involved. *Again for Phase 2, we are improving our equipment by addressing parts with design weaknesses.*

Phase 3: Periodically restore deterioration with a decision diagram: Once the parts with design weaknesses had been addressed, their lifespan will change; hence, we need to thoroughly review how we will maintain them. This is accomplished with a maintenance algorithm; RCM refers to this as the decision diagram or algorithm. A thorough study of the maintenance tasks must be done to establish the correct maintenance. Since the basics had already been established, we can perform this phase through Reliability-Centered Maintenance. Not all parts need to belong to Time-Based maintenance. This can be done by understanding how to maintain our equipment and the tasks to be performed. Each failure has its own unique pattern. The key in this activity is to understand the 6 failure patterns so the team can derive the correct maintenance tasks for each function using an Algorithm or Decision Diagram. We are no longer dealing with the equipment in this activity, but we are creating a sustainable process. What parts must undergo time-based, what parts can be predicted, what parts need an inspection, and what parts do not have any consequences and can be allowed to fail? *In this Phase, we are no longer improving the equipment but rather how the equipment is being maintained.*

Phase 4: Predict Equipment Life: In Phase 4, we introduce Predictive Maintenance or Condition Based Monitoring Techniques. This is similar in using the human senses in a much higher perspective specialized instruments' with precision accuracy. Some failures can be predicted through the use of these specialized equipment techniques. This is done by checking the condition of the equipment. The key to using this technique is to determine the P-F Interval. This means that if a part shows failure symptoms, these failures will be under the watch of Predictive Maintenance. For example, a bearing may produce noise, increase temperature, increase vibration, or temperature. These are symptoms or potential failures indicating that a failure is imminent or is on the verge of occurring. Therefore, maintenance can schedule the equipment for replacement. The advantage of using Predictive Maintenance is that the failure can be forecasted before it happens.

To conclude, these two powerful strategies are not meant to contradict one another, but rather, they complement each other regardless of their origin. There is no silver bullet solution for every single problem we face in our industry. Perhaps an independent consultant on RCM with little background or knowledge on TPM will be biased on which strategy to implement and vice versa. Both strategies have their limitations and strengths. I think I am just one of the lucky people who know these strategies by heart and mind. So my advice is before establishing any RCM efforts in your plant, try to address the basics first since by addressing the basics, some of these failure modes can be addressed.

2.10: October 2008: Maintenance, The Last Man Standing

Twenty-two years ago, I can still vividly recall my first work as an oiler in a 1500 gross tonnage cargo ship we call MV Sea Raider. During those times, employment was scarce in my country. I was a fresh graduate from Mechanical Engineering and took the Mechanical Engineering Licensure Board Examination, which I did pass. This ship was owned by my close friend in high school. Knowing my situation that I don't have work since I find it most difficult to find work, he asked me if I was interested in working in one of their ships to gain experience, and I said **"okay,"** which I gladly accepted. Today, this ship had already been totally scrapped due to its age.

I belong to the engine room, and this is where I first got my taste of work as a maintenance person, I mean as a mechanic. I was later promoted to the position of Preventive Maintenance Administrator. I have no idea what it meant or what it's all about. All I know is that it has a cool name. Most of our work includes scheduled maintenance, inspections, gauge readings, and partial overhauls of the twin 750 hp diesel engines. We also have a couple of generators to maintain. The main generator was run by a diesel engine generator and another backup generator in the form of a dynamo coupled to one diesel engine, which we regularly inspect daily. We used this backup generator when the main diesel engine generator was out of service or under scheduled overhauls, or simply when it was down. I just can't recall if the dynamo was positioned on the port side (left) or starboard (right) side of the engine.

Here was also where I first experience some sort of love and hate relationship between operations and maintenance. Although there were no operators on the ship except for the crane operator. I am referring here to the people on the navigation side and the engine room people. The Chief Engineer was always in a debatable and argumentative mood with the captain of the ship. When the officers on deck want more speed, the Chief Engineer always said, do you really want to destroy the engine? Well, I guess you can understand the rest of the story. It seems that they just can't seem to get along most of the time. While the officers on navigation, together with the captain, were highly responsible for meeting the cargo's schedule, we at the engine room and the Chief Engineer were responsible for taking care of the equipment and ensuring that the equipment is available and running whenever we need it.

We compose around 20 crew members, myself included. Six from the engine, six people from navigation, and two cooks while the rest under the supervision of a boatswain were on the ship's deck most of the time. This ship's route is to carry and disembark cargo from Manila to Singapore and vice versa. Singapore is one of the busiest ports globally, and you can just see the beauty of the city at night when we watch at a distance from our ship. I believe it took us around 6 to 7 days of travel time at a speed of between 18 to 21 knots. I spend a year or more on this ship, and it was the first time I celebrated my birthday at sea. I also spend Christmas and New Year and all of the holidays on this ship where at night you can only see the moon and the stars while during the daytime all you see is the sea and a thin line at the end. We don't even bother and think about Christmas time or holidays and go on as usual with our routine work, because thinking about the holidays will make you sad and think of your family. That was the advice of most crew members with me, and I believe I got along with it pretty well. There were no cell phones or internet during those times, and the only means of communication was a 2-way radio.

Figure 2.10: My First Work as an Oiler on Board MV Sea Raider

There were several occasions where the Chief Engineer visits us during our routine shift not only to inspect the engines himself if operations were normal but also to inspect our hands, which should be dirty and full of stain, dirt, grease, and so on since his thinking was having a clean hand was a sign of being lazy during your shift. As a sort of punishment, you will be spending around a couple of hours with him after your shift ends and do whatever he wants you to do. Our mindset during those days is that if you want to be a good mechanic like him, then you need to keep your hands dirty **"all the time."** I have no idea what Root Cause Failure Analysis was about during those days, and I never heard about it, so I don't even bother to argue about analyzing problems.

Funny to think that as time flies by and we age, there are many things I tend to forget easily, yet there are quite a few things that remain stuck in my skull that I never seemed to forget. After a year or so, I decided to step down and work in the land industry. It's time to see what it was for me on land. The machines, equipment, operations remain different, but most of the things I had experienced when I was on board that ship was not that different, after all. For land industries, a feud exists between operations and maintenance. In some industries, maintenance is considered a stuntman and a scapegoat for all the operation's problems. When something is wrong, or the output is not reached, all fingers point to the mechanic. Most industries depict maintenance as a form of something evil and a cost center. There is some mild discrimination, and most operations see maintenance as some form of the plant repair business. I am not certain if this sort of misunderstanding can finally be put to an end. I struggle to convince industries of the value and importance of the maintenance function. All I know is that it takes a great amount of time to "**change people**," as it takes a lot of time where black Americans were treated like slaves during the time of Abraham Lincoln to a point in time where a Black African American (Obama) is now the leader of the most powerful country in the world.

I have always thought that all men should be free, but if any should be slaves, it should be first those who desire it for themselves, and secondly, those who desire it for others. When I hear anyone arguing for slavery, I feel a strong impulse to see it tried on him personally," From Abraham Lincoln, 16th President of the United States, March 4, 1861, to April 15, 1865

Maintenance will always have a place in any organization. I believe the best positions in any industry belong to the maintenance function. But we need to understand that maintenance is not a repair business. The goal of the maintenance function is not to create repairmen. Maintenance is not about how fast we repair and put out fires but understanding why a certain part fails. When we understand these failures, then new doors would open up for the maintenance function. Lubrication, spare parts, failure analyst, tribologist, fractography, reliability people, CBM or predictive maintenance group, CMMS, vibration analysis, thermographer, oil analysis, oil contamination control group are just some of the many positions we can have in our maintenance function.

With Christmas and New Year holidays are just around the corner, we take the time to break off with our work to spend some time with our families to celebrate this festive season. There will be people that will celebrate this season not with their families, not with their loved ones but with their equipment. Despite these holidays, they need to light up the plant, provide power through their substations, cool some portions of the plant with their Air Handling Units. And as the New Year begins, they peek outside the open to see the beautiful fireworks, yet they remain vigilant and try their best to remain awake so they can monitor their equipment. These are the maintenance people from your plant's facilities and utilities. These people will always be "The Last Man Standing."

2.11: November 2008: Is There Still Value on Training

Most industries I worked with had some form of training programs that regularly schedule their people to attend from time to time. Others perform some sort of measurement per employee to complete a certain number of hours before the year ends. Other industries send lists of their people to offsite training, conferences, and conventions. And at the end of the day, we asked if we **"can"** apply these thoughts to our workplace, or is it just something that will end up nicely in our **cv and resume**? Is training just a place where we can relieve our pressures temporarily and escape from work? Is it a venue for a free cup of coffee or donut? Or is it something that our plant should take on very seriously?

I believe in the value of training if we use it for the right reasons. Most industries have some sort of Training Needs Analysis, which is used to assess their industry training needs from time to time. The training needs analysis determines the gap between their people's knowledge and the skills needed to do their job. This is a good start, but we also need to ask if this training will be used by their people. Remember that each organization is bounded by its own culture or way of doing things in its workplace. If what we have learned from the training differs from how they do things around, then the default is not to use what was learned. Just join the crowd for good. This is where I hear such remarks and feedback that their management should be the once attending this training, not me, and when you revert back to their management, they simply say, that's why I'm sending my people because they are the ones doing the hands-on and not me. Again everything ends up in a Merry-go-round, which makes me dizzy.

For example, a plant just completed their training needs analysis assessment on their maintenance people. One of them is the need to analyze the root cause of equipment failures. So a list of courses had been developed by their in-house local trainers such as the following:

• Pareto Analysis
• Ishikawa or Fishbone Diagram
• Why-why analysis
• Fault Tree Analysis and Failure Mode and Effect Analysis

But neither of these tools is really designed to understand the root cause of their equipment failures since these tools are only meant to address the most probable or likely cause of the failure. Because to understand its Root Cause, we need to perform and understand the following:

• We need to slowly take photographs and preserve the part that failed before fixing or repairing the equipment.
• We need to know the truth behind this present failure and not the past failure
• We want to know the root cause and not the probable cause or failure mode of this failure
• We need to understand how deep should our probe be on root cause or when to stop
• We need to allow time for the team to interview people and gather shreds of evidence

And when we asked the people if they will be allowed to do these things, the reply will surely be the opposite, especially with operations if maintenance delays their operations just to analyze the equipment before fixing it. Their remarks will be, I think you need to talk first to my boss.

Another plant finds value in using Predictive Maintenance instruments to aid in their maintenance programs and sends their Maintenance Managers to training and conferences. They set up a budget and finally decided to purchase some NDT instruments such as vibration monitoring and thermography initially. But when asked, the people who will use these instruments will be trained and certified because, from my knowledge, you need training and certification to use these instruments. They would not allow their people to be certified as this will again cost them money. They will just rely on the vendor's training; perhaps how to click this and that button or perhaps just an overview would be enough because they thought these instruments will just be used to monitor incipient failures in their equipment. My thoughts, *how very wrong they are!* These instruments are not just for monitoring incipient failures but also for allowing the users to decide whether to continue operating the equipment or finally stopping it. Let me be more specific, a level two thermographer can grasp a full understanding of how to avoid or correct errors on emissivity, reflection, transmittance, and spot size. Let me explain emissivity in a layman's term. It is the relative power of a surface to emit heat by radiation. Emissivity is the ability of radiant heat to leave the surface of an object. If we go back to our high school days during science class, we learned that heat is transferred in three ways: conduction, convection, and radiation. Conduction and convection will transfer heat via direct contact, while radiation is the only heat transfer that will require no direct contact. We feel the heat from the sun even if we are not touching it. Therefore, if we have two cars parked under the sun, the other car's bumper is black, and the other is chrome plated. We ask which one is hotter. The answer is black since black absorbs heat faster. But over some time, if you touch both the black bumper and the chrome-plated bumper, the chrome-plated is simply much hotter. Why? Because as black absorbs heat faster and chrome tends to reflect heat, their emissivity is simply the opposite. Meaning chrome has an emissivity of 0.4 while black is close to 1.

The black bumper will heat up much faster, but 95% of the absorbed heat is free to be emitted or discharged at a rapid rate. The chrome bumper reflects 96% of the heat and only absorbs 4%, but this heat is trapped in the bumper as only 4% can be emitted from the surface. When a thermographer goes out in the field to measure the emitted heat using infra-red thermography, the user must consider the existing environment, whether he is in a hot place, under the sun, in a humid place, etc. The users must have an indebted and thorough understanding of the principles of heat transfer and the different materials' emissivity. These people must understand that the instrument contains errors resulting from emissivity, so decisions can firmly be made. Hence, an experienced thermographer must understand that an object can emit heat, reflect heat, or transmit heat from other sources through the process of heat transfer. Hence, when deciding to purchase these instruments, think again if we need to cut costs on these training because it is quite obviously expensive.

If we asked why industries are reluctant to send their people to training or why is training the subject of many cost reduction measures? The answer is quite obvious: they fail to provide an ROI (Return on Investment). Most industries think about training as non-value. While it is true that it is hard to measure the ROI on soft skills training and that measurements are intangible. On the other side, technical training is much easier to quantify. A safety training program can be felt through a reduction in plant accidents. Training on maintenance can yield results in lowering the cost of doing maintenance and improving its reliability. Perhaps a team-building course can probably be measured on the number of turn-over for employees. So here are a few points I would like to share to find value in your training:

1) Consider training not as a cost but as an investment: Industries invest heavily in equipment, modifications. They invest in TV ads to promote and market their products, top-of-the-line software, equipment spare parts, overtime pay for repair work, and ensure that the part is always available if something fails. If we asked if they are willing to invest in their people's training, well, the answer is maybe later since we have no time right now or worst, they will frankly say no, and the closest excuse they can give is we have no budget for training right now. To me, it doesn't make sense, nor does it? Yet when you're in the lobby of their plant, you can see how romantic their vision and mission are, stating something like being the best and delivering world-class products that surpass their customer's expectations. Sometimes I wonder how they can achieve this without investing in their people. They believe that people can be easily replaced. No wonder some industries have a high rate of turnover these days. And when we think that training is a non-value added cost, I hope we also think about the cost of mistakes, ignorance, repairs, downtime, and breakdowns. Many say that people are the company's greatest asset. Am I right? Wrong, people can only become an asset if they are equipped with the right tools and knowledge to perform their jobs right the first time around; otherwise, they become liabilities.

2) Start your training from the Top Management: Let me share my personal experience with the reader that every time I went to India to facilitate some training and seminars on World Class Maintenance and Reliability-Centered Maintenance, I observed that most of the people who attended these training are from the Top Management level, such as General Managers, Deputy General Managers, and Executive Directors. I also experience a couple of times that one of the people seating in my class was a senior vice-president or a president of a large corporation. This is the opposite of what I totally experience in other countries like the Philippines. When I conduct offsite training, they send their engineers, technicians, and people from the ranks. I guess the difference is quite obvious. These c-level people will have the authority and decision of whether their people need this training or not. If they do, they can arrange for an in-house training with their people. On the latter, when we send engineers or non-decision makers to offsite and public training, you'll always hear them say that my boss should be attending this stuff too. Suppose management people from India can find the time to attend training; in that case, others will always have an insurmountable amount of excuses that they can't leave their plant or are just too busy with their work; I am so sick with this excuse that it makes me puke. Maybe it's the other way around; maybe the plant will run better if Top people are not around since no one is there to intimidate the workforce.

I would strongly suggest that Top Management should be trained by outside consultants and not from people in their training department itself to avoid the trainer being intimidated most especially if they are teaching people on a higher level as they can be dictated by the participants, which of course is your Top Management as well as minimize the chances of being biased or one-sided. If one of the Top Management says, we heard this stuff before, can we finish at 12:00 pm and not 5:00 pm as most of us have things to attend to? Most likely, the trainer would get intimidated and cut their training short.

3) Management must set expectations with their people that they send to training: Before sending your people to training, talk with them and explain to them the reason why you are sending them to this training. Talk to them again when they return to their work after the training on how they can start applying what they have learned in their workplace. If there are some hindrances or obstacles in implementing their learnings, then the boss will be your ally. This way, people attending the training are aware that something is expected from them when they return to work. When they know that something is expected from them, they will be attentive and ask many questions during the session. There is a greater chance that whatever knowledge learned from the training will be implemented and that management is serious about how they can address once and for all the problems they have in their operations and maintenance. Participants from the training are guaranteed that their management will support them in their initiative.

4) Define the needs and the time to use it: One of the questions that need to be raised in the Training Needs Analysis is that we need this training, but are we going to use it anyway? Since the training needs are mostly based on a survey generated with their people, it is important that once the needs are summarized, set up a meeting with Key people and discuss the outcome of the training lists generated. Ask them if they are going to provide their people the resources and time to adopt the learnings. If yes, then the plant needs to conduct the training; if not, then we might as well not go through with the training since you'll just end up spending money on nothing. Classroom training can only be effective if the attendees will be given the right amount of time to practice what they have learned. I often tell my students that this is like teaching you all the basic functions of a car, what an engine does, how the brakes function, the road signs, air-fuel mixture, combustion process, and so on, but after explaining this, I just can't be confident enough to give you the car key and drive me where I want to go, I must be with you, or someone who knows how to drive should be with you initially.

5) Determine where you are and where you want to go: As the year close to an end, it is best to take time off and reflect on how we performed and where we want to go next year and in the years to come. What trainings will our people be needing? This is best done with a strategic maintenance planning. It is good to plan for the upcoming years to discuss what we have accomplished and missed and what we do in the following year.

6) Consider training as a long-term and not a short-term investment: I just do not want to overemphasize this anymore, but better people will yield better results. Invest in your people's knowledge as this is the starting point of building their skills. A highly motivated workforce will result in a low turnover of their employees since they can grow with their company. I know of one industry with an exit interview for people who wish to resign and even provide exit training on what to do after resigning, guiding them on handling whatever separation pay they got. This company cares for their people up to the very end, where they permanently walk out of their plant.

7) Communicate the results with your training department: Usually, for large organizations with a vast amount of departments, the training department can have a way of measuring the number of hours every single employee had attended their training by updating their attendance sheets on their logs and computers. Yet, I doubt if they can record or monitor every single training used by many people in their department that yield success. I am pretty much sure there is. The problem is that the loop and feedback end after the participant finished the training program. Hence, if your organization really values their training, each department in their organization must always provide a loop and feedback to their training department regarding the benefits when applying their learning to their workplace. These remarks, messages, feedback, testimonies should be highlighted and credited to their training department.

Well, I hope you find this newsletter of value, and "yes," there is always value in training if we use it for the right reasons. Training may always not be the answer for every problem we have but what I believe is that having the right people will be one of the keys to an industry's survival, and these people must always be equipped not only with the right tools but with the right knowledge so they can develop their skills in doing their job better.

2.12: December 2008: A Typical Maintenance Culture

I would like to touch on this topic on "**culture.**" First, let me define it in its simplest terms; it means how people perceive and do these things in the plant. It refers to common values and shared beliefs, while others refer to them as shared thoughts and feelings. According to Schein, culture is the pattern of basic assumptions that a given group had invented, discovered, or developed in learning to cope with its problems of external adaptation and internal integration that had worked well enough to be considered valid and therefore taught to new members as the correct way to perceive, think and feed concerning their problems.

I had worked in many industries, and from the very first time I stepped in their plant and sat in their lobby waiting for an interview, you just can't resist the temptation to look at the walls and read their vision, mission, values, beliefs and thought to yourself, what a great place to work. You can read such words as Teamwork, Effective Communication, People first, and many good things. They speak very highly about their employees in the kindest-hearted manner. The words are fashioned together, making you think for a moment that this is the best industry on the planet to work with, and so you do your best in the interview. With a little prayer, finally, you got hired, a dream come true.

On your first day at work, since you are new to the plant, they treat you a little bit mildly, but the longer you stay, the more you understand what their culture is all about. I hope this sounds familiar. Through the years, I've worked with different industries; I was privileged enough (I think that is not the right word), I was "APPALLED" enough to have worked with people who have Masters and Ph.D. degrees in Anger Management and Reactive 101, or perhaps Bosses who never get the hang of being satisfied. As I reminisce about the past, I vividly remember some of my unforgettable recollections. As time passed by, I just laughed at them out loud (LOL) as it fills out my experience path. I recall one time when I was just a couple of weeks in this industry X, the Big Boss told me and asked me where I was last night since he was looking for me. He asked me what time I went home, and I said 5:30 pm. (This company has a free bus shuttle as part of its benefits to their employees, so I thought to myself to use it.) The Big Boss told me that if I'm here, you should be here, and then he shouted to everyone in the room, and everyone should be here. From that day forward, I only went home when ELVIS had left the building.

As an engineer in a plant, part of my work that you can't find in your job description is to prepare a weekly report on what you had accomplished for the week, every week, or the rest of your time while you are still in this industry. We need to write this report. During my time, we also need to present this to my boss to give you an idea of what this Boss was like; I have never seen this person smile in my lifetime. He is the sort of creature that never gets satisfied and usually has a built-in loud voice; some say he came from the military. He always yells at us. I guess it's what makes him happy. Our regular staff meeting is set every Monday from 1:00 to 2:00 pm to give us around 10 to 15 minutes to present, but believe it or not, sometimes the meeting ends up in the evening since he spends a great amount of time on each one of us scrutinizing what we presented and interrogating us most of the time. He will always look for the tiniest thing he could think of to criticize, whatever that is. This "Creature" never gets pleased even if you have accomplished something good for the week. I recall one of the engineers who smilingly told me, Rolly, my time with Boss won't be long this time. He showed me his report, and I was amazed by what he had accomplished for the week. So the time had come for this person to report. There was silence as he spoke, and the Boss was listening most of the time. As he ended his presentation, again, a moment of silence was observed.

Then I recall the exact words uttered by this "Creature," "Are you insulting me?" And the engineer, to his surprise, said No, Sir! Why? You dare come to my office and present your report in front of me without even combing your hair and shaving, and your shoes are dirty, that's why we have lots of problems on this planet, because of people like you, even though it's too early for church, the next hour is spent on the sermon, telling us during his days these was how they do things and that sort of stuff. I am beginning to think that some industries hired these kinds of "Creatures" just to get mad all the time since they are being paid well for that. I think Eric Clapton's song is meant for these people to do nasty things with their people. *I must be strong and carry on cause I know I don't belong here in "heaven."* by Eric Clapton, Tears in heaven. I believe that people like this won't go to heaven

I am now inclined to think that the real culture is not what is really written and hanging on the walls of each plant that can be read as People's Values and Beliefs, but rather, the real culture is confined to how the highest person in your department runs the show. If he is somewhat the tough type of person that doesn't know how to smile, then this is what he wants to project in each of his people. If you can mirror yourself like him, you have a pretty good chance of being promoted and becoming one of his disciples. Still, if you don't, you'll be considered an outcast and probably belong to the resistance group. The pressure will soon build on you to transfer to other departments or perhaps just leave the plant for good. These are the type of people who wants you to know that they make the decisions here, so leave the thinking with them and just do your job. If you can't do your job, just tell me, and I'll find someone who can do the job for you, understand? What an a%*&^le.

In my training, I often joke a lot to keep the delegates awake, build rapport, and ease interacting more freely. I often say that, if you want to stay long in your industry, here's my advice, just apply "Newton's" law of gravity, which states that what comes up must come down; similarly, what goes in must come out. My point is what comes in (whatever you hear even though it's unpleasant) must come out of the other ear. If not, you will just go nuts and mix it with your emotions, making you crazy. I failed in this subject; that's why I retired early in the industry.

The 5 Gorillas in a Cage

To understand how culture works is to understand "The Gorilla Story" if you have not yet heard about it. This is a story about culture. This story starts with a cage containing five gorillas and many bananas hanging above the cage's center. To start with the experiment, the gorillas were given some tranquilizers to sleep, then they put a ladder in the middle of the cage. Before long, a gorilla climbs the ladder and starts to grab the bananas. As soon as one touches the bananas hanging, all the gorillas are sprayed with cold water, and gorillas hate cold water. After a while, another gorilla attempts and get the bananas, and the same result; all the gorillas are sprayed with cold water. Every time a gorilla attempts to get the bananas, the others are sprayed. Eventually, the gorillas come to the point of quitting and leave the bananas alone.

One day, one of the original gorillas was removed from the cage and replaced with a new one. The new gorilla sees the bananas and starts to climb the ladder. To his horror, all the other gorillas attack him. After another attempt, the same thing happened, and he was attacked. Now the new gorilla knows that he will be assaulted if he tries to climb the ladder. After a few attempts, the new gorilla finally succumbed and left the bananas alone. Next, the second of the original five gorillas was replaced with a new one. The newcomer climbed the ladder and was severely attacked. The previous newcomer was called by the other gorillas and took part in the beatings with joy and enthusiasm. Next, the third original gorilla was replaced with a new one. The new one went for the ladder and was attacked as well. Two of the four gorillas that beat him have no idea why they were not permitted to climb the ladder nor the reason why they are being punished for getting the bananas, but I guess they enjoy it. After the fourth and fifth original gorillas have been replaced, all the original gorillas that experienced being sprayed with cold water were now gone, but nevertheless, no gorilla ever again climbed the ladder. If you asked why? Because that's the way it had always been done here.

Each industry has its own distinct culture. It's like a fingerprint. No industry has exactly the same culture, and in every industry, it will even boil down to each department having its own set of cultures. Many industries rely on the don't reinvent the wheel concept since this is how they do it, and they are successful with it, so this is how we should do it. In my thoughts, "how very wrong they are." They forgot the most important part of their organization, which is their "**culture**." If you want something to work for you in your plant, it must be accepted by the culture; otherwise, you will just end up wasting a lot of money and effort. Company Y wanted to initiate a TPM Program, and so they hired people with extensive knowledge on this subject; after a couple of years, the people they hired resigned because there was no commitment and involvement, most especially from management. If your plant is in a reactive mode and still on the status quo, or you still in the initial phase of struggling, you will understand what I mean.

Is it possible to change the maintenance culture?

Many articles and books are written regarding this subject on culture change; some are nice to read but rather difficult to implement. Some suggest having a catalyst inside your plant that can influence other people. All I can say is that there is no easy answer to this question. It depends on many factors, but I know that changing culture isn't going to be done overnight. It will take a lot of time. But the most important part is that change must come from the Man in the Mirror. I think Michael Jackson sang that song.

The industry is like a piece of equipment, and equipment has its subsystems. Each subsystem comprises hundreds or even thousands of parts, and each of these subsystems serves a specific function. Like an organization, each function must blend in harmony to deliver a common good, service, or product. Likewise, an industry has different departments; and that each department serves a specific purpose. Whatever purpose or mission that department has must always synchronize with the overall company's goal and targets. The main difference between a man and his machine is that people think while machines don't.

From a maintenance point of view, everyone agrees about the importance of reliability. The reliability of our equipment depends highly on how we maintain or sustain it, yet there is no doubt that most managers will think about cutting costs on every corner before initiating any reliability improvement in their plant. I think this story obviously fits well with the majority. On one typical day, the Plant Manager walked into the plant floor and found an oil leak on the floor. He immediately called the maintenance in charge and asked him why an oil leak was on the floor. The maintenance indicated that it was due to a leaky gasket in the pipe joint above. The Plant Manager then asked when the gasket had been replaced. The maintenance responded that they had already installed five gaskets over the past few weeks but that each one seemed to leak. The maintenance also indicated that they had already talked to the Purchasing people about the gaskets because it seemed they were all bad. The Plant Manager left and went to talk with the Purchasing Manager about the problem on the gaskets, and the Purchasing Manager indicated that they had been trying for the past two months to try to get the supplier to make good on the last order of 5,000 gaskets that were all seemed to be bad. The Plant Manager asked the Purchasing Manager why they had purchased from this supplier if they were disreputable. The Purchasing Manager said they were the lowest bidder when the quotes were received from the various vendors. The Plant Manager then asked why they went with the lowest bidder, and the Purchasing manager indicated that it was the direction he received from the Vice President of Finance. The Plant Manager left the Purchasing Manager and went to talk to the VP of Finance. The Plant Manager asked why he set up that direction? The VP of Finance said because you indicated that we had to be as cost-conscious as possible, and purchasing from the lowest vendor saves us a lot of money. The Plant Manager was horrified when he realized that he was the root cause of why there was an oil leak on the floor. The end of the Story.

Rather than telling you how to change your culture, take the time to reflect and ponder upon these couple of thoughts:

Learn from your own failures and not from other people's success: Leo Tolstoy once said, everyone wants to change the world, yet nobody wants to change themselves. Whether from the lessons of life or from industry, it tells us one thing. A lot can be learned from our own failures and mistakes. Pain and sufferings are an unpleasant but necessary part of our lives. The most successful people who lived in this world learn from their own failures. During my TPM days, many industries visited our plant to benchmark our success story; the benchmark includes a couple of hours of presentation and a line tour. We only show them the good stuff, and we keep the problems to ourselves. When we first hired a JIPM (Japan Institute of Plant Maintenance) consultant to assess us in our TPM journey, one week before his arrival, the TPM office was busy preparing production areas to clean their areas for any unnecessary mess. We taught people what to say and so on, and when the JIPM consultant arrived, he surveyed the plant and smilingly said to us, this is "TPM one night." We were shocked. In our minds, we asked, how the heck did he knew that. Later on, we learned that it was always his initial remarks to his clients.

The same is true for industries; with countries feeling the effects of the "Global Financial Crisis," industries tighten their belts as much as they can. Cost-cutting is the name of the game industries play. I tried to send an email recently to a couple of clients. Should there be any need for some future training on maintenance for 2009, let me know. Their response was they are cost-cutting on training due to the "Global Financial Crisis." I just nodded my head and smiled as I read their emails. All I can say is just be careful about what you wish for with your cost-cutting program, especially when it comes to reliability. These two don't mix. A low maintenance cost is always a good maintenance practice, and it cannot be the other way around. Some say that changing culture should start from the top; others say it's a bottom-up approach. There is no easy and conclusive answer to this, but all I can say is that change must start from within us.

Small problems matter most: Big problems are caused by small things, yet we tend to focus most of our attention when the problem had already erupted itself. These small things and not the obvious failures pose the most significant threat to our industry. History tells us that the worst accidents and disasters in the past are caused by small things that are neglected. In 1986, the Shuttle Challenger exploded not because of the O-ring on the right solid rocket booster but because the solid rocket boosters' management decided not to listen to their engineer's concern regarding the O-ring erosion. Many times when we dig up the cause of big problems, we'll end up on small things that should have been done in the first place. A Facility Manager attended my training and asked me what maintenance I can perform on their equipment UPS. I said that you need to have redundancy. I can read his thoughts when I said that, which states that you're crazy. Do you know how much it costs? A couple of years later, sad to say, there was a fire in their plant, and the UPS and their equipment were affected, which halted their operations entirely for a day. I learned that now they have not one, not two, but three new UPS, the other 2 serving as a backup. Let us not wait for these things to happen on our plant and act intelligently to address these small problems. And so until our next edition, I think that's all I have to say about that and wish you all good maintenance people a Merry Christmas out there.

2009 RSA Reliability Newsletter Vault Archive

> *In any maintenance, reliability or continuous improvement journey, we do not need management support. What we need is both management commitment and ownership. Remember that management can support but never commit and own the initiative.*

3.1: January 2009: Is Reliability Just a Numbers Game?

Every single industry has its own priorities and focus. The focus can either be on Safety, Quality and the not-so-popular one, "Reliability." For the latter, I say this is unpopular not because they do not know what it means but rather only a few and a handful actually focus on it. Mining, oil and gas, refineries, metal industry, shipping, construction, all these industries will prioritize safety as their focus of the discipline. On the other hand, the manufacturing, automotive, pharmaceutical, food, beverage industries, just to name a few, will prioritize satisfying customers' needs and focus their discipline on the quality of their products. But if we think of it for a while, which companies actually focus on **"reliability"**? Can you think of one? But isn't it true that if we focus on reliability, then the equipment is safe to operate and will likely produce good quality products? First, let us define what reliability is all about:

- It is a probability that a system will not fail.
- Reliability of an item is the probability that the item will perform its specified function under specified operational and environmental conditions under a given and specified time.
- Reliability is the probability that no failure will occur throughout a prescribed operating time or period.
- According to Bazovsky, the modern concept of reliability in popular language is simply stated as the capability of equipment not to break down in operation. When equipment works well and performs to do its job for which it was designed and intended to do, such equipment is reliable.

In its simplest term, reliability is a survival probability. On the other hand, those that do not reach their period had actually failed in the first place. When we speak about failures, these are any parts or items that cannot fulfill their designated function. With an adequate amount of samples, we can calculate a part's reliability performance.

As time passed by from one generation to another, equipment's tends to become much more complex and sophisticated, and as equipment becomes more complicated and automated, the way we do maintenance still remain the same as it was 50 years ago or even to the day the caveman invented fire, and that is, still firefighting? Am I right? The practice of allowing equipment to fail will bear different consequences compared today to 50 years ago when the equipment was made much simpler in nature. Today, the software is now available to calculate critical parts reliability, such as "Weibull," derived from its author Wallodi Weibull. So there is no need to calculate it manually using exponential equations and mathematical formulas; it is good to understand its principles before applying them.

A friend of mine had installed a tool algo on all their tooling equipment to monitor the number of cycles the tool will eventually wear out. Since we all know that tooling will eventually degrade or wear out after n-number of cycles, the equipment hooked up with the tool algo will provide an alarm if the tooling is nearing its life and automatically stop the machine from operating once it reached its calculated number of strokes and cycles. Pretty cool stuff.

My experience dictates that there are cases when focusing too much on quality can sometimes conflict and probably jeopardize reliability. I mean no disrespect to the quality people, but we need to understand reliability first before initiating any quality policy in our organization. Let me try to give an example in a manufacturing plant. This was my source of bread and butter during my employment days. Every manufacturing plant have their own Quality Department. This department comprises several people assigned to different locations within the plant, incoming, outgoing, QC inspectors, production inspectors, or when a product is moved from one department to another, quality is inspected. Almost anywhere in the plant, a group of quality people is deployed. They have quality audits and provide non-conformance tickets to those that simply do not comply and adhere to the processor specification that had been generated by the plant, just like the police. Hence, if the equipment is scheduled for a monthly Preventive Maintenance activities for some checks and replacement and some activities had not been performed, a non-conformance ticket will be issued on maintenance. Am I right here? Or worst, the Preventive Maintenance team will need to submit a formal report on why they missed their PM activities on one of their equipment even if the reason for not doing the replacement was because their Predictive Maintenance group found that the parts due for replacement are actually still fit to perform their function and so they decided not to replace it but to continue using it.

Another classic case was when I was still employed in "Company X," I recall teaching my first batch of Reliability-Centered Maintenance in 1999 to around 18 people, 2 groups from both the production line composing of operations and maintenance and one group from facilities and utility section of the plant. There were around 6 people per group. After the training, the group was highly motivated and filled with enthusiasm to perform the RCM analysis. They were convinced that they had found a better way to improve their equipment's reliability by improving the way they maintain their equipment. We prepared a roadmap, and all three groups completed their documentation on the RCM analysis. They submitted the final report to me. I checked them for a few errors. I think that they actually absorbed the training's message. One group had overall inspection lists of about 350 items in their equipment that needs to be done every year. It was trimmed down to 150 items. So when I told them to implement the RCM tasks and replaced their current Preventive Maintenance tasks, and adopt this newly derived RCM list, they were hesitant. Their leader said to me, Rolly, you better talk to my boss. When I talked to their supervisor, the supervisor told me he can't decide on this matter and advise me to talk to their Department Manager. When I talked to their department manager, he gave me an option and said, Rolly, we cannot change our PM tasks to RCM.

Suppose you insist on complying with the RCM requirements; in that case, we will do them both in which they will both perform their usual PM routine and the newly derived RCM analysis. As I was listening, I was just imagining some rubbish on the floor and sweeping it to keep the floor clean. After sweeping it again, garbage is put back on the floor. I told the Department Manager "no" and explained to him that your team had successfully re-engineered your lists of items to be checked and had removed all intrusive activities in the tasks lists. The boss explained that the accumulation resulted from both the customer and Quality Group audits that added most of the items on the checklists. He said, if you really want to change it, you better need to talk to them. Now, I understand what he meant. Both 3 groups have successfully finished their RCM documentation, but only one group had implemented the RCM process with good results. It was from the plant's facilities and utility section.

Our RCM Analysis led us to understand that the activities we are performing on our regular monthly and quarterly PM on our AHU and sub-stations were not really meant to prevent the failures, and those activities which must be performed to address the failure were not even included in our PM activities. By Arman Jusay, RCM Team Leader

It was my first RCM case study. Two had failed, but one pushed through. It is not my intent to blame quality for this. In fact, blaming itself will not help but rather, I reflected and figured out what mistakes I made in the past. I think one of them was to exclude the quality group in the RCM training. Or perhaps if only they know our intentions and the need to adopt a more robust process in our current PM structure, they will leave the maintenance alone and stop issuing any non-conforming tickets when they missed out on an activity.

Reliability as a Business Strategy

How I wish industries understand that improving reliability affects the way they do business. Last night as I was watching the news on TV, around 1500 people were laid off on a semiconductor firm I used to work with, and the reason stated by their spokesperson had something to do with Global Financial Crisis. A not so unique excuse. As you see, there is a period of peak demand and fall-out in semiconductor industries as well. When the demand is low, they use skeletal force. There were mass lay-offs as well. When I was still employed, I saw it happen a couple of times. The only difference is that we do not call it the Global Financial Crisis. When I try to knock on some industry doors if they have some maintenance training needs, I get the same answer, sorry Rolly; we are cutting costs due to the Global Financial Crisis. Every time they say no, the reason is the same. When matters get worst, it is all because of the Global Financial Crisis. Industries now have an easy excuse and way to explain to people regarding lay-offs and cost reduction schemes, just like the "Gorilla Story" in my December 2008 reliability newsletter issue.

But on the other hand, industries that take reliability seriously by improving their maintenance and adopting proven best practices such as Reliability-Centered Maintenance, Total Productive Maintenance, Condition-Based Maintenance, Oil Contamination Control, and other strategies have a better chance of truly surviving this crisis. Although I have said this many times, trying to cut costs on maintenance is not improving reliability.

Maintenance people have a common sentiment; they always feel under staff because of too much crisis and emergency works. Hence, a company initiating cost-cutting as their ultimate strategy will have its downfall and perhaps cut maintenance overtime, only to realize that nothing changed and matters just get worst. On the other hand, one industry had reduced their overtime rate simply because their RCM analysis led them to eliminate many intrusive replacements, overhauls, and inspection on their equipment that was deemed unnecessary.

Traditional Belief on Maintenance

When I am conducting training outside my country, my training sponsor booked me in a hotel. I need to adjust to the time from my country to their country. When you are in a hotel, you can either place a "Do Not Disturb or Clean My Room" sign outside the door. If you place a do not disturb sign on your door, they know that you don't want to be bothered simply because you want to rest or sleep. If you place a, do not disturb sign and people come knocking at your door, just imagine how irritating this will be to you. I recall placing a do not disturb sign, and as I fall asleep easily since it was already late. A few minutes, someone kept on knocking and asked if I feel comfortable in my room. I arrived at this hotel at 12:00 am (2:30 am in my country), went to my room, and have no time to change and just slept. 30 minutes later, I woke up because of a knock on my door to ask me if my room was ok. Just imagine how I felt. This is no different from our assets, parts, and items inside our equipment. Some of them are better off undisturbed because replacing the part or overhaul will likely cause infant mortality failures.

Maintenance people believe that "**all**" parts after consistent use will reach a point of wear and tear. Hence, overhauling or replacing the part before it fails on a specific fixed schedule will ensure the reliability of the equipment. Therefore, the concept of Preventive Maintenance will solve the problem of unexpected failures, right or wrong? Suppose point A is the assumed life and the start of the wear-out mode; in that case, this is actually when maintenance intervenes on the asset to perform their Preventive Maintenance tasks. This is a very wrong concept. Many cases exist when schedule overhauls or Preventive Maintenance can increase overall failures by introducing Infant Mortality failures into otherwise stable systems. If you want to know more about this, you can read my previous newsletter: An Inconvenient Truth about Preventive Maintenance.

In my humble opinion, reliability is not just a mere numbers game. Reliability is something that we can definitely improve. Maintenance people should know about this. Decision-makers must also participate in their plant's reliability improvement strategy. This needs simple support from them and, more importantly, making it a commitment from them. In fact, the truth is simply stated that our equipment's reliability has a lot to do with maintaining it. We must look into this matter very seriously.

I have often been denied training merely because it has something to do with my skin color or the country I lived in. It makes me feel sad as a human being, but it does not stop me from preaching what I know. It took me more than 10 rejections before I can actually teach people in

Malaysia. I thank people from other countries for vesting their trust and allowing me to share whatever knowledge with them. I thank them from the bottom of my heart for judging the book "not" by its cover but rather by its interior. Like reliability, it is not just merely a numbers game. In fact, anyone from whatever race, color, country, or industry can improve their equipment's reliability by understanding its principles and concept since, as always, a reliability-based culture will always start with its basic foundation, and that is through education.

3.2: February 2009: TPM Planned Maintenance Pillar

The first book I wrote, World Class Maintenance Management, the 12 disciplines, depicts the life and struggles of maintenance in seeking better ways to manage and maintain their equipment and assets. The author of this book shares his passion and experience about the day-to-day struggles in the life of maintenance. What is interesting about the author and this book is that he hails from the Philippines, but the problems, issues, and struggles we face in maintenance are generic and can be felt by any industry from whatever location, place, and race. This book contains real-life stories, situations, and many actual experiences by the Author in his career in maintenance and currently as a Reliability and Maintenance Consultant.

The WCM book is much easier to absorb as it is structured into three parts: the Basics, the Strategies, and the Advance Disciplines. The Twelve Disciplines are grouped according to these three parts. Maintenance often time seeks advanced ways to deal with their everyday problems and issues. This book's message is straightforward: there is no better way to start by going back to the basics and addressing these very small problems in our plant. Big problems, unplanned breakdowns, and catastrophic failures are just an accumulation of small problems that had always been ignored and mostly neglected in the first place. The author strongly emphasizes the importance operators' play in addressing the basic equipment condition and is considered partners with maintenance on this shared responsibility towards their equipment. It is difficult or impossible for the maintenance people to transcend from a reactive to a proactive mode if operators are not involved. When the Basics had been set and well established, maintenance can move on with the different maintenance and reliability strategies. Each chapter covers a specific maintenance discipline. Chapter 14 of the WCM book covers an implementation plan on how to proceed with these disciplines by understanding where the current maintenance is and bringing the maintaining herd in the direction they want to go. Finally, the author would like to share his message on the Conclusion Part of this book and share his actual experience on what it takes for industries to reach World Class Maintenance Management level. This book is not only about a book on reliability and maintenance. It is a book that will make each maintenance feel proud that they belong to the maintenance function.

IBIDEN Philippines Corporation passed TPM Excellence Awards

First, I would like to congratulate, Ibiden, Philippines Corporation, for achieving their TPM Excellence Awards Category (A) last year. It was privileged to train them on TPM training such as Understanding the Relationship between Equipment Losses and OEE and Planned Maintenance Four Phases to Zero Unplanned Breakdown. Their leader, Bong Lalatag, truly has the heart of a TPM Leader. As I recalled asking Bong how confident he was in achieving the awards despite all the hurdles, resistance, and obstacles he is facing, Bong said, *we will die aiming for it.* Truly his leadership, and efforts had sparked a change in their people. They finally succeeded in their TPM Journey in reaping out the fruits of their TPM endeavor. This year, their company will be part of a traditional ceremony given by JIPM to accept the awards on this prestigious TPM Excellence Awards. Once again, congratulations to Bong and Team IBIDEN on your success, and I wish you more power in your TPM Journey.

TPM Planned Maintenance Pillar

Spending many years in TPM, I have learned that while Autonomous Maintenance is the pillar of TPM with most members, Planned Maintenance should always be the strongest pillar in any TPM implementation. Why? It will be Planned Maintenance that will be responsible for developing and supporting the pillar of Autonomous Maintenance. Planned Maintenance will also be responsible for guiding and coaching operators regarding their equipment to perform the seven basic steps of Autonomous Maintenance. Planned Maintenance will play a crucial role in any Autonomous Maintenance. Maintenance will serve as the coach and mentor of operators. The role of operators involved in Autonomous Maintenance is to help in establishing basic equipment conditions. This means that if the Planned Maintenance pillar is weak, so will the Autonomous Maintenance pillar.

Figure 3.1: Myself at IBIDEN Corporation (Philippines) Teaching TPM

During our TPM days, when I asked some of my Planned Maintenance members what support they have provided to their operators, what should we do to be mutual partners with our operators to have a shared responsibility for their equipment? One member said that he makes sure that the operator is beside him while he is repairing when he repairs the equipment. This way, he can communicate freely and tell the operator what particular part had failed and its function. I said that is good. He said that he was not finished and told me that he makes sure that they eat together during lunch breaks. As they eat, he told her all about what she needs to know about her equipment. Again, I said, that is good. He told me that he still has something to say and that during her shift, he told the operator how to perform light repairs in the equipment that he believes she can handle. But that is not all; at the end of their shift, since they are both on the same shift, they go home together with the bus shuttle service (since the company offered a free bus shuttle as part of its employee's benefits). It's an hour-long journey, most especially during rush hour due to traffic, so instead of being bored sitting on the bus, he told her stories about his actual experiences in the equipment, and from time to time, he still discusses with her what he can think of regarding the equipment. He said that he just never ran out of words. Today, they are now married and have three kids. This person had taken my word literally when I said that operators and maintenance should be partners with their equipment since performing maintenance is always a shared responsibility for the two. Perhaps if Autonomous Maintenance is done in 10 steps, the last step will be for Maintenance to marry the operators if both are from the opposite sex. Anyway, I think that's about it.

Likewise, in my experience, Planned Maintenance will be the most difficult pillar in TPM, and the leadership of both the Planned Maintenance Facilitator and the Planned Maintenance core team must be strong. Planned Maintenance itself has its own eight pillars. Two of the most important pillars of Planned Maintenance will be Planned Maintenance support for Jishu-Hozen or Autonomous Maintenance and the Planned Maintenance Steps or Phases. Some of the Planned Maintenance pillars will be improved directly due to implementing the Planned Maintenance Steps or Phases. The following are the eight pillars of Planned Maintenance in no particular order.

1) Support for Autonomous Maintenance: Planned Maintenance must fully understand its role and responsibility to provide support and guidance to Autonomous Maintenance teams. Maintenance will be indirectly involved with Autonomous Maintenance by providing them with the training operators need about their equipment, safety, lubricating points, and other equipment facets. Planned Maintenance will likewise provide technical assistance in eliminating sources of contamination on their equipment. When Autonomous Maintenance reached step 4 of their roadmap of activities, Planned Maintenance will provide operators with theoretical knowledge concerning lubrication, bolting, and conducting a general inspection on their equipment. Operators must know that bolts should be completely tightened based on the correct level of torque required.

2) Planned Maintenance Steps/Phases: Like Autonomous Maintenance, Planned Maintenance is performed by performing its different steps or phases. This will depend on the type of industry and the directions of the JIPM consultant itself. However, whatever steps are being performed will all boil down to the Planned Maintenance Four Phases to Zero Unplanned Breakdown. The first two Phases of Planned Maintenance aim to reduce the number of unplanned breakdowns in the equipment, but the last two phases of Planned Maintenance aim to improve the system, or the way maintenance is being performed in the equipment. This is where RCM will perfectly fit into the TPM implementation. For detailed and variations of Steps and Phases in the Planned Maintenance implementation, you may want to refer to my previous newsletters

3) Spare Parts Management: A good Spare Parts Management system must provide the right part at the right time when maintenance needed it most. Movements of parts in the storeroom must be monitored and controlled at all times. Having a good Spare Parts Management System is not only about housekeeping but maintaining an accurate inventory of parts from the system and its actual physical location. When the quantity of parts in the system does not match the physical inventory, maintenance loses its confidence in the storeroom. Maintenance should decide what to do with non-moving parts sitting inside their storeroom since the equipment was decommissioned or retired. The parts affected must also be disposed of. Some of the Fast Moving parts in the storeroom will be subject to Planned Maintenance Phase 2, where the parts will be analyzed to increase their lifespan by addressing the part's design weaknesses.

4) Lubrication Management: As Autonomous Maintenance helps check the lubricating points in their equipment, maintenance seeks to understand how oil is being contaminated and how to analyze the oil for contaminations. Many failures are attributed as a direct result of poor lubrication. Maintenance must also understand if the different lubricants used in their equipment are sufficient to reduce friction. It is important to understand how oil is being contaminated and what contamination can do to their equipment. Having a thorough understanding of this can aid maintenance in developing strategies such as Oil Contamination Control to reduce the number of harmful contaminants in their oil. Maintenance must likewise understand the different oil properties and what Oil Analysis test is best suited to their equipment.

Figure 3.2: TPM Planned Maintenance Pillar (Myself seated 3rd from Left)

5) Maintenance Skills and Knowledge Upgrade: TPM believes that the equipment must be improved, but much more important is improving the skills and knowledge of the people doing TPM itself. It will be the people that will improve the equipment, and the people can only improve if they are equipped with the right skills and knowledge. Specialized training in different reliability and maintenance strategies and Predictive Maintenance technologies must be provided to the maintenance people. Both theoretical and practical applications of these skills and knowledge must be provided to maintenance. A maintenance training curriculum must be drafted at the start of the maintenance journey into the Planned Maintenance implementation.

6) Predictive Maintenance: As the maintenance people's knowledge is upgraded, maintenance realizes the necessity of using non-destructive and diagnostic instruments to monitor and check their equipment condition from time to time. The most common Predictive Maintenance instruments are vibration analysis, infrared thermography, oil analysis, and ultrasonic monitoring. Predictive Maintenance must work hand in hand with the Preventive Maintenance group in addressing failure modes in their equipment. These instruments will monitor failure modes that will provide signs or symptoms that they are on the verge of failing.

7) Maintenance Management Budget Control: It is expected that through the initial stages of Planned Maintenance activities, maintenance budget and cost will grow due to restoration activities and training its people. However, as maintenance pursues its activities on Planned Maintenance, the cost will definitely be reduced due to improving equipment's reliability and reducing maintenance cost. The maintenance budget will likewise be reduced as maintenance performs the other pillars of Planned Maintenance, such as Spare Parts Management, Lubrication Management, and Predictive Maintenance. Planned Maintenance begins to mature, and maintenance costs will definitely go down as reliability starts to improve.

8) Building an Information Maintenance Management System: An information management system or CMMS must be in a place and should include all relevant and important information maintenance needs from spare parts, lubrication, parts life monitoring, maintenance people 201 files, criticality analysis ranking, maintenance indices and KPI's, breakdown history records, machine 201 file, predictive maintenance monitoring and so on. The goal of building an information management system is to automate and streamline the maintenance process. Remember that CMMS can only be useful if we put in useful information to aid maintenance in performing their jobs well.

3.3: March 2009: TPM and RCM Do They Contradict or Complement Each Other?

In a sequel to our June 2008 RSA Reliability Newsletter: TPM vs. RCM, which is the best lb. for lb. improvement strategy, I would like to discuss this subject more and provide a detailed account and experience whether these two methodologies contradict or complement one other. If you want to read our June 2008 RSA Reliability Newsletter: I would like to point out the facts between these two methodologies to serve as a basis and as a guide.

These two methodologies originate from different parts of the world; TPM started from the east while RCM from the west. Origins of RCM can be traced back to the works done by Stanley Nowlan and Howard Heap from the United Airlines industries, which later on was adapted to land industries by the late John Moubrey.

About Overall Equipment Effectiveness (OEE)

First, let us speak about the primary measure of TPM, which is OEE or Overall Equipment Effectiveness. In John Moubray's book, on page 302, he quotes;

- *There is often a tendency to focus too heavily on primary functions when assessing maintenance effectiveness. This is a mistake because trivial secondary functions embody bigger threats to the organization if they fail than secondary functions. As a result, every function must be considered when setting up maintenance effectiveness and measures. The OEE, as defined above, only relates to the primary functions of the asset.*

- *And again, on page 304 on RCMII book, Moubrey quote, the OEE defined above only relates to any asset's primary function. This is misleading because every asset, machine tools included, has many more functions than the primary function, as in the case of a gasoline storage system. Each of these will have its own unique performance expectations. Consequently, the OEE is not a measure of the overall" effectiveness but only a measure of the effectiveness with which the asset's primary function is fulfilled.*

My point about John Moubray's statement is that this is true. OEE is just a measure of the primary function and will never measure the effectiveness of secondary functions, but the activities of TPM, most especially the pillars of Autonomous Maintenance and Planned Maintenance, will take care of the secondary functions of the equipment. While Autonomous Maintenance will address the equipment's abnormalities, Planned Maintenance will mostly address the equipment's deteriorations. These two pillars, when joined together, will address the failed secondary functions of the equipment. The purpose of these two activities is to bring the equipment back to its original condition when the equipment was previously commissioned in the plant where all secondary functions of the equipment were working. In hindsight, if we address the secondary functions of the equipment which are not working, then we can expect fewer failure modes to maintain.

Which Comes First, To Maintain or To Modify?

There is no easy answer to this question, but allow me to explain this the best way I can. TPM is an improvement methodology, and its concept is about incremental step by step or drip by drip improvement. Again from John Moubray, on page 188, he quotes, which comes first, redesign or maintenance? Reliability, design, and maintenance are inextricably linked. This can lead to a temptation to start reviewing existing equipment design before considering its maintenance requirements. In fact, the RCM process considers maintenance first before redesign for two valid reasons. First, most modifications take six months to three years from conception to commissioning, depending on the new design's complexity. Secondly, most organizations are faced with many more apparently desirable design improvement opportunities than are physically or economically feasible. RCM does much to develop a rational set of priorities for these projects by focusing on failure consequences. In other words, RCM simply tells us to first look at the consequences of failure so as we can set our priorities.

While I agree with the concept of RCM at this point, we must also look at these two strategies from a different perspective. RCM is more flexible since it can be adapted to almost all industries, including manufacturing, while TPM is designed mostly for manufacturing industries. I recall having lost an opportunity here in my country, just for being honest. A Power Plant industry wants me to conduct a TPM training. I told them that TPM is not design for their plant, and I never heard from them again. Imagine having an OEE of 90% for a Power Plant. Certainly, the city will suffer many blackouts, but for manufacturing, an OEE of 90% will be considered world-class if secondary functions are likewise working at this stage. Again OEE is not design for the Oil and Gas sector unless you measure every contaminant or bad molecule the oil has. TPM has its own territories and boundaries. The duties and responsibilities of operators in a Power Plant or Oil and Gas industry differ much from a manufacturing plant. Manufacturing plant operations can have hundreds or thousands of equipment inside its operations. Performing modifications or redesigning one piece of equipment to improve its design flaws and weaknesses will definitely take a much shorter time than other plants, if you understand my point. There are pillars of TPM that will definitely not be 100% applicable for an Oil or Gas Sector, and one of them will be Autonomous Maintenance since the duties and responsibility of operators differs from these types of industries as well as the type of equipment they handle, but nevertheless, there will be TPM activities for Autonomous maintenance that can be adopted by a power plant such as performing the basics and minor repairs on their equipment. There will be TPM pillars that oil and gas and power plants can adopt, such as

Planned Maintenance.

Modifications and redesign for manufacturing equipment will not take 6 months to 3 years but a much shorter time. A design flaw spotted on a punch in which it does not reach its designed lifespan can be subject to a Planned Maintenance modification and improvement by studying the cause of its failure and identifying changes in its strength or shape to lengthen the part's lifespan. But one thing I learned from RCM that I include in the selection of TPM's pilot machine is to reconsider the time the equipment will remain in operations since if the equipment will no longer stay or will be decommissioned in a few months from now, then it will be pointless to perform any modification or improvement in this equipment.

How Both RCM and TPM is Performed

RCM is performed by first writing the asset's operating context, which will undergo a thorough RCM process. This is followed by answering the seven basic questions of RCM. The first part will be to conduct a simple FMEA on the asset and understand each failure's consequences by subjecting it to an algorithm or decision diagram to derive the most feasible maintenance task to address each failure mode. On the other hand, TPM consists of eight pillars. The pillar that is closely related to RCM will be Planned Maintenance. There is a wide variety of approaches to Planned Maintenance, depending on the consultant the plant hires. However, in my experience with TPM, whatever steps or phases Planned Maintenance will undertake will all boil down to the generic 4 Phases of Planned Maintenance.

- **Phase 1: Stabilized MTBF:** This initial phase of Planned Maintenance deals with restoring the equipment back to its original condition, which is done parallel with Autonomous Maintenance, which will address equipment abnormalities and deteriorations for Planned Maintenance.
- **Phase 2: Lengthen Equipment Lifetime:** The second Phase of Planned Maintenance is to identify parts with inherent design weaknesses by analyzing them and modifying them to improve their lifespan.
- **Phase 3: Periodically Restore Deterioration:** Once the parts with design weaknesses had been addressed, their lifespan will change. Hence, we need to have a thorough review of how we will maintain them. This is accomplished with a maintenance algorithm; a thorough study of the maintenance tasks must be done to establish the correct maintenance tasks and establish the correct maintenance frequency.
- **Phase 4: Predict Equipment Lifetime:** In Phase 4, we introduce the concept of Predictive Maintenance or Condition Based Monitoring Techniques. Although this is similar in using the human senses in a much higher perspective with specialized instruments. Several parts of the equipment can be predicted using these specialized equipment techniques, which are done by checking the equipment's condition.

While TPM will aim for zero breakdowns, RCM will deal with the consequences of failure. This statement is true, but there is also another point we need to understand. In TPM, the first two Phases deal with improving the equipment's overall performance in breakdown and MTBF. TPM classifies breakdowns into two parts Planned and Unplanned Breakdowns. What TPM aims for is zeroing out all unplanned breakdowns. TPM will not or can never eliminate Planned Breakdowns. An example of Planned Breakdowns is Scheduled Preventive Maintenance over the equipment. While the last two Phases of Planned Maintenance are no longer design to improve the equipment's performance. The last 2 Phases are done to improve how maintenance is performed on the equipment. In short, we are dealing with improving the system and not the performance of the equipment.

Phase 4 is about Predictive Maintenance; it is my initial thought that Planned Maintenance aims to transition from a Preventive to a Predictive Maintenance structure. How very wrong I am. Phase 3 is done to identify which parts are more suitable to undergo a Predictive Maintenance stage. There is no transitioning from Preventive to Predictive since it is unlikely that all failure modes are suitable for a Predictive Maintenance task. The second point is that TPM believes that it is more feasible and effective to use Predictive Maintenance if the asset's basic equipment conditions are addressed in the first place.

Suppose you understand the concept of Reliability-Centered Maintenance. Isn't it that the goals of Planned Maintenance Phase 3 and 4 are similar to what RCM wants to do? In fact, I truly believe that RCM will fit perfectly into Phase 3 and 4 of Planned Maintenance. When I was still with the TPM Office, I asked our JIPM Consultant to add RCM into the Planned Maintenance structure. The consultant just smiled and said, do not be in a hurry. You can do that in the later steps of Planned Maintenance. At first, I do not understand what he meant, but as we progress into the Planned Maintenance implementation, I understand his point completely.

Figure 3.3: Can You Perform Step 1, Autonomous Maintenance Cleaning here?

No disrespect to other TPM and RCM consultants. If an independent RCM consultant with little knowledge of TPM will be a bit biased, and that is what I had experienced before with an RCM consultant. He wants us to drop TPM and implement RCM. That is why, instead of following his advice and dropping TPM off, we drop him off. When I was in India last year and conducted a series of offsite training on RCM, I received many negative reactions from some delegates when I explained the RCM process's principles. Some say, is that it? Others said, where does cleaning and doing the basics be done on RCM? You see, India's industries are heavy on TPM, and for most industries, the RCM concept is quite new to them. Most of them are familiar with TPM, and some of the companies there have already been awarded on TPM Excellence Awards. However, what I was teaching was RCM and not TPM. Some of the people complained to the sponsors and wanted to quit the training; I remain patient. On the last day, I explained to them where does RCM fit in the TPM process. I told them that it fits perfectly into the last Phases on Planned Maintenance and the delegates began to smile. A couple of them even apologized for their behavior, while others said, why didn't you explain that initially? And so they went home with a smile on their faces.

This newsletter's message is straightforward. These two methodologies do not contradict each other if you understand their principles. In fact, they even complement each other. RCM fits perfectly into the last phases or steps of Planned Maintenance. I think only the independent RCM and TPM consultants do the contradiction but not the methodologies themselves, and I think that's all I have to say about that. I hope that you enjoy reading this newsletter as much as I do enjoy writing it.

3.4: April 2009: How Much of Industries Problems are Man-made?

I do not know if you agree with me, but based on my experience, the bigger the plant's population, the more problems it has. And entirely all of their problems are, in a way, self-induced or simply means man-made. I have asked a few of my sources from a percentage of 1 to 100, how many industries' problems are man-made? Most of them answered 100%, to which I am not surprised. In my humble personal opinion, around 98 to 99% are man-made problems or induced by man. I just leave the 1 to 2% as unfathomable. Below are some classic cases of maintenance problems that simply stated are man-made itself.

CASE 1: Let me give you a classic example. Last May 2009 this year, I went to one of the plants here in my country to teach a one-day course on Total Productive Maintenance. The course was supposed to be attended by supervisors and superintendents, and the class was supposed to start at 8:30 am, which around 4 people arrived at that time. At 9:00 am, around 4 more people joined the class at 9:30 am, some more joined, and the class was finally completed. When the class was complete, I stopped the lecture for a moment and asked the class why they were late. Admittedly, almost all said they went to their line of operations and gave the operators some instructions. I asked one person how long he worked in this industry? He said that he had worked here on this plant for the past 30 years, 20 years in the supervisory position. I asked them how long he has been doing this. The supervisor told me he had been doing it for the past 20 years. And I asked them why do you give the operator's instructions? He told me that it was his job in the first place. I further explained that I am not disappointed by the class being late, but I have one major concern with you guys. It is very much related to the course that we were covering. The goal of TPM is about empowering operators; it is about trusting operators that they too themselves are capable of thinking and doing something without being told. You just cannot implement TPM if you continue to do this every time, all the time. If your plant is serious about adopting TPM, you must learn to change what you have been doing for the past 20 to 30 years.

CASE 2: If we take a look at our Preventive Maintenance Checklists. Every PM Checklist has a form of lubrication performed in a timely and scheduled fashion, indicating something like this.

• What: Apply lubrication on a roller bearing,

• When: Frequency monthly
• Who: Preventive Maintenance

If you think that this is the right thing to do, then you are so wrong. This is actually where **CHAOS** and **CONFUSION** start. If I ask one of the maintenance to perform lubrication, open up the cabinet, and see a dozen types of grease with different brands, some mineral oil, and a couple of synthetic oil, what will he choose? Since all of them are lubricants. As to what specific lubricant to be used is unknown. Would it be grease or oil? For example, if grease would be applied, what corresponding NLGI number must be used? What specific brand and type of grease would be used? What is the specific temperature of grease that will be required for this application? How many shots or pump is needed for the grease gun or when do we know when to stop pumping the grease gun? How many grams of grease is required for this application, and do you have any means of measuring it? These are just a few questions to start with your lubrication, and often we blame it on the lubricant, but much of the lubricant failures we experience in our plant are caused not by the lubricant itself but by us. It is due to how we perform our lubricating practices in our plant. Applying the wrong lubricant or mixing lubricants of different brands can cause incompatibility problems with the base oil and additives. This would likewise compromise and shorten the lifespan of the part it lubricates as well as the life of the lubricant itself. If we audit how many grease guns we have in our plant or how many grease types are being used over the years. I think you already know the answer, and each of these grease guns has a different output or volume of grease per pump. Suppose we study the cause of this problem. In that case, people from different departments, stations, or areas simply confine their responsibilities to their departments. In short, they do not talk or communicate at all.

CASE 3: When production is low, or the target output was not hit for the day, all fingers point to the maintenance people for a long downtime. What caused the downtime to increase was that the spare was not readily available in the storeroom. Maintenance had to purchase it manually outside the plant. When everything was fixed and done. Our boss often tells us how come we did not check the part's inventory in the first place. But when you check the inventory of every part you need for the equipment and have it requisitioned, you would be questioned why you need to have this in the first place. I believe the right word is interrogated as to why you need to have every single spare part requisitioned since we are on a really tight budget right now. Or worst, if you have some bosses that often told you that if you can't do the job right, just let me know, and I can find someone better. These people are always sarcastic and cynical in the way they speak, and believe me when I say that they are very good at this. It sometimes makes me wonder that these kinds of people are paid to do these nasty things to their people. I call this the maintenance trap. Your only choice is to select from the lesser evil. If you go to one point in the line, you will be bitten by snakes, while if you go to the other end, you will be eaten up alive by crocodiles. The worst part is we end up cannibalizing other idle equipment just to make our machine running.

CASE 4: A facility manager once asked me the best maintenance tasks to perform on their UPS equipment. (Note: UPS unit is used to protect the critical loads from power disturbances and used as a stand-by power supply during interruption of regular power supply due to load shedding, power failure, power fluctuations. It provides reliable and stable power to the equipment's power variations and interruptions. It functions as a voltage stabilizer and, at the same time, isolates the equipment and systems from the power lines.) I told the facilities manager that I cannot think of any besides having redundancy for the UPS. The facility manager just laughed at me, and a couple of years later, this plant went on fire, which stopped its operations for more than a day. The fire started from the UPS. Later on that year, I learned that they bought not one, not two, but three UPS as a means of redundancy for their equipment.

Same as true hold for January 28, 1986, STS-51L Challenger Disaster. It was not the O-ring that killed the 6 astronauts and 1 civilian teacher but rather it was the decision by both NASA and Morton Thiokol Management and their disregard to the engineers who warned them of the grave danger the O-rings would induce in launching Challenger at a cold temperature of 31 degrees Fahrenheit. The engineers insisted on a Root Cause Analysis investigation on the O-ring erosion problem from the previous flights, which never came. As Roger Boisjoly Quote:

- One month, one lousy month later, we all realized that we had no power, no authority, no resources, and management support; then how the heck can you get anything done if you don't have anything of those items I have mentioned? You can't. And the frustrations just absolutely go up minute by minute, day-by-day.

Because of the tragedy and the Rogers Commissions' findings, NASA and Morton make design changes to the space shuttle and added a 3rd O-ring. The previous O-ring was redesign by including a heated layer to prevent its effects from cold. But I wonder if these design changes would occur if NASA did not suffer any casualties from that flight. I think we need something really hard to hit us in the head to make those changes, or sometimes we need to experience a disaster before doing the right thing. Am I right?

I am sure that there is much more, but we all leave it to our experiences simply because this is how we do things here in our plant since the beginning of time. We only act on problems if they finally haunt us. I hope that we do not wait for the consequences to happen due to the decision we have made today.

- Mistakes are painful when they happen, but years later, a collection of mistakes is what we called experience." _Denis Waitley_

CASE 5: I always preach that the best way to lower your maintenance and operating costs is by improving your equipment's reliability. And improving the reliability of your equipment has much to do with how we perform maintenance in our equipment. Any good maintenance manager always knows these things, but even if he wants to do the right thing, external forces are beyond their control. When maintenance tries to requisition parts, their procurement department or purchasing will always opt for the lowest cost without realizing what havoc will cause in their equipment. The reason is simple. These people from Purchasing do not realize what Life-Cycle Cost is all about since they are only interested in the initial cost of the maintenance requisitioned part. Worst, some industries do not even allow the maintenance people to talk to the vendors directly as this is part of their policies in their plant, or it had been part of the system where purchasing are the only ones allowed to discuss these matters to their vendors.

I could give you more, but I will stop at these five cases. And I think you may have experienced any one or two of these cases. My message to the readers is that it is never too late if we act on these problems. People create problems, but I also believe that people can solve their problems, but sometimes we need to make these changes ourselves. Change must always come not from other people, not from the top, not from the bottom, not from consultants but from within ourselves. If we have the guts to finally put our fingers down and start off by looking ourselves in the mirror so that we can learn from the problem. As long as I see maintenance smile, then it is always a sign of hope, and I see a reason to hang on to this business and continue doing what I'm doing, which is teaching people the value of their work as maintenance. And I think that's all I have to say about that.

3.5: May 2009: 13 Grave Mistakes in TPM Implementation

Mistakes in TPM implementation are costly; hence it is important to know its basic fundamentals and roots before initiating it. These things are non-negotiable if we speak about implementing TPM. Many have tried to short-cut the TPM process, which eventually causes their failure and blames the TPM process itself. TPM is a very good improvement methodology; I have seen it worked. Both people and equipment are improved. This can only be achieved if we follow the process accordingly. I have listed some of these common mistakes in no particular order of importance. Most of these are not written in TPM books and are based on my own account and TPM experience. If your plant is initiating any TPM initiative or has been implementing TPM for a long time with little or no success at all, then perhaps you need to take time out, read this edition of our newsletter, share it with your management team, and give it some thoughts.

1) Never Use Contractual Employees in your TPM Implementation: Never attempt to perform Autonomous Maintenance on your plant if the operators are on contract. It will simply not work as they will leave the plant in a short period, and we need to retrain your operators again. TPM aims to empower operators, and I do not see any form of empowerment or enthusiasm when the operator's contract to work is nearing an end. During their final days of work, what is inside the operator's mind is not about getting to Step 2 or Step 3, but where will I work tomorrow. What happens to my family, or where will I find money to enroll my kids in school and that sort of stuff. If you tell me that this is our plant's practice, I only have one logical piece of advice for you. Stop doing TPM, whatever you do, and how extensive your efforts are in driving TPM; it simply will not work. It's that simple. How can we motivate people that are about to leave the plant for good? You simply cannot. The goal of TPM is not just to improve the equipment but for both operators and equipment to improve together in parallel with each other.

2) Too Many Improvement Initiatives in the Plant: Many people are somewhat numb or too naïve to understand the implication of too many improvement initiatives and strategies in their plant. All improvement initiatives are good if the plant focuses on them, but if there are too many improvement initiatives such as six-sigma, TPM, Lean Manufacturing, Just in Time, TQM, RCM, and the rest is not being consolidated, your plant will end up with each of these initiatives having their own what you called "Champions" or "Facilitators." You will just end up with too many meetings and action items to be done. Perhaps your day at work is consumed by endless meetings. How much of your time is spend on meetings and action items where you have a 9:00 am to 11:00 am TPM meeting review followed by 1:00 to 2:00 pm meeting with six-sigma, then a 2:00 to 4:00 meeting with Safety, and finally a 5:00 to 6:00 Operations Review Meeting with your boss. The next day will be no different, and so are the rest of the days. All plant improvements and strategies must be consolidated, integrated, and aligned. When champions of these initiatives do not communicate with other champions of other initiatives, there will just be redundancy of activities, people are confused about which strategy to prioritize.

3) Assigning Part-time People to do TPM: TPM is done in a step-by-step cookbook format. One of the important aspects of its preparation is assigning full-time people assigned to the TPM Office. Each of these people will play a very important role in the TPM implementation in the plant and the respective pillars they are handling. These people act as your plant's TPM internal ambassadors responsible for providing directions and roadmaps in their team's TPM journey. The number of full-time staff on TPM will depend on the population of the plant. Many industries who want TPM implemented in their plant merely assign a part-time staff or engineer. Again I will be frank with you; it simply will not work.

4) TPM Office Reporting to Operations: TPM Office should directly report to the President or CEO of the organization and must not report to any part of the department such as Operations or Quality Control. The TPM office should never be dictated and should be an independent body. Let me put it this way, during the initial process of implementing TPM, expect a lot of resistance initially. When TPM is under the control of operations, what happens is that operations can delay TPM activities to push through with production unless the Operations Manager sees value in the TPM process; otherwise, it would be the other way around. In this case, TPM activities will just be deferred and delayed until it is completely forgotten. I recall during my time when our TPM Office was reporting to one of the high echelons of operations, he completely changes the TPM direction and dictates us to act as police, provide photos of any untidy, messy, or anything unusual either from offices, equipment, an environment that we can spot and highlight them to their management. One of the engineers handling Office TPM reported a messy office of his boss; the engineer said, Sir, this is your office; how can we expect them to follow and the rest was history.

5) No Budget for TPM: TPM is an investment; you can only reap the fruits when you understand what is at stake here. Although many readers, especially Top Management, would think twice about this item yet, I would rather be honest about this issue. TPM will require heavy investment during the early stages of implementing training for their people, correcting abnormalities, and restoring activities done by the Planned Maintenance teams. The problem if their requisition had been disapproved and the machine is scheduled for an Autonomous or Planned Maintenance Certification or Audit, the teams will have no resort but to cannibalize parts on other equipment. By doing this, then we are just fooling ourselves with this practice. Like any other department, TPM must have a budget to allocate, which should be approved by the decision-makers of the plant.

6) Trying to Shortcut the TPM Process: There are two kinds of trains: the fast train and the slow train. Most managers simply want the fast train. They want everything done fast. When equipment fails, operations and management wants the equipment to be repaired in the quickest possible time. If I say wait, let's perform a Root Cause Failure Analysis first, they will say that they have no time for that. The problem with the people riding the fast train is that when your equipment fails and you repair it, the evidence is washed out. Most of the time, the part that fails is thrown away where much information can be obtained from that failed part. I believe that to be fast, we need to take things slowly. Like the TPM process, this is a very slow process; everything is done step by step, phase by phase, and one step at a time. Plants initiating TPM must fully understand that we are changing the culture, and we are not simply improving the equipment, but rather we are making a paradigm shift in the minds of the people. Many will be tempted to perform TPM, set up an Autonomous Maintenance Team, and perform Initial Cleaning at once on the machine. WHOA! Stop, you are way so wrong. That is not how it's being done. Remember, when one of your operators gets injured during the Initial Cleaning process, I believe that will be the last time you will hear about TPM. Initial Cleaning can only occur when the team understands their equipment's safety, which is being initiated by the Planned Maintenance team. All machines should be ranked so that the team can finally select their model machine to undergo the TPM process. Only Rank A or Worst Machines must be targeted first in selecting their pilot model machine. For TPM to work effectively, do not try to shortcut the process.

7) TPM and Company Goals are Not Aligned: Each company has its own indices, measurement, yardstick, and KPI's that they measure from time to time to determine where they are and how well they are doing, which is a good thing. Likewise, TPM has its own indices and measurements as well. The problem begins when the company and TPM goals are different. The effect will be that TPM will be considered a separate program and not part of the plant's corporate strategy because, in industries, we only focus on things that we measure. When a plant is measuring Availability and TPM is measuring OEE, it will create a separate direction for TPM.

8) TPM Office Insufficient Knowledge on their Respective Pillars: Again, it is not enough that a plant has a full-time staff on TPM, but the TPM staff must be equipped with sufficient knowledge on TPM or, more importantly, on the TPM Pillars they are driving. References, books, and training must be accessible to the staff of the TPM Office. For example, when a TPM Facilitator driving Autonomous Maintenance is only knowledgeable on Step 1 of the 7 Steps of Autonomous Maintenance, the teams that have completed Step 1 of Autonomous Maintenance will have to wait until the training on Step 2 are finally ready. The team's momentum and enthusiasm simply fade away, and the machine gets worst once again. The team needs to undergo and repeat the process of Step 1 once again. Before initiating TPM, roadmaps should be completed; training for each step and phase in the TPM pillar should be available so that TPM activities will run smoothly.

9) TPM is thought of as a Short Term Approach: Many people, mostly managers, are disappointed when they cannot gain results from TPM in a short period, completely abandoning the process and blaming it once again on TPM as another fad or flavor of the month. Industries must realize that TPM is a slow process. It takes time to realize the benefits of TPM, both from the equipment and from the people. TPM is done on a Step by Step process. Doing TPM

Preparatory Step can take 6 months or even more for a large plant. These things are crucial in any TPM implementation and must be understood by the plant or industry before initiating any TPM activities. The benefits that can be reaped from TPM can be felt first on a small scale and will increase gradually. Remember that the most difficult part of any improvement initiative is about changing culture. Adopting a TPM culture cannot be done overnight, if you know what I mean by that.

10) Frequent Reorganization in the Plant: It is quite difficult to implement TPM if there is frequent reorganization in the plant. One thing for sure that will happen is that team members will be affected. Just to tell you a bit of a story, one team has completed the 7 Steps of Autonomous Maintenance; the team was highly motivated and empowered. From time to time, the team had been invited by other plants to present their success stories. Three of its members were transferred to other departments, and they have been replaced by new operators; after some time, another two original members were again transferred, leaving one original operator in that equipment. What happens, their model machine deteriorated rapidly, and the basic equipment condition had been forgotten. How can you simply push through with a Focused Improvement Project if your members keep changing all the time? Perhaps we need to change the name of this pillar from Focused Improvement to a Not So Focused Improvement Team.

11) Wrong Selection of Pilot or Model Machine: Almost all TPM books will recommend that the worst machine be selected as the model machine to start their TPM activities. While this is true, we must not take the word literally. Suppose you go to your plant to identify the worst equipment. In that case, you might be selecting equipment that had been abandoned for a long time or equipment that had not been used because of too many problems in the past, or you might be tempted to select equipment that will take more than the cost of the equipment to restore and rebuild. Likewise is selecting a machine that will be decommissioned and phased out in a short period. To me, this is complete nonsense. When selecting your model equipment, make sure that the equipment is being used regularly in your operations.

12) No Management Review on the TPM Process: One of the important parts of TPM is having a regular review with Management people to determine their TPM progress on how the teams are doing in their implementation. The TPM Master Plan serves as a guide if the team is behind or ahead of schedule. What Steps are we in the Autonomous or Planned Maintenance process? Which department is fast and slow in their TPM roadmap implementation? When will we complete this step and that sort of stuff? Management also asked what changes that take place when doing TPM on both the people and the machines. When a regular review occurs, the people involved in TPM know that something is expected from them. The review process also creates a feedback forum between management and the TPM teams involved. The review process also creates a venue for management to show their support and commitment to the TPM process.

13) Ignoring the Real Message of TPM: Industries that value their people have a great chance of implementing TPM successfully. Remember that TPM is about changing people because it believes that they will also change how they do things in the plant if people change. When equipment improves, it did not improve by itself. Someone did something, and it was the

TPM team. TPM is about people. When people become empowered, there is no limitation on the benefits it can bring to the plant. Therefore, remember that in any TPM initiative, the focus must be on the people since they will be the ones to improve their equipment, and it is not the other way around. And I think that is all I have to say about that; until the next reliability newsletter, keep on reading.

3.6: June 2009: Why are Operators Important in the Reliability Strategy?

If there is one thing that TPM and RCM both agree on, it involves operators in the reliability strategy of improving the equipment. As I keep on saying, the equipment is always a shared responsibility for both operations and maintenance. Maintenance can only overcome the vicious cycle of reactive maintenance if operators join them in sustaining and improving their equipment. TPM has a thorough and detailed plan for operators. The journey on empowering operators is done through the Seven Steps of Autonomous Maintenance, while on the other hand, operators are strongly recommended to be part of the team covering the RCM Analysis because both strategies believe in the importance of operators.

In TPM, operators learn from Maintenance: From the traditional mindset of "I operate (which is the operator), you fix (which is the maintenance) to a paradigm that the equipment is a shared responsibility for both operations and maintenance. They both must take care of the equipment. Operators must understand their role in the equipment by maintaining the asset's basic equipment condition. But to make this happen, maintenance must take the time to teach operators about their equipment from time to time. Here are some of the things that maintenance must teach their operators:

• Basic Machine Function
• Safety on the Equipment
• Importance of Establishing the Basic Equipment Condition
• Preparing Inspection Standards
• Basic Lubrication Standards and Techniques
• Correct Tightening of Bolts
• Minor Repairs and Troubleshooting
• How to Use their Senses to Detect Machine Problems
• Importance of Addressing Sources of Contamination and many more

Autonomous Maintenance is a slow process. Do not expect results in a day or a week. It takes years to transcend and empower operators in this journey, but it will be very much worth the time. Autonomous Maintenance is not just a mere transfer of maintenance responsibilities to the operators. It is changing the mindset of operators that they too had a role in taking care of their assets by helping maintenance establish the basic equipment condition in their machines by keeping the equipment clean, having the correct amount of lubrication, and completing all the bolts and nuts in their assets as these small things are often the cause of sporadic and catastrophic problems in the equipment. Most industries' problem is that they do not allow their operator to think but often tell them to focus on one thing, and that is "output." Those industries that are now making their journey to World Class Maintenance Management are finally learning that maintenance cannot do this alone; they need to be partners with their operators to move forward. This is the missing link in any reliability improvement strategy. It is like having your own car. If you care for your car, you just don't drive your car to wherever you want to go. You need to do some basic things in your car, such as opening the hood and checking the radiator if it still has water, checking the tires if they are inflated, cleaning the car, checking the dipstick to check the oil level, and so on. These small things lengthen the life of the car. Take note that it is not the mechanic nor the technician who will tell you that the car is starting to overheat. The car's temperature will indicate that. It is the car's driver who should be the one monitoring this symptom. Once the operator changes, expect that they will do the same to their equipment, which is one of the benefits of having Autonomous Maintenance.

In RCM, Maintenance Learn from Operators: Reliability-Centered Maintenance strongly recommends that an operator be part of the team to compose the RCM analysis. When equipment fails, the first line of defense will always be the operator. Kindly note that when something fails on the equipment, it is not the maintenance who will be in direct contact with the failure; rather, it is the operators. Therefore, operators must have experienced something on the equipment before it fails. This information is vital and critical in the RCM Analysis. By allowing operators to relay their experiences on failure, maintenance learns a great deal about the failure itself. Remember that the operators will always be the first line of defense in any breakdowns or failures encountered on the equipment. It is usually the operators that will sense the symptoms of the failure first and not the maintenance. Maintenance usually goes to the equipment after the failure has happened because the operator called them. Suppose there is a symptom such as noise, a change in vibration, some changes in the gauge's reading; it is the operators who will experience this and not the maintenance. One problem with most industries is that there is little or no communication between operators and maintenance. When the equipment fails, and the maintenance is around, the operators break and leave the maintenance alone to fix the equipment. The operator only comes back to the equipment when maintenance calls them that their equipment is ok. When both operators and maintenance communicate, maintenance can have a deeper understanding of the failure. Operators might not be technical people like the maintenance, but they know something for sure about the failure that happened in their equipment. RCM believes that the best source of failure modes are the operator.

Tips on How to Involve Operators in Maintenance

Involving operators in maintenance is not easy, especially if operations management is not committed to the process. I recall a dear friend who once told me that he strongly recommended building an operator-maintenance relationship in the line in one of their operations meetings. As he was midway in his talk, a higher ranking operations manager shut him down right on his face and told him, "Boy, you watch too many movies." So how in the heck can you start a relationship with operations? You can't. You simply can't. Perhaps this is where 3rd party consultants would fit in. They cannot talk to us in that fashion simply because we are not employees from the plant. Hence, you may start with these basic simple tips for those industries without a clear and detailed Autonomous Maintenance in their plant.

When the equipment fails, ask the operator to stay for a while when you are working with the equipment. When you have experienced the failure before, tell the operator stories about your experience with that failure. Create some bonding with the operator. This way, the operator learns something about their equipment and the failure as well.

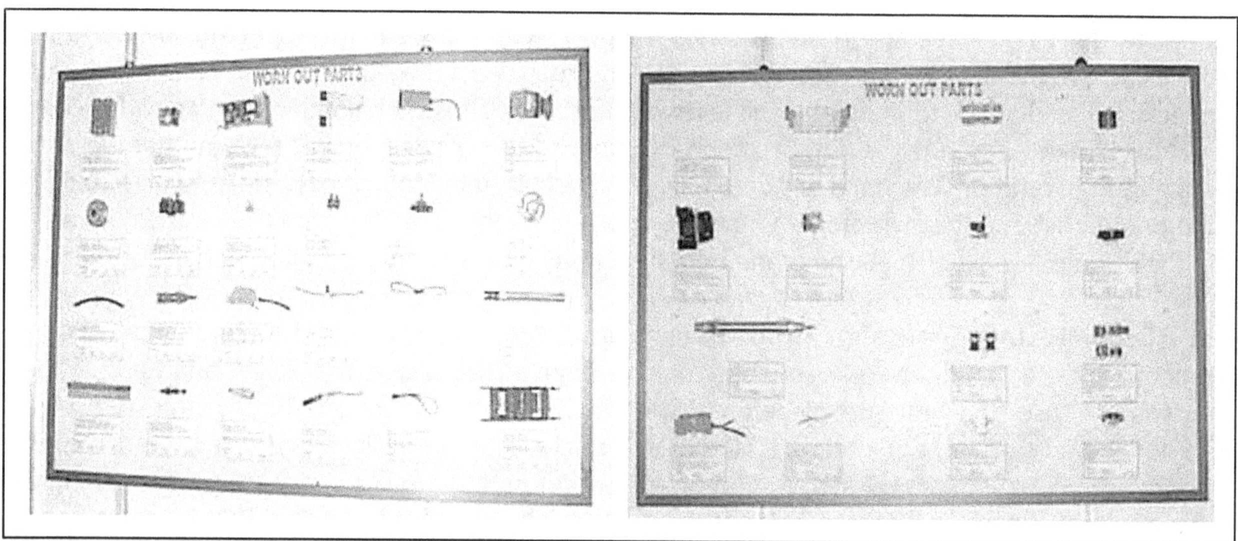

Figure 3.4: Teaching Operators Broken and Worn-Out Equipment Parts

Second, always keep the part that fails. Make it a habit to teach the operator one part a day. Spend around 10 to 15 minutes per shift. Discuss the function of the part and why this part is important. By doing this, the operator can learn a great deal about the part that failed unless your operators are engineers themselves. Most of the time, operators are non-technical people. They find it difficult to tell what went wrong because they do not even know the name of the part that failed. They can only describe it to some extent. By teaching operators just one part a day, it can facilitate better communication between operators and maintenance. We mount these failed parts on a 4 x 8 whiteboard shown in figure 3.4 and discuss one part, item, or spare each day with the operator during my time.   It creates better communication with operators.

Third, most people complain that maintenance prepares operators to check on the equipment, yet operators do not actually perform these inspections. Remember that operators will only perform the checking religiously if they are the ones who made the checklists themselves. Maintenance must understand that operators are capable of thinking. One of the roles of maintenance is to coach and guide operators in allowing them to make decisions on the critical things to inspect in their equipment. If this is done, then there is a greater chance that operators will perform the inspection themselves since they were the ones who made the checklists.

To conclude, operators' involvement will play a vital role in any reliability improvement strategy if operators learn from maintenance or the maintenance itself learning from operators. The equipment is always a shared responsibility for both operators and maintenance. Maintenance people can only advance to other relevant tasks on maintenance if operators participate in this initiative.   I believe that this is the missing link in any reliability and maintenance strategy.   It is difficult to succeed in any Vibration Analysis initiative if the equipment lacks a couple of bolts. Operators will play a major role in establishing the basic equipment conditions on their equipment and assets. Always remember that when a failure strikes, it is the operator that encounters the failure first and not the maintenance; therefore, your operators shall remain the first line of defense on equipment-related failures.

3.7: July 2009: The Classical RCM Approach, Should There an Eight Question - 1?

This edition of our newsletter is a series of two parts covering the 7 Classical questions. In the process, I basically asked if there should be an eight-question on the RCM process. The first part will explain the seven basic questions, which I think some of you might already know, and the second part of this newsletter will explain the need to have an eight-question.

Classical RCM Seven Basic Questions

To begin with, first, let us define what Reliability-Centered Maintenance is all about. RCM is a process used to determine any physical asset's maintenance requirements in its present or current operating context. It is also a process used to determine what must be done to ensure that any physical asset continues to do whatever its users want it to do in its present operating context. In its simplest definition, RCM is used to sustain the reliability of equipment, asset, sub-system, or process by improving the way we perform maintenance. The Classical RCM Approach conforms to SAE JA1011 by answering the 7 basic classical questions, which are as follows:

1. What are the functions and associated performance standards of the asset in its present operating context?
2. In what ways does it fail to fulfill its functions?
3. What causes each functional failure?
4. What happens when each functional failure occurs?
5. In what ways does each failure matter?
6. What can be done to predict or prevent each failure from occurring?
7. What should be done if a suitable proactive task cannot be found?

RCM Question 1: What are the asset's functions and associated performance standards in its present operating context? The first question on RCM asks us to define the functions of the asset being analyzed to have a better grasp and understanding of what functions of the equipment or asset are being used currently and need to be maintained. To take special care of defining the asset's function is distinguishing between a function from a feature. A function summarizes why the asset was acquired in the first place. A feature is a nice to have or additional attractions to the asset. If we want to define one of the car's main functions, it allows us to go from one point to another. But there are other features a car can have, like having a hi-fi sensurround radio with a built-in DVD component. Although one feature can be considered a function to another. There are cases and confusion on whether to state it as a function or a feature. For example, I asked one colleague what his car's most important function was. He told me that it was his radio when I asked why. He told me that he is on the night shift, and the radio keeps him awake during driving. Hence, in this case, the feature now becomes a function. It means that his radio should be available whenever he is using his car. Functions can also be classified as Primary Functions, which answers why the asset was acquired in the first place, and secondary Functions, where an asset is required to fulfill other functions rather than the primary function itself. Also, note that there will cases where the secondary function's failure poses more threat and danger to the organization than the primary function's failure. Hence, the first step in any RCM Analysis is to derive the asset's functions in its present operating context or in the conditions that it is currently being operated.

RCM Question 2: In what ways does it fulfill its functions? The second question on RCM asks us when do we consider that the asset has actually failed. It is important to define what is meant by fail, especially for everyone involved in the asset, such as operations, maintenance, safety, quality, third party contractors, etc., to agree or disagree on when to declare the asset failed. A safety officer may declare the equipment in a failed state from a safety standpoint if it is leaking. It might cause someone to slip, resulting in injuries. A maintenance point of view of failure is that if a certain part or spare had worn out rapidly, such as the seal which is causing the leak, then this is when he will declare it failed. But perhaps from an operations standpoint, as long as the equipment runs and can produce what they want, the equipment is not in a failed state. Hence they will declare a "No Failure Situation." Perhaps the only time they will declare it fail is when the equipment stops because it runs out of lubrication. In this case, we have three different people who have three versions of failure. If these people are in disagreement as to when to declare the asset failed, the problem is, what the heck are we trying to avoid in the first place.

RCM Question 3: What causes each functional failure? The third RCM question deals with the probabilities that cause the failure to occur. When we speak about each functional failure causes, we ask the hypothesis or, most likely, the failure modes. Failure modes should be defined in enough detail for it to be possible to select a suitable failure management policy. It is likewise important to list every single failure mode that is likely to affect the asset. For example, when a pump discontinues discharging, one possibility is that the valve was closed. Another possibility is the motor burned out. Still, another possibility is that the bearing seizes, or the strainer screen is blocked by contaminants. Another failure mode is the impeller shaft is broken, or the shaft keyway disengages with the impeller causing the pump not to discharge any fluid at all. We need to list down all the possible failure modes or probable causes of failure at this point. When we speak about failure modes, we refer to all the probability of failures and not the Root Cause of the failure itself.

RCM Question 4: What happens when each functional failure occurs? This question asks us if this failure happens, what will be the effect of each failure mode on the asset. In short, in this question, we ask about the failure effect. Will some alarm sound as a result of the failure or stop production? What is the estimated time the machine will be down? Usually, how long will the maintenance be working on the equipment? Are spares needed to place the equipment back in operation? Are they readily available to the maintenance crew in the storeroom? Knowing the effects of this failure mode will make us understand what will occur after the failure had happened on its own and how maintenance will respond to this situation. Failure effects usually are written in paragraph form and should be written in enough detail so that maintenance can better grasp and understand what to do if the failure takes its toll.

RCM Question 5: In what ways does each failure matter? Every failure has its own unique consequences. This question asks us that in the event of a failure, what will be the consequences of the failure itself. Will the failure be hidden or evident? Will it affect the environment? Does it have safety consequences? Will the failure affect the operations, or will it not matter since the equipment is equipped with a backup or standby unit? RCM simply tells us that if the consequence of failure is not acceptable, then maintenance should exhaust all efforts to reduce or eliminate the risks associated with the failure. When the failure has minimal consequences, then RCM allows failure to occur on its own. RCM tells us that it is more important to understand the failure's consequences than completely eliminating the failure itself. Understanding the consequence of failure is critical to the RCM process. This will spell the difference in having a sound maintenance task and understand which failure modes are unacceptable.

In December 1984, a methyl isocyanate gas leaked in one of the tanks in Bhopal, India. 40 tons of highly poisonous methyl isocyanate gas escaped the pesticide factory in the Union Carbide plant in Bhopal. As a result, more than 6,400 died, many from the effects several years after the disaster, while 30,000-40,000 were seriously injured in the world's worst chemical disaster. Over 500,000 men, women, and children have been exposed to the poison clouds, and at least 6000 people died within the first week of the disaster. The current death toll is still rising and is over 16,000. Pregnant women were miscarried. Those who survived have their babies stillborn. The poison affected some people's lungs and nervous systems. The consequences of this single failure were simply unfathomable and affected so many lives in Bhopal.

RCM Question 6: What can be done to predict or prevent each failure? The sixth question asks us if the tasks performed by Preventive or Predictive maintenance can address the failure mode when it occurs on its own. Suppose the failure mode occurs on a wear-out pattern; in this case, it can be anticipated using Preventive Maintenance replacements and overhauls. Simultaneously, suppose the failure mode provides some signs or symptoms that it is on the verge of failing, then these failures can be captured using Predictive Maintenance instruments. For example, infra-red thermography can provide users with an increase in heat or temperature in one of their motors. Monitoring can signal them that the motor has already reached its potential failure stage. The maintenance is being alerted that something must be done so that they can intervene to avoid further damage.

RCM Question 7: What should be done if a suitable proactive task cannot be found? This is the last question in the RCM process, which states that if both Predictive and Preventive Maintenance will be unlikely to address the problem, what options in the maintenance tasks are left to address the failure mode itself. The remaining maintenance tasks left according to RCM will be failure finding tasks or functionality inspection for protective devices and standby components, a run to fail situation for lesser critical failures, or allowing the part to fail as long as the only consequences of failure will be the direct costs of repair and the last option we have is to redesign and modify which means that if doing maintenance cannot help solve the problem then the only option left for maintenance will be to modify or redesign the system. There will be cases in which maintenance cannot help solve the problem, and if this is the case, then maintenance needs to walk the thin line and opt for a redesign or modification to eliminate or reduce the consequences of the failure itself. Redesign and modification are applicable if the consequences of failure would simply be not acceptable at all in the plant.

These seven basic questions complete the Classical RCM process. The first four questions comprise the first part of the RCM process, usually written in the RCM Information Worksheet. The remaining three questions refer to the RCM Decision Worksheet, which is being performed using an Algorithm or Decision Diagram. SAE JA1011 "Evaluation Criteria for Reliability-Centered Maintenance (RCM) Processes is a standard for the RCM process to distinguish the Classical RCM from Streamlined RCM Versions. But to carry out a comprehensive RCM approach, I believe that there should be an eight questions. The eight-question would spell out the difference between the success or failure of the RCM analysis. If there will not be an eight-question, then the first seven basic RCM questions will be deemed useless. Our next Newsletter will answer the eight-question in the RCM process that needs to be answered.

3.8: August 2009: Classical RCM Approach, Should There an Eight Question – 2?

Our previous July 2009 Newsletter covers the first part of the RCM's seven classical questions. It discussed the 7 basic classical questions on RCM. This second part will unveil if an eight-question is required in the RCM process.

First, let me start on how I started my involvement with RCM. 10 years ago, in September 1999, two of my colleagues at work, myself included, attended a three-day Public RCM Class in my country in the Philippines. After the training, I spend a great amount of time studying and researching this topic. My interest in RCM was overwhelming, and finally, I develop my own training materials on RCM. I created three prototype teams, two from operations and one team from the Facilities and utility section, to undergo the RCM analysis. The team was carefully selected with an incumbent operator from each team except for the facilities since all the maintenance staff are also the operators. Each team also comprises a maintenance manager to support this initiative. Finally, in December 1999, I completed my materials. I delivered a three-day RCM Class for these three teams. The people that attended the training were convinced beyond a reasonable doubt that their current way of doing Preventive Maintenance was not solving their day-to-day maintenance problems and that RCM should be implemented. An RCM roadmap was developed for the teams to go through the RCM process step by step, and the RCM analysis began. They provided me a schedule of their meeting twice a week, in which, in most cases, I facilitated the meeting. Each meeting lasted from 2 and sometimes up to 4 hours. We follow the RCM process very carefully and finally completed the RCM Analysis. Each of the 3 teams submitted the RCM Analysis Report, including the RCM Information Worksheet, RCM Decision Worksheet, and the operating context statement. The report also compares how much failure mode their current Preventive Maintenance tasks addressed compared to the derived RCM Analysis tasks.

I spend around a couple of weeks checking each document averaging around 1 week per team to check how the process was done. I returned their documents with few corrections in which they immediately corrected. Finally, I gave the signal to them to carry out the implementation of the RCM Analysis. The two teams from operations were hesitant to change their PM process simply because to change something, you also need to revise the current specs they are using, subject to several approvals from different people. You see, everything we do in this plant is heavily documented, and each has a corresponding specification number which includes PM. There is always an SOP for everything we do inside that plant.

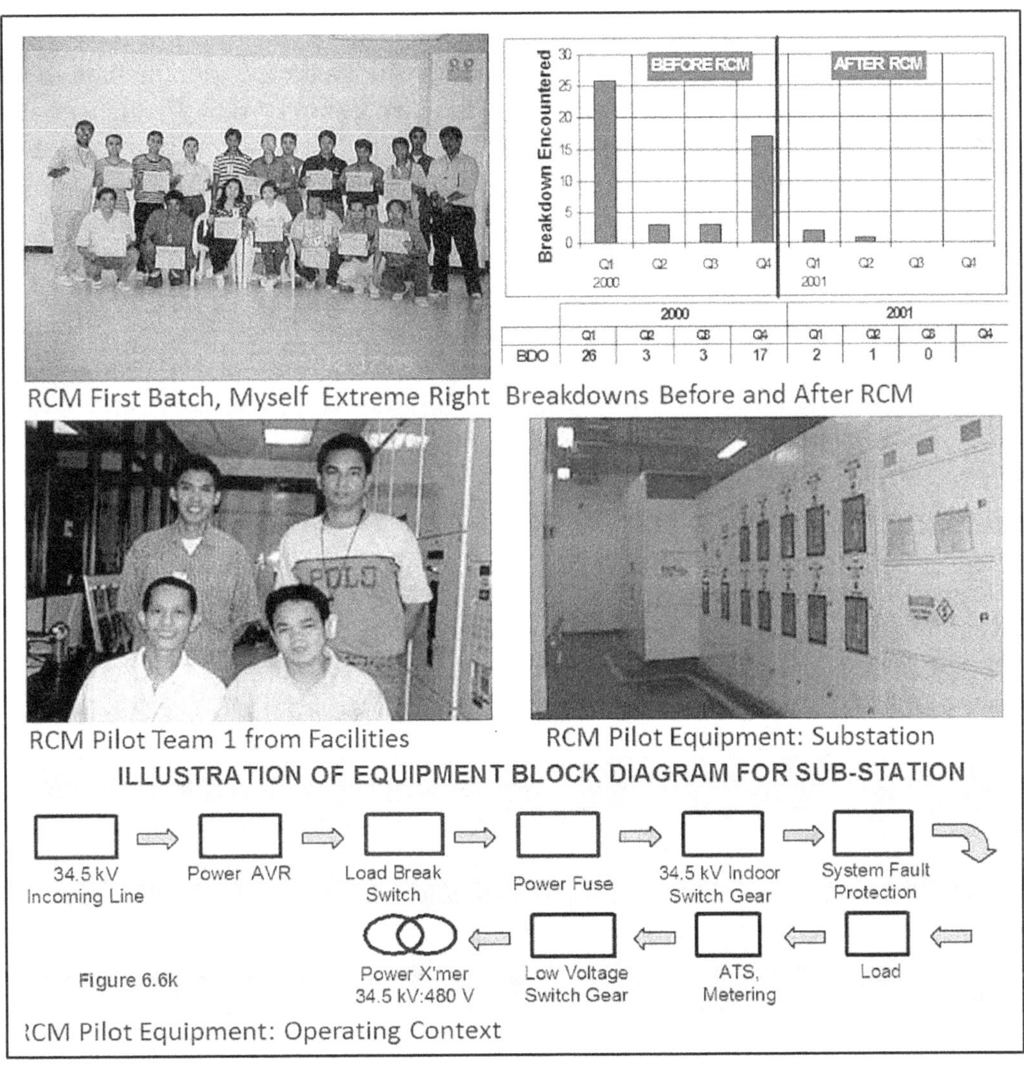

Figure 3.5: RCM Implementation Two Failed, One Succeeded

Finally, the team who was hesitant in implementing the RCM told me to talk to their supervisor. I went on and talk to their supervisor and discuss the implementation of the RCM Analysis. The supervisor shook his head and said that he cannot decide on this matter. He instructed me to talk to their manager. These were the same managers who attended my RCM Class, and so after talking to them, they said that they cannot decide on this matter and told me to talk to their Department Manager. And so I went on and talked to their Department Manager. I told him that his team had completed the RCM analysis and dramatically trimmed down all activities on the maintenance tasks from 350 to 150, which the team has accomplished. The Department Manager told me that to comply with your RCM, he told me that he will do both the RCM and their current PM. He said that they still need to perform their current Preventive Maintenance activities to conform with their specs. My jaws dropped, and I told him that if we do them simultaneously, it is like sweeping the floor to get rid of dirt, and after cleaning it, we put dirt and trash on the floor once again. I told him that if you implement RCM, you must drop whatever PM activities you are currently doing since everything is already covered in the RCM Analysis. The Department Manager disagreed with me to drop their current PM tasks. I left the room dismayed. The same thing happened with the other team, which is also from operations. Finally, their leader approached me and told me that the added activities in their PM resulted from action items from meetings, Quality, and Customer Audits throughout the time the machine was operating. In short, the RCM Analysis of the two teams from production was never implemented, and I think the documents remain in their drawers. I just cannot tell what happened with the documents.

The third team from Facilities and Utilities implemented the RCM Analysis, and as the months passed by, they were all shocked with the results. After completing the RCM Analysis, they implement it on their 10 substations. Their breakdowns from their substation were reduced dramatically just by implementing the RCM Process. A failure or breakdown on any of their 10 substations can paralyze more than a couple of hundred pieces of equipment in operations. Here is the operating context statement of the RCM team from the Facilities section.

Operations' Overview: Substation operates without interruption. It is stand-alone equipment that operates without human intervention. The substation is equipped with monitoring instruments and protective devices that shut off the unit and transfer the power source to the generator automatically in case of circuit faults and power failure. Operator intervention is being introduced intentionally during scheduled power transfer, and Preventive Maintenance activities are performed based on the schedule. Raw power is obtained from MERALCO (Manila Electric Company), the NPC's power distributor (National Power Corporation). An AVR (Auto Voltage Regulator) transformer with a variable ratio was used to maintain the line's transmission voltage at 34.5 kilovolts (kV), 3 phase, 60 Hz. This voltage is transmitted through a 3 phase high tension wires, # 4/0, bared, referred to as the transmission line. A Meralco power fuse type S and C SM -300 E rating at 300 amperes is installed before the substation's plant load or primary side. It disconnects any portion of a primary circuit during overload and a broad range of primary side electrical faults. The fuse is the weakest spot in the circuit made up of a fusible metal with a low melting characteristic, which will melt at a rated temperature. Should there be an overload, the fuse melts as temperature increases from the overload current. This will disconnect the supplied power.

Just recently, I conducted an in-house training on RCM. After the training ended, they organized a short management presentation the following day in which their CEO, VP for Operations, and all high-ranking people in their organization were present. I delivered the RCM presentation for a couple of hours and answered questions from the management team from time to time. Finally, when I ended the presentation, the CEO stood up and said, "Just do it." But up to this point in time, they have not started with their RCM Analysis even if their CEO had given the nod. Perhaps what he meant was Just do it after finishing priority no. 75 or just do it next year or just do it when you want to or if you have the time. Am I missing something here?

Finally, I am convinced that with these experiences, which I have shared with you, I hope you agree that there should be an eight-question in the RCM process. And this must be answered even before doing any RCM Analysis, and the eight-question is "Are We Ready for the Change?" because if the plant or industry is not ready with the change, then questions 1 to 7 in the RCM Analysis will be deemed useless even if you hire the best RCM consultant in the planet since it will not be implemented anyway. Remember that conducting the RCM analysis is the easy part; implementing it will be the challenging part. And so to wrap it up, the Classical RCM Eight Questions should be:

1. What are the functions and associated performance standards of the asset in its present operating context?
2. In what ways does it fail to fulfill its functions?
3. What causes each functional failure?
4. What happens when each functional failure occurs?
5. In what ways does each failure matter?
6. What can be done to predict or prevent each failure?
7. What should be done if a suitable proactive task cannot be found?
8. Are we ready for the change?

3.9: September 2009: Is Root Cause Failure Analysis Reactive or Proactive?

Many are asking whether Root Cause Failure Analysis is reactive or proactive. This is like answering the question of which comes first, the chicken or the egg. Although there is no scientific proof that the chicken emerges from a single cell and just came out of the farm a couple of thousand years ago. All we know is that the chicken comes from the egg and the egg comes from the chicken, but whichever comes first, I think, I leave it up to the reader.

Although Root Cause is both reactive and proactive, I believe that in Root Cause, we need to be reactive first to be proactive. Root Cause Failure Analysis can only be performed if there is an actual problem or a failure. Nobody can perform a Root Cause Failure Analysis if no failure actually happens. A failure must first occur before a Root Cause Failure Analysis investigation can be performed. Without a failure, then how in the heck can we perform a Root Cause Failure Analysis? Likewise, the failure must be fresh, meaning it just happened. It is highly unlikely to perform a Root Cause Failure Analysis if the failure happened a year ago or 6 months ago. Why? Because to perform a Root Cause Failure Analysis, the evidence must be frozen.

One of the most difficult parts of doing Root Cause Failure Analysis is that eventually when something fails on the equipment, two teams will respond to the failure. The first team is the restoration team, or simply known as the fast team. These people are fast; they need to restore the equipment as fast as possible since productivity will always be the utmost priority. The longer the equipment is restored, rest assured that operations people will be watching you at the back, whispering how much more time is needed to fix the equipment. On the other hand, we have a slow team. These people are the ones who will investigate and probe how and why the problem or failure occurs on the equipment in the first place. The problem is that when the slow team arrived. Everything is restored, and the part that eventually failed had been thrown away for good, and every bit of evidence had been washed out and destroyed by the fast team or restoration team. So my question is, how in the world can you perform a Root Cause Failure

Analysis in this situation? You just simply can't.

When Is Root Cause Reactive? Root Cause will always be reactive at the start since a failure needs to be in place before anyone can perform an investigation. For example, a ball bearing had failed that caused the equipment to stop functioning. There was no stock in the stockroom for the part; hence downtime was enormous. Management warranted an analysis of why the bearing had failed, so an investigation team composed of a Principal Investigator and Evidence Gathering Team had been formed to investigate the failed phenomenon. We can only perform an analysis when something had failed or a problem erupts that causes the operation to halt. The team is interested in understanding the underlying causes of the problem so that the people involved can learn from the things that go wrong in their industry.

I recall one time, I was called to a plant to provide a brief presentation on their management team about Root Cause Failure Analysis since one of their leading customers got very irritated, which almost pulled out their business in that plant. Equipment failed that eventually caused a delay in their shipment. When the customer asked the reason for the delay, the director said that a bearing failed in the equipment processing the units, which eventually stopped the equipment. When the customer asked what did they do to address the problem? The director said that they replaced the bearing with a new one. The customer was so horrified and almost pulled out their business with this plant because they do not have an answer to why the bearing failed in the first place. This plant, which I worked previously, called me and asked me to present the courses I teach. As I was presenting, someone interrupted me since what they want is a total one-time approach to capture all the failures on this equipment. When I told them that Root Cause was not designed to handle that and provided my recommendation to do RCM, I lost that opportunity. In my mind, these people do not know what Root Cause Failure Analysis is all about. It is not something that will address all the equipment's problems, but rather it is used to understand the cause of a particular failure that actually occurred in the equipment. You see, the people above wanted to address every bit of probable cause that can occur on the equipment; I think what they need was some sort of FMEA or RCM and not RCFA. One manager even proposed to have a why-why analysis to the 10th level. In my mind, this guy is a freak. The difference between the two is that FMEA is designed to capture the probable causes and not the root cause of the failure and prioritize the failure according to its severity, occurrence, and detection, while RCFA is designed to address the true cause of failure based on the evidence unfolded so that we can learn from the things that go wrong.

When is Root Cause Proactive? First, Root Cause Failure Analysis can only be Proactive if we can finally learn and understand that big problems are just an accumulation of small problems that had been neglected in our equipment. Most Root Cause initiatives and efforts seem to address catastrophic and big failures. The important thing that matters is why not address the problem when it is still small. One of the leading Root Cause providers in the US, where I am affiliated with Failsafe Network, greatly considers the importance of small problems. One of the most memorable things I've learned from Failsafe Latent Cause Analysis Experience training is that big things go wrong because we do not act on small things. Waiting to do a Root Cause Failure Analysis on big problems will only assure us a continuance of bigger problems. Hence, Failsafe recommends performing root cause on MAXI or big events and on MINI and

MIDI or small-scale events. Their training, "The Latent Cause Analysis Experience," believes that our focus should be on small problems since big problems can only be taken care of if we take care of the small problems. Isn't it that TPM also wants the same thing in mind that when we take care of the basic equipment condition, then the equipment's life can be prolonged?

Second, most people say that performing Root Cause Failure Analysis will prevent a recurrence of the problem. Let me rephrase this. It is much more likely that performing a root cause failure analysis will help us prevent the same failure from occurring. When we have taken care of a single or couple of causes, there is still a probability that the failure can occur again in the future due to a different cause. Root Cause is not designed to address every possible cause in a single investigation unless the failure has only one cause, but rather we are only interested in the cause of the problem based on the evidence unfolded.

Let us assume that the RCFA analysis shows traces of silicon contaminants and metal debris on the bearing's raceway, causing the bearing to suffer from fatigue and spalling, and the investigating team indicates that the oil was too contaminated with metal contaminants. A particle count indicates that the ISO range was way out of the standard. Good filtration practices were provided, and contamination control practices were adopted, not only on this equipment but also on the rest of the assets with lubrication. How lubricants were stored was addressed, and so on. But all these efforts were done as a result of finding out the cause of the failed bearing. After initiating these corrective measures, they experience smooth operations due to their proactive efforts. They addressed that single equipment that failed and the other equipment could probably fail due to a similar cause.

Third, Root Cause Failure Analysis can only be proactive if we can learn from the failure itself. This is easy to say, but it is much more difficult to do in the real world. There is always a deeper underlying cause behind the physical and human cause, which is the latent cause. Many industries think that when they find out the physical cause of the problem, they stop, only realizing that the failure will revert back to them in the least unexpected time. Leo Tolstoy quote, everyone thinks of changing the world, but no one thinks of changing himself. The late "King of Pop," Michael Jackson, in his song "Man in the Mirror," sang, "If you want to make the world a better place, take a look at yourself and make that change." Latencies are hidden causes that need to be exposed. It is not just about system causes, rules, and procedures that we ought to follow, but it is about how each of us has contributed to the problem. To address the latent cause of the problem, we need to answer the following questions:

• What is it about the way I am that contributed to the problem? (About you)
• What is it about the way we are that contributed to the problem? (Organization)

• Space Shuttle Challenger exploded because the O-ring on the right solid rocket booster leaked, which contacted the flammable external tank. The Roger Commission concludes that the explosion was caused by the solid rocket booster joints' faulty design. Morton Thiokol engineers already knew this problem and had informed their management about this. However, Morton Thiokol's management team also knows that their billion-dollar contract with NASA was nearing an end and needed to be renewed soon so they can once again be in business with NASA. Richard Feynman, a novel prize winner, physicist, and part of the Roger

Commission that probes the Challenger Disaster investigation said that both engineers and managers were not communicating effectively. NASA was not communicating well with its suppliers.

• An electrician rewind the motor backward. The motor was set up in a critical location on the plant. Management decided to discipline and suspend the electrician without pay for a month, which eventually wrecked the turbine. Later on, a deeper probe into the investigation revealed that the electrician had already been working for 24 hours straight because the person who should relieve him was on sick leave. The operations manager requested the electrician to extend his shift as they badly needed the motor, to which the electrician agreed. Was it really the fault of the electrician?

• The team probing the failure of the pump found out that misalignment caused the pump to fail. Likewise, they disciplined the person who performs the misalignment. When the investigative team probed deeper into the cause of the problem, they found out that this person only used his eyesight to perform an alignment. No instruments were being used; no training was ever provided to this guy. They learned that the instrument had already been requisitioned 5x but was disapproved by higher management 5x due to cost-cutting schemes. Was it really the fault of the person who performed the alignment, or was there a much deeper cause?

I can go on and on and give more cases, but let me stop at this point. Whether we accept it or not, latent causes exist, and they must be exposed. Change can only occur if we have the guts and courage to look ourselves in the mirror and ask ourselves, what is it about the way I am that contributed to this problem, and what is it about the way we are as an organization that contributed to the problem. Therefore to answer the question, is Root Cause Failure Analysis Reactive or Proactive. Root Cause Failure Analysis is both Reactive and Proactive. Still, Root Cause will start on a reactive scale. Once we learn from the problem and provide the necessary corrective measures in which the problem no longer recur, then this is when Root Cause becomes proactive. Root Cause is proactive if we can address the small things and expose the latencies of the problem, and I think that's all I have to say about that.

3.10: October 2009: Operations and Maintenance - Will the Feud Ever Stop?

This is often a typical conversation between operations and maintenance people in the plant.

• Operations: How many times have I need to remind your thick skull that you cannot perform your PM this month since we have many shipment schedules this week?
• Maintenance: But this is the 5th time you have waived the PM?

When capacity is high, and production is at its peak operations people focus on just one thing, and that is "output," but when production people cannot deliver the required productivity or output for the day or week, they will get all the necessary details you have never imagined why they were not able to deliver. Only one thing pops up in the minds of the operations people, and their most lame excuse will be maintenance. During their operations meeting and review, operations will always find a clever excuse and blame maintenance for the downtime. Since maintenance people are just human too, they will fight back and accuse the operations people of flooding their equipment to death by waiving the equipment for their monthly Preventive Maintenance. Everything ends up in a merry-go-round.

Sounds familiar! When operation people are present in my training class (since most of my delegates come from the maintenance function), I ask them if they are in good and harmonious relationship with the maintenance guys in their plant, and most of them will just laugh at it at loud (lol), and I believe I got their message.

Although I would like to write this newsletter in an unbiased fashion, which I would try to do my best, both have valid points about their situation. Yet, most of the time, they seemed not to get along pretty well. When productivity for the day had been delivered for the day or week, every bit of glory and recognition goes to operations and production people. On the other end, when production had not been up to speed, then all fingers point to the maintenance people. I even recalled a meeting when I was still working in the mining industry where the maintenance was being bombarded by operations where the maintenance manager finally stood his ground and said, if you really want your output, I can place all my maintenance people to operate the equipment, and I will provide the output you want. And so the battle continues in which both parties have lost the war. In the end, the industry lost. Hence, instead of admitting that they are part of the problem, operations people will always find a scapegoat, and blame is always given on the maintenance side. In short, maintenance is treated like a stuntman (just like in the movies). Let's say we are filming an action film and the leading actor (which of course is the operation guy) is in a fistfight with the villain, both were given instructions by the director, and as the villain is about to punch the leading actor in the face in which he needs to fall as he needs to absorb the punch, the director will shout "and cut," the fists of the villain are inches away from the face of the leading actor. The movie director tells the leading actor to take a seat and drink his orange juice. At the same time, all the makeup artist is busy retouching the face of the actor. The movie director shouts, where's the stuntman (which, of course, is the maintenance)? The maintenance will come along, take the leading actor's position, and the director shouts, "action." The maintenance gets punch right in the face and falls down, and that's the way it goes.

During my last employment in the mining industry, they have a room they called "the war room" when I was still new around; I used to ask why they call it the war room, to which I already knew the answer. One of the maintenance managers asked me to sit in during one of their operations review meetings. The meeting was presided by their Resident Manager, the highest person in the mine. The meeting started very calmly, but when the Resident Manager asked why the week's productivity was not met, they started what they called a war. Operations were accusing the maintenance, and the maintenance are likewise accusing the operations of waiving the PM of their equipment. I wondered why these people kept shouting with each other, or perhaps their oratorical hearing ear was not that clear. I have seen cups of coffee spilled on the floor, perhaps due to the excitement of one another. I recalled the maintenance manager defending his tribe and said that your operators just don't seem to care about their equipment and operate them to destruction. We have been providing these checks for over a year for your operators to perform, yet nobody seemed to follow them. The story continues with what needs to be done the next time, and some action items were generated. But looking at it from a personal point of view, nothing had actually been accomplished, and they will just end up the same way they have started. One of the problems, in this case, is that sometimes it can get ugly and personal. When it gets personal, good people resign or transfer to another industry ending in the same merry-go-round situation.

There was an industry which I taught last year. I learned from one of the delegates that one of their policies is that if a breakdown occurs on their equipment, and then the breakdown must be charge to a specific department (accountability issue). Still, there are departments with common equipment being used. When this common equipment between the two departments failed, each group will find a way to charge the opposite department why the equipment failed. I have learned from this plant that people from these two departments have not communicated for several years due to this policy. The communication gap had become personal.

From an Operations Point of View

In my book on World Class Maintenance Management, The 12 Disciplines, I wrote, out of order again? No wonder they PM it. Maybe if the PM group did not touch this equipment, I'm pretty sure this will still be running smoothly! I believe operations have a valid point here. If we performed Preventive Maintenance in our equipment, then it should be the other way around. Are we missing something? These things simply are called Infant Mortality Failures, and many factors contribute to them, but most likely, Infant Mortality Failures are caused by human errors committed during the process of performing scheduled overhauling and replacement during Preventive Maintenance shutdown. Perhaps this is one of the reasons that operations waive their PM. Most maintenance people believe that the more often equipment is overhauled, the fewer failures will be encountered. In reality, this is just the opposite. The truth is that PM overhauls and replacements increase failures by introducing what we call Infant Mortality Failures into otherwise stable parts of the equipment. This means that if we are not equipped to disassemble and assemble systems in our equipment or lack some training or tools to perform these tasks, I believe that the best thing to do is leave the equipment alone. I think there was a rapper in the 1990s, MC Hammer, who sang the song "U Can't Touch This." Just don't disturb stable systems unless we are really equipped with the skills, tools, and knowledge to do it. The bottom line is that if operations people always complain about having some difficulty starting up the equipment right after an exhaustive Preventive Maintenance, then expect this as one of the reasons for deferring or waiving the PM on the equipment. Maintenance should review the lists of activities they performed on the equipment, perform only the necessary maintenance needed on their equipment, and apply Precision Maintenance.

From a Maintenance Point of View

Operations people must realize that their equipment is not a plug-and-play instrument, just like your television set or laptop we have in our home. For as long as parts are moving inside your equipment, they will always be subject to stress. Some parts eventually need to be maintained. There are perhaps more than a hundred ways your equipment can fail. What maintenance can actually do is prevent, predict, anticipate, control, manage, or prolong the failure itself. But maintenance cannot do this alone since I firmly believe that operators play a vital role in maintaining their equipment. Let me put it this way, when we drive our car, it is actually the driver who checks his car if there is still fuel in it. The driver also has to monitor the temperature of the car from the dashboard. When something is wrong with your car, it is the driver that actually feels and experiences it. Suppose we simply drive the car and ignore these symptoms and signs or avoid performing the simple basics; in that case, we expect the car to fail sooner than expected. Same true holds to our equipment. It is actually the operator who will witness the failure itself, and whatever information the operator can provide maintenance is of critically vital importance. In fact, RCM believes that the first line of defense against failure will be its operators. The sad thing is that some operations people do not merely understand these things and tell their operators to focus only on one thing: output. What a shame!

Figure 3.6: Operators and Maintenance are Just like Puma and Adidas

The bottom line is that the feud between operations and maintenance is pretty much alive in most industries. I just can't tell when it will stop; perhaps it is part of their subconscious culture. Suppose both groups can just set aside their politics, red tapes, ambitions, pride, position, ego, and that sort of stuff, start working out things in a civilized fashion; in that case, that will be a good start. Operations and production people must understand that no industry can exist without maintenance, and maintenance people must also understand that they cannot escape the vicious cycle of firefighting and will always be trapped in a reactive mode if operators will not accept the fact that maintenance is always shared responsibility for both of them. Nobody can exist without each other. Hence, instead of blaming one another for the failure or downtime, why not work together and not against each other. Whenever we blame the maintenance or operators for the downtime, the problem has still not been resolved. Blaming and finger-pointing are just an excuse for our problems. Everything will just get worst over time. Being cynical and sarcastic is not the way to improve productivity and reliability. In my class, I frequently say that the next time these kinds of things happen in your operations meeting, just bring two things with you, a stone and a mirror. If operations people accuse you again of the problem, give the mirror and stone to them and tell them to look themselves up in the mirror and ask them that if they think that maintenance is the only problem, then let them throw the stone to us. Well, of course, if operations will throw the stone, you need to avoid it. My point is instead of blaming each other, why not sit down resolve this matter together. When the Mafia bosses have their disputes, they have a rule which they call a "sit-down" in which the two bosses will talk and settle their disputes, but of course, after the sit-down, they kill each other. Operations must realize that their equipment needs to be maintained and that maintenance should understand their role in maintaining their equipment rather than dismantling all the parts that later cause problems. Both operations and maintenance simply cannot exist without each other; these two are just like Puma and Adidas. Did you know that these two are actually brothers born with the same parents in real life? And I think that is all I have to say about that.

3.11: November 2009: Cost Cutting - The Wrong Way to Save Maintenance Costs

Suppose we conduct a survey and ask each and every reliability and maintenance manager in every industry which of these two items is more important, reducing costs or improving reliability. In that case, almost everyone will agree that improving reliability is more important than reducing costs simply because if the reliability of the equipment starts to improve, then the cost will definitely go down; however, it is not the other way around. In fact, I know and experienced many cases that reducing costs on maintenance will create problems with reliability. But suppose we conduct an actual survey in the real world on what most organizations and industries focus on. I think it is more on cost-cutting or having some form of cost reduction program. This is the name of the game most industries play. Maintenance often realized a cut or slash in their maintenance budget by 10, 20%, or even more, and the figures increased every year. Headcounts are reduced as a result of a study on a man-to-machine ratio. And what happens to those affected or classified as excess, sad to note that most of these good people will be laid-off. When old maintenance people are forced to retire, industries replaced them with a new breed of younger people with no experience maintaining the asset. Spares and budgets have to be cut, and some need to be localized to cope with the plant's cost-cutting scheme program.

Purchasing people are busy looking on the yellow pages or sourcing the internet for vendors that can provide the lowest possible cost. Maintenance people are always in a never-ending battle against production people in getting the equipment for scheduled PM work. The budget for maintenance training is often cut short, and maintenance can barely attend the training since there are too many fires to put out in the battle. In short, maintenance is left in a never-ending war against being reactive with sudden breakdowns all around. Maintenance is definitely going to lose this war. The pressure is building on the maintenance side. As pressure increases, maintenance people are in a state of chaos and confusion. When maintenance is confused, it only boils down to one thing, "under pressure" (I think the late Freddie Mercury sang that song), and that is the life of a maintenance person. It doesn't have to be this way. There is always a better way. If you are dead serious about saving costs on maintenance, why don't you consider the following?

Take a Look and Study Your Spare Parts Management: Instead of laying off people in your maintenance organization, why not let 2 to 3 people improve your storeroom and organize a Spare Parts Management System in your plant. Most industries have a storeroom to stock parts, but not all stores know how to manage their spares and items. Let just give you some hint. Suppose you have a form of computerization for your spare parts. Does the physical inventory match the system's quantity on the computer? If they don't match, then you got a big problem. Just imagine going to your storeroom to get some parts you needed for a repair only to realize that the part has no more physical quantity left even if the system says that there are at least a dozen more of them. How would you feel? What do you do with the spare parts whose equipment had already been decommissioned or in the process of decommissioning? What do you do with those parts that are classified or deemed obsolete? How about the same parts with different part numbers because they came from different vendors and are used by different departments? How do you control them? These are just some of the questions that can be improved inside the storeroom. I am pretty much sure that if we can utilize these people to improve your spare parts, the savings they can generate will go far and out weight the time they spend in improving their storeroom. Remember that the Spare Parts Management goal is to provide the right part at the right time when both maintenance and operations needed it most. Remember that the longer the spare part is acquired from the storeroom, the longer the equipment's downtime can be realized. For those industries who are playing a dangerous game on cost-cutting, perhaps I can recommend the following in this picture:

Consider the Study of Life Cycle Costs of your Spares and Equipment: There is always a temptation to purchase equipment or spares based on the lowest or cheapest source for industries. This might not be such a good idea since purchasing based on the initial cost only tells us one side of the story. The true costs can be seen based on their performance. In my experience, this is the problem with most procurement and purchasing people since they make their decision to purchase parts based on the lowest bidder or lowest possible costs. The savings these departments claim are insignificant since both operations and maintenance can encounter many failures. They always look at the part's initial cost and not the cost of problems the part may give the user in the long run. This is the problem with cheaper parts and equipment that fails repeatedly and will likely yield a much larger amount of cost in a period compared to slightly higher equipment. The equipment's true cost can be felt when the equipment was commissioned to the plant until decommissioned or disposed of.

Let me give you an example of two oil filters; one is rated as a nominal filter with a Beta Rating of 2. This filter has an efficiency of 50%, while the other filter is absolute and has a Beta Rating of 1000. The efficiency of this oil filter is 99.9%. The filter's efficiency refers to the number of solid particles or contaminants that the filter element can remove. Let us say that we have a 5-micron size filter with an efficiency of 50%; this means that if there are around 100,000 contaminants that will pass through the upstream of the filter of 5 microns, the filter can only capture 50% of the contaminants or around 50,000 out of 100,000. The remaining half will travel downstream of the filter together with the oil and back again into your system.

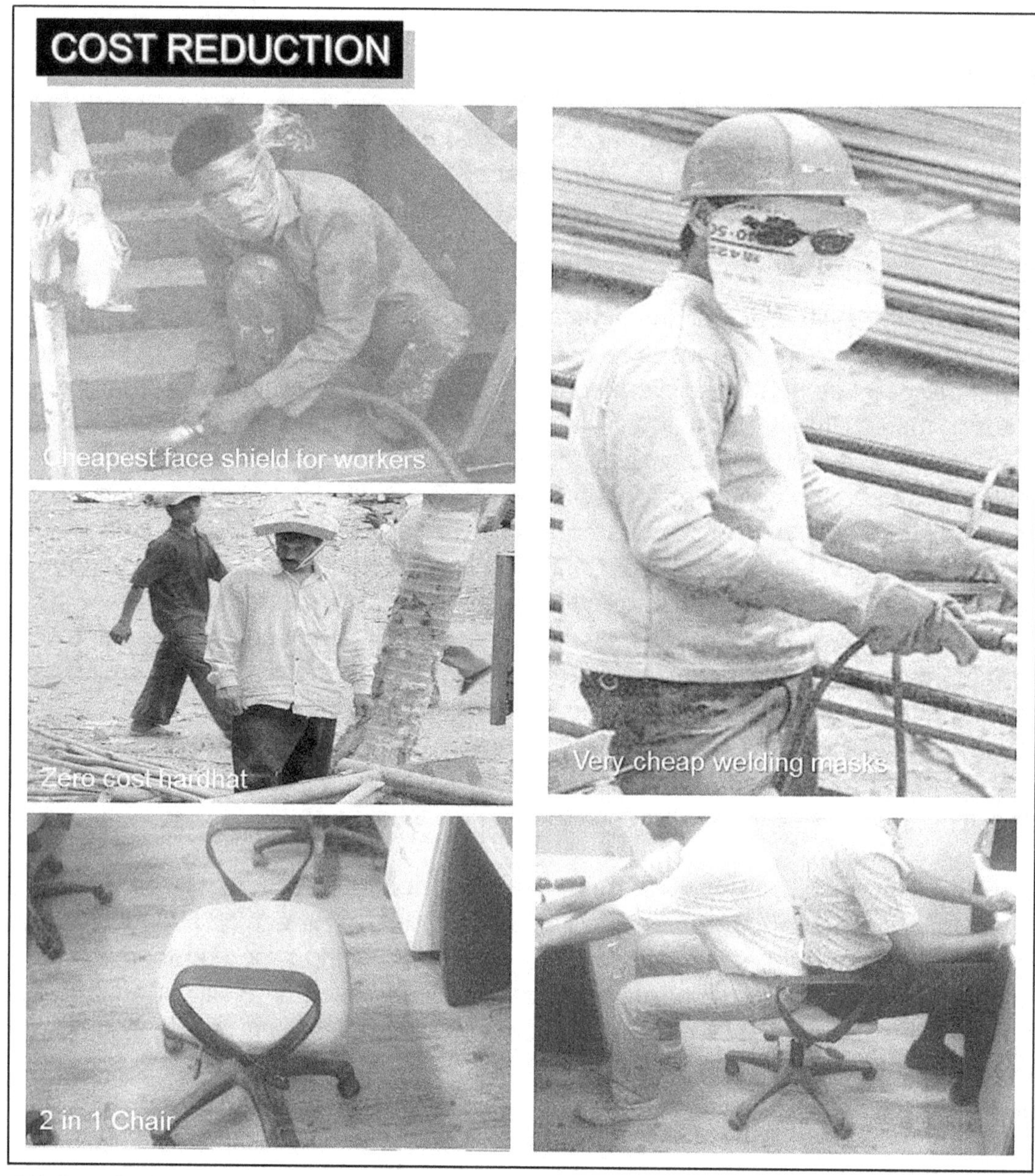

Figure 3.7: Cost-Cutting, the Wrong Way to Save Cost

On the other hand, if you have an absolute filter of 5 microns and a rating of 99.9%, this means that the oil filter is guaranteed to trap all five-micron size contaminants and larger. This means that 99,500 out of 100,000 will be trapped by the filter. Obviously, the filter that can trap more contaminants will provide you more reliability and smooth operation than a nominal rating of the filter. Now you must be wondering about the cost of this highly efficient filter. If I told you that the cost of both filters is 450. Definitely, you would like to purchase a filter that has an absolute rating, isn't it? But there is one big difference in the costs; although both filters cost 450, the absolute filter will cost 450 USD while the nominal filter costs 450 Philippine pesos (1

USD = 46 Pesos). This means that the other filter costs are actually 10.65 USD compared to the absolute filter, 450 USD. But suppose you can compare the other costs, including downtime costs, replacement costs, spare parts costs, overtime costs, and other costs, then this is where the more expensive filter will be cheaper in the long run. Again if we ask the maintenance people which filter they prefer, I am pretty much sure that they will choose the oil filter with a higher efficiency rating, but perhaps the sentiments of the procurement or purchasing people will be different since they will select the nominal filter since it is cheaper in cost compared to the absolute rating filter. Although they will be saving around 430 USD (if they buy the nominal filter). What happens is that they will leave the maintenance people to suffer. See what I mean.

Establish a Lubrication Strategy in your Plant: Most of the time, lubricants are just taken for granted. Oil is just oil; if the level is low, we add oil or grease to our equipment. If there are just too many equipment types around, then maintenance decides which equipment gets the oil. From the maintenance point of view, that is all we do because that is all we know. Putting too much lubricant or less lubricant will not be good for your equipment, so the question raised here is how much lubricant quantity is sufficient to be placed on each piece of equipment. Suppose we are greasing a bearing on the motor, for instance. Does maintenance know what type of grease to place in the bearing's raceway? How much quantity is required? How do they know that the grease is reaching the raceway of the bearing? What type of grease gun to use, and all that sort of stuff? These are some of the important questions that should be answered because, in most PM activities, the level of information is insufficient. Most PM checklists state what equipment is to be re-greased, how frequent and who should be responsible, but as to what type of grease, the correct amount in grams, type of grease gun to be used, special accessories such as high-pressure digital meters, and the hearing aid is not explicitly known. There are many more ways to reduce your maintenance costs, and listed above are only a few of them. Hence, instead of laying off maintenance people in your industry, why not utilize them in improvement projects or permanently improve your spare parts, life-cycle management, lubrication strategy, oil contamination control, predictive maintenance, to just to name a few.

Another important aspect to consider in your Lubrication Strategy is Oil Contamination Control. How are your lubricants stored? What are your dispensing methods, how do we transfer oil to the equipment, disposal procedures, oil analysis tests performed, addressing oil leaks, and so on? The smaller the contaminant, the more abrasive it can turn out to be. These tiny contaminants, often invincible to the naked eye, are the main cause of problems that affect our equipment, and they can be controlled if we know the ways and means to address them.

Address the Basics in your Equipment: If there is one thing I believe that industries should do is to take care of the basic equipment condition. Big problems and catastrophic failures are just an accumulation of small things frequently neglected. The process of addressing the basic equipment condition simply means keeping the equipment clean, applying the correct lubrication, and tightening bolts. These are very simple tasks to perform, yet the impact will be felt on having equipment with lesser failures. It is useless to conduct vibration monitoring on the equipment if it lacks bolts, nuts, and fasteners. Remember that it will only take one loose or missing bolt to create a chain of destruction and havoc in our equipment.

So the next time your industry will lay off some maintenance people in your organization for cost-cutting schemes, why not convince the decision-makers in your organization to permanently place some of your maintenance people on areas that will reduce your maintenance most. This is a good deal; the savings generated will go far out weight their salaries. Some of them are explained above. Maintenance is not only a department; it is just more than an activity or task. These are your very own people that look upon the welfare of your assets and equipment. Maintenance is a business, and in any business, there is always an investment. Maintenance is more than a repair function but sad to note that most industries perceive maintenance to be performing repairs. Worst are industries that don't value their maintenance people because they are inclined to think that they can be replaced anytime. I tend to agree to some extent but kindly note that the skills, knowledge, experience, and camaraderie will take a much longer time to replace. All I would like to share in this newsletter is that if we value our maintenance people, then provide them with the right tools, the right knowledge, training, skills, and most especially the respect they deserve, and I am pretty much sure that these people will deliver with all humility and I think that is all I have to say about that.

3.12: December 2009: Why Performing More PM Will Just End Up to More Problem?

In a typical plant or industry, the equipment is scheduled on a time-dominated or calendar-based pattern with a list of activities to comply with. Some of these activities will include the following:

• Cleaning and lubrication
• Replacement of parts
• Overhauling and scheduled restorations
• Routine and Functionality inspections
• Performing rebuild and repairs

The activities performed vary from one type of industry to another and depend upon the equipment. By definition Preventive Maintenance (PM for short) is also a series of tasks performed at a defined frequency dictated by the passage of time either in the number of machine-hours, mileage, or running hours that either extend the life of the asset or detect that an asset had a critical wear and is about to fail or break-in operation. PM is also a basic maintenance performed on the equipment and facilities. The main goal of performing the PM task on a scheduled basis is to extend the equipment's life and assure its capacity to support its plant's goals and targets. In doing Preventive Maintenance, the thing to consider is that the cost of doing PM should always be less than the cost of failure it is meant to prevent. If the cost of performing Preventive Maintenance is 1000 USD and the cost of failure it is meant to prevent is 100 USD, then performing PM is not feasible to use. The intentions of performing PM are, without a doubt, very noble, but most industries simply cannot maximize the benefits derived from performing their PM; let us explain some of the reasons behind it.

Problem Number 1: Not All Parts Will Eventually Wear Out: The problem begins when most people who perform PM assume that the machine's condition is deteriorating and that the need to perform maintenance is correlated with the passage of time or age. This means that the item can be expected to operate reliably for an amount of time and is expected to eventually wear out concerning its operating age. This is true if the part, spare, or component to be overhauled or replaced has a clear wear-out or age-related pattern. It means that the part is said to survive this specific age. The truth behind all of this is that not all parts of your equipment will eventually wear out. Others will fail randomly, while others will fail prematurely right after an extensive PM overhaul or replacement is done. Let me put it this way, if you are driving your car and a rock hit your windshield that created a small fracture, the question to raise is when will the second rock hit your windshield. Nobody knows. Preventive Maintenance cannot tell. This is what random failures are all about. They can happen anytime. A different type of maintenance task or strategy is needed, but not PM. This means that if the failure you are experiencing in some parts of your equipment is random, then Preventive Maintenance will be useless and ineffective. Suppose most of the failures you experience are from electronic boards; the best thing to do is duplicate the PCB that fails frequently or perform a thorough Root Cause Failure Analysis Investigation to get to the bottom of the problem. Do not overdo or overuse Preventive Maintenance tasks. Performing intrusive or forced maintenance on an asset will just do you more harm than good.

One of the most important discoveries in the reliability and maintenance field that all maintenance people should know and understand is that there are six patterns of failure in the real world. The six failure patterns will provide us with an indebted understanding of how certain equipment parts behave and how it fails. The six failure patterns were discovered by Stanley Nowlan and Howard Heap of United Airlines way back 50 years ago to improve commercial aircraft's reliability by improving the way they do their maintenance. Their discovery "shocked" the maintenance function because only 11% of failures can be prevented by performing scheduled Preventive Maintenance, while 89% of equipment failures cannot be Prevented by doing any form of Preventive Maintenance.

Problem Number 2: Similar Machines Will Not Require the Same Degree of Maintenance Requirements: Another problem with performing Preventive Maintenance is that people who plan and perform their PM activities will assume that similar equipment will have the same maintenance requirements regardless of how the equipment was operated. This is not true. The only time that the activities will remain the same for similar equipment types is that both equipment types operate under the same conditions. Suppose I have a pump and the pump is discharging water, I cannot perform the same amount of maintenance to a similar pump pumping out acid or slurry. In this case, the impeller will erode much faster than the other pump pumping water. Likewise, for manufacturing plants, you simply cannot perform the same degree of maintenance requirements for a machine running 24 hours a day continuously for weeks against a machine that is only running 1 shift for 5 consecutive days.

Another case that I can share is that if a team performs an improvement or modification on a piece of equipment, there is a temptation to replicating this improvement or modification on similar machines. Their presumption is that they do not want the problem to exist in other machines. This is valid if the other machines that will be replicated suffer the same problem as the original machine in which the modification or improvement had been generated. But suppose some machine similar in nature simply emits no form of a problem and is considered stable; it might as well leave that equipment alone. Don't replicate the improvement in this machine. There is a probability of inducing new problems on that piece of equipment.

Problem Number 3: Introduction of Infant Mortality Failures: Suppose your operator encounters many problems getting the machine running smoothly right after an extensive PM overhaul; it is good to revisit how we perform our Preventive Maintenance tasks on that piece of equipment. Infant Mortality failures exist. The most likely cause is overdoing your maintenance on the equipment.

Our equipment is like a hotel room. Inside the hotel are many rooms. If someone is staying in room 502, you have the option to place either a "Do Not Disturb" or "Make My Room" sign. If you place a do not disturb sign only to be disturbed, then you get pissed off. But on the other end, if you place a Make My Room sign on the door, then expect someone to knock on your door to clean your room. My message here is very simple if you have all the tools, knowledge, and skill to dismantle and bring the equipment back in one piece as it was before dismantling it, then go ahead and overhaul it for all I care. But if you have the slightest doubt in bringing it back together in one piece, then you might as well think twice or thrice or even more before trying to dismantle the machine. What I am driving at is that in maintenance, the more we touch, the more it fail, the less we touch, the less it fails.

One of the best examples I can provide the readers is the discovery of Stanley Nowlan and Howard Heap from the civil aircraft, which showed that 68% of items that failed conformed to pattern F, considered Infant Mortality Failures. An example of some of the benefits achieved from this learning on infant mortality failures by the civil aircraft industry was a dramatic reduction in their scheduled overhauls in their DC-8 aircraft from 339 items which are previously subject to an overhaul which they dramatically reduced to only 7 items for their DC-10 aircraft, where DC-8 is much smaller compared to a DC-10. But in DC-10, they have dramatically reduced the number of overhauls done on this bigger and more complex aircraft. One of the items no longer subject to overhaul was their turbine engines. Likewise, on the initial program developed for Boeing 747, which took the United Airlines 66,000 man-hours on major structural inspections to reach an inspection interval of 20,000 hours. Compared to their traditional 4 million man-hours inspection on smaller and less complex aircraft such as DC-8. Today, it is simply much safer to ride aircraft than 40 or 50 years ago because the rate of accidents per million take-offs during those times was high, which was around 40 to 60. This means that for every 1 million planes that take-off, expect around 40 to 60 planes to go down. Was it magic? No, of course, it was just about learning from the lessons on infant mortality failures.

Infant Mortality Failures are failures that actually occur during the beginning of life. You might have experienced this where operators have a hard time starting the equipment, especially when the equipment was newly commissioned, overhauled, or right after performing a shutdown or scheduled outage. Others might refer to this as starting-up failures, debugging, or commissioning failures. When operators complain to maintenance that they find it difficult to run the machine, you have been a victim of Infant Mortality Failure. Many factors can induce infant mortality failure, such as lapse or slip on the maintenance in putting the equipment back together in one piece with the same condition before dismantling. Human errors can occur during reassembling or putting the equipment back together to realize that a small spring, bolt, nut, or item was not placed back in the equipment. Again it will take us a couple of hours or more just to correct this kind of thing. Most of us just pretend that no one saw that and just keep this small part away in the PM bay for good, or perhaps you have dismantled some stable systems inside your equipment that should have been left alone.

To conclude this newsletter, in adopting any maintenance strategy, it should always be done 3 folds. There should be maintenance tasks that must address not only age-related but also random failures. Remember that Preventive Maintenance is only designed to capture age-related failures, which is roughly around 20% of the overall equipment failures we experience, and that leaves us 80% of equipment-related failures that need to be addressed as well.

Year 4

2010 RSA Reliability Newsletter Vault Archive

> *Japanese industries do not believe in best practices. Why? Because they do not understand the word best. They only understand the word better because they know that the best can still be made better. This is the main concept of their continuous improvement or Kaizen strategy. They believe that the older the equipment gets, the better it runs, which is contradicting to others.*

4.1: January 2010: Why Do Most RCM Initiatives Fail?

Some articles written about the process of RCM yield very little or completely no success in implementing it. Others somehow come up with a streamlined version in which their selling point is to arrive with the same results an RCM Analysis can achieve with less time in performing it. While most people who perform their RCM implementation blame their failure on RCM as another 3 letter acronym and another flavor of the month. My question that always plays in my mind is RCM really to blame for their failure, or have they missed out on some of the very important aspects in the RCM Analysis that lead to their failure?

So far, I consider RCM to be the most important document and piece of research ever to emerge to determine the best maintenance requirements that can be done to ensure that our equipment continues to do whatever the users want them to do. In light of this, there are very important factors that we need to consider before initiating an RCM Analysis in our equipment. Failure to realize these factors can eventually result in failure. I have listed some of these in no particular order.

1) Wrong selection of the asset or process to be analyzed: The type of equipment, asset, process system, or sub-system to undergo the RCM Analysis depends on the type of plant and industry itself. Suppose you are working in the Oil and Gas industry; the RCM Analysis can be a series of assets and components that make up a system or sub-system process which can include a motor, pump, valves, supply tanks, pipes, discharge tanks, protective devices, and so forth. One may consider some equipment that will be most critical to their operations for a manufacturing plant. Equipment from their Facilities or Utility section will be a good candidate. One can consider a specific sub-assembly (not the whole machine) that fails a lot during operations if it is a highly complex asset from stand-alone manufacturing equipment. One of the mistakes in performing an RCM Analysis is to perform the analysis on a highly new piece of equipment since the outcome of the analysis will provide little or no benefit at all. The team might have been in the plant for many years, but the failure modes that can occur on this new equipment might be entirely new. Besides, you also have to address the equipment's warranty issue since anything that will not follow the OEM tasks would provide some problems with the manufacturer. RCM Analysis should be carried out only on critical assets or equipment in the plant. One should also consider the horizontal replication or templating of the RCM Analysis to similar equipment with the same operating context or if the asset has other similar equipment or processes in the plant with an almost identical operating context or the conditions that these assets operate exactly the same or almost identical.

2) RCM is not for everyone: RCM is carried out by a cross-functional group from both the maintenance and operators. RCM is always a team-based approached and is never intended to be performed by a single individual. One of the most critical factors will be the team's experience that will compose the RCM Analysis. Consider these as the Rolling Stones people in your plant. If you don't understand what I mean, these are the people with white hair. For no hair, maybe 50/50 as I know many bald people who are still young. The RCM analysis should be carried out by maintenance and operators with the most extensive knowledge of the asset they are about to analyze. The experience will play a very important factor in this. You simply cannot set up an RCM Analysis team if the people doing the analysis are new to their work or are simply novices. The analysis they will be performing will be inappropriate or will either be shallow. One thing for sure is that they will miss a lot of failure modes. The team should be experienced enough to detail all the possible failure modes and effects that they have experienced and witnessed throughout their lifetime in maintaining their asset. This is one very important lesson that must be known for conducting an RCM Analysis of their assets. If we speak about the military, I think the RCM team can be compared to some special force on a specific mission such as Delta-force, Black Ops, Commandos, and the like. Therefore, in selecting the team to carry out the RCM Analysis, the team must compose the most experienced people from both the maintenance and operations sides. Yes, you read it right; operators should also be included in the RCM team.

3) RCM team must include operators: Although any RCM Analysis goal is to improve the way we maintain our equipment and assets, maintenance must realize that their equipment is always a shared responsibility for both operators and maintenance. An operator should be part of the RCM Analysis team because these people are the best source of failure modes. When the equipment fails or breakdown in operation, the operators usually have witnessed the failure and not the maintenance. The operators will simply call the maintenance to fix their equipment after the failure occurs. Operators must have some knowledge about the failure that needs to be shared with the RCM Analysis team. Second, some maintenance tasks will default to the equipment operators, such as inspections, minor adjustments, cleaning, and other light maintenance tasks. Experience tells me that operators will only perform these tasks if they understand the importance and criticality of not performing such tasks. Suppose we exclude operators in the RCM Analysis and maintenance, provide them with a list of things to be checked and monitored on the equipment, expect them to be forged, or, worst, expect them not to perform these routine checks.

4) RCM facilitator must be present in every meeting: The Facilitator can be someone from your plant who understands the RCM process very well or can be an outside consultant who will play a vital role in the initial process of conducting the RCM Analysis. Let me put it this way. If you want to know how to drive, you search the net, and you will find a great deal of information about cars, engines, cooling systems, fuel systems, braking systems, and even down to the different road signs. By knowing all of this information, you simply cannot be confident to drive to any place you wish to go thru by just reading. In the initial stages of your learning curve on how to drive, there must be someone sitting beside you who can coach you when driving. This is one important facet in the RCM Analysis, and I believe that most teams undermine its importance. One of the RCM facilitators' roles is to decide when the group does not know or ignorant and when the group is uncertain. Meaning, the group cannot decide; perhaps a couple of members from the team experienced the same failure in a different situation. This is where an outside expert's opinion may be needed. Do not focus on one failure mode for too long. If the team is uncertain or has no knowledge or very little knowledge of the failure mode being discussed, skip this failure mode and proceed to the next. You might have to invite someone knowledgeable in one of your RCM meetings to cover this.

5) Biased outcome of the maintenance tasks: Almost every plant has some form of Preventive Maintenance that provides a list of maintenance tasks on their assets on a specific fixed period and interval. Some RCM teams perform RCM Analysis only to realize that they end up doing the same Preventive Maintenance tasks that they have been doing previously. The team just cut and paste the tasks on their current PM to speed up the process. Performing the RCM Analysis will be composed of two parts. The first one is deriving a simple FMEA or Failure Mode and Effects Analysis. The second part is deriving the RCM Decision Worksheet based on an Algorithm or Decision Diagram. Our goal in the RCM Analysis is to make sound decisions on what to do about the failure mode before it eventually occurs on its own. What can we do about the failure mode? Can we prevent it, predict it, anticipate it, prolong it, control it, or simply allow the failure to happen? Remember that Preventive Maintenance or Scheduled Maintenance is feasible to use when the failure mode has an age-related pattern (Patterns A, B, and C) and the Airline Industry found out that only around 11% of all the failure modes fit this pattern, which leaves them around 89% of the failure modes that need to be addressed too. Therefore, if almost all the maintenance tasks generated through the RCM Analysis result in the same Preventive Maintenance tasks they are currently doing, nothing will change. One question that needs to be raised will be if the current Preventive Maintenance tasks that are being carried out will likely prevent the failure mode from occurring or not?

6) The team's operating context statement should be well understood: Before performing an RCM Analysis, the team must write and understand the operating context. What makes the RCM unique is that it will attempt to provide a holistic maintenance task based on how it is operated independently. Some of the questions that need to be addressed and answered in writing the operating context will be as follows:

• Are machines independent of each other, or will they affect the entire process? Is there a redundancy or stand-by for this asset we are analyzing? Take note that the degree of maintenance requirements for a stand-alone and stand-by unit will be different
• Are there any failures from the past that have resulted in any known breach in environmental regulation or have affected society? Are there failures that have caused injury or death in the past?

- Will the asset run continuously for 24 hours a day, 7 days a week, or will the asset run intermittently? Are there any peak months during operations that the asset is expected to deliver continuously?
- How about the speed of response and repair of failure, are there any critical spares? Do we stock them at the warehouse? If not, what is the lead time in procuring these spares?

7) Management must understand that RCM requires a budget: Management must understand the RCM process and why it needs to be done. If some parts of the equipment need to be requisitioned, such as broken gauges sensors, malfunctioned protective devices to track some failure modes. This will be requisitioned, and the manager needs to approve it. In any improvement initiative effort, management support and commitment are very important. You simply cannot perform an RCM Analysis without the approval of your management people. Management must take their time to attend formal RCM training and understand their role in the RCM process. One of the team's roles is to formally present their actual RCM Analysis to the senior managers and decision-makers in the plant. The RCM team needs to defend why the RCM Analysis must be done on their equipment. Every plant and industry have their own set of SOPs', procedures and specifications to follow, and their current Preventive Maintenance is no exemption. Hence, before changing your current Preventive Maintenance procedure with the RCM Analysis, senior managers must understand the reason for the change itself since they will be the ones who will approve or disapprove it.

8) RCM is not about software: Although there are many RCM software you can purchase to ease documenting the RCM Analysis, this software should only be used if the team understands the principles of the RCM itself. An RCM software might likely to include all the possible failure mode that may eventually occur to the equipment, but what if there are other failure modes in its current operating context which is not included in the software package, can we simply add them or not? If not, then you have a big problem. Remember that making decisions in the RCM Analysis is not about just clicking here and there or pushing some buttons but understanding that each failure mode has different consequences. The software is designed for a particular plant and may not work perfectly in your plant as the failure modes may not be exactly and 100% identical. That is the main reason why RCM specifies the operating context of the equipment.

9) RCM requires a paradigm shift: It is very difficult or impossible to change how we perform maintenance in our plant if we do not adhere to or believe in the RCM process's principles. Performing the RCM Analysis is easy; implementing the RCM Analysis is a different story, which is the most difficult part. In implementing the RCM Analysis, the current Preventive Maintenance procedures the plant is initiating should be stopped and superseded with the RCM Analysis. The team initiating the RCM Analysis needs to understand the principles themselves, but this goes to every plant and people outside the plant who have affected how they perform maintenance on the equipment. Suppose intrusive maintenance tasks are continuously being added to the PM checklists due to quality or safety audits on the equipment, then Quality and Safety people should also be trained and understand the RCM Analysis itself. To conclude, RCM might be the best strategy to adopt to improve the way we maintain our equipment and assets, but RCM itself is not a perfect methodology, and the output of the RCM analysis will always depend on the team doing the analysis. The team should conduct a review every 6 months to 1 year after implementing the RCM process to determine if other failure modes need to be added or deleted in the process and if the equipment's operating context has been altered or changed in any way. I hope you consider the following points above before conducting any RCM Analysis in your plant.

4.2: February 2010: What Does It Take to Change the Maintenance Culture?

Perhaps the title of this newsletter requires no easy and conclusive answer, but what I can say in most cases is that one of the main reasons why maintenance remains reactive can be rooted down to the entire organization as well. I mean that there are many cases where the organization itself is the root cause why maintenance remains reactive. Slashing the maintenance budget to cope with cost-cutting schemes as well as cutting in the training budget for the maintenance function are just some of a few to start with hence maintenance is left with no option but to do what they think is best, which is to put out fires all the time at the least possible time. But first, let me define what culture is. Culture means how people perceived and do these things around here on their plant. It refers to common values and beliefs, while others prefer them as shared thoughts and feelings. This is how people do things in this plant because this is how it was being done, and these things have existed here for a very long time. According to Schein, culture is a pattern of basic assumptions that a given group had invented, discovered, or developed in learning to cope with its problems of external adaptation and internal integration that had worked well enough to be considered valid and therefore taught to new members as the correct way to perceive, think and feed concerning their problems.

Everything that we do in our plant has something to do with the culture, how we communicate with our people, how we do things around the plant, how we deal with failures, etc. These things that we do have been in existence for a very long time since the people who started doing them thought that it was the right thing to do. Remember the Gorilla Story in my previous newsletter? But what most of us left here in the plant failed to miss out, is that as time passed by, our equipment and assets become more complicated and automated, but the way we do maintenance remains the same as it was before since the caveman invented fire, which is fire-fighting.

Changing the maintenance culture is not an easy job and will not happen overnight. It will require a great amount of time, patience, and dedication. Each industry has its own distinct way of doing things. A culture is like a fingerprint. No industry has exactly the same culture, and in every industry, it will even boil down to each department having its own unique set of cultures. Culture is invincible, but as human beings, we can feel and sense it. It is pretty much alive and felt in the way people do things in their plants. Many industries relied on the do not reinvent the wheel concept and just copy and paste syndrome, since this is how others tend to do it and seemed to be successful, so we should just do the same. *These people are so wrong!* They forget the most important part of their organization, which is their "culture." If you want something to work for you in your plant, it must be accepted by your culture; otherwise, you will just end up wasting a lot of time, money, and effort.

The More We Blame, the Less We Understand:

I got this message from one of the threads on a Root Cause Forum facilitated by my good old friend Bob Nelms. I truly believe in this statement *that the more we blame, the less we learn and understand things. The less we understand then, the less we learn from the problem itself.* If we asked ourselves, why do people blame people? It is simply a case of human nature. I believe Michael Jackson sang that song "Human Nature." To give you an example, just recently, I hired a carpenter to make an office table, he completed his work in just a couple of days after which I told him that I wanted to place a glass cover on the surface of the table and so I have him measured the top of the table. He gave me a measurement on a piece of paper. I told my daughter to find a glass shop and have this dimension cut out. That evening a truck went to my house carrying the glass. I told them that it might have been a mistake because that is not the size I have ordered. When they look into the piece of paper for the dimension, it measured as:

L = 124 cm and W= 47 (the person who cut the glass thought that the length was in inches and not in centimeters)

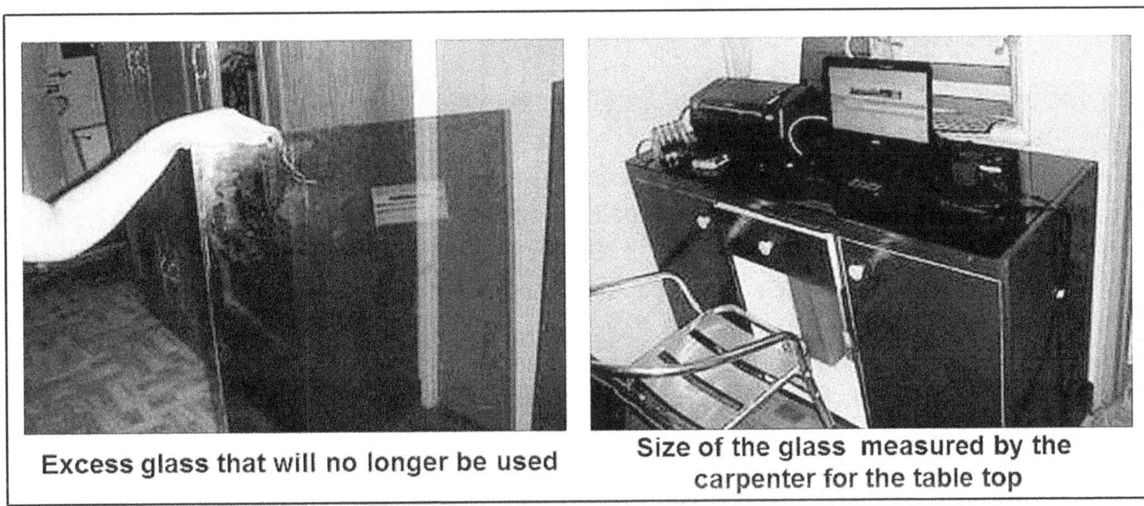

Excess glass that will no longer be used
Size of the glass measured by the
carpenter for the table top

Figure 4.1: Miscommunication, a Common Problem on Industries

The length was correct, but the width was done in inches. I called the carpenter, who was still working and explained to him the discrepancy. The carpenter explained to the driver, who at the same time did the cutting, that if there was no unit of measurement, then it should be understood to be in centimeters. Still, the driver who cut the glass assumed that it was in inches, so he cut it in inches. I told the driver to cut it according to the correct measurement to fit the table, but the driver said that I cannot return the excess glass since the mistake will be charged to him. In short, I let them cut the glass to the dimension I wanted (both in centimeters) and paid for the excess glass, which is still in my home. The carpenter said that if I don't place a unit here, then that is already understood to be in centimeters, but the driver who at the same time cut the glass said that the mistake can be avoided if you place the unit in both the length and the width. I learned from this simple situation that both the carpenter and the driver who cut the glass never learned from their mistake because they were up to speed in blaming each one for the failure. In industries, it is no different; when something goes wrong, each department involved is busy thinking of excuses rather than working together to solve the problem completely. Am I right?

In my years of experience conducting training and consulting, one of the industries I find most difficult to convince is **"POWER PLANTS."** The irony of this is that these plants are just some of the few industries that I know who believe and focus on the value of having good reliability in their equipment and assets, but most of the time, they just overdo it. During their Plant Outage and Shutdown, they will replace as many parts as possible, even those with no signs of wearing or aging. They often replace components with new ones even if some of them are still in working condition. Why do they do this? Because that is the only time on the planet they can perform these replacements and overhauls. The consequences of failure will somehow affect their environment. Personally, if you ask me if what they are doing is the right thing to do? My answer is definitely no, but why are they doing this? Suppose something bugs down or fails during operation, which will result in a trip on one of their systems which could lead to a power failure, then maintenance will be blamed. Power plants do this because they comply with government laws and regulations. If a failure happens on a power plant, then all eyes and ears will be on that plant, not to mention the media covering the news, the comments of people affected on social media, and the remarks of those crooked old politicians. This is what the "Blame Game" is all about, and nobody wants to be a part of it, but everyone is actually doing it.

It is sad to note that nobody in the plant really cares when maintenance does a good job. No one will thank the maintenance for doing the right thing simply because maintenance is a thankless job, but out there somewhere, operations are celebrating because the credit goes to them. Some will even tell you that you are paid to do that job, so just do it. On the other end, if production will lag behind due to an event of a failure or an unexpected breakdown, then this is where maintenance will be made known and highlighted. The more we blame, the less we understand, and the lesser we blame, the more we understand. When we put our fingers down and start working together on the problem, this will be a good start. I hope that when a problem arises, operations will come forward to the maintenance calmly and say, how can we help? You and I are partners, right?

As the "King of Pop" Michael Jackson sang in his song "HUMAN NATURE," If this town is just an apple, then let me take a bite, why, why tell them its human nature, why, why, I like doing it this way **(meaning if this is how they do things around, then let me just join the crowd)**

The Organization Must Understand What Maintenance Can and Cannot Do

In a nutshell, there are a hundred ways or more equipment can fail, but most of the time, we ended up being reactive and address breakdowns and failures after they had eventually occurred. A typical role for maintenance will be to fix the failure without understanding the underlying cause. Most industries perform their routine Preventive Maintenance activities in a scheduled fashion, but they cannot seem to get rid of the failure in the first place. This is because of the wrong belief that most maintenance simply refuses to accept that not all failures can be prevented by doing scheduled Preventive or Time-Based Maintenance. Only a fraction of these failures can be prevented after all. Being reactive is addressing each failure one at a time from time to time, but to be proactive, we must have a plan for each of "all" these failures that might eventually occur in the equipment. Hence, what maintenance can actually do is prevent, predict, anticipate, control, manage or prolong the failure's duration. This means that some failures can be:

• Prevented by doing Preventive Maintenance (usually if the failure has to wear out or age-related pattern)
• Predicted by doing Predictive Maintenance if some failures provide some signs or symptoms that they are on the verge of failing out.
• Anticipated by allowing them to occur on their own if the consequences of failure are minimal
• Control or prolong the duration of failure by performing modifications and redesign to reduce the consequences of failure acceptable to the user.

Every failure has its own unique pattern. Understanding the pattern can provide us some clues as to which maintenance tasks are the most feasible to use. This is what maintenance is all about. Contrary to what other books say, maintenance is not about transcending from a Preventive to a Predictive maintenance stage. There is no such thing as transitioning because not all failures can be predicted. Predictive maintenance can only be feasible to use if there is what you call a potential failure or a symptom. For example, before a bearing will fail, it can increase in vibration, noise, or temperature. We can monitor these signs or symptoms through non-destructive Predictive Maintenance instruments to make a sound decision on when to stop the equipment and have the bearing replaced.

It is the maintenance that should understand their role in their equipment, but the entire organization must also understand that indirectly they contribute to why maintenance is reactive. To give you some ideas, purchasing people will always go for the cheapest vendor to save a few bucks, making the maintenance suffer. Management slashes budgets on training the maintenance function because putting out fires is always the priority. Tightening belts and cost-cutting schemes are the game most industries play. As a result, the entire maintenance budget, together with their people's training budget, is often cut short. The benefits gained from these cost-cutting schemes are narrowly short-termed, but the bigger cost will show up somewhere in much larger proportions at an inconvenient time when nobody expects it to happen. The only way I know to reduce the cost of maintenance is to focus on reliability because if reliability starts to improve, whether you like it or not, automatically, the cost will definitely go down, and it cannot be the other way around. Another way of reducing cost on the maintenance function is to understand what Life Cycle Costing is all about. Remember that focusing on reducing costs will sometimes hurt reliability. I hope that organizations understand the role maintenance plays in their organization. These people can only perform their job well if they also understand what maintenance is all about. Before preparing for a battle, the organization provides its soldiers with slingshots instead of an M16 rifle because of cost-cutting schemes. So my question is, how it the planet can you expect the maintenance workforce can do its job with just a bunch of WD-40 and duct tapes.

How Can Maintenance Start with their Culture Change?

Therefore, from the book of Stephen Covey, Principle-Centered Leadership, a four-way approach and strategy can be applied.

- **1ˢᵗ: Determine where we are:** The first step in initiating change is to understand where we are right now and asks ourselves if we are happy with the way we do things here in our plant or we need to move forward. This defines the current situation as well as the baseline in understanding where the maintenance function currently is.
- **2ⁿᵈ: Determine where we want to go**: Once we understand where we are right now, we ask our maintenance where we want to go. This will be the direction in which we are headed forth in the future. It is important to know the destination and let the whole maintenance people know the direction of the entire maintenance workforce. (Similar to a Vision)
- **3ʳᵈ: How will we get there?** Once maintenance understands where they are right now and where they want to go, maintenance must understand what strategies they need to adapt to bridge the gap between where they are and where they want to go. This refers to a roadmap to bridge the gap between where they are and where they want to go.
- **4ᵗʰ: How will we know we have arrived?** The first and last step is to define and measure the KPI's and indices that not only tell us that we have arrived but whether or not we are headed forth in the right direction. It is important to measure our performance to determine if we have realized our goals and objectives or not.

Finally, to conclude this newsletter, it is not enough to adopt this four-way approach to change the maintenance culture. In my book, World Class Maintenance, The 12 Disciplines changing culture and achieving a World Class Maintenance stage requires three elements, .a plan, a team, and, most importantly, a big heart. If you got a plan and a team to do it, but you ain't got the heart to push it, then the change effort will just be short-lived. Real transformation takes time. Changing culture may take years depending upon the maturity and acceptance of the people involved, but what is important is developing both short-term and long-term goals along the way. Otherwise, if we let our people march on this long journey of changing culture, we might lose the people since most will join the resisting force. The best way to initiate change is to start with the Man in the Mirror. If there is one thing that I believe, maintenance is all about people; if the people are right, then the culture will be right. Industries and organizations that value their people provide a better chance of changing culture than industries that keep changing their people and always think their people are expendable. Most especially for the maintenance function, changing culture will never work if industries keep changing, and I think that is all I have to say about that.

4.3: March 2010: Why Does MTTR and Root Cause Failure Analysis Do Not Blend?

When a failure strikes, it is critical for maintenance to restore the equipment as soon as possible. If we talked about MTTR or Mean Time to Repair, this is the average time required to repair a given equipment, machine, or asset. Others termed this as Mean Time to Restore or Mean Time To Recover. The trend we want for MTTR will be the lower, or the shorter the time to repair, the better. MTTR may also be defined as the time it will take to bring a failed system back to its available or operating status once again. MTTR analyzes how long repairs and maintenance tasks will take in the event of a system failure. The unit mostly used for MTTR will be in hours.

The formula used for MTTR will be the total repair time divided by the number of repairs in a given time. On the other hand, Root Cause Failure Analysis simply tries to understand why something went wrong so people can finally learn from the things that go wrong. It will identify the cause or origin of the problem based on the evidence unfolded by separating the facts from the other probable causes. Root Cause Failure Analysis identifies the basic source or origin of the problem to prevent its recurrence. It provides a methodology for investigating, categorizing, and eliminating the cause of safety, quality, reliability, and equipment-related problems and consequences. Every system, spare, or component failure happens for a specific reason. There is a specific succession of events that will eventually lead to the failure. RCFA will follow the cause-and-effect path from the final failure back to its origin. Therefore, Root Cause Analysis is a tool to better explain what happened, determine how the failure happened, and better understand why it happens. If we add the word failure, then this is simply equipment-related failures.

I have several friends who used to work for industries, and one of them is Charlie. If I recall him right, he had been working in the plant for a good number of years, 25 years to be exact, and is still working on that same old plant up to this point of writing. There is a lot of rotating equipment in their plant, and Charlie is one of the maintenance guys that maintain them. He told me that when he was new at that plant, it took him several hours to fix the equipment experimenting on what to do because you are usually on your own. When a bearing fails, it usually took him a couple of hours or more to replace the failed bearing. 25 years had passed, and today when the same bearing fails, it will just take him around 20 to 30 minutes at the most to replace it. My question, is that a good thing or a bad thing? The answer here is both good and bad. If we speak about MTTR or the Mean Time to Repair, Charlie's MTTR improved a lot through the years he had worked in the plant by gaining experience. If we speak about the trend of MTTR, the lower, the better, then this is a good thing. On the other end, if we asked about Root Cause Failure Analysis, Charlie simply cannot find the cause as to why the bearing failed, since the failure keeps on repeating itself because he never bothers to identify the root cause of the problem, and I believe that this is not a good thing. Why did Charlie become good at fixing bearings? Because the bearing fails all the time, and that's about it.

Although we still have to consider the economic point of view in this situation, meaning if the consequences of failure will be limited to the cost of failure alone, or if the motor that failed has a standby motor or redundancy in place, then MTTR can either be prolonged to allow us the time to gather evidence on the part or component that failed. However, suppose this is a stand-alone motor where the failure can affect some critical parts or the entire operations of the plant, then the repair time should be in the minimum and shortest possible time. This is where performing root cause will have its stall.

The Fast Team and the Slow Team

As mentioned in my previous newsletters, when a failure occurs, there are actually two teams the will come to the equipment. First, we have a fast team. The fast team will always be present in the plant in the shortest possible time. They will arrive at the scene of the crime in a matter of seconds. They will be the restoration team, and they will be responsible for bringing the equipment up and running in the shortest possible time. They need to do things fast, and they know that they are under pressure to repair, even if they did not cause the equipment to fail since most of the time, operations people would be monitoring them. The fast team will be taking a ride on a fast train. The restoration team knows that time is always of the essence.

On the other end, we have a slow team. These people will be responsible for analyzing the root cause of the problem, and they will be taking a ride on the slow train. They will take things rather on a slow basis by picking up debris, taking pictures initially on the failed equipment, spending time interviewing operators and people nearest to the asset during the failure. However, when these people arrived at the failure scene, the fast team had restored the equipment. In this case, how in the world can you perform a Root Cause Failure Analysis when the equipment is already restored? You just simply can't because all the shreds of evidence and traces of failure had been washed out in the first place. One of the most important things that I've learned in performing a thorough root cause failure analysis investigation is to freeze out the evidence, but in this case, the evidence had completely vanished. And most of the time, their managers complain that their maintenance people lack root cause analysis skills. Come on, give me a break.

In my class on Root Cause Failure Analysis, one of the things I do in my class is to conduct some surveys where I asked each and every one of the people in the class a question something like, what if your plant is in the process of reducing your maintenance headcount in half and you have six people in your maintenance crew. Three of them are good at analyzing equipment-related problems (root cause). The other three are efficient in performing repairs (MTTR is low); which will you retain in your crew?

- A - Retain 3 of my maintenance who are efficient in performing repairs
- B - Retain 2 of my maintenance efficient in performing repairs and one good in analyzing failures
- C - Retain 2 of my maintenance efficient in analyzing problems and one good in repairing
- D - Retain 3 of my maintenance who are efficient in analyzing problems

The results were not surprising as the majority voted for A, which is quite expected where they will retain 3 of their maintenance who are efficient in repairing failures followed by B, in which case they will retain 2 of their maintenance efficient in performing repairs and retain one good person in analyzing failures. Very few voted for C, where they will retain 2 of their maintenance efficiency in analyzing problems and one good at fixing failures. Nobody voted for D. What a pity.

Suppose we perform a Root Cause Failure Analysis on the equipment or asset that failed, MTTR will be prolonged since we need to pick up whatever evidence can be found on the equipment before restoring the asset. I don't think operations people will like that very much.

Where MTTR Should Be Used

Some industries used MTTR as a way to provide a comparison among their maintenance craftspeople. Others even included MTTR as one of their key indices and even include them in their Performance Appraisal. Hence, if Joe has the lowest MTTR among all the maintenance people around, then he will get the highest increase in his next payroll. I think that this is a terrible mistake. Why? Many systems are repairable when the system is said to have failed in operation. We also need to understand that MTTR varies depending upon the difficulty of repairing a system failure. MTTR will be difficult to estimate because one must consider a variety of repairs. An engine repair will include tightening a drain plug bolt to overhauling an entire engine assembly. We cannot just simply compare these two. Suppose all the repairs done on the equipment are identical in every way; in that case, we can use this to compare the repair time for different people, but this is unlikely because, in the real world, the equipment can fail in many ways. If we compare Joe and Charlie, Joe repairs fast, but after a few minutes, the machine will fail again. Charlie is slow in repairing, but he made sure that he will not be called back again on the equipment to repair the same problem after he completes that repair. How can we compare these two people? If we follow the rule on MTTR, then Joe wins over Charlie. The repair time will vary depending upon the repair's difficulty and the person's experience performing the repair.

In most cases, when an experienced person retires or leaves the plant for good, his experience usually goes with him to the grave. Most industries have not captured nor documented the experience of this person who retired on how did he performed his repair effectively. What was his first step, then what, after that, what's next, and so on. When people like these exist in your plant, we need to know what they know and capture their experience before leaving for good. We can use these people to teach others who are new around the plant. I believe this is where the true value of MTTR should reflect upon. Most of us in maintenance admit that we still cannot get rid of the hero stuff where no one at that shift can repair the failure, called us late into the night, come back in the plant to fix something that nobody can fix, and with all the sweat you got, and experience dealing with this magnitude of failure, the equipment runs, and the boss gives you a pat in the back just like what we do with our dogs. Good job, boy. And a hero comes along, with the strength to carry on; I believe this time Mariah Carey sang that song. I hope you agree with me that, in a way, music is a universal language. Sometimes, when I teach, I expressed myself in music to better understand what the heck I am talking about.

Perhaps if MTTR is being monitored in your plant and in a certain group of 10 people, Charlie is said to have the lowest MTTR in performing repair and troubleshooting for hydraulic systems, while Joe has the highest MTTR in the same category, we can define proper procedures on repairs based on Charlie's practices that can be easily applied and followed by other people thereby avoiding trial and error in conducting repairs. These repair procedures can be encoded in the system if the plant has some form of CMMS or EAM software. If the same failure happens in B-shift and a failure occurs, the maintenance on duty can just check the system if a repair procedure for this type of failure exists and if yes, then he can have a printout to serve as a guide on how to conduct the repair especially for newbies in the plant. We can also tandem or partner Joe and Charlie so that Charlie can teach Joe from time to time. The goal will be to improve the repair time performed by the lesser skilled maintenance.

I would also like to clarify in the MTTR calculation that when a machine stops, the operator will find someone who will repair the failure. When the maintenance arrives at the failed equipment, he looks at the machine and tries to diagnose the fault. If a spare part is affected, he will check the spare part in the system if it is available or not. If the part is available, the maintenance will go to the storeroom to withdraw the part. Then he goes back to the machine and starts repairing it. The maintenance will provide trial runs until he finally endorses the machine back to the operator. Maintenance will finally log on to their system and determine the amount of time required for this machine to be fixed.

If the spare part required is not available and it took the Purchasing people 3 days to acquire the part and 1 hour to repair the failure, what will be the total repair time? Will it be 3 days and 1 hour or just one hour? When I pose this question to the class, I have a mix of answers. Some will say that it should be 1 hour because that is the real-time to repair the failure. Others will say that it will be 3 days and 1 hour. They will state that downtime due to no availability of spares should be included in the repair time. If you ask me about this, my answer is that the repair time should be 3 days and 1 hour. Why? Because Spare Parts is still the responsibility of the maintenance department. Is it not? When a part is not available, then let the maintenance be held accountable for that.

For failures that keep repeating themselves, the best strategy will be to finally address the root cause of the problem and prevent it from recurring on its own. The saying goes that when our people become really good at fixing failures, then something is terribly wrong with our people and maintenance organization. What is wrong? It seems that they cannot seem to let go of the problem. Try to think why many of our maintenance people become good at repairing as time goes by? It is because the failure keeps on repeating again and again, or the problem itself keeps on resurfacing. When failure keeps repeating itself, maintenance must take the time to analyze and address the problem's root cause. If you really want to understand why the failure keeps recurring, take things slowly, and start analyzing them. I believe that to be fast, we need to take things slowly. Perhaps, if I am still employed as maintenance, the next time the machine fails, I'll give the operators a repair menu. So whenever the operator calls me to repair something, I'll give them the menu. If the operator asks, what the heck is this? I'll tell the operator to choose what repair he wants, a quick one or a sure one where the failure would no longer repeat.

Finally, to conclude, Root Cause Failure Analysis and Mean Time to Repair do not actually blend. It is like singing an opera song when the beat or type of music being played is reggae. These are two separate strategies that maintenance can benefit from if we actually know when and where to use these strategies in our equipment and assets. Root Cause Failure Analysis is not recommended for every single failure, but all I can say is that if the failure keeps on repeating itself, then I believe that we can set aside MTTR for a while and take the time to take things slowly and perform a thorough root cause investigation. Having a low MTTR due to your years of experience is not good because the failure simply keeps on repeating itself. You just become good because you keep on practicing the same old things over and over again. However, having a low MTTR for most maintenance people due to repair or troubleshooting guides and procedures based on your people's experiences will be of value and worth to the company if knowledge and skills can be shared. When these good people finally retire, they know that they have left a legacy in their plant. And for now, I think that is all I have to say about that.

4.4: April 2010: What Does it Take to Initiate a Plant-wide Improvement?

Have you ever wondered what it takes to initiate or implement a plant-wide maintenance improvement initiative in your industry? Let me tell you that the best and most frustrating days of my life being employed were when I was tasked to handle the overall pillar of TPM Planned Maintenance plant-wide. The primary role and objective are to provide a plant-wide strategy to improve our equipment's reliability and maintenance. While many say that Autonomous Maintenance is the pillar with the most population, I believe that the TPM Planned Maintenance pillar should be the strongest in any TPM initiative. Why? Because if the structure of Planned Maintenance is weak, then the same goes for Autonomous Maintenance because Planned Maintenance will be the ones to guide the Autonomous Maintenance pillar in convincing operators of the importance of establishing basic equipment conditions in their equipment. The tasks were very noble, indeed, but the obstacles, pressures, limitations, hindrances, and frustrations of doing this so-called Planned Maintenance were simply overwhelming, if you know what I mean. And this frustration simply grows minute-by-minute, day-by-day. Just to provide you some idea of what I meant, to start with, I set up a plant-wide membership for Planned Maintenance, and most of my members are hard-knocks and experienced maintenance senior engineers and managers whose position is 3 times more than my current position was. Some have stayed in the plant for several years, and most of them longer than my term. When you provide them some initial tasks and activities, most of them will be hesitant. It is quite normal to hear things like:

• I don't believe that this is part of my job description
• We have our own priorities, and you are adding up work to my load, so which one should I prioritize?
• Rolly, I don't report to you, so why in the world should I follow you?
• We have done that before, and it did not work, so why do you think this will work?
• What's in it for me if we do what you want us to do?
• Will our salary increase if we join you and many more

And of course, there are more that adds up to my lists, and the longer you stayed in this field, the more frustration you get. How do you deal with these kinds of people? Remember that they do not report to you directly. Providing them some additional tasks at the beginning will be such a pain in the neck. If you don't get the patience to handle this, then perhaps you will just lose your cool and punch them right in the face, but of course, that is not the right thing to do. I don't recommend you to do that. This is the challenge of handling a reliability and improvement initiative in your plant. If you have been assigned to handle this, then you will understand what I meant. If you cannot handle the frustration, then either you request a transfer, or you will just leave the plant for good. In short, handling an improvement or reliability initiative is tough. The most difficult part will always be the start-up. This is where you need to transition from being a manager or engineer to being a leader. This is where you need to turn these negative remarks into positive thoughts. This is where you need to command respect regardless of your position at that time.

But most importantly, this is where you need to realize the virtue of extending your patience.

It is your job to motivate and inspire these people. I might not be in a position to write all my books if I quit, despite the negativity and frustrations. Nelson Mandela united his people through a sport known as rugby and motivated his players to win the world cup. I admired Mr. Mandela, despite serving 27 years in prison. He had no grudge with his captors. Another inspiring person was Mahatma Gandhi. He inspired his people through non-violence means. He was a stubborn man and who was willing to starve to death by fasting. Remember that your maintenance people are accustomed to doing the same old things every day, and if what you say is new to them, then expect a lot of pressure and resistance during the beginning. Hence, if you are assigned to my position, instead of quitting your post, I suggest being prepared for the worst. People assigned to handle this situation should have a strong conviction and passion. You need to believe that what you are doing is for the benefit of the whole maintenance force and the entire organization.

Initiating a reliability and maintenance improvement initiative is all about bringing the whole maintenance herd in one direction. This can never be an effort of a single individual. It can never be a one-man strategy since you need to bring the whole maintenance herd working together towards a common goal and objective. You need to have just one direction. So let me provide you with some guidelines on how to initiate a plant-wide maintenance improvement initiative.

1) Develop a Roadmap for your reliability and maintenance initiative: The first thing you need to do is develop a roadmap for your maintenance function. You may start off by conducting a survey and having a random group from across all divisions answer the survey. The survey should tell you what areas of maintenance that you need to focus on and improve. This will serve as the guidelines on how to start or kick off your activities. It is important that before trying to embark on any reliability and maintenance improvement initiative, you must have a specific and clear direction on where you want the people to go. Provide a time frame on how much it will take them to complete each set of roadmap. This is to provide them with a clear direction on where the maintenance is headed forth in the future. Provide specific indices and KPIs on what you want to measure.

2) Management Buy-in of your Reliability Initiative: After you have developed your initial roadmap. Provide a brief presentation with your Top Management Team. Explain to them what will be required initially in undertaking this effort. Ensure that they understand the benefits that can be derived from implementing this reliability initiative in your plant. Seek not only their support but their commitment as well. Once you got their nod, the next step will not be quite difficult to do since they will be the ones to provide it. Now you got the side of the decision-makers.

3) Create a Core Group for the Maintenance Function: Suppose you are a large plant with several departments; you need to have permanent members from each division or department in your plant. Do not exclude your facility or utility department as they play a specific and major role in maintenance. They will be included. If you belong to a manufacturing plant, the consequences of breakdowns and failures from Facilities and Utilities would be much devastating than the failure of a single manufacturing equipment. It can have a major impact on your plant. Since you have initially got the nod from the top management, this activity will not be a problem. Each member of the core group represents their division or department. They will initiate this initiative in their respective division or department. You need to provide the roles of the core team, the Planned Maintenance teams, and every single entity that will be involved in the Planned Maintenance pillar.

4) Conduct Strategic Planning Together with the Maintenance Core Team: Organize a strategic planning with your selected core team. Set up your maintenance vision and mission together, which should be aligned with your corporate vision and mission. Conduct some gap analysis between the current status and what you want to achieve in your maintenance journey. A good starting point can be derived from the book of Stephen Covey on Principle-Centered Leadership by asking the following questions:

• Where are we currently right now?
• Where do we want to go?
• How will we get there?
• How do we know we have arrived at our destination?

5) Determine the Training Initially Needed by the Team: Once the improvement areas had been determined, define what initial training is required to start up the reliability initiative. For example, suppose your focal point of improvement will be on lubrication; in that case, you and your people need to take a full-blown course on Lubrication Strategy. Suppose you want to improve your current Preventive Maintenance practices; you need to have your people trained on Reliability-Centered Maintenance. Remember that training is a long-term approach, which means that this will be required by your people from time to time and not just a one-time deal. My second book, on Maintenance-Roadmap to Reliability, has an assessment questionnaire for the 12 disciplines on World-Class Maintenance you can use to guide your team where your strengths and points for maintenance improvement are.

6) Set-up Pilot Teams from Each Division/Department: Once the training had been deployed to members and the core team, it is time to set up a team per division/department. A detailed roadmap should be provided to the Planned Maintenance teams. For your start-up activity, they need to accomplish the following:

• Determine machine inventory for all assets to determine how much equipment they have
• Conduct a machine ranking to classify which are the worst equipment in your plant
• Determine your pilot machine or which machine to start with (should be included in the worst machine category)

Once you have completed your initial roadmap, move on to the next phase or level of your reliability and improvement roadmap. This should always be a continuous process. Review the indices together with your core team regularly. Once these people realize the value and impact of this initiative, resistance will be much less at this point. Now you got your allies. It is important to let your maintenance people realize that this is not a once-a-week meeting and that whatever activities should always be aligned to their day-to-day activities.

7) Provide Simple Recognition and Reward to Successful Teams: People want to be recognized. Provide some simple yet memorable and lasting recognition and rewards to teams that successfully implemented Planned Maintenance. Rewards and recognition may not necessarily be in the form of money. Perhaps a plaque of appreciation, dinner with the team together with your Top Management, highlights your plant's bulletin board where everyone can see. A free movie ticket with your family, or if you really got the time, you can organize an in-house symposium or friendly competition among the teams and solicit judges from other industries to make it more exciting. This will be appreciated much by your members.

I have been in the business of training, consulting, and facilitating reliability and maintenance courses for five consecutive years now and have been privilege to have walked through many different plants and industries from near and from a distance. It has allowed me to meet different people from maintenance working in different industries and countries. But despite all of this, I see a common trend in all these industries. The way we do maintenance has something to do with the people doing it. Maintenance is all about people. If we provide these people with the right knowledge, they can improve the reliability of their equipment. It is always the people that actually determine the culture of their maintenance. And this aligns with my mission in helping industries improve the knowledge and wisdom of the entire maintenance human race. And all I can say is that despite all the day-to-day pressures they have at work, maintenance people are simply humble and down-to-earth human beings. (I speak about the majority.) On a typical day, these people mostly work in shifts, and when equipment is down, they attend to it even if their shift is over. Like a good soldier, these people will never abandon their post and see that their equipment is running before calling it a day's work. Maintenance people are like soldiers; they will never leave a good man behind. Whatever heroic deeds they have done for the day, they know that nobody will ever thank them for doing a good day's work because they know that maintenance is simply a thankless job. They know that you are either paid to do the job, or it is simply your job to do it.

I have trained many people from different industries, and when some of these people tell me that I am one of the best trainers in the world, it flatters me; most especially, it humbles me. Honestly speaking, in my mind, I simply cannot accept that title since there are a lot of dedicated people out there on reliability and maintenance who have stayed in the park longer than I have, and I owe these good people a great deal. Some of them have not only been my friends but have also been my mentors, as well as the likes of C. Robert (Bob) Nelms of Failsafe Network and Vee Narayan. And suppose there is something worth fighting for, is having our maintenance people finally being respected, and I believe that is all I have to say about that.

4.5: May 2010: The Golden Rule on Root Cause Failure Analysis

I had just recently concluded a three-day Public Training and Master Class on Root Cause Failure Analysis last July 14 to 16, 2010 of this year. The first day was really tough because of a recent typhoon that hit the country. Day 1 of Root Cause Failure Analysis was also signal number one in the Philippines, and three delegates did not make it on the first day; however, two of them were able to finally come on the last two days of the RCFA training class.

I would just like to share some important points and highlights that we have just uncovered in this RCFA Master Class. First, before applying or initiating any Root Cause Failure Analysis investigation in your plant, kindly remember the "Golden Rule" of Root Cause Failure Analysis, and that is "**AMNESTY.**" Nobody should be punished or disciplined due to any Root Cause Failure Analysis probe or investigation effort unless this is a clear case of sabotage. If the reason for conducting a Root Cause Failure Analysis is to find out the guilty person and to punish them outright or pin them down, then all your collective efforts on having the chance to learn from the problem or from the things that go wrong had been deemed unsuccessful and useless. Or perhaps if your intention in performing a Root Cause Failure Analysis is to put the blame or punish someone, then you need to call your analysis **"WHO-WHO ANALYSIS"** instead of **WHY-WHY.**

Quote from my good old friend and mentor of Latent Cause Analysis, C. Robert Nelms: The issues of blame, discipline, fear, punishment and accountability are near the root of why industries don't learn from the things that go wrong. There is most certainly a need for rules, regulations, policies, and procedures in our businesses. Likewise, there is most certainly a need to enforce them by applying discipline to those who do not abide. However, it is vitally important to acknowledge the reason for our rules. They exist to help **PREVENT** incidents. The time for discipline should be before an incident, not after it happened. Unfortunately, most organizations do it backward. They do not consistently apply their disciplinary policies ahead of time. Instead, they will wait for something awful to happen and then go after the person. This is outrageous, illogical, and immoral.

From the book of Keith Mobley on Root Cause Failure Analysis (page 4): The purpose of RCFA is to resolve problems that affect a plant's performance. It should not be an attempt to fix the blame for the incident. This must be clearly understood by the investigating team, and those involve in the process. Understanding that the investigation is not an attempt to fix blame is important for two reasons. First, the investigating team must understand that the real benefit of the analytical methodology is plant improvement. Second, those involved in the incident generally adopt a self-preservation attitude and assume that the investigation is intended to find and punish the person or persons responsible for the incident. Therefore the investigators need to alleviate this fear and replace it with the positive team effort required to resolve it.

But in most industries, when the impact of the failure is high, the insurance companies, lawyers, the media frenzy, and those fat-crooked old politicians are involved. These crooked old politicians will always give some reactive remarks after an incident of major interest happens and not before. They will be interviewed by the media, and they will state a remark that they should have done this and that. Why are people so dumb in voting these? No more comments, and back to maintenance. Anyway, management and everyone wants a name. They want a culprit. They want someone to be blamed for the incident to appear that the failure had been resolved. An investigation was conducted, and the probe indicates that John Doe closed the valve that subsequently wrecked the turbine. John is severely discipline and is lucky to still be employed. The management, together with the lawyers and insurance people, are satisfied because they thought that the root cause of the problem was finally resolved. One year later, the same problem occurred once again, and this time by a different person named Johnny Thor. When Johnny was interviewed about why he committed that mistake and closed the valve, he said that he just used to sweep the dirt in the plant, and the supervisor asked him to closed the valve, and so I closed all the valves I can find. Was it really the fault of Johnny Thorr? Therefore, when performing RCFA, it is important to determine why people do what they do. A deeper Analysis tells us that John Doe does not possess the skills or knowledge to perform his job safely and correctly. In the case of Johnny Thor, it was not even his job in the first place to do that.

Therefore, before applying the RCFA process, it must be explicit that nobody should be blamed or punished for the incident unless a crime is committed or simply this is a clear case of sabotage. In short, for those who want to perform Root Cause Failure Analysis, there must be an amnesty policy. Before we start putting fingers or placing blame on someone else, kindly answer one simple and honest question, are we convinced beyond a reasonable doubt that if we were the person who committed the mistake have we done the same thing, or have we done it differently given the same circumstances? If your answer is doing the same thing, you need to probe further beyond the human cause because there is a much deeper cause which is the latent cause of the problem.

I had a participant from one of my RCFA class who said that previously when an equipment failure or problem hit their plant, all maintenance people would come down running as fast as they could towards the failure or failed equipment to extend their help to the maintenance on duty which is a very noble gesture indeed on the part of maintenance. That was their culture. Nobody was left behind; they will always help because they wanted to involve themselves and resolve the problem together. These people understand the word team. They work as a group to resolve the problem completely, even if they belong to different parts of the plant. One day, a new management team was placed in the plant and started changing SOPs and policies. After management starts changing their policies and rules, they started suspending people involved in the failure and incident. Things changed dramatically in this plant because, unlike before, instead of involving themselves and extending a helping hand as they did previously, these very same people started walking away from the failure. They do not want to involve themselves anymore. It's not my problem anyway, so I do not want to get involved since I might end up getting caught in the problem too, or management might accuse me of the problem. Perhaps if the person who did this policy himself is perfect in every way and has not yet committed an error in his entire lifespan, then I will succumb and submit to that policy.

When we accuse people or blame them, people's lips begin to seal. You will unlikely get to the root cause of the problem because people will practice their right to remain silent. They rather not talk about it, and they start to become more defensive and later on defiant when you interviewed them. Ask me why; it is simply a case of human nature. Try to do something good all the time, and nobody will talk about it, but when you do just the opposite, everybody in the plant will know about it, and you will be the talk of the town until the next incident comes. Am I right here?

While it is true that people err and commit errors, a lot of cases had been stated so far that people commit errors and mistakes way beyond their control. If this is the case, then we must dig deeper into the real reason why the person committed that error so we can finally learn from the things that go wrong. There underlies a deeper reason why people commit mistakes and errors. A solid Root Cause Failure Analysis will uncover that fact. In RCFA, all physical failures are triggered by humans. But humans are negatively influenced by latent forces. Therefore, in any Root Cause Failure Analysis investigation, the goal must be to expose and identify these latent causes.

First, let us define what human error is. Human error is an inappropriate action or intention to act, given a goal and the context in which one is trying to reach that goal. It can also be defined as an action planned but not executed according to the plan. Human error is the by-product of poor planned tasking or execution of a task resulting in a goal's failure. Humans make errors. The most common human errors are what you called "slip and lapses." Slip is when you carry out planned tasks incorrectly or in the wrong sequence. If you need to perform steps 1, 2, 3, 4, 5, and you did steps 1, 3, 2, 4, 5, then you committed a slip. If the person needs to do a task in sequence and does the wrong sequence, you commit a lapse. This is when you need to do steps 1, 2, 3, 4, and 5, but what you did is steps 1, 2, 3, and 5. You forget to do step 4. Slips are human actions that take place, which were not intended in the first place. The occurrence of a lack of one's train of thought, which is derived from unconscious behavior.

On the other hand, lapse is when you missed out on a step in a planned sequence of events. How many times have you went to work and went back home because you forgot something, or perhaps you missed out on a very important occasion because you were busy with work? Lapses occur when someone misses out on a key step in a sequence of events or activities. Slips and Lapses happen because the person was distracted, preoccupied, or simply was absent-minded.

The difference between humans and machines is that people can sense, change, hear, smell, see, feel, or taste something different and take the necessary actions to correct the anomaly. Humans are flexible, while machines are frequently not. Research by Dr. James Reasons of the University of Manchester in England found that humans commit 6 errors per week. The latest official current world population estimates that for mid-year 2009, the world's total population was estimated to reach about 6,790,062,216. This means that, on average, around 5,839,453,506 errors are being committed daily. And with all the errors occurring and all the ways we could destroy ourselves, life is still preserved. Why? Because humans can sense change and break the error chain. In short, we cannot eliminate human errors; but we can only reduce them or learn from them. I would prefer the latter and learn from the failure itself because it makes us a better person.

To conclude, if any industry wants to perform a thorough Root Cause Failure Analysis, do it for the right reasons. If the reason why you want to perform a Root Cause Failure Analysis is to put the blame on someone and find a culprit, then we are not learning from anything. There must be complete "amnesty" at the very beginning when performing a Root Cause Failure Analysis. The people involved in the investigation should understand that there is a much deeper cause why people commit mistakes, which is the latent cause of the problem. The only way that we can learn from our own failures, whether from our equipment failures or from the lessons of life, is when we set aside our position, pride, ego, sarcasm, pessimism, negativity, and once in for all, try to look ourselves in the mirror and ask ourselves that we too are part of the problem. Whether we like it or not, humans will continue to make errors and mistakes; that's why they need to analyze them is a must simply because failures will allow us to begin again more intelligently.

4.6: June 2010: The Importance of Life Cycle Cost in the Reliability Strategy

Reducing cost has always been the focus of most maintenance managers, and that perhaps we need to learn from the lessons of history. The cost must be studied thoroughly, not just based on its initial cost but also on the equipment's entire life cycle cost. Considering the initial cost is like seeing the tip of the iceberg, but underneath the iceberg lays the true cost far greater than the equipment's initial cost. If one has seen an iceberg, multiply the area 3 to 10 times, which will be the area or size underneath the iceberg.

One of the more important aspects of improving our equipment's reliability is Life Cycle Costing or LCC study. Life Cycle Costing refers to the total cost of equipment throughout its entire lifespan. The US Management and Budget defines LCC as the sum of the direct, indirect, recurring, non-recurring, and other related costs of a large-scale system during its period of effectiveness. In terms of production equipment, LCC can be described more simply as the design and fabrication cost, which is the initial or acquisition cost, plus the operation and maintenance cost, which is the running costs. The initial cost will always be easy to see, but the running cost is not. Failure to consider the running cost can lead to many problems in the long run. At least 80% of equipment LCC can be conceptualized at the design stage. Hence, Life Cycle Costs are the sum of the initial costs and its running costs. The message here is very simple: when purchasing parts and equipment, one must not only consider its initial costs but its running costs as well.

Let us consider this equipment, for example, in figure 4.2. When this equipment was conceptualized and designed, there are already costs involved from the manufacturers and vendors, such as evaluation costs, design costs, revision costs, engineering costs, designer costs, etc. Once the equipment is fabricated, you have a new set of costs such as fabrication, material costs, procurement costs, assembly costs, spares costs, etc. When an industry purchase the equipment, the vendors and its technical people will commission the equipment in your plant and install the utilities required to run the equipment. Hence, the industry will have to shoulder the procurement costs, installation costs, transportation costs, contractor costs, debugging costs, and other costs until the equipment can finally be endorsed to the plant's operation and maintenance people. From this onwards, the plant now operates the equipment and encounters costs on operating it, labor costs for both maintenance and operations, energy costs utilized by the equipment, cost of downtime both planned and unplanned, spare parts costs involved, schedule shutdown costs, some modifications that have been done on the equipment over time, facilities, repair costs and so on. Finally, when the equipment is decommissioned after (N) number of years, we still have to go with the burden of shouldering the decommissioning cost such as the costs of disposal, transportation costs, spare parts cost that will become obsolete and will go together with the decommissioned equipment, and so on. As a human being, there are still costs involved when one dies: the coffin, burial cost. In the Philippines, you can hire crying ladies, in which all they will do is cry, but you need to pay them.

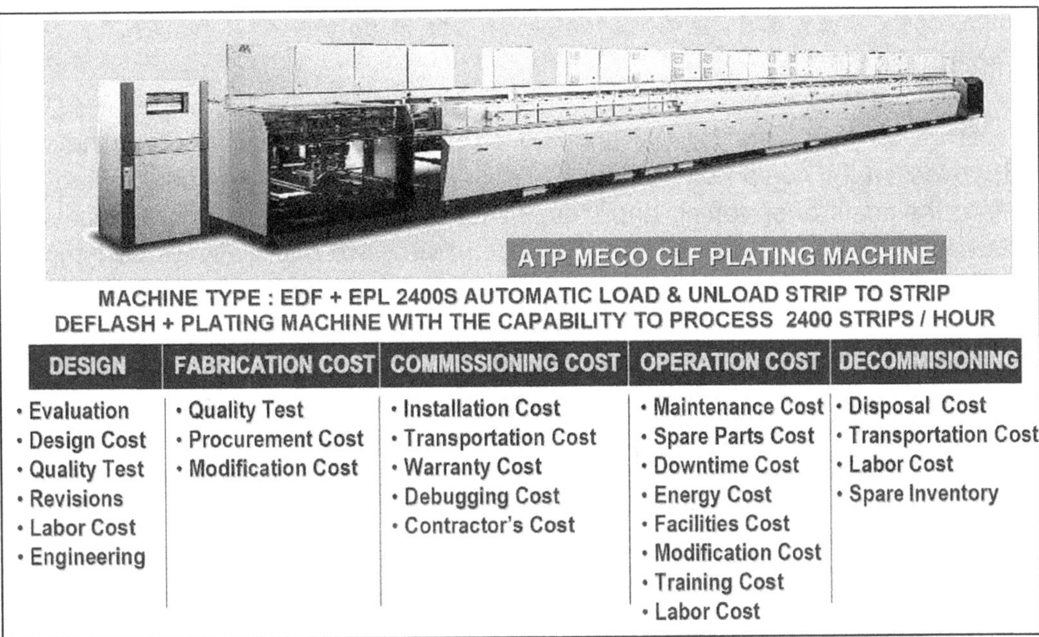

DESIGN	FABRICATION COST	COMMISSIONING COST	OPERATION COST	DECOMMISIONING
• Evaluation	• Quality Test	• Installation Cost	• Maintenance Cost	• Disposal Cost
• Design Cost	• Procurement Cost	• Transportation Cost	• Spare Parts Cost	• Transportation Cost
• Quality Test	• Modification Cost	• Warranty Cost	• Downtime Cost	• Labor Cost
• Revisions		• Debugging Cost	• Energy Cost	• Spare Inventory
• Labor Cost		• Contractor's Cost	• Facilities Cost	
• Engineering			• Modification Cost	
			• Training Cost	
			• Labor Cost	

Figure 4.2: Equipment's Life Cycle Costs: Design to Decommissioning

This means that when an industry makes decisions about purchasing equipment for their plant's use, there are actually two choices. The first choice will be to consider only the equipment's initial costs: This is the equipment's tag price or procurement costs. In this regard, the decision will be based on the vendor with the lowest equipment cost. But a better choice will be to consider the cost of actually owning the equipment, where one has to actually understand the costs of maintaining and operating the equipment. Decisions about purchasing or acquiring new equipment and spares should not only be based upon the tag price or procurement cost of the part or equipment itself but rather on its entire Life Cycle costs. Some of the questions that need to be considered are:

• How will the equipment or part perform in the long run?
• What will be the estimated cost of operating and maintaining this equipment?
• What are the critical parts of this equipment?
• How frequently should this part or equipment be replaced?
• How much is the cost of spares and downtime?
• How long will the equipment be down in the event of a breakdown?

Purchasing cheaper parts and equipment may actually mean spending more in the long run. A true and meaningful cost reduction effort can be achieved not by purchasing the cheapest costs but by having a thorough understanding of what Life Cycle Cost is all about. Many industries initiate a cost reduction program in their plant. The best way to reduce costs is to understand that it is more important to look into the equipment or part's running cost than the equipment or initial procurement cost.

Sometimes a higher performance product costs less than a commodity type product, even though it is slightly higher. To gauge whether this will be the case, one should look at the product life cycle cost rather than its purchase price or initial cost. Life Cycle Costing is a way of analyzing equipment purchase choices. If the decision was based on several factors rather than its initial costs, we would make our selection based on the least amount to own over its entire life span.

If this will be the first time for the plant to order the equipment, perhaps not much can industry do to lessen the costs from the design up to the commissioning costs, but much can be done to lower the costs of operating and maintaining the equipment and partly this is the study of life cycle costing or LCC. From a layman's point of view, we might always be tempted to buy items on sale because the store offers a huge discount on a specific product, but again, buying cheap products will eventually be expensive in the long run. Many years ago, I decided to repaint my roof, so I hired a painter. He advised me to purchase the more expensive paint brand. I wanted him to paint it with primer paint since some roof parts are starting to rust. When I went to the shop to buy some primer paint, there are many paints, and the costs vary. A well-known paint would cost around 450 Pesos or around 10 dollars per gallon during those times, while a not well-known brand would cost around 100 to 200 Pesos or roundly around 2 to 4 dollars per gallon. I needed around 5 to 6 gallons, so I decided to go for the lower costs and not follow the painter's advice. In our country, the Philippines, there are two seasons: the summer and the rainy season. After merely one year, when I went up to the roof to correct my aerial antenna's orientation since it was disoriented by the wind and storm that recently passed, the paint vanished completely. I learned my lesson the hard way this time, but I learned it for a very long time (1 year) and decided to repaint the roof once more and bought the well-known paint brand. The painter knows much more about what life cycle cost is all about much more than myself, which really matters most.

In industries, the experience I have shared with you regarding the paint is no different. Most purchasing or requisitioning departments are more inclined to always go for the lowest bidder or vendor when ordering parts. The awarding of spares, raw materials, or anything the plant needs will go to the vendor who can provide the lowest possible costs, which will prove disastrous to both operation and maintenance people in the long run. Perhaps one might think that they have saved a few bucks by switching or awarding them to the lowest bidder, but it is both the operations and maintenance that suffer as a result. Worst, they blame each other because of this. Some people might argue that the part or product specs are the same, but can we say the same for its lifespan? Hence, one of the key points in this situation is not just to award the part to the lowest bidder but to vendors that can provide the highest MTBF or MTTF on their product or spare parts. Perhaps a part may be slightly expensive in its initial costs but will last much longer than a cheaper brand and will actually save us a great amount of cost on downtime and replacement in the long run.

On Purchasing New Equipment

What is important in purchasing new and additional equipment for your plant is considering how the previous asset had been decommissioned and reducing or lessening the time to commission and install the equipment. Remember that the longer the equipment is being commissioned in your plant, the more the plant will not generate revenue. As your equipment stays with you in your plant, some modifications, redesign, and improvements had been done from time to time by some departments from operations, maintenance, engineering on some parts and assemblies for it to run more smoothly and to lengthen the life of some parts that fail frequently and randomly during operation. These modifications and improvements should be properly documented and prioritized so that when the plant will decide to order additional equipment in the future, these modifications and improvements should be discussed with the equipment manufacturers so that when the new equipment is installed, we can expect fewer problems since we have already modified some of them.

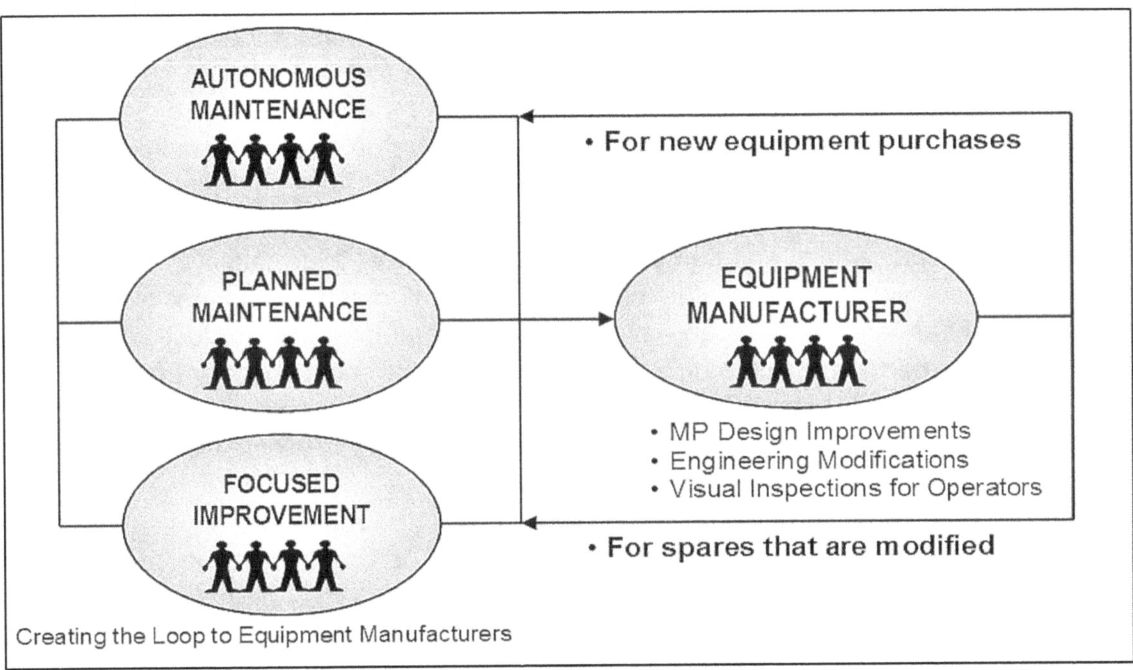

Figure 4.3: TPM's Early Equipment Management Pillar (EEM)

If the industry really wants to improve on cost, one should focus not on the Initial cost but on the equipment's life cycle cost. This is where a true and meaningful savings can be realized on maintenance. Every maintenance should focus on improving reliability and not reducing cost. Why? Because if reliability starts to improve, then the cost will definitely go down; it cannot be the other way around. Remember that there will be times that focusing on reducing cost will affect reliability, a lesson we should reflect upon. Having a low maintenance cost is always a consequence of good maintenance practice.

4.7: July 2010: Understanding Human Errors

It is much easier to blame people for their mistakes and errors rather than learn from the things that go wrong, whether in industries or about lessons of life itself. If you ask why? It is because of human nature. Let us admit it, myself included. This is how people react in the first place. On the other hand, people that committed the error would always have tons of excuse for not doing the right job rather than taking their time to analyze the root cause of their errors. If possible, they will always look for ways to pass the error to someone else rather than admit them. If the error or mistake was done by one of the c-level people or decision-makers, then no one will admit to the fault as if nothing happens. As a result, most industries have some policies and rules for people that commit mistakes. If an operator commits an error during operation or when a technician poured the wrong oil into the equipment, we punish or reprimand them instead of understanding the reason behind their failures. Why? Because it is much easier, faster, and most importantly, this is how we do things in our plant since the beginning of time. In short, it had been part of our culture for many years. If you are an engineer and your management people commit an error, we have limitations on what we can do. Instead of correcting them, we just avoid them as we do not want to be involved. Rogier Biosjoly of Morton Thiokol warned his management of the O-ring problem on the solid rocket booster they supplied to NASA and recommended to delay the launching, but he was ignored, and the shuttle exploded, killing the six astronauts and the teacher Christa McAullife. What would you do if you were in the management's shoes?

As defined in the previous newsletter, human error can be defined as an action planned but not executed according to the plan. Errors are committed daily by physicians, doctors, engineers, designers, carpenters, nurses, and you name it. For people involved in the medical field, some errors can be fatal and can compromise the lives of their patients. You have road accidents reported almost daily by the news. Errors from management decisions, flaws in systems, procedures, and policies can have a detrimental effect on its product and operations and affect how we perform maintenance in our plant. But the most important factor in human error is learning from our failures to become a better person. If people create problems, then people must also be capable of correcting them, but this can only happen if we look ourselves in the mirror and admit that we, too, are part of the problem.

Humans play an important role for industries during the design, installation, production, and maintenance phases of a product. In maintenance, human error may be defined as the failure to perform a specified task that could disrupt scheduled operations or result in damage to property and equipment. While human error had existed since the beginning of mankind, only in the last 50 years has it been the subject of scientific inquiry. There are various reasons for human errors such as stress, fatigue, working too long without adequate break time, incorrect tools, forgetfulness, complacency, pressure, distractions, insufficient knowledge, lack of training, lack of communication, misinterpretation, ignorance, lack of standards, lack of awareness, and the list goes on. However, with every error that we make, there is typically an associated change or something out of the ordinary occurring in our environment. The difference between humans and machines is that people can sense, change, hear, smell, see, feel, or taste something different and take the necessary actions to correct the anomaly. In industries, human error exists either because we tolerate them or we are just ignorant. Let me state some cases.

Case 1: Try to loosen some bolts in your equipment and place a toolbox near it. Let us say that the toolbox will contain a hammer, pipe wrench, adjustable wrench, pliers, socket wrench, double-ended open type wrench, and a torque wrench. Ask one of your technicians in the plant to tighten the loose bolts and observe how he tightens them. After he has tightened the bolts, ask him to loosen them again. Try to call 4 to 5 more technicians, one at a time, and give them the same instructions to tighten the bolt and write down how they perform the tightening process. Perhaps you will observe that some have similar ways of tightening the bolt and have used the same tools, while others simply use other methods to tighten the bolts. Why? Because you just ask them to tighten the bolts, but if we are explicit in our instruction and procedure, we can instruct them to use a torque wrench with this amount of torque, or perhaps a warning to indicate not to use an adjustable wrench, hammer, or pipe wrench so that each one of them will do the exactly the same procedure on tightening the bolts.

<u>**Case 2:**</u> Every single PM Checklists has a form of lubrication activity that is being performed in a timely fashion, which states something like this:

• What: Perform lubrication on bearing
• When: Monthly
• Who: Preventive Maintenance Group
• As to what specific brand and type of grease or oil are used, is unknown?
• If grease is to be used, what specific temperature of grease is required for this application?
• How many shots or pump is needed or when do I know when to stop pumping the grease gun?
• How many grams of grease is required for this application, and how do we measure it?
• How many pumps or strokes are required before stopping?
• Is this the correct interval to perform lubrication?

These are just a few questions to raise because lack of lubrication or over-lubricating will not be good for the bearing. Frequently we blame it on the lubricant, but most lubricant failures are caused not by the lubricant we use but by the way we perform our lubricating practices in our plant. Applying the wrong lubricant or mixing lubricants of different brands can cause incompatibility problems with the base and additives. This would likewise shorten the lifespan of the part it lubricates, not to mention the life of the lubricant itself.

Categories of Human Errors

When considering people and machines' interactions, human errors can be classified into four categories: Not all human errors are necessarily the fault of the person who made the error. In many cases, the error is either forced by external circumstances or other factors; therefore, if blame is allocated for any error, care must be taken to identify the real root cause of the problem.

A) Anthropometric Factors: These factors rely on the person's size, shape, and strength. Errors occur because the person simply cannot fit into the space provided, cannot reach something, or is not strong enough to move or lift something. Industries aiming for multiskilling should also consider anthropometric factors to avoid human errors. The human resource department hires people with special built, height, or perfect eyesight vision because it is mainly required in their work. Hence when a taller person is required to operate certain equipment, we simply cannot replace him with a shorter person because the height requirement is needed to perform their job. Many industries that practice multiskilling should also note its limitations, and one of them is anthropometric factors.

B) Human Sensory Perceptions: These are factors that people can see, hear, touch, feel, and even smell what is going on around them. Some failures will give some sort of symptom or warning that they are in the process of occurring, which can be captured by the human senses, such as excessive vibration, loud noise, heat, and foul smell. Therefore if errors are occurring or thought to be likely to occur for any of these reasons, then the human error is not the root cause of the failure, and we need to dig deeper into the real cause of the problem. When maintenance cannot smell a burnt motor because he has a terrible cold and was not allowed to go home by his boss, he should not be blamed for the failure.

C) Physiological Factors: The term physiological factors refer to the environmental stresses which affect human performance. These stresses include high or low temperatures, loud or irritating noise, excessive humidity, high vibration, exposure to toxic chemicals, radiation, or working too long without adequate break time. Exposure to these stresses leads to reduced sensory capacity, fatigue, and reduced mental alertness. These are all manifestations of human fatigue, and all greatly increase the chances that the people concerned will either make a slip, lapse, or mistake. If errors occur or are thought to occur for any of these reasons, human error is not the root cause of the problem. For example, reducing temperature or providing hearing protection should minimize exposure to noise and stress of the workers.

D) Psychological Factors: Psychological factors can be grouped according to those that are unintended and intended. Unintended actions can be grouped into slips and lapses, while intended action is grouped into mistakes and violations.

Slip occurs when somebody does something incorrectly or does something in the wrong sequence. If a person performs something in sequences such as step 1, step 2, step 3, step 4, and step 5, and the person does step 1, step 2, step 4, step 3, and step 5, he committed a slip. These are human actions that take place which was not intended. The occurrence of a lack of one's train of thought, which is derived from unconscious behavior.

Lapses occur when someone misses out on a key step in a sequence of events or activities. For example, a mechanic leaves a tool behind after working in a machine or simply forgets to fit a key component while reassembling it. If a person will do step 1, step 2, step 3, step 4, and step 5, and the person missed out on step 4 in the process, he committed a lapse. As humans age, our memory will not be as sharp as our high school or college days. People tend to forget many things as they grow old.

Mistake is a planning failure, where actions go as planned, but the plan was simply bad. These are errors in judgment. Mistakes are the real challenges in the analysis of human errors. Mistakes are influenced by external factors and can be group as knowledge-based, ruled-based, and violations.

Knowledge-Based Mistakes are types of mistakes that occur when someone is confronted with a situation that had not occurred before and had not been anticipated. In other words, there are no rules to follow. In situations like this, the person has to decide quickly about an appropriate course of action, and a mistake occurs due to the wrong decision. This suggests that the first and foremost way to avoid knowledge-based mistakes is to improve people's knowledge to make these decisions.

Ruled-Based Mistakes usually occur when people believe that they are following the correct course of action when doing the tasks based on a specific rule, but the course of action is not appropriate. An example will be replacing parts in equipment to comply with their PM procedure even if the pattern of failure is random as Preventive Maintenance is only feasible for age-related failures.

Violation occurs when someone knowingly and deliberately commits an error intentionally. Violations fall into three categories, routine violation, exceptional violation, and acts of sabotage.

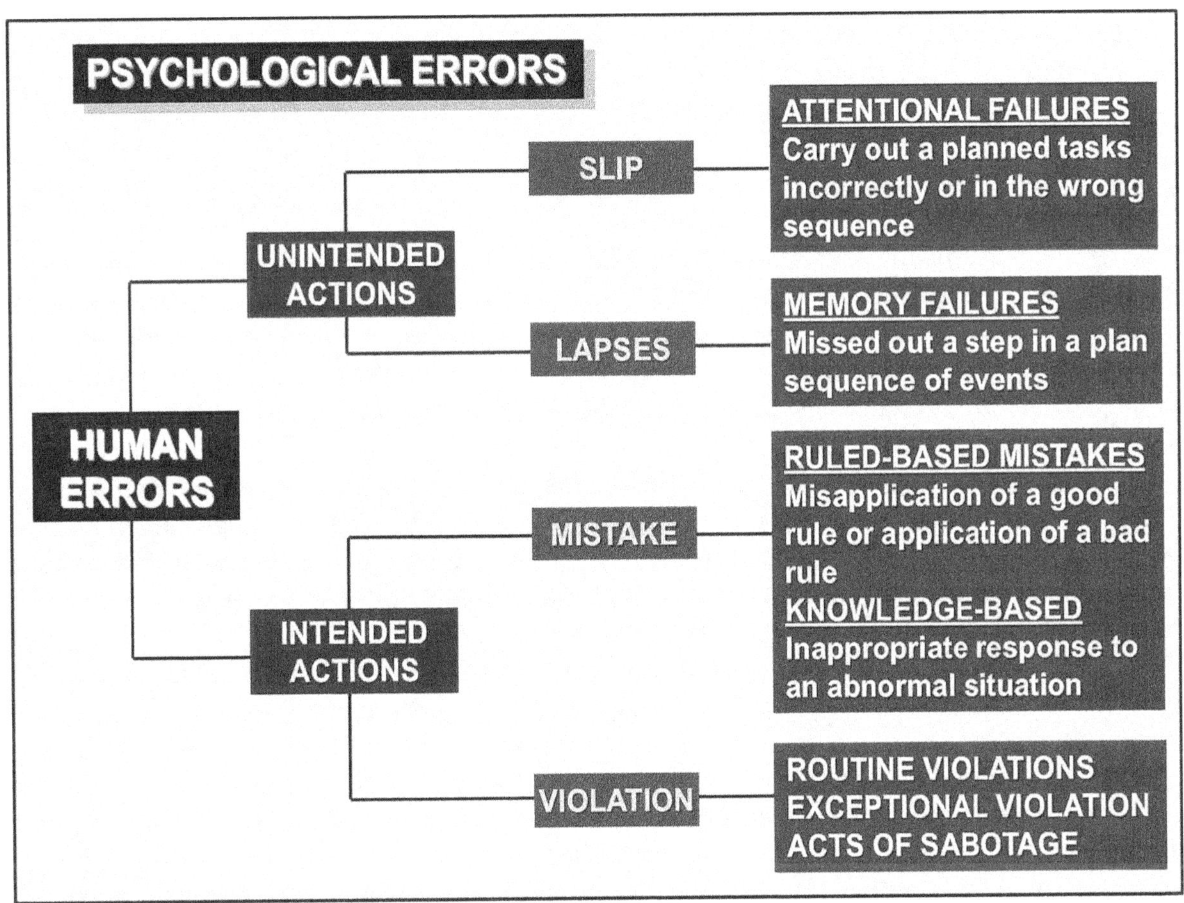

Figure 4.4: Understanding Human Errors

Therefore whether we like it or not, humans will continue to make errors and mistakes. That is why the need to perform a thorough Root Cause Failure Analysis is a must. Failures are not really bad, after all, if we can learn from them. Failures allow us to begin more intelligently. We can only benefit from failure if we can learn from it. Perhaps we cannot change the human condition, but we can change how humans perform their work. When people commit errors, we must steer away from blaming or punishing the person. We must understand the reason behind the error or failure. Changing policy, procedures, and systems is not enough. It also requires a change in oneself. The person who commits the error and the people accusing them should also change because the more we blame, the less we understand the failure. When human errors occur before we blame or point fingers at someone, it is more important to understand the reason behind the failure. All people commit mistakes and errors. We must realize that most of these errors happen beyond the person's control. Remember that the most successful people in this world failed miserably in their lifetime before they become successful. One thing these people have in common is that they learn from their mistakes and become better, and I think that's all I have to say about that.

4.8: August 2010: What Maintenance is All About

In a typical industry, maintenance is often faced with an eternal battle of dealing with their day-to-day pressure to resolve unexpected breakdowns, not to mention the ever-increasing cost of doing maintenance. When production for the day or week is met, all credit goes to the production and operations people. On the other hand, when the plant's capacity or output has not been met, all finger points to the maintenance people and are blamed for all failure even though they did not cause the equipment to fail.

The Evolution of Maintenance

If we trace the history, equipment was simple in design and not that complicated during World War II. The downtime did not matter much; hence, equipment downtime was not a real big issue. The demand for the products and services was low. When the equipment is down, they repair it; thus, there was no need for systematic maintenance; simple cleaning and lubrication routines were the best course of action when maintaining equipment.

After World War II, things change. There was an increase in demand for products and goods. This time, downtime was a real big issue for industries. This eventually led to the idea that equipment failures should be prevented, so the concept of Preventive Maintenance emerges. The traditional concept was thought that a direct relationship between the rate of wear and the equipment's operating age has some relationship. To cope with this issue, scheduled repairs, restoration, and overhauls were done to ensure that equipment is available whenever operations needed them. The expectation on maintenance was to have equipment with higher availability and lower cost, but even if the schedule has been complied, the opposite happens, and the cost of performing repair and maintenance also begins to rise sharply and take its toll. During the 1980s, greater expectations are now being realized on the maintenance function. As equipment nowadays are not only becoming more highly complex, diversified, and automated, more and more industries now realize that not only should they provide equipment with higher availability but also with higher reliability, integrity, greater safety, higher product quality, and most importantly no damage whatsoever to the environment. New techniques, strategies, and software on the maintenance function have emerged throughout the years, where failures can likewise be monitored by checking the equipment's condition and having it part of every single maintenance strategy. However, reality speaks that today, maintenance is still trapped with the vicious cycle of being reactive and industries focus too heavily on their Preventive Maintenance system dreaming that one day, the reliability of their equipment will start to improve by providing 100% compliance to their Preventive Maintenance activities. To be honest with the reader, that dream will never come true. The question to raise is, what exactly will fail in my equipment, and second, when will that failure happen?

If we know the exact answer to these two questions, then we're good, and that's about it. However, these two questions do not have a clear, concise, and definite answer, but rather to cope with this situation, some maintenance people are deployed to perform routine maintenance work on their equipment and assets. Most industries have some form of Preventive Maintenance group that sort of schedules their equipment for maintenance work. Some overhauling and replacements are being done in a scheduled and timely fashion. Some inspections are performed on the equipment regularly. Some deploy people do some walk-around inspection and functionality checks on the equipment. But despite all these noble efforts on the maintenance part, equipment still fails, and maintenance always gets the blame from operations people. I believe the first and foremost way to start a reliability improvement strategy is to change the way we think about failures, meaning before anyone can attempt to improve the reliability of their equipment, we need to change our paradigm first. Here are a few thoughts for the reader to think and ponder.

MAINTENANCE GENERATION

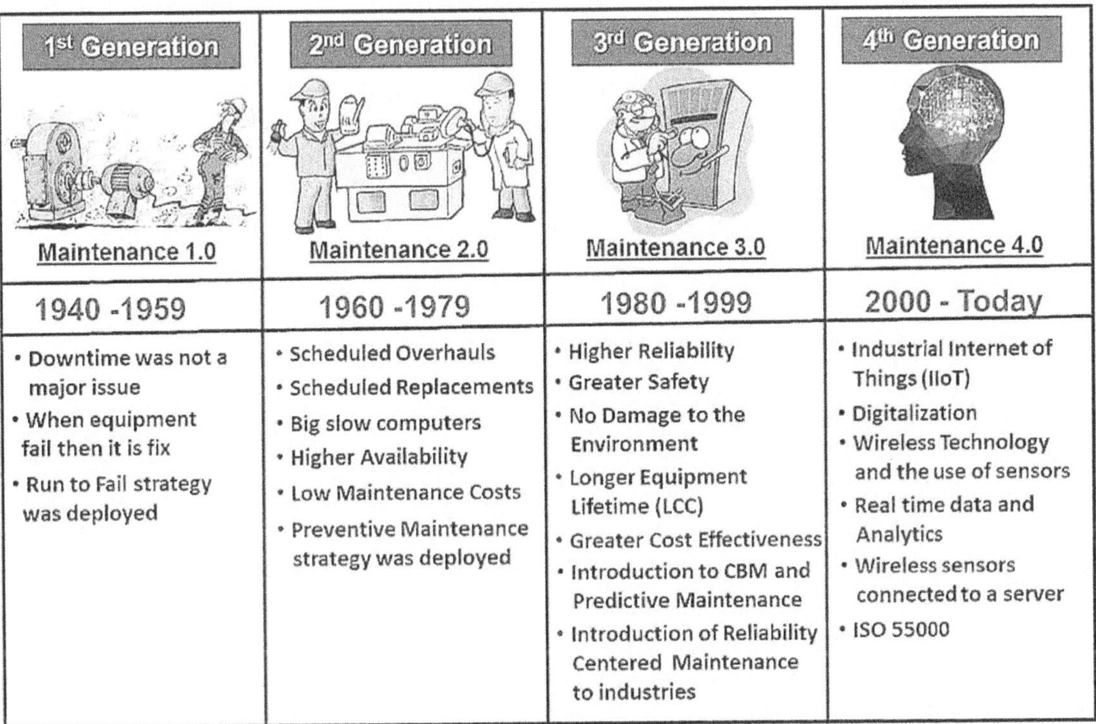

1st Generation	2nd Generation	3rd Generation	4th Generation
Maintenance 1.0	Maintenance 2.0	Maintenance 3.0	Maintenance 4.0
1940 -1959	1960 -1979	1980 -1999	2000 - Today
• Downtime was not a major issue • When equipment fail then it is fix • Run to Fail strategy was deployed	• Scheduled Overhauls • Scheduled Replacements • Big slow computers • Higher Availability • Low Maintenance Costs • Preventive Maintenance strategy was deployed	• Higher Reliability • Greater Safety • No Damage to the Environment • Longer Equipment Lifetime (LCC) • Greater Cost Effectiveness • Introduction to CBM and Predictive Maintenance • Introduction of Reliability Centered Maintenance to industries	• Industrial Internet of Things (IIoT) • Digitalization • Wireless Technology and the use of sensors • Real time data and Analytics • Wireless sensors connected to a server • ISO 55000

Figure 4.5: The Evolution of Maintenance

Maintenance is not about maintaining the whole equipment

When we speak about the word maintain, it means preserving or sustaining something. It also means keeping something in its existing state. Hence, the word maintenance is the act of maintaining or preserving something. It simply means ensuring that our physical assets and equipment continue to do what the users want them to do. In hindsight, maintenance is not about maintaining the whole equipment, but rather maintenance is about maintaining the functions of the equipment which the operators are currently using. Let me give a simple example of what I am pointing out. Today, almost everyone has their own mobile phone. It also comes to a point where it becomes a necessity. One can easily purchase it. I can recall vividly in the 1990s, where I used to work with Sharp Philippines Corporation, not all can afford one.

We called it a cellular phone, and it was expensive to own one. To use a cellular phone, you need to carry with you a cellular antenna and a battery, but nowadays, it becomes handier, and you can place it in your pocket. As the years passed by, more and more features are being added to the mobile phone.

Comparing the mobile phone today, you can do almost anything with it. You can watch your favorite TV program, email or chat with someone, locate an address using a GPRS, surf the net, listen to music, selfies, post on Instagram and other social media, and so on. The features are now becoming endless. Yet, I use my mobile phone to SMS, to call, check the time, and set the alarm if I needed to wake up early, and I think that's about it. I have never use the Bluetooth, email portion, or use it to surf the net. Therefore, if these features on my mobile phone fail, I don't mind about it, and it does not stop me from using my mobile phone anyway. In short, we maintain only what operators are currently using and not everything about the equipment. I cannot put it any simpler than that.

If you live somewhere in the middle east, let us say in the United Arab Emirates where the temperature can reach 45 to 50-degree Centigrade and your car air-conditioning fails, then almost immediately you will have it prioritized and have your ac (air-condition unit) fixed right away, but if your car's heater fails, then I don't think anyone with the right frame of mind will prioritize and find someone to fix it. On the other hand, if you live in Alaska and your car heater fails, I think you will have it fixed immediately.

Maintenance is not about eliminating breakdowns and failures

Most maintenance thinks that they can completely eliminate failures and zero out breakdowns. When I posed a question to my class and asked them if it is possible to eliminate breakdowns, some say yes, and when I asked them where they heard about that. They said that this is what TPM tells us. I have breathed, ate, and lived TPM for many years, and believe me when I tell you that this simply isn't possible. TPM has an ideal dream. It imagines an ideal factory with zero breakdowns, defects, and accidents. In reality, that factory does not exist. The goal of TPM is to bring the industry closer and closer to it, but eventually, up to this point of writing, it is still a dream. If I tell you that I have found a way for a person to become immortal and sell my formula for 100 dollars, then you know it's a ripped-off, but if I tell you that if you eat the right food, take a proper exercise, and take care of your health, then there is a greater probability for you to live a longer life. But even if you eat the right food, exercise regularly, and take care of your health, you can still get a bump from a speeding car and die. TPM tells us that when we encounter fewer breakdowns in the equipment, you actually are increasing the Mean Time Between Failure of the equipment.

Maintenance people must understand that performing maintenance is all about understanding that it is more important to reduce or eliminate the consequences of failure rather than eliminate the failure itself. Hence, instead of focusing on eliminating failures, the focus must be on understanding every single failure mode's consequences that can highly affect our equipment and assets. Like people, not all failures are created equal. Some failures can be confined to the equipment itself without affecting the other systems or equipment. Other failures can shut down the whole plant, while other failures may not only shut down the whole plant but might cause a threat to the safety of other people or perhaps damage to the entire environment itself. Therefore, when the consequences of failure are low, then it might be a good idea to allow failure to occur, but when the consequences of failure are simply not acceptable, then we need to do whatever it takes to reduce or eliminate the consequences of failure.

Preventive maintenance cannot prevent "ALL" failures

Maintenance must realize that adding intrusive or forced activities on Preventive Maintenance can only make your maintenance worst. Preventive maintenance cannot capture all failures; in fact, only around 20 to 30% of equipment failures can be addressed by doing Preventive Maintenance, leaving you with the other 70 to 80% of equipment failures that you need to address and deal with. In a nutshell, there may be around 100 ways your equipment can fail. Not all of this can be prevented but what maintenance can actually do is that some of these failures can be prevented by doing Preventive Maintenance, some of these failures can be predictive by using Predictive Maintenance and Condition Monitoring activities, some of these failures can be anticipated and allow to happen hence, there will be times that allowing the failure to fail can be a better option. Some failures can be studied so we can extend its life through modification. The message here is quite simple, do not overdo your Preventive Maintenance. Rest assured that the more PM activities you perform on your equipment, the more likely your equipment will be prone to infant mortality failures. Here are some excerpts from books by different authors, and they will tell you the same thing.

- In summary, we should be very careful about overhauling PM tasks because our equipment may not have an age-reliability pattern that justifies such tasks. Due to human errors, overhauls are likely to cause more problems than they prevent if aging regions are not present. *From page 61, RCM – Gateway to World Class Maintenance by Anthony Smith*

- One of the underlying assumptions of maintenance theory has always been that there is a fundamental cause-and-effect relationship between scheduled maintenance and operating reliability. This assumption was based on the intuitive belief that any equipment's reliability is directly related to its operating age. Because mechanical parts wear out, it, therefore, follows that the more frequent equipment is overhauled, the better protected it was against the likelihood of failure. Over the years, however, it was found that many types of failures could not be prevented, no matter how intensive the maintenance activities. *From the Original RCM Manuscript by Stanley Nowlan and Howard Heap, Chapter 1, page 2*
- In Tagalog: Sira na Naman ba? OO, kasi Bagong PM, Ginalaw pa kasi eh, malamang kung hindi ginalaw yan, maayos ang takbo nitong makina ngayon. Translated in English: Out of order again? No wonder they PM it. Maybe if the PM group did not touch this equipment, I'm pretty sure this will still be running smoothly! *From page 115 on World Class Maintenance Management - The 12 Disciplines by Rolly Angeles*

I have been mostly criticized by operations people who were present in my training if I make such a statement based on my actual observation and experience that maintenance people are humble and down-to-earth people while, on the other hand, operations people are arrogant people. (Of course, I am not mentioning all but only the majority I have dealt with.) When equipment fails, people will not accept why their equipment failed and always blame maintenance for not doing their jobs correctly. They always treat maintenance as a scapegoat on all equipment-related problems and issues. These people think that they seemed to be right all the time. I hope that sometimes operations people understand just a little of what they expect from the maintenance function.

I understand the pressures, sentiments, and frustration of the maintenance people, but change can only happen if we can accept the truth about what maintenance is. What it can do and what it cannot do. As the great leader Martin Luther King once said, I dream that someday all men are created equal. Like Martin Luther King, I also dream that one day, operations and maintenance will be treated equally and, most importantly, that one-day maintenance will finally be respected.

As the late Michael Jackson sang if you want to make this world a better place, take a look at yourself and make that change. (From the song "Man in the Mirror) R.I.P.

4.9: September 2010: Most Common Mean Time Indicators Explained (Part 1)

Many Mean Time Indicators are currently being used by our reliability and maintenance people to measure their equipment performance. The most common are MTBF, MTTF, MTTR, and MTBA, so we will cover this 3 part series of newsletters.

Mean Time Between Failure (MTBF) Explained

Failure simply means the inability of equipment to perform its required function. If we speak about a spare or component's failure, this is terminating its life. On the other hand, reliability is the probability that no failure will occur throughout its prescribed operating period.

Failure of a component indicates that it had become completely or partially unusable or has deteriorated to the point that it is undependable or is unsafe for normal sustained service. It may also be defined as an event in which any part of the equipment or machine does not perform according to its operational specification. In short, failure can also be synonymous with the term breakdown. As Bazovsky states, the modern concept of reliability in popular language simply means the equipment's capability not to fail or break down in operations. When equipment works well and performs to do its job for which it was designed to do, such equipment is reliable.

MTBF is an average measure of reliability that indicates a device will run without failing. The most common units used for MTBF are in hours. MTBF originated from the US Military Standards (MIL-STD-217) and has been widely used in other industries. The original reliability prediction handbook was MIL-HDBK-217. The Military Handbook for the Reliability Prediction of Electronic Equipment was published by the Department of Defense, based on the Reliability Analysis Center's work.

MTBF is a reliability engineering term that means the average operating time between breakdowns that require repair. The MTBF trend will be the higher the value, the more reliable the machine or part will be. MTBF simply means the average amount of time between failures. It is based on historical data or estimated by vendors and is used as a benchmark for reliability. The formula for MTBF is:

MTBF = Operating Time / Number of Failures

Where: Loading Time = Available Time – Non-Machine Related Downtime

Operating Time = Loading Time – Machine Related Downtime due to Breakdowns

- Available Time is given as the total time in a given period, 24 hours for one day, 168 hours for one week, and 720 hours for 30 days or 744 hours for 31 days.
- Machine-Related Downtime is the time the equipment is not operating due to machine-related downtime. In manufacturing, failure or breakdown is not the only cause of machine-related downtime. Other causes of machine-related downtime include short stoppages, change in cutting tools, quality-related problems on the product, set-up and conversion, changing equipment to produce one product to another product, start-up speed reduction losses. The only downtime due to equipment breakdowns should be included in the calculation of the operating time. Downtime due to set-up, conversion, short machine stoppages, and assists must be excluded in the MTBF calculation. This must be made clear to calculate a true and meaningful MTBF calculation; other related machine downtimes should be excluded.
- Non-Machine Related Downtime is the time the equipment is not operating due to non-machine-related factors. TPM refers to this as Planned Downtime. Factors reflecting non-machine-related downtime include operators' break time, meetings where equipment is stopped, planned shutdowns, Preventive Maintenance overhauls, and replacements, planned scheduled outages, holidays when the machine or equipment is not required to work in the absence of operators.
- Operating time is the time in which the equipment and assets are operating. It is also the time during which the equipment and asset are working without failure. Other terms to designate operating time are uptime, productive time, running time, or time the equipment was utilized. Let us provide an example here.

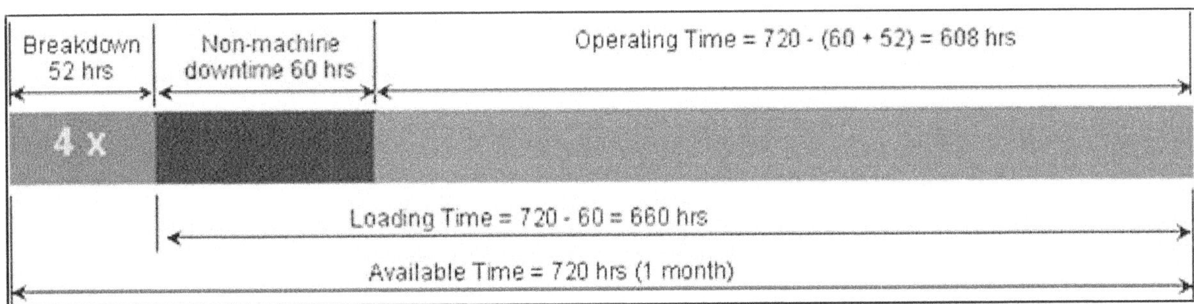

Figure 4.6: Case Study - Calculating MTBF

Given:
- Machine Related Downtime due to breakdown = 52 hrs.
- Non-Machine Downtime due to PM, Break time = 60 hrs.
- Available Time in one month = 720 hrs.
- Loading Time = Available Time - Loading Time = 660 hrs.

• Operating Time = 608 hrs.
• Number of times the equipment had failed = 4 times
MTBF = [(720-60) – 52] / 4
MTBF = 608 / 4 = 152 hrs.

So if the MTBF of this equipment for a specific month is 152 hrs. What does it mean? Most will say that the machine will fail every 152 hours. That assumption is truly wrong because the word means simply states an average. Hence, the correct way of stating this is that this particular equipment's average failure is every 152 hours. So don't expect your equipment to fail every 152 hours because, for all I care, the equipment can fail more or less than 152 hours since we are just basing this on an average.

If there is no breakdown or failure, the denominator will be zero, and an MTBF of infinity will be obtained. This simply indicates that there is nothing wrong with the equation. So in the case where there are no actual breakdowns recorded. Either we prolong the duration of MTBF until a breakdown happens or just assume a denominator of 1 to obtain a given value. Meaning in a month, when the MTBF is 720 hrs. Then no failure occurred during that particular period.

But before anyone can calculate and use MTBF, they must have a crystal clear understanding of what will eventually constitute a failure in the first place. Let me give you an example. I hope you take the time to answer these cases if this will be part of the MTBF calculation or not. The answer will be provided in the 2nd part of our newsletter on Most Common Mean Time Indicators Part 2.

Case 1: In one of the plants, half of the machines running on the production floor stopped because the main compressor in the utilities and facilities section failed. Most of this equipment are pneumatics. My question, is there a failure on the pneumatic equipment on the production floor? Will the amount of time the equipment was down be included in the MTBF calculation or not?

Case 2: If you have a pump and the pump has a standby mode, the standby mode will operate automatically if the duty pump failed. If the duty pump failed and the standby pump was automatically activated, there was no downtime. Operations will not be affected, but there is evidence that the duty pump had failed. Will this be included in the MTBF calculation or not?

Case 3: Some manufacturing equipment encounters what you call short stoppages or assists; TPM term this as Chokotei or Minor Stoppages, where the equipment stops, the operator resets the machine, and then runs again. Usually, it takes seconds or minutes to address this kind of problem. Will Minor Stops or short stoppages be included in the MTBF calculation or not?

Case 4: The Predictive Maintenance or CBM group spotted a spike or increase in vibration on one of the equipment types, which is a clear sign of potential failure. Finally, the group decided to schedule the equipment for overhaul and replacement. It took 4 hours of downtime. Will the downtime be included as unplanned downtime in the MTBF calculation or not?

Case 5: A Safety Officer audited one of your equipment with a severe oil leak and provided a non-conformance ticket. The machine was not allowed to be used for safety purposes. The machine was finally repaired for leaks, which took maintenance three hours to fix. Will this be included in the MTBF calculation or not?

Case 6: In some manufacturing plants, some equipment is used to process more than 1 product. If the equipment is used to run a different product, some parts will need to be replaced for the next product to run. This is called the conversion time or set-up time. Suppose some of the tooling was not fitted correctly, which cause breakage on the equipment, which occurred right after the conversion. Will this be included in the MTBF calculation or not?

Before attempting to measure MTBF, your people must agree about what will eventually constitute a failure and that consistency will always play a key to a credible and meaningful measurement. Otherwise, our data will just be compromised, and we are just fooling ourselves with the numbers.

Limitations of MTBF

Collecting failure data to calculate MTBF to determine the maintenance interval for replacement or overhauling is wrong. This should not be done since MTBF is just a measure of the average. Failures fail predominantly into three categories. Age-related makes up to 20% of all failures, and the bigger portion will be both infant and random failures. This will constitute around 80% of all failures. And for age-related failures, it is not MTBF but rather the remaining useful life that is significant when determining the best replacement period. In short, PM replacements and overhauls should be based on the useful life and not on the average life of the part being replaced or overhauled.

Another limitation of MTBF is that both failure and assists are not distinguished properly. Some industries make their own set of rules and assume that unattended assists for six minutes or beyond will eventually become a failure because the waiting time is far becoming long. An assist or error will still be an error and must be separated from the breakdown to get an accurate MTBF value despite of the waiting time.

MTBF may vary up to ten times, depending on the installation site, environment, and, most especially, how it is operated. For example, an environment with a high sulfur compound can provide a shorter life for electronic components. Field conditions vary enormously, from the power supply voltage to corrosive atmospheres, ambient temperatures, relative humidity, moisture, air cleanliness, all of which will affect the MTBF of the same component depending upon the conditions in which the part is being operated. Hence, vendors' MTBF may not be the same if the operating conditions will be different.

MTBF for Group of Machines

If you plan to use MTBF as one of your indices, we need to understand how your plant intends to measure MTBF in the first place. The formula can be quite confusing and depend a lot on the industry you are currently working in. MTBF can be used to compute the following.

• MTBF by critical part (usually called MTTF to be explained in Part 2 of this newsletter)
• MTBF by sub-assembly - so we can pinpoint the worst sub-assembly for a particular machine
• MTBF by machine - to determine the MTBF of a particular machine regardless of what part fails
• MTBF by process or system so we can determine the component in the process that fails frequently
• MTBF by a group of machines which will determine the total MTBF for a group of machines.

In calculating MTBF for several machines, you have the following three options to get the total. We can get the total MTBF, the MTBF in terms of the percentage of the average MTBF. If you want to get the total MTBF for these 4 machines, you can add the total MTBF, say for a given week or period.

• MTBF = (36 + 45 + 74 + 90) hrs. = 245 hours

Second, you may get the average MTBF and divide it by the total number of machines.
• MTBF = (36 + 45 + 74 + 90) / 4 = 61.25 hours

Third, you may get the percentage MTBF in which a perfect MTBF in a week is 168 hrs.
• MTBF = [(36 + 45 + 74 + 90) / (4 x 168 hrs.)] x 100%
• MTBF = (0.3646) (100 %) = 36.46%

When to use MTBF

MTBF can be used to compare two or more identical parts from different vendors. When we sub-contract some repairs of our equipment to some of our local vendors, one of the justifications in awarding the part will be its MTBF. A part or spare with a higher MTBF may require a higher initial cost, but lower life cycle costs can be achieved in the long run if we consider the cost of doing repairs, spare parts, overtime costs, overhead costs, downtime costs, just to name a few. The part with the higher MTBF will definitely have a lower cost in the end. A part or component with a higher MTBF will yield a better return in cost savings in the future. Although it is not recommended to use MTBF in determining the frequency of overhauls and replacement for our Preventive Maintenance schedule, MTBF can be used to determine the frequency of functionality inspection or failure. Failure Finding tasks for protective devices to eliminate the chance of multiple failures. Multiple failures are when the protective device is in a failed state simply because the protected function has already failed. Failure-finding tasks are done only on hidden failures. MTBF can also be used for routine inspections and lubrication as long as nothing will be dismantled on the equipment.

MTBF is used to determine the main contributor to why equipment keeps on failing. A group of maintenance people can conduct an MTBF study to determine what specific parts of their equipment keeps on failing and analyze the root cause of failure to modify or redesign the part with inherent design weakness. Continued.

4.10: October 2010: Most Common Mean Time Indicators Explained - Part 2

Mean Time Between Failure (MTBF) Explained: Before anyone can attempt to calculate and use MTBF, they must have a clear understanding of what will eventually constitute a failure and breakdown. Therefore, let us answer the following cases from Part 1 of this newsletter if they are included in the MTBF calculation.

Case 1: In one of the plants, half of the machines running on the production floor stopped because the main compressor in the utilities and facilities section failed. Most of the equipment is pneumatics and is operated by air. My question, was there a failure on the pneumatic equipment on the production floor? Will the amount of time the equipment was down be included in the MTBF calculation or not?

Answer: No, there was no failure or breakdown encountered on the production floor consisting of the pneumatic equipment, but there was evidence of downtime on the pneumatic equipment caused by the compressor on the plant's utilities or facilities. In this case, this should not be included in the MTBF calculation on the machine downtime, but if we are going to calculate the MTBF of the compressor, then the breakdown will be included in the MTBF calculation but not for the pneumatic equipment.

Case 2: If you have a pump and the pump has a standby mode in which if the duty pump failed, the standby mode will operate automatically, the question to raise is, if the duty pump actually failed and the standby pump was automatically activated, then there is eventually no downtime and operations was not even affected. In this case, there is evidence that the duty pump had failed. Will this be included in the MTBF calculation or not?

Answer: Yes, for the duty pump's failure, it should be included in the MTBF calculation whether we are getting the MTBF for the 2 pumps or for the single pump that failed since the duty pump eventually failed to have to be actually repaired. On the other end, if we calculate the MTBF of the entire system, the MTBF will still be high as there is no downtime.

Case 3: Some manufacturing equipment encounters what you call short stoppages or assists. TPM term this as chokotei or minor stoppages, where the equipment stops, the operator resets, and the equipment runs again. Usually, it takes seconds or minutes to address this kind of problem. Will minor stoppages or short stoppages be included in the MTBF calculation or not?

Answer: No, assist are different from breakdowns; some industries include them as breakdowns when waiting time exceeds six minutes and above. This should not be included in the MTBF calculation since what should be included in the MTBF calculation will only be downtime caused by breakdowns and failures. Suppose you insist on including them in the MTBF calculation; in that case, they might as well change its name from MTBF (Mean Time Between Failure) to MTBFAA or Mean Time Between Failure and Assists. Again, assists or short stoppages should not be included in the MTBF calculation.

Case 4: The Predictive Maintenance or CBM group spotted a spike or increased vibration on one of the equipment, which is a clear sign of a potential failure. Finally, the group decided to schedule the equipment for overhaul and replacement, which took 4 hours of downtime. Will the downtime be included in the MTBF calculation or not?
Answer: Yes, if we follow the formula that MTBF is equal to operating time divided by the

number of failures, operating time is equal to loading time minus machine downtime caused by breakdowns and failures. And loading time is equal to the available time minus planned downtime. Hence, the equipment's maintenance repair will be included in the planned downtime on the MTBF calculation and not in the machine downtime since no failure actually occurred yet, since it was predicted by the Predictive Maintenance group.

Case 5: A safety officer audited one of your equipment to have a severe leak and was provided a non-conformance ticket for safety reasons. The machine was not allowed to be used for safety purposes. The machine was finally repaired for leaks, which took the maintenance 3 hours to fix. Will this be included in the MTBF calculation or not?

Answer: This is quite debatable. In my opinion, yes, in this case, it should be included in the MTBF calculation, although there was actually no breakdown that occurred in the equipment. This will be included as a planned downtime of the MTBF Calculation. This situation is actually the same as in Case number 4. Let me give you a concrete example so you can follow it.

Given:

• Time to fix the leak is 3 hours, which is considered as a planned downtime
• Available time in one day is 24 hrs.
• Actual breakdown in this situation is zero

• MTBF = Operating Time / Breakdown Occurrences
• MTBF = (Loading Time - Machine Breakdown) / Breakdown Occurrences
• But Loading Time = Available Time - Planned Downtime = 24 - 3 = 21 hrs.
• MTBF = (21 - 0) hrs. / 1 = 21 hrs.

In this situation, we compute MTBF for that given day, and in the case where there is no actual breakdown, two options will be required to calculate the MTBF value. First, lengthen the time to get the MTBF until an actual breakdown is encountered, or second, assume a denominator of 1 to obtain the numerator; otherwise, an answer of infinity will be achieved.

Case 6: In some manufacturing plants, some equipment is used to process more than one product. If the equipment is used to run a different product, some parts will need to be replaced for the next product to run. This is called the conversion time or set-up time. Suppose some of the tooling was not fitted correctly, which cause breakage on the equipment, which occurred right after the conversion. Will this be included in the MTBF calculation or not?

Answer: In this case, an actual breakdown had occurred on the equipment caused by poor set-up and conversion. Therefore, this will be included in the actual MTBF calculation. Note that only when the equipment was down caused by the breakdown will be included, but the downtime caused by the conversion and set-up time will not be included in the MTBF calculation.

What is important before attempting to measure MTBF is that your people must be in agreement as to what will actually constitute a failure and that consistency will always play a key to a credible and meaningful measurement; otherwise, our data will just be compromised, and we are just fooling ourselves with the numbers.

Mean Time to Fail (MTTF) Explained

Case Study: The life of an incandescent bulb reached a lifespan of 3 years. Hence, the bulb was used for 3 years before It was declared fail and incapable of providing illumination. In its simplest sense, the bulb was busted. Therefore, if the bulb's life reached 3 years before it was replaced, what is the Mean Time to Fail of that bulb? The answer is 3 years. If the MTTF of the bulb is 3 years, then what is the Mean Time Between Failure of the bulb? The answer is also for 3 years. Therefore, if MTTF (Mean Time to Fail) and MTBF (Mean Time Between Failure) is three years, is MTTF the same as MTBF, or are they different?

Answer: The MTBF and MTTF of the incandescent bulb are 3 years, but the two mean indicators are not the same. They are different. MTBF (Mean Time Between Failures) is the expected time between two successive failures of a system. Therefore, MTBF is a key reliability metric for systems that can be repaired or restored, while MTTF (Mean Time To Failure) is the expected time to failure of a system. Non-repairable systems can fail only once. Therefore, for a non-repairable system, MTTF is equivalent to the mean of its failure time distribution. For MTBF, the actual time to repair should be included. Therefore, MTBF is equal to MTTR + MTTF.

If you buy a laptop, and after several months of use, it fails to function, then what is the MTBF in this case? Actually, this happened to me in the past. I went to a service center, and they diagnosed the failure to be on the motherboard. The technicians told me that the motherboard cannot be repaired, but it can be replaced. Hence, if we speak about the motherboard, which cannot be restored or repaired, we speak about its MTTF. If we consider the laptop which is repairable, then we speak about the MTBF. Similarly, if we speak about the equipment which can be repaired or restored, then we speak about its MTBF, but if we speak about the actual part that had failed, which cannot be restored but replaced, then we speak about the MTTF of that particular part.

MTTF is a basic measure of reliability for non-repairable systems. It is the meantime expected until the first failure of a piece of equipment. For constant failure rate systems, MTBF and MTTF are also the reciprocal of failure rate. Technically speaking, MTBF should be used only for repairable items, while MTTF should be used for non-repairable items. However, in today's modern language, MTBF is more commonly used for repairable and non-repairable items.

Mean Time to Repair (MTTR) Explained

When a failure occurs, it is critical to restoring the equipment as soon as possible. Typically much repair time is spent in determining the cause of the problem. Repair time should be performed in the shortest possible time. Our goal will be to put back the equipment in its current operating state. For failures that keep on repeating themselves, the best strategy will be to address the root cause of the problem to prevent it from recurring on its own.

MTTR (Mean Time to Repair) is the average time required to repair a component. It may be defined as the time it will take to bring a failed system back to its available or operating status. It is also the average time required to perform corrective maintenance or repair for all removable items in a product or system. MTTR analyzes how long repairs and maintenance tasks will take in the event of a system failure. Other terms used are Mean Time to Restore or Mean Time To Recover but the most common term used for this indicator is Mean Time To Repair. MTTR trend will be the lower, the better. The shorter the time to repair it better. Improving MTTR means shortening the time to repair the machine.

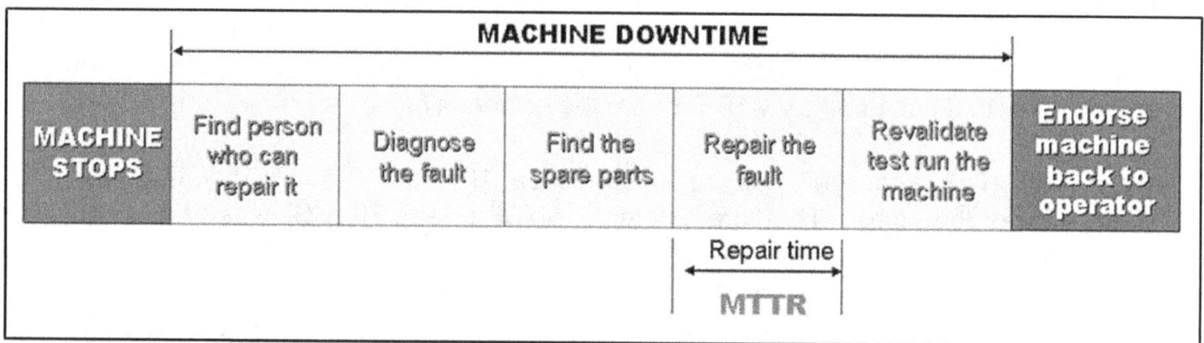

Figure 4.7: Case Study – Determine the Mean Time to Repair (MTTR)

Typically, when a machine fails, the operator will find someone who can repair it. Once the maintenance arrives at the failed equipment, he will diagnose the fault. If a part or spare is affected, he will leave the equipment temporarily to check the system and acquire the part in their stockroom. Then the actual repair will take place. After the part had been replaced, the maintenance will revalidate and make some test runs, then finally endorse the equipment back to the operator to continue production.

Question: If the part is unavailable in the stockroom and it took 1 week to acquire the part where the machine was not used and idle for that period, and once the part arrived, it took maintenance only one hour to repair and replace the part; in this case, what is the total repair time? Is it 1 week and 1 hour or only one hour? I have made a mini-survey to a small list, and as expected, their response differs from one industry to another. Here are some of the answers they provided and how they responded. The question, in this case, what is the total repair time? Will it be 1 hour or 1 week plus 1 hour?

a) 1 week + 1 hour (Total repair time is equal to the total downtime)
b) Only 1 hour (Total repair time is equal to the actual time the machine is repaired regardless if the part is available or not)

Response 1: It is on a case-to-case basis. When equipment fails, downtime for us is when the machine does not give a good product or no output. When we install an alternative part vs. OEM, it is a risk we have to decide, equating the cost of additional repair to the profit we can make to produce a good product. *From Ryan Ben R. Sabilala, Philippine Overseas Working in the Middle East*

Response 2: Answer (B) only 1 hour. For record purposes, we usually put the actual time to repair, which is 1 hour. The waiting time for spares to arrive can be devoted to other activities. If materials and spares are not available, this will automatically be considered pending work orders and will be active upon arrival of the required materials. Hope this would help. *From Rusty Peralta working in a Power Plant, Philippines*

Response 3: Good afternoon Rolly, we would consider the following conditions. If the machine was restored to the following conditions: the speed before breakdown is the same and the quality before breakdown is the same, we consider only 1 hour. If we could not operate the equipment, then repair time will be equal to the total downtime. I hope this will help. *From Albert Floresca, working in an Automotive Plant, Philippines*

Response 4: Greetings! My answer to your question is letter (B), only 1 hour. The reason for that is the maintenance will not consider the downtime due to the procurement of the parts considering that procurement is not under the maintenance function. We only consider the time upon the arrival of the said parts. *From, Kenn R. Rio, Head, Maintenance Group, Philippines*

Response 5: Rolly, good morning. My opinion on your question is the following, the repair time is only 1 hour. Since the rest of the machine downtime is just waiting time for the part to arrive. However, the machine downtime is 1 week + 1 hour. Thank you for considering me as one of your respondents. *From Fortunato G. Dequit, working in a power plant in the Philippines*

Response 6: Hi Rolly, my answer to your question is a letter (B). The total repair time is only one hour, which is equivalent to the actual repair time executed. *From Joemar Apolinario, working as a semi-conductor in the Philippines*

Response 7: Rolly, my answer is (A). Once you receive the permit to work, that is the time that the clock of downtime started. Once you receive the PTW (Permit to Work), they cannot use the equipment that has already been isolated. *From Joseleo Vicaldo, working in a power plant in, Philippines*

Response 8: Hi Rolly, one-hour downtime plus one-week waiting time are two different scenarios. When we are talking about maintenance, particularly skills, the computation should refer to letter B. The 1-week waiting time will become an assignable cause when trending the downtime. On the other hand, when we talk about OEE, I believe (A) is more appropriate. *From Anthony Bathan working in the semiconductor industry, Philippines*

Response 9: Rolly, I would choose (A). For me, repair time means exactly what it is because I consider the equipment under repair starting when it broke down until it is restored, regardless of delays in shipment, parts, etc. *From Mac Chan Lee, working in a power plant in, Philippines*

Response 10: Hi Rolly, Good day; we measure repair time as the actual time in doing the repair. If the reason is the non-availability of parts, then the machine done is for different reasons. So the answer is (B). *From Charles D. Mendoza working in a semiconductor plant, Philippines*

Here are the responses of consultants who have participated in this survey question, and their choice is almost unanimous except for number 14.

Response 11: Rolly, answer (A), Repair time = Active repair time + Preparation time. The reason the machine is there is to produce and not sit idle. An airplane only makes money when it is in the air. A machine can only be in one of two states, working or not working. If we don't need it to produce in the second case, it is still available to work to treat it as work. All the rest of the time, it cannot work, so it is down for maintenance, so no $$$. An efficient maintenance system will minimize the preparation time and scheduled work, so we minimize the loss of $$$. Finally, maintenance is a Business Process, not a Department. *From Vee Narayan, Lead Author, 100 Years of Maintenance: Practical Lessons from Three Lifetimes, Industrial Press, U.K.*

Response 12: Hello Rolly, answer (A), a failure event must be seen regarding the total loss value and costs incurred to the business. Total repair time is a measure that means exactly what it says: The time from the second the machine stopped being available for production to the time the machine came back up to its required production rate. That includes commissioning time and hand-back time, removing tags, and completing permits. What is most important to the business is the full impact of a machine outage on production uptime. From the business, the repair took 1 week and 1 hour. That is the correct business measure to truly understand the impact of the business's outage. *From Mike Sondalini, http://www.lifetime-reliability.com, Australia*

Response 13: Rolly, my selection is (A) if the maintenance function objective is "equipment ready to run," and the maintenance function has authority over resources required, diagnostics, preventive, and spare parts, etc. If the maintenance function is only "repair failure," then "b" could be my answer. I like the maintenance goal defined as "equipment ready to run." However, the resource consumption required needs to be logical to the cost of downtime. There is no need to own an expensive spare part and pay interest $ when the part can be obtained in a short relative lost production time $. So, the maintenance function needs to be closely aligned with the value-added activity of production. *From Greg Peitz, Affiliate, Failsafe Network, USA*

Response 14: Interesting question Rolly, I am not an expert on maintenance metrics, but I always believe that you measure what you want to hold to a certain performance standard or to improve, and the measures need to reflect the actual problem. In this case, the downtime needs to be recorded, but what bucket do you put it in. Most places have reduced spare inventory, but if it causes a week delay, it must be visible that spares or inventory was the issue, not maintenance working on the repair. So my answer is total repair time is 1 hour, total downtime is 1week + 1 hour, and the cause is inventory. *From Virginia Edley of SBK Consulting, LLC, USA*

Response 15: Rolly, anyway, I really don't have anything to add to either Greg or Virginia's response. I do like Virginia's inference that we measure what we're interested in improving. So, suppose we're interested in improving total availability. In that case, I'd opt for a. And as Greg said, if I'm only interested in maintenance response time, then I'd opt for b. Most people and organizations might be a bit too myopic and don't measure from a broad enough perspective. So, from the two options, I personally would opt for "A." Thanks for asking. *From C. Robert (Bob) Nelms, President, Failsafe Network*

My Response: In my own opinion, the total repair time will be 1 week and 1 hour, which includes the time the spare had to be on hand up to the time the equipment had actually been repaired. Hence, I choose (A) because spare parts management is one of the responsibilities of the maintenance function. If the part is not available when needed, then maintenance should be held accountable since we are the ones who control our spares. When the equipment is idle because a part or spare is not available, then the industry is not making any profit; consideration of the repair time should include the time the part or spare is acquired. This simple survey question is that if one industry opts to choose a or b, MTTR computation will vary greatly. Considering the total repair time is only 1 hour, then we can have the following computation.

MTTR = Repair Time / Breakdown Occurrences.

If we select (B) as our answer as the total repair time, then the actual repair time in one week is just 1 hour. In one week, the MTTR value will be 1/1 or 1 hour, while the rest of the week, it is idle and not working since the spare is unavailable. The MTTR, in this case, is one hour in a week, but the rest of the days, the machine sits idle. Although the MTTR value is giving us a good figure since repair time is low, but in this case, we are just fooling ourselves with the numbers since the rest of the time, the machine is not functioning and sits idle. On the other hand, if we consider both the waiting time for the spare, which is 1 week and 1 hour, to be the total repair time, then the MTTR value will be 168 hours in one week, giving us a more truthful and realistic value. A true and correct MTTR starts at the time of failure and continues until the item is operational once again, regardless if a system part or component will be available or not. To be continued.

4.11: November. 2010: Most Common Mean Time Indicators Explained - Part 3

Mean Time to Assists (MTBA) Explained

Assists or errors are any unplanned interruption or variance from equipment specification that requires human intervention on the equipment. If the time fixes an assist goes more than 6 minutes or more, many manufacturing industries I know will consider this a breakdown. Causes of prolonged assists may include waiting time for the maintenance to arrive at the equipment, or it takes too much time to troubleshoot the assists. In my opinion, an assist is always an assist and is different from a breakdown. When a part breaks down caused by the assists, this is the time to consider it a breakdown.

Assists are often encountered when the equipment is fully automated and full of electronic parts. The frequency of assists encountered will be much more frequent than the breakdowns occurring on that particular equipment. Meaning in a week, if you provide me with a data of 5 assists and 40 breakdowns, this is quite unlikely realistic as assists occur more frequently than breakdown. The second point I would like to point out is that if breakdowns are not clearly distinguished from assists, there is a temptation to include this downtime in the MTBF calculation.

For industries implementing TPM or Total Productive Maintenance, they are more familiar with the term minor stoppages or chokotei. A minor stoppage is an equipment stoppage due to a failure or an error in automatic handling, processing, or assembly of parts and workpieces. It sometimes occurs due to quality-related abnormalities. These are errors in automated processes where the workpiece flow stops, the operator resets, and the machine runs again.

In most manufacturing plants that encounter this type of problem on their equipment, it is quite tedious for the operator to record every single assist and error that can occur on a particular day. With the absence of software and system, it is unlikely that the operator can capture every assist that can occur on a particular piece of equipment; in this case, this is where an MTBA Snapshot can be taken down on that particular equipment.

Meantime Between Assists (MTBA) is the average time the equipment performs its intended function between assists. It is also the product or the operating time divided by the number of assists; hence MTBA is the average time between assist on any stoppage of a machine caused by any unwanted operation. The formula for MTBA is as follows:

• MTBA = Operating Time / Frequency of Assists or
• MTBA = Productive Time / Frequency of Assists where
• Operating Time = Loading Time - Machine Downtime due to assists

Procedure for Preparing MTBA Snapshots

• Step 1: Prepare MTBA Snapshot Form
• Step 2: Select equipment that has a high frequency of assists and errors
• Step 3: Sectionalize the equipment per station or sub-assembly and record all possible assists that can occur on each sub-assembly or station
• Step 4: Perform a Snapshot (min of 2 hours) and write the duration, frequency, and type of assists that occur on the machine
• Step 5: Generate corrective measures and perform modifications
• Step 6: Horizontal Replication of the MTBA improvement to same equipment similar problems

Figure 4.8: Actual MTBA Analysis Conducted on Meco Plating Machine

Below is an actual MTBA Snapshot Case Study we performed in the paste

• Area: Central Lead Finish Station
• Machine: MECO Plating Machine).
• Machine type: EDF + EPL 2400S automatic load and unload strip to strip deflash and plating machine to process 2400 strips per hour.

Central Lead Finish Meco Plating Machine

- Package or lead count: ABC / 52 L
- Area: Central Lead Finish (Plating)
- Machine Number: Meco 2
- Performed by CLF Planned Maintenance Team
- Date Performed: January 15, 1998
- Time Started: 10:00 am
- Time Ended: 12:00 pm
- Total Observation Time: 2 hours / 7200 seconds

Computing for the MTBA

Given from the snapshot

- Time Start (TS) = 10:00 am
- Time Finish (TF) = 12:00 pm
- Total Assists (TA) = 25 times
- Total Downtime (TDT) = 845 sec

MTBA SNAPSHOT FORM												
DETAILS OF ASSISTS	Frequency (Time in Seconds)										TOTAL ASSISTS	TOTAL TIME
	1	2	3	4	5	6	7	8	9	10		
Misloading on load	11	34	55	31	23	48	18	50	11	15	16	429
	33	58	35	17	20	15						
Mispick	17	21									2	38
Magazine stops	34										1	34
Strip jamming	60	19									2	79
Elevator jam	52	113	49								3	214
Magazine stop	51										1	51
Total Assists and Time in Seconds											25	845

Figure 4.9: Actual MTBA Analysis Snapshot Conducted on Meco Plating Machine

Total Productive Time (TPT)

- TPT = Loading Time - Machine Downtime
- TPT = [(2 hrs. x 60 min. / hr. x 60 sec / min.) - 845 sec.
- TPT = 7200 sec. - 845 sec
- TPT = 6,355 seconds

Mean Time between Assist (MTBA)

- MTBA = Productive Time / Total Assists
- MTBA = 6,355 seconds / 25
- MTBA = 254.2 sec
- MTBA = 4.24 minutes

PROCESS	CURRENT UNFORSEEN ISSUES	CORRECTIVE ACTIONS	MP DESIGN NO.
LOADING	- Frequent turn-over of strips during change of magazine that cause mispick - Frequent alarm of mispick due to undefined setting of vacuum pressure - Early replacement of gripper plate holder for rotating cylinder	- Modify y-aligner arm by extending its arm to 3mm to cater upcoming strips. - Convert analog pressure to digital pressure switch - Change material from aluminum to ss plate	- CLF-00SP-MP-006 - CLF-00SP-MP-007 - CLF-00SP-MP-005
DEFLASH to DRYING	- Early deterioration of conveyor clips that cause strip jamming to processed cells	- Modify flywheel to non opening type that prolongs the pressure switch clip life span to reduce jamming	- CLF-00SP-MP-008
UNLOADING	- Frequent alarm and mispick due to undefined setting of vacuum pressure - Frequent turn over of strips during off loading	- Convert analog pressure switch to digital pressure switch. - Modify y-aligner arm by extending its arm to 3mm to cater upcoming strips.	- CLF-00SP-MP-007 - CLF-00SP-MP-005

Figure 4.10: Modifications Done on Meco 2 Machine to Increase MTBA

PHASE 0	PHASE 1		PHASE 2	
START	BEFORE	AFTER	BEFORE	AFTER
January 1998	February 1998	April 1998	May 1998	February 1999
4.24 min.	16.61 min.	22.54 min.	45.84 min.	115.74 min.

Figure 4.11: Before and After Data After Conducting MTBA Analysis

This means that MECO 2 has a probability of assists occurring every 4.24 minutes. The following figure indicates the modifications done on the CLF Planned Maintenance team's plating machine increased the MTBA of the MECO equipment. The following were the actual data taken from CLF Plating Meco Machine as part of their improvement activities during their TPM Planned Maintenance activities. Based on actual records, all MECO machines passed the Planned Maintenance Phase 2 audit and certification. The group was able to improve the MTBA of their equipment from 4.24 minutes to 115.74 minutes after conducting several modifications on the equipment that affected these errors. Before and after data derived from CLF Planned Maintenance activities 1998-1999.

4.12: December 2010: Autonomous Maintenance for Industries Part 1

During the first 3 months of this year, I was quite busy refining and completing my training materials on Autonomous Maintenance, which was recently requested by one of my clients here in the Philippines. This client was not from the manufacturing sector but from a power plant. Autonomous Maintenance is one of the TPM pillars with the biggest population, which involves operating their equipment. TPM is originally designed for manufacturing plants, and although not all industries can implement Autonomous Maintenance, there are non-manufacturing plants such as power plants that can benefit from its implementation, but it cannot be designed the way we want it to work for a manufacturing plant since the set-up of operators is different. If Step 1 of Autonomous Maintenance can be performed, rest assured that the remaining steps can also be done. So let me share this two-part series of newsletters on Autonomous Maintenance for Manufacturing and Non-Manufacturing Plants.

In my training on World Class Maintenance Management - The 12 Disciplines, I asked people to group themselves into a team and select one of the 12 Disciplines that they would like to improve most in their plant. Top of the list is Autonomous Maintenance, meaning that the operator should not just be operating the equipment but should accept that maintenance is a shared responsibility for both parties. When equipment is about to fail, the operator will sense and witness the failure first and not the maintenance because these people are the ones who are closest to their asset during the time of failure or breakdown. The role of Autonomous Maintenance is to enhance the senses of operators so that they can easily detect problems, irregularities, and abnormalities in their equipment by addressing the basics and accepting maintenance responsibilities, which is keeping the equipment clean, completing the bolts, nuts, fasteners and applying the correct lubrication on the equipment.

If we trace back history on why operations and maintenance are separated, it all started with Fredrick Taylor's concept. According to Fredrick Taylor, the best way to manage an organization was to standardize the activity into simple repetitive tasks and closely supervise them into doing it. In effect, management people do all the thinking and decisions, while the supervisors act as the watchers making sure that the decision is followed to the letter. Workers are focused on doing what they are told to do and just follow instructions until they get bored to death because of their routine work. While the western countries focused more on producing big volumes, capacity, and production, the Japanese people learned that the best way to run an organization is to focus more on the voices of their people by allowing them to make decisions to perform their work better.

The western countries slowly realized that they are being beaten badly by their Japanese competitors, but many industries are still stubborn and remain trapped in the old Taylor paradigm. Thus, most American and western style factory management clearly separated the production and maintenance departments' roles. Managers were convinced that this style was the most effective way to utilize human resources. Operators concentrate on production with little or no knowledge of the structure and function of their equipment while maintenance received work orders and performs repairs on their equipment. As a result, both operations and maintenance went their own separate ways instead of following the path to mutual cooperation and shared responsibility. That is why today's feud and friction are pretty much alive on both sides. Much worst is the lack of communication between operators and maintenance. When a breakdown occurs, operators tend to disappear, just like Harry Houdini after calling the maintenance.

On the other hand, traditional mechanics would love the smell of breakdown. They know that they have become indispensable specialists of the trade. Maintenance thinks that they are assured of a stable job every time a repair is needed on their machines, not to mention that maintenance love to work overtime because of the added pay. And this vicious cycle goes on in which the aftermath is an immense amount of waste in man-hours, production time, loss of opportunity, and ballooning of spares and maintenance expenses.

Operators must understand that maintenance is a shared responsibility for both parties. There are maintenance activities that must be done by the operators themselves. Maintenance must understand that Autonomous Maintenance can only work if they teach operators about their equipment. Maintenance should be the ones to make the first move. But most importantly, for Autonomous Maintenance to work, the industry needs to change from the traditional mindset that operators are hired to operate, while maintenance will fix every time there is a failure. We need to transition into a paradigm in which both operators and maintenance are responsible for taking care of our equipment. Blaming and finger-pointing have to be stopped, and culture needs to be changed, otherwise forget implementing Autonomous Maintenance because this simply cannot be forced on operators.

Autonomous Maintenance is the activities in which each worker performs daily inspection, lubrication and cleaning, minor repairs and troubleshooting, accuracy checks, and so forth on the equipment, aiming to keep their equipment healthy and in good operating condition. Big problems are just simply an accumulation of small problems that are frequently left unattended. Training and nurturing operators with strong equipment skills can develop operators to detect early signs of problems before becoming serious enough to cause catastrophic failures. One feature of Autonomous Maintenance is to develop a checklist for operators, which will include cleaning, lubrication, and inspection to be done on the equipment. In short, operators will develop their own standards and checklist.

Autonomous Maintenance in 7 Steps

Autonomous Maintenance is done in a standard 7 Step approach and might differ in detail from one industry to another depending on the type of industry and how TPM consultants approach them. Hence, let me discuss the standard and generic 7 Step approach on Autonomous Maintenance, which will begin with Step 0. Autonomous Maintenance aims to develop a self-directed and empowered operator to improve their equipment and assets' performance.

Step 0: Preparatory Stage: Before going through Step 1 of Autonomous Maintenance, it is important to start with this preparatory step, Step 0. The basic concept of Step 0 is to organize a small group of operators who will compose the pilot team. Selection is critical and should compose 100% pure operators. Maintenance should not be part of the team as they have their own TPM pillar to cover with. The role of maintenance is to act as a coach and training partner for operators. If your plant works on a shifting basis, decide if the team will be a vertical set-up (members will include operators from A shift, B shift, and C shift) or the team will be a horizontal set-up (members will include operators from A shift only, B shift or C shift only to start with). Both will have their pros and cons but based on experience, vertical set-up will be more difficult but will provide an edge as all people within the shift are fully aware of the activities to be done by the team. Select a pilot equipment, process, component, or system to start with. Selection should be made on the plant's worst and critical equipment currently being used by operations.

Note: Do not select good or brand new equipment since the team's output will be useless as there will be very little or nothing to improve on the equipment selected. In this preparatory step, maintenance should train operators on their equipment's basic function and, more importantly, the equipment's safety. Never perform Step 1, Initial Cleaning, if operators are not trained on the equipment's safety. If someone accidentally gets injured during the cleaning process, that will be the end of your Autonomous Maintenance. Important activities to cover on this step are requisitioning all the cleaning materials and tools needed for Step 1 and scheduling the time to perform Step 1.

Step 1: Perform Initial Cleaning: During Step 1, operators clean their equipment thoroughly as planned from the preparatory step. Each member is assigned to a specific place on the equipment. When they clean their equipment, they are actually inspecting it. The inspection will allow operators to touch their equipment and will reveal problems, irregularities, and abnormalities. Operators use their senses to detect looseness, wear, cracks, or leaks on their equipment and tag them. These tags are placed near as possible on the abnormality found. Tags will only be removed once they had been addressed completely. Take all necessary safety precautions before starting your cleaning activities.

Review the cleaning map from Step 0. Frame a cleaning plan as to what parts need to be clean. Clean thoroughly to remove dirt and dust accumulated over many years. Open covers and lids that have never been seen and clean them thoroughly. Clean dirt on the equipment itself and transfer equipment, electrical boxes, fluid tanks, and other auxiliary equipment. Operators need to understand why they need to keep their machines clean and the harm of not doing it. Abnormalities found by operators should be corrected by the operators themselves. For those who will require job orders to restore, they will be passed on to the maintenance. But the golden rule in this step is that if the operator can correct the slight abnormality found, they should correct it themselves. At this step, the operators will experience areas on the equipment that are hard to access or hard to clean and take note of this. It will be addressed in the next step of Autonomous Maintenance. Initial cleaning time to be done on the equipment may vary on the equipment's size to be cleaned and the amount of dirt, dust, grime, scales, and debris removed. Hence, do not rush this activity. This will be the start of the learning process for the operators.

Step 2: Eliminate Sources of Contamination and Hard to Clean Areas: After completing Step 1, the Autonomous Maintenance team realizes that dirt accumulates after some time on their equipment. The team also discusses problems on areas and parts of the equipment they find difficult to access and clean. AM team brainstorm on eliminating or reducing contamination sources and thinking of simple specialized cleaning tools required for hard to reach or difficult to clean areas. The Autonomous Maintenance team shares their experiences in performing Step 1 Initial Cleaning with the group and initiates simple improvement activities to address their problems during the initial cleaning process. The AM team begins to eliminate or reduce contamination sources that make the equipment or critical parts of the equipment dirty once again. Another important activity on this step is to reduce the cleaning time by preparing cleaning charts, reducing contamination sources, making it hard to clean places more accessible, and introducing visual control on the equipment. Teams also become aware of hard-to-reach or clean areas and feel obliged to improve their accessibility. Step 2 aims to reduce the time it takes for the team to clean their equipment. Measures to control hard-to-access areas are taken by improving places where cleaning and inspecting are difficult and time-consuming so that cleaning and inspection can be performed easily.

Step 3: Establish Tentative Equipment Standards: This step aims to lock in place the gains and benefits from Steps 1 and 2 to ensure the Basic Equipment Condition and keep the equipment in peak operating condition. During Step 1 Initial Cleaning activities, the team put a tremendous effort into cleaning, correcting minor flaws, and establishing Basic Equipment Standards. In Step 2, they reduce the time required for these tasks by addressing contamination sources, cleaning, and lubrication which are easier to perform. During Step 3, the team prepares tentative standards and establishes checkpoints for cleaning, inspection, and lubrication. With the help of maintenance, operators try to understand why these tasks are important and what harm they can do to their equipment if it is neglected. These standards specify what needs to be done. What parts of the equipment need to be clean, lubricated, and inspected? Many industries already have these standards in place. Still, very few operators follow them thoroughly because the people who set the standards are seldom the ones who apply them. Operators might not understand the importance of doing it in the first place. Before operators can perform these standards and inspections, maintenance must fully train operators on the importance of doing them. Remember that it is the operators who will witness the failure first and not the maintenance people. It is important to let operators identify what needs to be inspected on the equipment.

Step 4: Develop General Equipment Procedures and Training: In Steps 1 to 3, the operator experienced performing initial cleaning, detecting abnormalities, identifying hard-to-clean areas, generating lubricating, cleaning standards, and basic knowledge about safety and the machine function. In Step 4, the operators must conduct maintenance to understand the fundamental theory and its principles. For example, an operator may detect that a bolt is not properly tightened or loose, but the operator does not know the correct amount of torque needed to apply on the bolt and what happens when we over-apply the torque in the bolt. An operator may know that the oil level is below the minimum, so the operator decides to tap oil into the equipment, but what is the correct viscosity for this application? If we speak about grease, what specific type of grease quantity and interval is needed. What corresponding temperatures can they be used? In this step, fundamentals of machine elements, hydraulics, pneumatics, lubrication, bolts, nuts, fasteners, electrical, power transmission, rotating machine elements are learned by operators to find confidence in themselves in performing the standards generated on lubrication, cleaning, and inspection. Step 4 focused more on training that the operators will need to advance in the Steps of Autonomous Maintenance. This will be the longest step in the AM journey. Planned Maintenance must be prepared for this step as they will be the ones to conduct this training.

2011 RSA Reliability Newsletter Vault Archive

> *Many say that maintenance cannot improve equipment reliability but can only sustain it. All I can say is that maintenance, as a task cannot improve equipment's design reliability, but maintenance as a human being can definitely improve the equipment's reliability through improvements and modifications.*

5.1: January 2011: Autonomous Maintenance for Industries Part 2

Step 4: Develop General Equipment Inspection Procedures: As the first three steps are being accomplished, their involvement in their equipment becomes more evident. Maintenance, on the other hand, understands the role operators play in this Autonomous Maintenance transformation. In Step 4, Autonomous Maintenance is taught about the theoretical knowledge they need to know to perform their inspection thoroughly. Maintenance should be responsible for transferring this knowledge to operators. Training is an essential part of this step and should be prepared by Maintenance in advance or when the Autonomous Maintenance is still in Steps 1 to 3. This will be a continuous process. Maintenance prepares training materials, cut-out models, and one-point lessons. It provides one-on-one coaching and other relevant materials needed for this step. Training will not only be limited to classroom training. In the first three steps of Autonomous Maintenance, Maintenance educates the operators about their equipment's basic function and safety. In this step, Maintenance teaches operators about nuts, bolts, lubrication, pneumatics, hydraulics, and other important things they need to know about their equipment. As the operators' senses are enhanced in the first three steps to capture equipment abnormality and irregularities, operators learn their way through experience by touching the parts on their equipment, but the operators also need to know the theoretical aspect. Although in Step 3, a tentative standard for cleaning, inspecting, and lubrication has been established, operators try to enhance these standards since operators do not only have experienced the process but have also have known the theory and principles. The general inspection goal is to raise the operators' skills by understanding both the practical side and the theoretical side of their equipment.

While in Steps 1 to 3, the operator experienced performing initial cleaning, detecting abnormalities, identifying hard-to-clean areas, generating lubricating and cleaning standards, and basic knowledge about safety and machine function. In Step 4, the focus will be on training to clearly grasp the basic theory and its principles. An example of this is that an operator may detect that a bolt is not properly tightened or loose, so the operator will tighten it, but what the operator does not know exactly is what is the correct amount of torque needed to apply on the bolt and what happens when we over apply the torque in the bolt. Another example is that an operator may know that the oil level is below the minimum, so the operator decides to tap oil into the equipment, but what is the correct viscosity suited for this application. If we speak about grease, there are different types of grease, and what corresponding temperatures can they be used. Since the operators have to experience the actual inspections, cleaning, and performing lubrication on their equipment and had been taught the basic principles on lubrication, bolting, and other relevant topics they should know, AM teams with the guide of the PM group finalize the best frequency and interval to perform these activities on their equipment.

This step aims to understand the structures, functions, and principles to learn optimal conditions to achieve the equipment's maximum efficiency. In this step, fundamentals of machine elements, hydraulics, pneumatics, lubrication, bolts, nuts, fasteners, electrical, power transmission, rotating machine elements are learned by operators to find confidence in themselves in performing the standards generated on lubrication, cleaning, and inspection. Unlike the first three steps, Step 4 is probably the longest time to accomplish, so what is important is not to rush this activity since we are building the operators' skills. If done correctly, the company should see around 60 to 80% reduction in their breakdowns and other downtime forms at the end of this step, such as errors and assists.

Step 5: Conduct General Equipment Inspection Autonomously: Once Step 4 is completed by the operators, the standards generated in Step 3 are thoroughly analyzed and finalized to enhance the inspection process. These inspections that are carried out by operators and Maintenance are split to avoid redundancy in the process. The general inspection process is organized so that the correct frequency and interval can be derived. Operators prepare an inspection process swiftly, daily, weekly, monthly, or even quarterly. Step 5 is done to generate a sustaining process in the previous steps of Autonomous Maintenance. Once this step is performed, the equipment becomes a shared responsibility for both operators and Maintenance. Operators have grasped a full understanding of why they need to perform these activities on their equipment. The inspection is carried out regularly, and equipment failures are minimized since operators can detect early failures. As operators become full partners in this shared responsibility, maintenance focuses on carrying out the higher phases/steps of their Planned Maintenance activities on their equipment.

The cleaning and lubrication standards set in Step 3 and the tentative inspection standards prepared for each given category in Step 4 are now combined in Step 5, where they are combined into a unified Autonomous Maintenance Standard. Cleaning, inspection, and lubrication standards should be reexamined to accomplish a higher check efficiency and eliminate the possibility of human errors. If problems occur, operators work with Maintenance to develop inspection points to prevent the problem from recurring. At this step, operators continue to think of ways to shorten the time to perform cleaning, lubrication, and inspection on their assets. Because of the training gained from the previous steps of Autonomous Maintenance, operators' skills are truly enhanced. At this step, operators can detect abnormalities and irregularities in their equipment and what to do if they spot one. It is important to consolidate the Maintenance and operators' activities to avoid duplication and redundancy on the maintenance side. Suppose such activities as cleaning, lubrication, and tightening activities are included in the current PM lists; CMMS is revised, and a responsible person is changed from Maintenance to the operator since this will be carried out by the AM teams. The inspection being finalized and generated is very much different from the traditional checklist provided by Maintenance since the operator understands its value.

Step 6: Manage Systematic Autonomous Maintenance in the Workplace: As the operator completes the first five steps of Autonomous Maintenance activities, equipment starts to improve, and failures are reduced dramatically. Operators now focus on how to improve their workplace by using simple principles of 5's, visual control management, and housekeeping. Tools are properly organized, cleaning materials are easily accessible, jigs tools and fixtures for conversion are properly stored, etc. In simple layman's terms, let us say that the first 5 steps refer to cleaning and inspecting inside your house; in this step, the owner would like to maintain his backyard and clean up unnecessary things in his garage. These things are done to provide orderliness in the workplace. People working for quite several hours can stress and fatigue themselves easily. Human stress is accelerated if the environment they are exposed to is untidy. Also, as Autonomous Maintenance is a continuous effort, the operator's device simple yet innovative ways to improve inspection, lubrication, and cleaning in their equipment with the effective use of visual controls.

All activities geared toward completing Steps 1 to 5 focused on reducing equipment breakdown. In Step 6, AM activities are focused on aiming for Zero Waste and Zero Defects. This can be achieved by improving equipment and work methods. Operators' human senses, visual controls, and Poka-Yoke or Mistake Proofing devices to eliminate defects are used. Operators should be keen on determining abnormalities in the equipment and learn to distinguish materials that can contribute to waste and quality defects. Step 6 aims to foster operators' knowledge in quality, equipment, and procedures for fully implementing autonomous supervision without fully detailed instruction by supervisors and managers. Although most industries want equipment with minimum breakdowns, more industries are interested in quality matters since achieving Zero Defects in quality is much more difficult to achieve than Zero Unplanned Breakdowns.

Step 7: Achieve Empowered Autonomous Maintenance Workforce: This is the last step of Autonomous Maintenance activities where the operators become self-sustaining. As operators become empowered, minimal assistance is required. The operators have already achieved a stage of being fully competent in their equipment to make their own decisions and conduct improvement initiatives with the other pillars of TPM, such as Planned Maintenance and Focused Improvement in eliminating other losses in their equipment. The operator has fully been transformed into this final stage as an empowered and highly motivated workforce. They challenge themselves to further enhance the inspection process, generate further improvements in their equipment, and challenge themselves to improve their equipment indices. Autonomous Maintenance becomes full-pledged partners with Maintenance and joins them in further improving their equipment by analyzing the causes of failures and other related product quality issues. You just cannot imagine what the power of this workforce can achieve when you reach this stage of Autonomous Maintenance. Both the human and equipment side are improved. Better operators can yield better performance. Operators know what is best for them and their equipment. The best part is that operators join Maintenance in teaching other operators about this transformation.

In this last step, operators become independent, skilled, confident, and well-motivated and can perform simple repairs and equipment restoration by training them in repair skills.

Operators' improvement skills are raised as they join improvement initiatives and projects in eliminating the 6 big equipment losses on their equipment and assets. In Step 7, knowledgeable operators conduct autonomous supervision and follow the standards set by themselves on the orderly shop floor, where any deviation from the normal or optimal operating condition is detected at a glance by the AM team. Completing the steps from Step 0 to Step 7 is a very long journey. At an average speed, this will take around 5 years or more to reach this step. Maintenance should have patience since if this is done correctly, breakdowns and quality defects will dramatically be reduced. Still, if done incorrectly, failures will revert back, operators become saturated. The team's enthusiasm diminished as frontline people, engineers, and Maintenance, including plant managers, are being replaced every year. It is important to sustain the team's enthusiasm at every step that the team will undertake.

TPM believes that constant improvement has to come from the individual and that empowerment is the only way to get the people to adopt constant improvement as a way of life in the individual business. In other words, the way to empowerment is to create a working system based on principles that will enable people to take charge of the situation

TPM believes that the people closest to the asset know how to perform and improve their day-to-day operations. Frequently, operators do not perform any inspection of their equipment. Maintenance develops the checklist, but operators simply don't perform the checks. This seems to be the main complaint of the maintenance people. Operators do not resist change. They just do not want to be dictated by anyone. The secret is to let the Maintenance guide them into understanding their own equipment. Once they learn, they will take care of their own equipment. The key is to let the Maintenance understand that operators are capable of developing their own checklist. If operators develop their own checklist, it is guaranteed that they will perform it themselves. Remember that Autonomous Maintenance cannot be forced on operators; it will require a tremendous amount of initiative, guts, patience, and humility to develop self-directed and empowered operators. It will take 3 to 5 years to achieve this stage but the time spent will be truly worth the time.

5.2: February 2011: Is Zero Breakdown Really Possible?

To answer this question directly and bluntly, the answer is **"NO."** Breakdowns and failures are simply not possible to eliminate in your equipment and machines. Those of you who are reading this and practicing TPM or Total Productive Maintenance claim that one of the goals of TPM is to eliminate breakdowns or zero them out in your equipment. The only time you can actually have a zero breakdown, and zero downtime is when your equipment will no longer be used or decommissioned.

I was invited to Malaysia and participated in a maintenance conference where I was one of the resource speakers during their morning session. My topic was about comparing "TPM vs. RCM, which is the Best Pound for Pound Maintenance Strategy." I discussed the importance of shifting our maintenance focus from trying to eliminate breakdowns to eliminating or reducing the consequences of failure since, in reality, breakdowns simply cannot be eliminated in the first place. During the afternoon session, a speaker from **TPM Club India** presented about TPM and discussed the possibility of eliminating breakdowns through the implementation of Total Productive Maintenance and was contradicting what I was discussing previously. If you were one of the delegates who attended that conference, there would be confusion in the first place on which side to believe.

First, let me clear them out to you when TPM calls them zero breakdowns; it does not necessarily mean zero downtime. Preventive Maintenance is included as one of their Planned Downtime. This means that when a part is about to break or fail in operation and Maintenance, replace them in advance before the actual failure and took them 30 minutes to replace; then, no breakdown is recorded, but there is an actual downtime of 30 minutes. Downtime caused by parts replacement, PM shutdown and overhauling, and other maintenance activities will be included in Planned Breakdown or Planned Downtime. If we speak about manufacturing plants, breakdowns are not the only cause of downtime. Downtime can also be caused by assist or errors (TPM called these minor stoppages), set-up or conversion, start-up losses, and cutting tool changes.

Second, the word Zero Breakdown is quite confusing and, in fact, very intriguing. Suppose I have equipment with a severe oil leak; a safety officer can declare the equipment fails for safety purposes even if it is still running. From a maintenance point of view, when the oil level is always below the minimum, they may declare the equipment in a failed state. Operations people may declare the equipment fails when it overheats due to the failure of the lubricant. Therefore, before one can declare a zero breakdown, your Maintenance, and operations, people must first understand what failures and breakdowns are all about and understand its diversity in the first place. Kindly click this link for a more thorough description of each failure and breakdown.

A) Patterns of Failure
- Infant Mortality Failures are failures occurring at the beginning happens mostly right after a PM overhaul
- Random Failures are failures the occur at any given period
- Age-Related or Wear Out Failures are failures that will survive to the age or period specified

B) Classification of Failures
- Hidden Failures are failures that will not be immediately felt by the operator unless a second failure happens
- Evident Failures are failures that will be known and become evident to the operator

C) Types of Failures and Breakdowns
- Function Loss Breakdown are failures in which the equipment will totally stop
- Function Reduction Breakdown are failures in which the equipment can still run and produce output

D) Occurrences of Breakdowns and Failures
- Sporadic Breakdowns are breakdowns that result from a single cause that is easy to rectify
- Chronic Breakdowns are breakdowns that have a wide array of probable causes

E) Categories for Failures
- Maintenance Induced Failures are failures caused by maintenance people themselves

• Non-Maintenance Induced Failures are failures caused by other parties aside from Maintenance

F) Stages of Failure
• Potential Failure indicates that a functional failure is on the verge of occurring
• Functional Failure is the inability of an item to meet a specific performance standard

G) Groupings of Breakdowns for TPM
• Planned Breakdowns are non - machine-related downtime due to PM, meetings, lunch break
• Unplanned Breakdowns are machine-related downtime due to the 6 equipment losses such as breakdowns, assists, and conversions

Third, in TPM Planned Maintenance, whether you apply it in Steps or Phases, this will always be done two folds. The first few Steps/Phases of Planned Maintenance will reduce breakdowns to a minimum or eventually zero them out but expect the breakdown to occur again during the later phases/steps of Planned Maintenance. Once the breakdown repeats itself, a system for determining the best frequency will be established. Meaning in the first few steps of Planned Maintenance, the focus will be on eliminating breakdowns, but as the team matures on their journey, the higher steps/phases of Planned Maintenance focus more on developing a robust strategy for maintaining the equipment and achieve a Predictive Maintenance stage. This is where the different maintenance tasks such as Preventive Maintenance, Predictive Maintenance, Maintenance Prevention, Corrective Maintenance, Time-based Maintenance will be introduced, and a decision diagram or algorithm will be developed to fully understand when these tasks are feasible to use.

Similarly, much like RCM. Suppose zero breakdowns are really possible in the technical sense; the team cannot go forward on the higher phases of Planned Maintenance. What TPM is actually doing is not really about eliminating the failure but rather prolonging the life or Mean Time Between Failure of the equipment.

Fourth, breakdowns and failures may be easy for mechanical parts as most of these parts will wear out or deteriorate when they reached a given time or period, but we cannot say the same for electronic parts. Failures of electronic parts and components are random in nature, which means that they can fail at any given period. Electronic parts do not wear out. The failures and breakdowns we encounter for electronics always happen when we least expect them most.

Fifth, typical equipment can fail in a hundred ways or more. What Maintenance can actually do is to understand every single failure and adopt a feasible task before it happens. What Maintenance can do is prolong or actually control the duration of the failure.

• Understand which failures can be prevented and apply Preventive Maintenance
• Understand which failures can be predicted and apply Predictive Maintenance
• Understand which failures have minimum consequences and just allow them to fail
• Understand which failures have high impact and consequences but cannot be prevented nor predicted, so the last remaining option for the Maintenance is to modify or redesign the system

I have nothing against TPM. In fact, I have lived, eat, and breathe TPM when I was handling the pillars of Planned Maintenance and Initial Flow Control Activities during my time with Amkor for 8 years, so I definitely know what I am saying. There are no contradictions in what I am pointing at. Sometimes you need to experience TPM before really understanding it. There are many things I have experienced that are not written in any TPM books. Let me give you an example; during my time handling the Planned Maintenance, our goal was to zero out all breakdowns, and so I asked each of the 22 departments involved in our Planned Maintenance to provide me the breakdown data of their pilot equipment for the last 6 months to create a benchmark. One department (let us call this Department A) provided data for their pilot machine with an average of around 10 breakdowns a month; the other department (let us call this Department B) gave data of a consistent 1 breakdown a month. If I ask you which department is performing better, obviously you will answer Department B, since it is only experiencing 1 breakdown a month. Suppose you answer that Department B is the better department; this is where you are terribly wrong because that department belongs to the Facilities/Utility group. Their machine undergoing the Planned Maintenance was a sub-station. Meaning a failure of one sub-station in a manufacturing plant can paralyze all the production equipment.

But before even thinking about a strategy on totally zeroing out all the breakdowns, Maintenance must understand precisely what will eventually constitute a breakdown, which is where I think every industry will have its own definition. Let me give you this quiz as an example and kindly answer the following questions. (Answer to be revealed in our next newsletter)

1) A Function Loss Breakdown where machine totally stop
   a) Considered as a Breakdown
   b) Considered not a Breakdown

2) A Function Reduction Breakdown where the machine still runs
   a) Considered as a Breakdown
   b) Considered not a Breakdown

3) A Scheduled Preventive Maintenance performed on the equipment
   a) Considered as a Breakdown
   b) Considered not a Breakdown

4) Failure of a pump with a standby unit which runs automatically if we speak about the failure
   of the pump
   a) Considered as a Breakdown
   b) Considered not a Breakdown

5) In item 4, if we speak about the whole system in which the pump is just a part of the system
   a) Considered as a Breakdown
   b) Considered not a Breakdown

6) An assists or error which is consuming 10 minutes due to no technician performing the
correction
   a) Considered as a Breakdown
   b) Considered not a Breakdown

7) Random breakdown of an electronic part that stopped the equipment
   a) Considered as a Breakdown
   b) Considered not a Breakdown

8) Infant mortality failures or failures encountered right after a major overhaul or outage
   a) Considered as a Breakdown
   b) Considered not a Breakdown

9) Unscheduled repair, replacement, or overhauling of parts
   a) Considered as a Breakdown
   b) Considered not a Breakdown

10) Run to fail components that have operational consequences
   a) Considered as a Breakdown
   b) Considered not a Breakdown

11) Detection of a potential failure such as noise on the machine detected through vibration monitoring
   a) Considered as a Breakdown
   b) Considered not a Breakdown

12) Run to fail components with great possibility of secondary damages
   a) Considered as a Breakdown
   b) Considered not a Breakdown

13) Software problems on the equipment where machine runs intermittently and then it suddenly stopped
   a) Considered as a Breakdown
   b) Considered not a Breakdown

14) Actual monitoring of Predictive Maintenance in the equipment
   a) Considered as a Breakdown
   b) Considered not a Breakdown

15) Failure of secondary functions such as sensors and bulbs
   a) Considered as a Breakdown
   b) Considered not a Breakdown

16) A compressor fails in such a way that it was not able to provide compressed air to 100 pneumatic equipment; if we speak about the compressor, then
   a) Considered as a Breakdown
   b) Considered not a Breakdown

17) In item 16, if we speak about the 100 pneumatic equipment that encountered downtime due to the failure of the compressor
   a) Considered as a Breakdown
   b) Considered not a Breakdown

18) Waiting time due to no available parts and spares
   a) Considered as a Breakdown

   b) Considered not a Breakdown
19) A Failure in which a component will be involved
   a) Considered as a Breakdown
   b) Considered not a Breakdown

20) Downtime due to no inventory and no operator at hand
   a) Considered as a Breakdown
   b) Considered not a Breakdown

To conclude, not all failures are created equal. There will be failures that will have very minor consequences when they occur and will be limited to the cost of the part that fails. At the same time, there will be failures and breakdowns that will have a high impact. The consequences of these failures may not be acceptable to both operators and Maintenance. If this is the case, then we simply do not allow that failure to happen, but when the failure has minor or light consequences, then we tolerate that failure to occur. That is why most power plants, oil and gas sector, airline industries, or even your utilities and facilities group have redundant functions or standby equipment in the first place. Therefore, in my point of view, it is more important to understand the consequences of every failure rather than eliminating the failure.

When TPM says Zero Breakdowns, what it is actually stating is Zero Unplanned Breakdown. TPM grouped breakdowns into Planned and Unplanned, or more precisely, Zero Unplanned Breakdown for this period and that to be much more accurate, perhaps January 2011 to May 2011. If you encounter no breakdown in one year, then the correct way of saying it is Zero Unplanned Breakdown from January 2010 to January 2011 but stating a ZERO BREAKDOWN, come on, give me a break, and I think that's all I have to say about that.

5.3: March 2011: Top 10 Problems Experienced on PM Revisited

Last May 2008, I have written this newsletter based on a survey conducted with a few people from maintenance. 15 people responded to this survey, to be exact. Instructions were to select at least three problems from the list they think they experience most on Preventive Maintenance. However, the survey conducted was random, and the results did not establish some trend after all. I continued this survey on some of my courses, and I think at this point of writing, we can now see a trend as to the real problems we face right now on Preventive Maintenance, and I would like to share with the readers the results of that survey. First, let me again list the Top 10 problems on Preventive Maintenance with a brief explanation of what the problem is all about.

1) Add on PM Checklists Syndrome: When your equipment was new and your vendors and contractors had completed commissioning your asset on the plant, they left you with some maintenance activities that need to be performed regularly from time to time on the equipment. These activities include some basic inspection, parts replacements, periodic or scheduled overhauls that needs to be done in a scheduled and timely fashion. But even if you comply religiously with these activities, unexpected and unplanned breakdowns happen all the time, and when the Boss's attention was finally caught down by these breakdowns, his only question to you is, son, is this included in your Preventive Maintenance activities? And when you replied

"NO," then the Boss, make sure that you add them to your unending PM activities list. In short, when starting out with your maintenance activities when the machine was newly commissioned, there were just around 15 activities to be done, but as time passed by, more and more activities had been added up, and today, there are more than 100 activities added, and the list never seems to stop. The lamest excuse of them all is when a breakdown occurs, add this to your PM activities.

2) Introduction of Infant Mortality Failures Infant Mortality Failures are failures that simply occur right after a major maintenance intervention on the equipment. Others refer to commissioning failures, starting-up failures, and so on. This problem is mostly experienced right after a maintenance intervention or Preventive Maintenance is performed on the equipment or asset. In most cases, extensive overhauls were performed on the equipment.

3) Replacement of Parts Just to Conform to PM Specs: Preventive Maintenance utilizes the concept of JIC. JIT means just in time; hence, JIC simply means "Just in Case." Most of the sentiments of maintenance people are that when a machine is open or being overhauled, that is the only time they have on this world to replace the parts they think is on the verge of failing, and that needs to replace because their thinking is that if this part fails and the boss question you if you replaced this part last outage and you answer no, then you are in a lot of trouble. Maintenance mostly assumed that the parts they replaced are nearing their rupture, but most of the time, the parts are still working and in good condition. This is partly what makes our maintenance costs because most of the assumptions made are not right. When a part is still in working condition, then it should remain in service.

4) The Case of Random Failures: Random Failures are failures that occur in any given period. It simply means that the probability that an item will fail in any one period is the same as in any other given period. In short, the part eventually has no life and can fail at any given time. One characteristic of random failure is that a wear-outage is not identifiable and that the failure can occur at any given time. When failures that are occurring are random in nature, Preventive Maintenance is at its weakest point. In simple terms, PM is not a recommended option, and other tasks to use will either be a Run to fail only when the consequences of failure are low or Condition-Based Maintenance or Modification. Samples of random failures include electronic boards and parts, bulbs, ball-bearings, seals, hydraulics, etc. Most industries think that Preventive Maintenance can address random failures by performing maintenance activities, such as regular replacement and overhauls. Doing these activities will just induce more infant mortality failures on the equipment.

5) Ageing workforce nearing retirement: When good maintenance, people retire, or simply said when the industry retires these people, their experience goes with them most of the time. The industry will now hire newly fresh graduate people with no experience on the asset because they will be paid much less in salary and benefits. Because these people's experiences were not captured or documented, these new people will experiment on how to deal with the failure. The cycle goes on for generations. These older people have experienced most of the equipment failures since they have stayed with the plant long enough to witness the failures. Most of the time, industries do little or nothing to document how they repair or troubleshoot the failures.

6) Lack of training on the maintenance function: Many say that people are the companies' greatest asset. I totally disagree with this statement because the correct way of stating this is that the right people are the company's greatest asset; the wrong people are simply called liabilities. And we can only have the right people in maintenance if they have the right skills and knowledge to perform their jobs right. Training will provide the foundation and backbone of any cultural change or initiative because this is where we acquire knowledge. But sad to note that when cost reduction is the name of the game industry play, then the number one department to be affected is their training. In my small firm, I often tell people that maintenance is not just a verb (action word). It is also a noun. Meaning maintenance are also people. When we invest in the people, then these people will do the right thing in their work. It is sad to say that maintenance is always busy catching up on repairing and troubleshooting their assets and equipment. I am also saddened to hear that when I conduct in-house training on their plants, many say that this is the first taste of training they have ever had on maintenance. These people have stayed with the plant for more than 15 to 20 years.

7) Still Reactive and a lot of corrective maintenance even with a sound PM Program:
When I ask maintenance people from different industries if they are satisfied with their current Preventive Maintenance activities, almost all will say "NO." Despite their most noble efforts and compliance with every single piece of PM activity listed, breakdowns and failures still happen. Every industry has its own way of doing Preventive Maintenance, yet most of them are not satisfied with their PM outcome. First, industries must understand that Preventive Maintenance includes risks because there is always an assumption that everything dismantled will be put back together in place.   Hence, despite the best efforts and structure on Preventive Maintenance, failures are inevitable and will not be captured solely by PM. Zeroing out all breakdowns is like catching lightning with a Polaroid Camera, and why used a Polaroid camera if you can capture the lighting with a digital camera? And why won't Preventive Maintenance capture all failures? Because typically, only around 20 to 30% of component failures will wear out or are directly related to the equipment's age, and around 70 to 80% or all failures will fit the random and infant mortality failures. And when the failure is random in nature, then no amount of PM can address this issue.

8) Frequent reorganization in the plant: When an organization is dynamic, so are the people. People are transferred from one department to another due to a direct order from the Top. I have seen projects and ongoing improvements and modifications abandoned because the person had been transferred. When a new boss heads your department, a new system takes place, and the culture also changes, but the validity of that system only lasts when the Boss is still around the corner. When the Boss is again transferred, a new system occurs because this is what the new Boss wants. No disrespect, mostly to those affected, but when maintenance reports to operations people, your plant has a big problem, most especially if the person you report to is a non-technical person since his frame of mind is focused only and only one thing, and that is output.

9) Lack or poor documentation in PM: Sometimes, documentation in PM is incomplete and faked. Suppose you have around ten people from maintenance in your department handling more than a couple of 100 machines where you have a lot of firefighting and troubleshooting going around then; maintenance has simply no time documenting every event happening around him. What is important in PM documentation is documenting what is relevant and not. Maintenance makes decisions based on these documents, and if these documents are flawed, half-baked, or faked, then our decision on PM will be jeopardized, and at times, there will be repercussions. Documentation is important; hence let us do this not because we will be audited but for the right reasons.

10) PM is waived: Operations won't give equipment for TPM to cope with production: When the equipment is subject to a maintenance intervention or scheduled PM, in short, operations will frequently be deferred or waived equipment because the output is always the top priority and supersedes everything. When the demand for your product is high, or there are months where production is at its peak, then maintenance will find it very difficult to get the equipment for a scheduled PM activity. As a result, more unexpected failures occur, and the feud between operations and maintenance is high. Maintenance always ending up on the losing end because operations decisions always prevail. Operations will give the equipment to maintenance only when they want it to.

	SURVEY ON TOP TEN PROBLEMS ON PM (Updated August 2011)				
No.	Problem on PM	As of May 2008	RANK	As of Aug. 2011	RANK
1	Add on PM Checklist Syndrome	9	1st	130	6th
2	Introduction of Infant Mortality Failures	8	2nd	159	4rth
3	Replacement of good parts to conform with PM specs	4	5th	125	7th
4	The Case of Random Failures	6	4rth	123	8th
5	Ageing Workforce, maintenance nearing retirement	1	last	73	last
6	Lack of training on the maintenance function	8	2nd	186	2nd
7	Still reactive even with a sound PM program in place	9	1st	156	5th
8	Frequent reorganization in the plant	2	7th	75	9th
9	Lack or poor documentation for PM	7	3rd	179	3rd
10	PM is waived by operations	3	6th	235	1st

Figure 5.1: Survey on Top Ten Problems on PM Revisited 2011

There you have it, the top 10 problems industry faced in doing Preventive Maintenance with a brief explanation of each problem.

5.4: April 2011: Why Operations are Not Maintenance Customer?

I often hear this statement from maintenance people that operators are their customers. And from the traditional saying, which states that the customer is always right and when the customer is wrong, you go back to the basic rule, which states that the customer is always right.

I would like to quote from the book of Anthony Smith, RCM - Gateway to World Class Maintenance, page 16. He quotes view maintenance as a Profit Center. This suggests that the maintenance organization must be treated as a key element in your business strategy and plans for achieving profit targets. Like any organization's function (design, manufacturing, marketing), maintenance costs perform their routine tasks. But recognizes that routine (i.e., scheduled) maintenance tasks, when properly performed, can dramatically affect the ability to achieve or exceed targeted production output. This means, among other things, that operations (Production) and maintenance must be treated as equals. No longer should operations dictate when maintenance can or cannot do his job. Rather, there must be decisions made for the common good, and each must respect the other's role in meeting the customer's demands (i.e., the "real" customer, the one who pays your salaries). No longer should maintenance feel that it's one, and the only customer is Operations; it is not. Unquote

I tend to agree 100% with what the author Anthony Smith is saying. Therefore, if your company's policy or culture follows the rule that operations are maintenance customers, then I believe they need to rethink and change course while there is still time; otherwise, that industry is in a lot of big trouble. Here are some reasons I believe why I consider operations are not maintenance customers:

1) Maintenance is a Shared Responsibility for Both Operators and Maintenance: First, operators must accept that maintenance is not only for maintenance alone to perform. There are minor maintenance activities such as cleaning, inspection, and basic lubrication done by the operators themselves and not by maintenance. TPM refers to this as addressing the basic equipment condition. The logic is as simple as driving your car because if you own a car, it is the operator or driver who checks if they still have fuel left to travel. The driver should also be monitoring if his engine temperature is normal or increasing when traveling. The driver needs to check if their tires have the right pressure or are inflated. The driver should also be the one to check if their car contains water on its radiator. I can go on and on with this, but all I am saying is that these activities I have just mentioned are **not** performed by the mechanics but rather by the operator or driver themselves. As I have said many times, the second point is that operators are the first line of defense on any failure that can occur in the equipment since they are the people closest to the asset during the time of failure. They will be the ones who will encounter the failure first before the Maintenance arrives at the scene. Maintenance can only advance to any continuous improvement effort and advancement if the operator accepts the responsibility to play a major part in establishing Basic Equipment Condition, which aims to keep the equipment clean, correct lubrication, and complete bolts with the correct torque of the machine. Operators must provide continuous communication with Maintenance because it is important to maintain people understanding what eventually happens before the failure. If there is some sort of noise, smell, or anything that the operator senses. One problem with traditional industries is that when the equipment breakdown and fails, operators call the Maintenance, and when Maintenance arrives on the equipment, the operators disappear or take a break; in short, communication is lost between the two. Both TPM and RCM agree that Maintenance is a shared responsibility for both operators and Maintenance. TPM do this by adopting a structured and detailed 7 step approach on Autonomous Maintenance, while RCM will require an operator as part of the RCM team since RCM believes that there are things to be learned from the operator most especially during the time of failure because these operators are the closest people to the asset during the time of failure.

2) Maintenance Should Report to Maintenance and Not to Operations: When we speak of any industry's organizational chart and hierarchy. Often, you cannot see the maintenance function, or it is often missing because they directly report to operations people. With this in mind, operations control and dictate what maintenance must-do on the equipment. The worst thing that I can think of is that if the highest person in operations is a non-technical person, all priority will end up on only one thing: output. This is one of the main reasons why maintenance people are treated slightly lower than operations. What I believe is that both operations and maintenance people should report to two different people. This means that the highest person on operations and the highest person on maintenance must be of equal level or with the same position, and neither one reports to the other but reports to a much higher person, who is perhaps the General Manager or Factory Manager. Although this is a sensitive issue to discuss, my experience tells me that if Maintenance reports to operations, then all hell breaks loose, and maintenance pressure is intense on keeping the equipment running despite operations waiving the equipment for a Preventive Maintenance schedule. And when the machine fails, then operations people are behind or at the back of maintenance repairing the fault, keeping an eye on Maintenance, and asking all the time when the machine will run. I think maintenance should report to the head of the maintenance and not to operations themselves. In one of my world-class maintenance classes, one guy from the maintenance function, a very vocal guy, raised a question about a simple case of an oil leak that he wanted to address. I asked him why he won't address such a small issue. He told me that if he repairs the leak, the whole plant will need to be shut down for a while. And I asked him, why don't you schedule it in your PM outage? It is because operations keep on waiving the equipment for PM. My last question to the person was to whom did he report, and he said to the operations manager. And I guess that answers the question. Perhaps from a business point of view, operations people might be thinking that they are doing the right thing, but from a maintenance point of view, big problems and catastrophic failures are just an accumulation of very small things.

3) Maintenance are Not Mechanics or Repair People: The role maintenance play in an organization is diversified in nature, perhaps belonging to the maintenance function allows me to understand that the best positions an industry can have always belong to the maintenance section, but in most organizations, maintenance is merely treated as repair people that troubleshoot breakdowns. Maintenance should always be in the line and not in their offices so that when a breakdown occurs, they can be called at once. Some even have a man-to-machine ratio that, for every one maintenance, his responsibility will be to look out for breakdowns on at least around 15 machines assigned to them. Others, even worst, provide a record of what these people actually repaired on an hourly basis. If we survey how many industries actually have some form of Predictive Maintenance or CMMS in place, perhaps not all or even half can say that they have one. Still, I am sure that each of these industries has plans to have them since the beginning of time. Still, the problem is that every time they requisition or provide some P.R. (purchase requisition) regarding these instruments and software, they have always been denied by their management. When maintenance people are not visible at the production site or walking down the production aisle, operations people need to be happy simply because it operates smoothly. Maintenance are not mechanic, and mechanics are simply not maintenance people. These two are different. Mechanics use their hands more often, while maintenance people use their brains most of the time and less their hands in addressing the problem. There are many doors on maintenance that still remain closed on industries because they want their maintenance to be just mere mechanics that troubleshoot and repair the problem all the time, and when maintenance get the hang of repairing the fault, they are less inclined to learn what the best practices on maintenance for fear of losing their job. Lubrication is one of them. And this subject is taken very lightly or shallow. Let's speak about hydraulics; for example, more than 80% of hydraulic system failures are caused by oil contamination or unwanted foreign object. If you have an exact activity like this, then this is the start of a bigger problem. This is actually where chaos and confusion begin. What specific lubricant to be used is unknown because the PM states only to apply lubrication? Note that both oil and grease are lubricants. Let us say that grease will be used, then the next question is what corresponding NLGI number must be used for this application? At what specific brand and type of grease to be used? How many shots or pumps are needed, or when do I know when to stop pumping the grease gun? How many grams of grease is required for this application, and do you have any means of measuring it? And most of the time, when failure struck us, we blame it on the lubricant and not on how we perform lubrication. This means that different grease guns contain different pumping capacities. Grease should never be mixed, which means that if a grease gun contains calcium grease and you place some silicon grease on the same grease gun, you are building up incompatibility issues on the grease. This means that you are shortening the grease's life and the component that you will apply the grease. When your people become good at fixing failures, your plant has a big problem since the problem keeps on repeating itself. In today's fierce competition among industries, we need people who can analyze and address the root cause of the problem, and there are only not even half of the maintenance population knows how to analyze problems. I hope that there comes a time where the industry begins to mature and understand the role maintenance play in their organization so that they can transition from a firefighting stage to a more robust and structured approach on Maintenance besides the ***Shout and Scare Tactic*** that most operations people are doing. Remember that there is a better way of doing maintenance rather than repairing the problem and pressuring Maintenance all the time.

4) Maintenance Should Not Be Blamed When Production is Not Met: Again, when we accept that operators are maintenance customers, operations will always have a lame excuse for why their production was not met for the day. Maintenance is always a scapegoat for every single problem that occurs on the equipment, and this seems to be the trend happening in most industries that when unexpected breakdowns and failures occur on their assets and the cost of doing maintenance is high, the blame virus prevails, and maintenance is the answer to all evil not to mention the relationship operations people have on maintenance. The feud between operations and maintenance is pretty much alive. When production for the day is met, all credit goes to operations, but on the other side, if production for the day will not be met, then all fingers point to the maintenance. Maintenance cannot simply follow orders from operations, which keep on waiving the equipment for a Preventive Maintenance schedule. Suppose the reason for delaying the PM is to reduce or minimize infant mortality failures experienced; then, I agree with operations for waiving the equipment. Still, on the other hand, if the reason for deferring PM is to give way to production, then maintenance should make a stand and not be dictated by operations. Maintenance should not be blamed for every problem and downtime on the equipment or simply being held accountable. Even if we point our fingers and blame maintenance for it, the equipment is still down. This is what happens when maintenance treats operations as its customers. We need both parties, the operations, and maintenance, to finally work together to find permanent solutions to problems. As quoted from the book of Keith Mobley, "An Introduction to Predictive Maintenance, 17% of equipment-related problems are maintenance induced, which means that 83% of equipment problems did not come from maintenance but from other sources.

To conclude this newsletter, both operations and Maintenance should work together to resolve problems instead of accusing each other all the time. When both parties sit down and lay their fingers down and start communicating instead of blaming, that is a good start. Remember that operations are not maintenance customers ***because operators are simply maintenance partners, and they can never be divorced, just like The Rolling Stones; oh yeah!*** Our customers are the ones who buy our products and services because they are the ones where we get our salaries.

Answer to February 2011 Issue: Are Zero Breakdowns Really Possible?

1) A (function loss means a failure in which the equipment totally stops)
2) A (this is when the machine is still running, but when listing the number of breakdowns, these can be omitted)
3) B (Scheduled PM or outage is not included as a breakdown)
4) A (although there is a redundancy in this case, the running pump eventually failed)
5) B (the system was not affected; hence, there is no breakdown on the system)
6) B (an assist is not a breakdown and cannot be considered as one)
7) A (breakdown of electronic parts that stop the equipment is considered as a breakdown)
8) A (infant mortality is a pattern of failure that occur after a PM overhaul is performed)
9) A (this is considered as a breakdown)
10) A (breakdowns usually have operational consequences)
11) B (potential failure is a stage where a functional failure is about to occur)
12) A (this is considered as a breakdown)
13) A (software problems that stop the machine from running will be included as a breakdown)
14) B (the activity of monitoring the equipment through Predictive Maintenance is not a breakdown)
15) B (these are the same as item 2 and called function reduction failures)
16) A (if we speak about the compressor, then there is a breakdown)
17) B (if we speak about the 100 pneumatic equipment, there was downtime but no breakdown)
18) A (I consider this part of the breakdown, which is waiting for the spare or part)
19) A (if a part or component is involved, then there is a breakdown)
20) B (this is not a breakdown)

5.5: May 2011: Different Maintenance Strategies Explained

There are four different maintenance strategies as far as I know that can be applied to the asset; however, the application and feasibility of adopting these strategies depend entirely on the failure consequences.   These strategies include Reactive, Preventive, Predictive, and Proactive Tasks. We shall explain each of these strategies in detail. Every industry has its own term they used to denote these strategies.

Reactive Maintenance: A strategy tells us that when a machine fails, then it is time for maintenance to fix it. Maintenance is done at a point when there is repair or actual breakdown. It occurs when repair action is taken on a problem only when the problem results in machine failure. Reactive maintenance results in unplanned downtime. In its simplest definition, reactive maintenance simply means fixing it when it fails. This strategy's limitations are that a run to fail will cause unplanned downtime and production delays, resulting in revenue losses and extending the maintenance to work overtime to fix the equipment. Repairing the equipment after it fails usually creates the possibility of secondary damage, which increases the maintenance costs. Allowing failures to occur can be applied to the asset if and only if the consequences of failure and the cost of repair is minimal and acceptable to both the user and maintenance. Other terms used to denote Reactive Maintenance includes: run-to fail, run-to destruction, firefighting maintenance, unplanned breakdown, power plants often called this corrective maintenance (note: TPM practitioners refers to corrective maintenance as performing improvement on the equipment), breakdown maintenance, RCM often called this as no-scheduled maintenance. I think that's all I can think of. If the failure is evident and does not affect safety nor the environment, or if it is hidden but does not affect the safety or the environment, then the default decision is to allow it to fail.

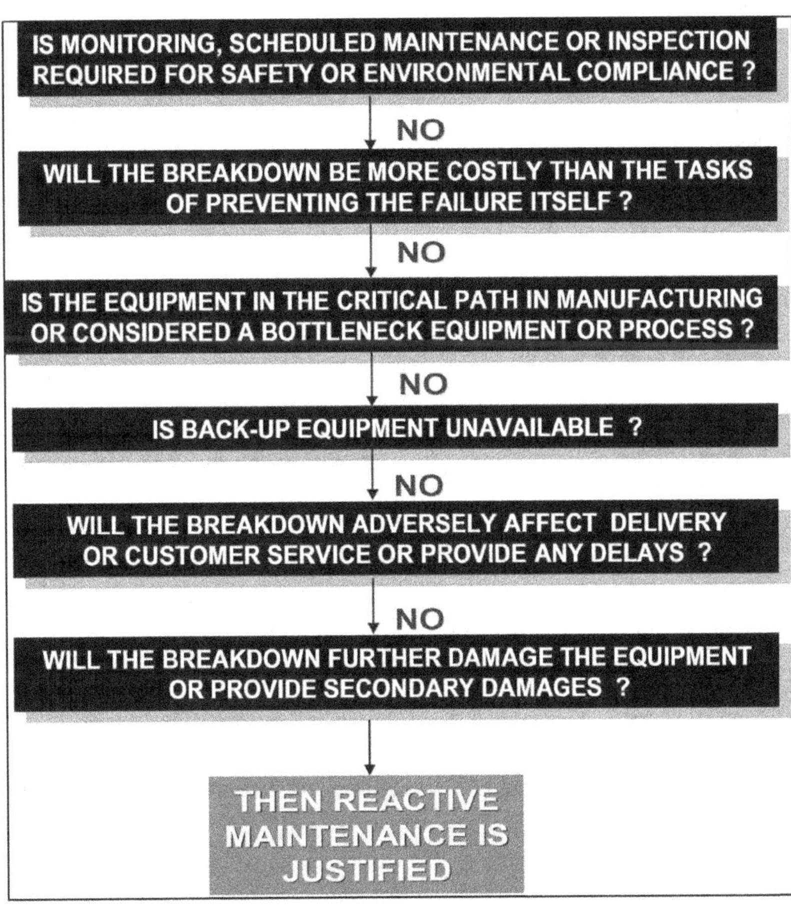

Figure 5.2: Justifying a Run to Fail Maintenance Task

Duplicating the system or component failures is allowed or tolerated through some redundant form or when the asset has some form of duplicated functions. The presence of standby or redundancy, an alternative means of production, is a feature of the operating context that must be considered in detail when defining the asset's functions in its present operating context. Often termed as standby-unit, and even with the same type of equipment, standby units can have different degrees of maintenance requirements as the duty unit, and most failures for standby units are considered to be hidden. Therefore, when the consequences of failure are limited to the direct cost of repair with minimum chances of secondary damage to the equipment, the default task will allow failure to occur. If we place a bearing under reactive maintenance, there will be no replacement on the bearing to be done; maintenance will just wait for the bearing to seize. When this happens, then the bearing will be replaced, production halts, downtime occurs, and maintenance performs repair and replacement on the bearing. The maintenance strategy will be to keep stock of the bearing on the stockroom to ensure that repair time will be minimal. The problem with this approach is that there is a possibility of having some secondary damage on the equipment or other parts that may be affected, downtime can be excessive, and maintenance people are caught by surprise on the bearing's failure.

Preventive Maintenance: This is a basic maintenance task performed on the equipment and facilities. The main goal of performing a task on a scheduled basis is to extend the equipment's life and ensure its capacity to support the plan's goals and targets. In Preventive Maintenance, the basic law to consider is that the cost of performing PM must always have to be lower than the cost of not doing it, which eventually will result in a failure. PM is also a series of tasks performed at a defined frequency dictated by the passage of time, the number of machine-hours, or mileage that either extend the asset's life or detect that an asset had a critical wear and is about to fail or break-in operation. PM, also called time-based maintenance, is founded on the belief that given a history of failures of a given component failing after a certain number of hours used or other measurements, most parts based on this approach are usually based on averages. One of Preventive Maintenance's problems is that this approach may carry out maintenance work at regular intervals. Since most tasks are based on the average maintenance, they may carry out work that is not required and the probability of replacing parts that are still in working condition. Since this is time-based maintenance, studies have shown that most equipment failures are not directly related to the number of hours or operating age. Preventive Maintenance activities include equipment inspection and checks, scheduled parts replacement, overhauls, routine cleaning, lubrication, planning, scheduling, preparing work orders, managing spares inventory, CMMS, and so on.

Besides, workers can record equipment deterioration, so they know to replace or repair worn parts before it causes a system failure. Applicability of Preventive Maintenance tasks is technically feasible if first, there is an identifiable age at which the item shows a rapid increase in the conditional probability of failure, which means that the part has a useful life and most of these items survive to that age, and they restore the original resistance to failure of the item. Hence when we place a bearing under Preventive Maintenance, it is assumed that 95% of bearings will reach their desired life, and the probability of failure is high. If this is the case, then it is good to place the bearing on a Time-Dominated Frequency or Preventive Maintenance. But in reality, or perhaps based on previous failure history records on bearings, the record shows that most bearings fail every 2 years or 24 months; therefore, maintenance decided to replace the bearing when it reached the 23rd month of continuous use placing it on Time-Based. The problem with this approach is that in reality, especially in a reactive world, we experience a lot of premature and unexpected failures encountered before the PM schedule is due, and maintenance is not guaranteed that even if the bearing reached its 23rd month if it is still fit for use or really needs to be replaced.

Predictive Maintenance: Condition-Based Maintenance or Predictive Maintenance checks the equipment condition using sophisticated measuring non-destructive instruments with precision accuracy. Predictive Maintenance instruments are just a higher form of the human senses. These instruments simply allow us to determine the problem much more than our senses can either hear or see. It allows us to make decisions once a potential failure is visible. Predictive Maintenance aids us in determining the potential failure or symptoms that equipment is in the process of failing. Changes in the condition of the equipment can denote a potential failure. Specialized diagnostic instruments can aid in detecting this potential failure on the equipment, such as an increase in heat or temperature, increase in vibration, changes in resistance, changes in conductivity, increase in noise, change in pressure and flow rate, lubrication contamination, wall thickness decrement, rate of corrosion, leak detection, crack detection and so on. Since our goal in maintenance is to keep our physical assets in an existing state. Prediction is a declaration in advance that something will happen. From the dictionary, Predictive Maintenance is a proclamation or declaration in advance based on observation to preserve something from failure or sustain it against danger. Predictive Maintenance is a maintenance activity geared toward indicating where a piece of equipment is on the critical wear curve and predicting its useful life.

CBM tasks entail checking for potential failures so that action can be taken to prevent a functional failure or avoid the consequences of a functional failure. Predictive Maintenance aids in detecting potential failures in equipment with the aid of specialized instruments. Maintenance is based on the condition of the equipment, which differentiates it from Preventive Maintenance. Many failure modes give some sort of warning that they are in the process of occurring; hence, the P-F curve shows how a failure starts, deteriorates to a point where it can be detected (Point "P"), and if it is not detected will deteriorate at an accelerating rate. Suppose we place a bearing under Predictive Maintenance, an ailing bearing will produce signs and symptoms or potential failure, that it is in the process of failing; therefore, the maintenance department will try to use CBM / PdM Techniques such as Oil Analysis and Vibration Analysis to detect a potential failure from the bearing such as an increase in vibration or contaminants. Although this is a good strategy, the problem with this approach is that since imminent failures can be predicted, production can be notified, but still, we need to replace the bearings, and the life of the bearing might not be actually reached. Although it is good enough that both operations and maintenance are not caught by surprise on the failure.

Proactive Maintenance: Proactive Maintenance is about analyzing why failures occur so that their recurrence can be finally eliminated, thereby extending the part or component's life. Proactive Maintenance is when maintenance or a group of cross-functional teams analyzes the failure with problem-solving tools and analytical techniques such as Root Cause Failure Analysis, FMEA, Ishikawa diagram, P-M Analysis, Fault-Tree Analysis are used to better understand why the failure occurred. In Preventive Maintenance, we replace the part that we think is in the process of wearing out. Our thinking is that replacing the part will bring the equipment back to its original condition; we have not considered the need to analyze further why a certain part keeps on failing. Proactive Maintenance is used to reduce the probability of failure mode occurring to a level, which is acceptable.

An example will be replacing a component with a stronger or more reliable replacement making the failure no longer a threat to safety and environment. Suppose we place this bearing under Proactive Maintenance; a cross-functional group will study and analyze the bearing's failure. Once they determine the cause of the problem, they will provide a permanent solution to prevent its recurrence and increase the bearing's life. Hence, if we place the bearing under Proactive Maintenance, after the bearing's analysis, the team found that the cause of premature failures was severe oil contamination. After adopting oil analysis and application of absolute filtration, finally, the bearing life improved. Studies made by different tribologists indicate a direct relationship between a bearing's life and the number of contaminants present in the oil. After the improvement, the bearing did not only reach its life, but it also tripled its lifespan.

	Fix it when it fails	Maintenance at Scheduled Intervals	Predict Equipment Lifetime	Equipment Improvement
TPM	• Breakdown Maintenance • Unplanned Maintenance	• Preventive Maintenance	• Predictive Maintenance	• Corrective Maintenance • Maintenance Prevention
RCM	• No Scheduled Maintenance	• Scheduled Restoration • Scheduled Discard	• On-Condition Tasks	• Redesign • Modification
OTHERS	• Run to Fail • Run to Destruct • Reactive Maintenance • Band-Aid Maintenance	• Time-Based Maintenance - Running Hours - Stroke-Based • Scheduled Outage	• Condition-Based Maintenance • Equipment Diagnostic Technique • Reliability-Based Maintenance	• Proactive Maintenance

Figure 5.3: Terms Industries Used for the Different Maintenance Tasks

Every failure has a specific set of consequences. Being PROACTIVE has something to do about reducing or eliminating the consequences of failure to a minimum rather than completely eliminating the failure itself. The best maintenance strategy to adopt will always have to be based upon the consequences of failure itself. The first thing to ask in the event of a failure will be the consequences of the failure, if it occurs on its own and whether it is acceptable to the user or simply not.

5.6: June 2011: The Two Sides of Failures

Many of the failures we encountered on our equipment are what we called MIF or Maintenance Induced Failures. By definition, in its simplest state, these are failures directly caused by the maintenance themselves. Although in my previous newsletter I had discussed in detail the grandeur of failure, I would like to add that there are 2 sides of failure, which is the left or proactive side of failure and the right side, which is the reactive side of failure, which we will try to explain in detail on this newsletter. Although all industries have a vision of what they want their maintenance to be, most are reluctant to achieve their vision and remain in the status quo or trapped in a reactive environment. Although there may be many reasons for this, I believe that there are two **major** reasons why industries remain reactive, and let me do my best to explain them in detail.

First, to change the way we do maintenance, we need to change its culture

A culture is like a fingerprint in which no human being has the same fingerprint, and every industry has its own unique culture; and the culture of a single industry varies and can even break down to every single department or function, which means that the culture of your department might entirely differ from other departments even if they work in the same industry. In fact, no industry has an exact identical culture.

First, let me define it in its simplest term what culture is all about. It means how people perceive and do these things around their plants. It refers to common values and beliefs, while others refer to them as shared thoughts and feelings. According to Schein, culture is the pattern of basic assumptions that a given group had invented, discovered, or developed in learning to cope with its problems of external adaptation and internal integration that had worked well enough to be considered valid and therefore taught to new members as the correct way to perceive, think and feed about their problems.

Many industries have a Preventive Maintenance program or strategy to plan and schedule their equipment for maintenance. Still, when a breakdown occurs or a failure strike without any warning, then people are deployed to put down the fire at all costs. The schedule is set aside for a while, and when the same thing happens, then the same thing will be done regarding this matter. As time goes by, people become so good at fixing and repairing failures, and the boss praises you for saving the day. These people are recognized by their top management for their efforts. They are promoted to a higher rank because management knows that they can count on them during breakdowns and emergencies. As people get promoted because of this, other maintenance sees the mode and follows the same path, and so the thinking goes who cares about going on training and using Predictive Maintenance or doing RCM (Reliability-Centered Maintenance), TPM (Total Productive Maintenance), Tribology / Oil Contamination Control or other reliability initiatives if people get promoted for being reactive. Henceforth, when a reliability initiative is being introduced to their plant, maintenance is reluctant to follow, or others are just in a wait and see mode, thinking that let us just wait and see since they know that this initiative won't be around the corner for the next few months. If you ask why, the answer is because their culture thinks that the "break and fix it tactic" is the one that is accepted by their culture, and these are what they teach to all levels of the maintenance workforce because this is how we get. Besides, having emergency breakdowns and failures keeps me working overtime, so I have additional pay.

To make matters worse, based on my experience, operations people always complain a lot if they cannot see any maintenance in the line and condemned them for being lazy and not doing their job. Because of this mentality, maintenance people prefer to remain visible in the line even if there is nothing wrong with the equipment. And as the incumbents on maintenance always say, besides, if you will follow that reliability improvement path, perhaps some of you may no longer be working around here because if the equipment improves and keeps running without failing, what do we need you for? The maintenance people perceive this as the right thing to do because this is simply accepted by their culture. ***Does this seem familiar to you?***

Second, if reliability improves, then this may mean reducing manpower in maintenance

As a reliability and maintenance consultant and having trained people regarding the value of reliability and maintenance, 75% of the delegates I teach or more belong to the maintenance side; I often hear the same sentiment repeatedly. Their sentiments are, I think my manager or boss should be in this training and not me since I only follow what they say and this is how I get paid and the second sentiment is, if we do what you tell us to do then I think that there would be no more need for overtime or what happens now to the rest of the people repairing, will they get fired or forced to retire? Hearing these sentiments from our brothers on maintenance is a very valid one which I cannot emphasize enough because in my small frame of mind inside my thick skull, what I am thinking is that your Top Management people and your industry do not really understand what maintenance is all about.

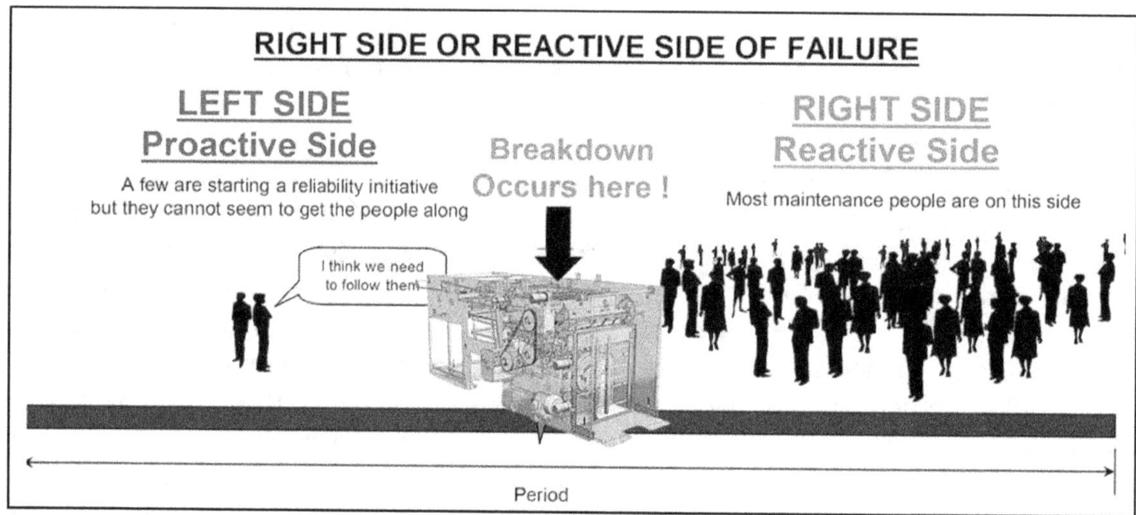

Figure 5.4: The Reactive Side to Failure

The sad is industries perceived maintenance as synonymous to repair and fixing breakdowns and failures. All I can say is that maintenance is much, much more than that. Suppose your line, department, or equipment reliability improves; one thing for sure is that the number of maintenance people performing repairs will definitely be reduced, and that is a fact. Overtime will be controlled and minimized, and that is another fact. And when this happens, **it does not** mean that the rest of the maintenance people will be left without work because what I believe is that once this happens then, this is the opportunity where new doors in the maintenance department will start to open. Having many repairs, breakdowns, and overtime is not a guarantee that your industry is stable and competitive. Industries aiming to better themselves and understand what it takes to achieve World Class Maintenance are finally moving this path, and if your industry still remains in the reactive mode, then you are simply being left behind in the race.

To give you some thoughts regarding what maintenance is and its scope, try to look at the figure above. These functions should be manned, and the best people that can fill up these positions are non-other than maintenance people. Because these areas should be governed and controlled by the maintenance department themselves. Definitely, these aspects of maintenance will not exist or are simply missing if your industry is in a reactive mode, or if some of them are in place, then you might not be benefiting much from them because being reactive is the accepted norm of your culture.

Just think about this situation, in one industry, all maintenance people are forced to submit an hourly report on what they do, how they spend their time, and what machines and activities they have fixed and repaired on an hourly basis. In another industry, the people analyze their failures and begin to adopt a robust oil contamination control and tribology group, and more than 80% of their hydraulic failures begin to disappear, and the breakdown was silent. Maintenance people are focused more on how to improve the reliability and integrity of their equipment. Which industry would you like to work with? I think the answer is pretty much obvious on this one.

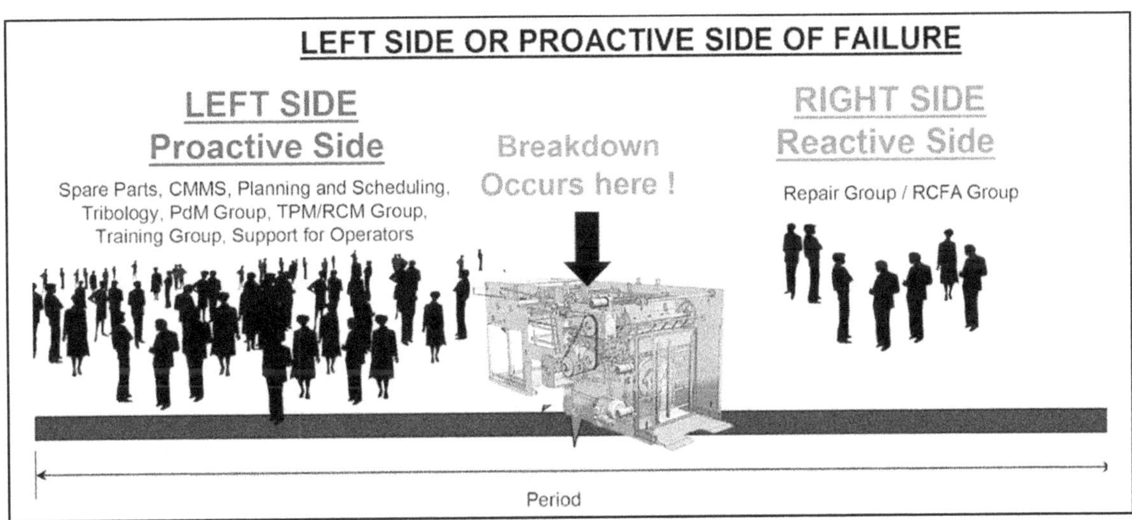

Figure 5.5: The Proactive Side to Failure

If most maintenance people from your industry are on the right side of the failure, they are reactive. A few will start out an initiative, but since the initiative is not accepted by their culture, things remain the same as it was since the beginning of time because fixing failures and repairing them is what they perceived to be the right way of doing things around here. On the other side, as reliability improves and is now being accepted by the culture of the plant and management knows the importance and value of doing maintenance, more functions are being opened in the maintenance side to plan for the equipment, and finally, people from all departments and ranks realize the true essence on what maintenance is all about and that it is not simply just merely troubleshooting breakdowns and failures.

5.7: July 2011:Tips in Performing Reliability Centered Maintenance

There are many negative articles on the internet I have read so far regarding RCM. Many users and consultants claim that the process of implementation is slow. Some analyses were abandoned midway through the process, while some claim that they have not achieved the desired results after implementing the RCM process.

If you search RCM on the net, there are way too many versions of doing RCM, which most are considered a streamlined approach or a faster version so that the analysis can be done at a much shorter phase of time. But in this article, I would like to talk about a version of RCM, which is in conformance with SAE JA1011 developed by John Moubray derived from its roots from Nowlan and Heap of the United Airlines industry and applied to land industries by answering the seven basic questions of RCM. Although in my previous article, I believe there should be an 8th question in which it states if you are ready for the change process. Therefore, before any plant or industry can implement the RCM process, there are things that you need to know so that the team implementing can derive the most benefit from the RCM process. Let me share with you some Tips on Performing the RCM Analysis:

Tip 1: Team Should Have Undergone the Training on RCM: Do not attempt to organize an RCM without having the team undergo basic training on RCM. You will likely have a greater chance of failure than success in the end. All team members should understand the principles of RCM and why existing assets and equipment should be analyzed using the RCM process. Let me share with you one learning from RCM. Most industries have their existing way of doing maintenance, which they referred to as Preventive Maintenance. This is done by scheduling equipment on either a time-based frequency, calendar-based, or running hours to undergo some replacements, overhauls, inspection, lubrication, and other activities on maintenance. But even after performing and complying with PM 100% of the time, there are still many breakdowns, failures, and emergency experiences on the equipment.

The reason behind this is that Preventive Maintenance cannot silence all breakdowns. In fact, complying with every single activity on PM can only guarantee around 20 to 30% of the overall failures encountered on the equipment. These will bring you to 70 to 80% of equipment breakdowns that cannot be totally addressed by doing PM alone; hence you need to divert to other maintenance tasks or strategies. In fact, doing too much PM on the equipment can induce infant mortality failures or problems during starting the equipment for operations. Why? Because in the real world of doing maintenance, there are actually 6 patterns of failure, and every single part or spare can actually fail in any of the 6 patterns. Preventive Maintenance is only applicable if the failure pattern is age-related or an actual wear-out process occurs on that part. The truth is Preventive Maintenance cannot capture all failures. If we speak about music and consider a band, for example, the beauty of the band's sound is because of the harmony played by each of the instruments performing the same note/chord. If we have 5 members in a band which include a bass player, 1 rhythm / lead guitarist, 1 drummer, 1 organist, 1 drummer, and a lead vocal, PM perhaps is the bass player, if you play bass 100% of the time without playing the rest of the instruments, then the audience will be bored to death and might throw some eggs on the band, but if we play them together in pure harmony and rhythm, then this is where we will enjoy the beat of the music. Meaning PM is not the only task on maintenance. You may also use Predictive Maintenance (the drummer perhaps), Failure Finding Tasks or Functionality Inspection for hidden devices (organist), run to fail / redundancies (rhythm and lead guitarist), and modification (vocalist). But this can only be done by understanding the RCM Logic Tree Diagram.

Tip 2: RCM Team Should Compose of the Most Experience People on the Asset: Unlike TPM, which will be done by everyone in the organization, RCM is not for everyone. One of the most critical processes in the RCM analysis starts from the beginning and has something to do about who will compose the RCM team. If we are in the army, the RCM team is the elite commando group, special forces, or delta force that had undergone a more advance, thorough and rigid training. As John Moubray points out in his book, RCMII, the team to compose the RCM process should be the people with the most extensive knowledge of the asset. This means that they are the people with an enormous amount of time and experience on the asset analyzed. Perhaps these are people with a great amount of white hair or no hair in layman's terms. (I have doubts on the 2nd one since there are young people who are already bald, just kidding!). Sometimes I tell my delegates that if I am the one to select one who will compose the RCM team, I will need a magnifying glass and examine if the candidate has any white or gray hair. You simply cannot include a newly hired person or people that had a brief experience on the asset that is about to be analyzed. The output of the RCM process will depend on the experience and integrity of the team. Remember that we are doing the RCM analysis. We are capturing these good people's experiences and documenting them because we know that these good people will go away and retire one day.

Tip 3: Operator Should Be Included in the RCM Analysis Team: There should be at least one incumbent or experience operator on the team. One of the mistakes I see in the RCM implementation is that 100% of the members are a cross-selection of maintenance people without any operator's presence. So what happens if there is no operator on the team? All I can say is that this is a big problem! RCM and TPM agree on the involvement of operators and the belief that maintenance is a shared responsibility for both operators and maintenance. Once the team reached the second part of the RCM analysis on the RCM decision worksheet, some of the team's tasks will default to operators such as easy inspection, creating visual controls, cleaning, etc. The best person to perform this would be the operator. Most of the time, the traditional approach to maintenance is to create a maintenance task and pass them to operators. And maintenance complains a lot that operators won't take part in performing the inspections or are being done half-baked. This is because operators do not understand the importance of doing them since maintenance did not teach them about the consequences of the failure it is meant to avoid. Operators learned a lot during the interaction of experience maintenance people on the RCM analysis, and the good point is that operators understand the importance and value of why a simple inspection should be done. So if you plan to organize an RCM team, always include 1 or 2 operators.

Tip 4: RCM Team Should Refrain From Being Biased On The Analysis: Another mistake I see in conducting the RCM process is that after completing the analysis and comparing them to their current Preventive Maintenance activities, almost everything is identical, especially the task derived and agreed upon. Meaning nothing changes, and the analysis is a complete waste of time. Why does this happen? Because the members of the team do not think outside of the box. They prefer to do the same old activities for the remainder of their lives. In the second part of the RCM Analysis, which derives the RCM Decision Worksheet, or when the team is going through to the last couple of questions on RCM, the question to raise is what we will do to prevent or predict the failure? If the failure cannot be prevented or predicted, what would be our remaining options for maintenance? These questions are replaced with the team's easier ones, which is what we are currently doing in our maintenance. That is why they revert to the same old tasks again. The members are biased or afraid to change their old routine work. In the final two questions on the RCM analysis, the team should challenge themselves if the current activities are acceptable or there is a better way of doing maintenance than they previously have. If there is, the team should agree and write them on the RCM Decision Worksheet. Remember that there may be more than one task required to address a certain failure mode in some situations, and there might be a combination of tasks needed to address a particular failure mode. You can be using both Predictive Maintenance and Preventive Maintenance at the same time.

Tip 5: Basic Equipment Condition Should Be Addressed Before the RCM Analysis: In my training, sometimes I overemphasize this statement. It sounded like a soundtrack that always itself repeatedly. Basic Equipment Condition includes having clean equipment, complete bolts, and having the correct amount and correct lubricants' viscosity. You can have the most sophisticated, top-of-the-line Predictive Maintenance instruments, such as vibration analyzer and thermography, to be carried out by a level 3 certified human being. Still, suppose your equipment lacks so many nuts, bolts, and fasteners in them, these predictive maintenance instruments will be plain useless. Your equipment will be more prone to frequent breakdowns and failures. It only takes one loose or missing bolt to create a chain of destruction on parts inside your equipment. Just one piece of a 20-micron dust is enough to destroy your hydraulic equipment. This is where TPM comes first. I think if your plant is doing a plant-wide TPM strategy and you want to implement RCM. Let TPM do their thing first and address the basics before applying RCM. Nothing much will change if the basic equipment condition is not in your equipment, and you will just end up being frustrated and blame it on the RCM process.

Tip 6: Initial Interval Is Only an Initial Interval: In the second part of the RCM Analysis, the team members will discuss an initial interval for every task. They will be asked what they think is the best frequency of performing this equipment's maintenance activity. If the team agreed every month, then they will indicate it on the task interval. This is what you called the initial interval. This means that the initial interval will be followed as the equipment functions well with no problems. But this is not likely the case for Predictive Maintenance activities; the initial interval or the frequency derived will be followed if there are no potential failures or early warning failures spotted on the equipment. Suppose a potential failure is spotted on the equipment; in that case, the monthly frequency will be shortened, perhaps every couple of weeks or weekly. If you are to follow the initial interval to the letter, there is a great chance that you will miss the failure. Therefore, remember that the initial interval as the word itself is only an initial interval and will be subject to change from time to time during the actual implementation of the RCM process on the asset.

Tip 7: Know Thy Boundaries: RCM applies to any industry as long as there are assets or equipment to maintain. This can be applied to mobile and stationary equipment and as far as I can say. Manufacturing industries can also apply RCM for their own benefit. But what is important is to understand the best way to use RCM for your equipment and assets. If you work in an Oil and Gas Plant, power plant, or plant where equipment is connected, then the best way to adopt RCM is to do this on a system level. This means that there will be a list of components that will be included in a system. You need to determine the starting point and where to end your analysis. Work in a process plant in which equipment is dependent on each other. You can select the bottleneck or constrained equipment as your initial or pilot RCM project. Suppose your equipment is a stand-alone and is independent of other equipment, meaning the failure of this equipment will not affect other equipment; you can perform your RCM on that piece of equipment. If your equipment is a complex one, perhaps one equipment contains around 100 PWB electronic boards such as a wire-bonder from a Semicon industry; the best way to perform RCM is to determine the worst sub-assembly of that equipment, and this is where you perform your RCM analysis and not on the whole asset itself. I would like to convey that RCM should be done only on your most critical system, bottleneck equipment, or worst subassembly. We do not recommend using RCM on new equipment or good running equipment because the gains and results will not be visible unless you want to perform RCM on newly commissioned equipment.

Tip 8: Dedication to the Completion: Members of the team should have the passion and dedication to completing the RCM Analysis. Once the team is organized, there will be a series of meetings that will be conducted. Typically every meeting will have a duration of 2 to 4 hrs. And the meeting will be once or twice a week. The leader and the members should agree on a commonly available time for all the team members to meet. There will be cases that members will be called out of the meeting to address some issues on their operations or for whatever reasons. It will be very discouraging for the team if a couple of members are always absent from the meeting or are being called out consistently.

Suppose this is an issue you are undergoing; I strongly recommend disbanding the team and returning to your old habits. This clearly indicates that your management people are not serious about the RCM implementation and whatever outcome of the analysis. The output of the RCM will be half-baked because you have not captured the experiences of other members. After all, they were not in the meeting initially, or worst, having just one member doing the RCM Analysis. This will simply not work because RCM is a team approach and not an individual approach. In one of my classes on RCM, a delegate approached me and said that this was one of their early problems on initiating RCM, so they requested all members to stay in one hotel until the completion of the analysis, which their management hesitate in the beginning but agree in the end. As a result, they have completed their RCM analysis to the best of their knowledge, and everyone was happy with the outcome and results of the RCM. I just do not know how much the bill for the hotel was. But I hope the message here is clear, let the team meet with as little interruption as possible, and management must provide support and commitment because support and commitment are 2 different words. Management can support but cannot commit to the RCM analysis.

Well, I hope that these tips help you ponder these messages for a while if you plan to conduct an RCM analysis in your plant. Remember that RCM is not a replacement for Preventive Maintenance; PM will always have a place and will always be included in the RCM analysis. The point is that we utilize Preventive Maintenance only if it is meant to prevent the failure, and if the failure cannot be prevented, then this is where we divert to other tasks. Maintenance is not about confining to Preventive Maintenance alone, but understanding that other tasks can be utilized simultaneously with our current Preventive Maintenance tasks.

To conclude this newsletter, RCM is not a perfect reliability improvement initiative, nor are other tools such as lean, 6 sigma, TPM, etc. The analysis will depend on the team's output together with their experience on the asset being analyzed. Nevertheless, all I can say is that perhaps this is the best-structured approach on the planet on how to derive your maintenance tasks on a piece of equipment up to this point in time. I just cannot say for other planets.

5.8: August 2011: Tips on Performing Total Productive Maintenance (TPM)

In my previous newsletter, I wrote about tips on conducting Reliability-Centered Maintenance, and I thought about providing some tips about implementing Total Productive Maintenance for the benefit of industries implementing them or planning to implement them. TPM is much tougher to implement than RCM because TPM involvement will be from everyone in the organization. Second, implementing TPM is always a Top-Down Approach, which cannot be the other way around. Perhaps your industry is thinking of having TPM as part of your strategies to achieve your vision and is planning to start this initiative, or you have been implementing TPM for a year or two, but it seems that you cannot seem to get moving at all, so the program takes its stall and the people in charge of the project where given other assignments/projects to undertake. Or your company had kicked off TPM and completed their initial cleaning on the equipment. Both management and operators are enthusiastic initially but lost their enthusiasm to continue because they do not know how to begin the next step or production supersede everything and is the number one and only priority. In 1998, I hope a got the year right, JIPM (Japan Institute of Plant Maintenance) published a book called TPM World Congress consisting of many case studies on TPM done on different plants around the world and states that in a survey conducted by them out of 3000 industries that implemented TPM, only 10% are considered successful and gained the results of TPM, the rest simply failed. Cited in the book as the number 1 reason for failure is lack of management support and commitment.

I have breathed, ate, and lived TPM for 8 years, working full-time where I was assigned as the overall facilitator for Planned Maintenance and Initial Flow Control Activities (others refer this to as Early Equipment Management), and honestly speaking, doing TPM is a very frustrating job, but nevertheless, I enjoyed every moment of it even when matters get tough. During my days in TPM, we experience many mistakes in its implementation; therefore, it is best to know TPM and traps to avoid doing TPM right the first time around. Mistakes in TPM implementation are costly and may have some repercussions, and the people involved might lose their enthusiasm for doing it, or you may reach what you called the saturation point of TPM. Hence, given this, let me share with you some of my experiences to make your TPM journey successful.

Tip 1: Deployment of TPM Training Will Be from Top to Bottom: As TPM is a Top-Down Approach, the first group to attend training should be the Top Management people. These people need to know what TPM is in the first place, the 8 different pillars, and what respective pillars they will be involved in, and their role on that specific pillar. These people need to understand that TPM is a long journey, and results cannot be achieved in a week or months. These people need to understand the basic requirements needed to implement TPM, such as having a full-time department/office staffed by full-time people assigned to handle the major pillars of TPM. Management must understand that TPM will require a budget, and those small details should be addressed at the very beginning. Let me cite an example, if Autonomous Maintenance is implemented, teams will be deployed, and if the team is composed of a vertical set-up or the members of the team composed of members of different shifts, this means that you have three shifts in a day, one or a couple of members will go on overtime to attend the meeting. These overtime issues should be addressed at the very beginning of implementation. These people need to decide whether to push through or not and be ready to implement it. I hate to say this, but these management people need to change to adopt the TPM process. It requires a culture change. As far as I have experienced it, TPM is 90% people and 10% equipment. If we focus on changing people, then results can be achieved. Therefore, management should make a buy-in decision on whether to implement TPM or not. If management says yes, then expect a series of training that will be done by levels:

• Level 1: Top Management People, Senior Management people
• Level 2: People to staff the TPM Office, those who will be assigned as TPM coordinators, supervisors, Maintenance Supervisors, Engineers, Quality people, those who will be involved in Focused Improvement Pillar
• Level 3: Those who will compose the pilot teams for the Autonomous and Planned Maintenance pillar. This will now include operators and maintenance people.

Tip 2: Organize the TPM Office Immediately: One of the most important things to accomplish at the beginning of the TPM journey will be to organize the TPM office and staff them with permanent and full-time people. These people will play a crucial role in the implementation. One of the most important criteria in selecting is that these people must possess two traits, passion and patience. They must have the dedication to accomplish their duties and responsibilities. How many people to staff the TPM office depends on the population of the employees in the plant. When I was still working with the TPM Office, we were a big plant, and we cater to two plants totaling about 10,000 employees. Overall, we had 8 people in the TPM office, myself included, with 1 TPM manager that reports directly to the CEO/President. Most of us were engineers at that time and handle one or two pillars at the most. As far as I can recall, I have around twenty members as the core team in the Planned Maintenance committee composed of maintenance leaders in different departments. We meet every week to monitor their progress. Overall, we have around 700 members in our Planned Maintenance pillar of TPM.

To give you some idea of what the TPM Office people will do, they will initially be responsible for their own respective pillars. In short, they will need to provide the specific training of the pillar they are assigned to. These people had undergone the very basic TPM Concept. Most TPM consultants will provide you the basic principles of TPM; I think that's about it. You need to break these principles into these nitty-gritty small details that each TPM team can comprehend and follow. In short, TPM Office people should be the ones responsible for developing the roadmap for each step or phase of their respective TPM pillars. Example Autonomous Maintenance has 7 Steps, and there are specific details to undergo each step. Planned Maintenance can be done in 4 Phases or Steps, and each step has its own details. And once the team completes each step or phase in Autonomous or Planned Maintenance, the team should undergo an audit, so again TPM Office will be the ones who will develop the audit process and perhaps be doing the audit themselves in the initial phases or steps of TPM implementation. Suppose a team completes Step 1 of Autonomous Maintenance, and the TPM Office had not yet completed the details and roadmap of the next step, the team loses its momentum and most probably will revert back to its old ways. One more thing, during my time, we at the TPM office were also responsible for consolidating all reports, improvements and providing rewards and recognition for every single team that completes each step or phase of their respective pillar.

Tip 3: TPM Office Staff Are "Think Tanks," Do Not Be Dictated by Anyone: Although this is a tough one and one of our biggest mistakes during my days at TPM. There were cases when we at the TPM Office were dictated by top management on what to do with TPM. For instance, since they finally realize that TPM is a very slow process, this person (a top senior management guy) who claims to know TPM force us to do police work by auditing every abnormality we can see in the plant and highlighting them to the manager in charge of that area. Even though we do not like what we are forced to do, we do not have any options. Either you stay around and follow or resign and work in another plant.   The problem is that since our plant is a very dynamic one, this senior management guy won't be around for the next couple of months, and we at the TPM office will end up facing a blank wall. Again our journey is back to zero. Hence, our TPM is just a merry-go-round, which makes me dizzy. If you are reading this newsletter and assigned to the TPM office working full time, remember that you are the "Think Tank. " You are considered as the plant's internal consultant or ambassador on TPM, so you need to know everything on your TPM pillar. Leaders and team members will ask for your guidance on what to do and how to do it. I cannot emphasize any further the magnitude of your role and responsibility on TPM, but one thing for sure is that you are not only changing people but their lives as well. Suppose this is one of your setbacks on TPM, I think you need to hire an external consultant on TPM, or perhaps if your plant is aiming for TPM Excellence Awards, you need to hire a TPM consultant from Japan itself, but I strongly recommend acquiring them if you are midway in your TPM journey.

Tip 4: Agree On What Indices You Want to Measure on Your TPM Pillar: TPM Office, together with their committee members, council, coordinators, or whatever you want to call them, should be specific and explicit on what indices they want to measure at the very beginning of your implementation. This can be done manually by each team and consolidated by the TPM Office, but as more and more teams emerge, you need a way to automate the data. You can measure those that are both tangible and non-tangible. For example, for the Autonomous Maintenance pillar, one non-tangible measurement will be the number of suggestions implemented by the team members, the number of abnormalities or irregularities corrected on the members' equipment, the number of one-point lessons generated, and so on.

These are considered non-tangible measurements, but these have a tremendous impact on the team. For Planned Maintenance, the measurements will be a bit more complicated as they focus more on tangible results such as MTBF, MTTF, MTTR, Number of Breakdowns, Spare Part Cost, Repair, and Maintenance Cost, Lubrication Cost, and so on. Just one point before we move on, measuring breakdowns is very tricky, and before you can measure the number of breakdowns on your equipment, maintenance should need to sit down and discuss what eventually will constitute a breakdown.

Let me give you an example if you work in a manufacturing plant, and your utility compressor failed to supply air, all pneumatic machines connected with this compressor in your operations will stop. Is this a breakdown or not if we speak about the machines in your production line? (Answer not a breakdown, but the compressor encountered a breakdown. What you have is a downtime) If you have a redundant motor parallel with the running one acting as a standby and the motor failed so that the standby motor automatically runs, was there a breakdown? (Answer is two folds if we speak about the whole system, then there is no breakdown, but if we speak about the component, which is the motor, then there is a breakdown). In manufacturing plants such as semiconductors, they have a downtime called short-stoppages. TPM termed these as minor stoppages; others call them assists. Most of the time, it will just take seconds or minutes to address them. Other plants consider that if an assist takes too long or more than 6 to 10 minutes, they convert that into a breakdown; perhaps there was no maintenance available at the time. My point is that whatever the time it takes to address an assist, this is still a short-stoppage and not a breakdown, so do not include them in your breakdown lists. Tooling had already worn out to the point that it is producing off-quality products, was there a breakdown at this point? The answer to this may either be yes or no. So again, if you consider breakdown to be one of your indices or KPI to measure, you and your members should sit down and agree on what will actually constitute a breakdown and be excluded from it.

Tip 5: AM, PM, FI Should Have the Same Pilot Equipment: Before selecting your pilot equipment to undergo the different phases or steps of TPM, the Planned Maintenance pillar should perform an inventory of all their assets/equipment and have a matrix for ranking each of them. Say Rank A will compose of Worst Equipment, Rank B will be Semi Worst, and Rank C will be considered as Good Equipment. Priority will be given to the worst equipment, and this is where each pillar will select its pilot equipment. To accelerate the TPM process results, it is good to focus on the same equipment as their model or pilot machine for the three TPM pillars, Autonomous Maintenance, Planned Maintenance, and Focused Improvement pillar. This approach's advantage is that there will be easy communication between the three groups, and redundancy can be eliminated in the initial steps or phases. Autonomous Maintenance corrects abnormalities on the equipment and focuses on the external part of the equipment. At the same time, Planned Maintenance performs their Phase 1 Restoration activities and focuses more on the interior part of the asset or equipment. At the same time, Focused Improvement identifies the equipment's major losses and performs their step-by-step activities. Also, Planned Maintenance can guide and coach Autonomous Maintenance on their activities more easily because they work on the same equipment but in different groups.

One of the mistakes I see on Autonomous Maintenance is that they will initially list all the abnormalities and irregularities they can see during their initial cleaning process, and after listing them, they will prepare a work order and pass them along to maintenance. The rule of thumb here is that whatever you list, let the team correct it themselves. If Autonomous Maintenance does not have the capacity or capability to correct them, then this is where Planned Maintenance will come into place; my point here is do not pass everything to the maintenance people because you are not learning anything from your Autonomous Maintenance journey.

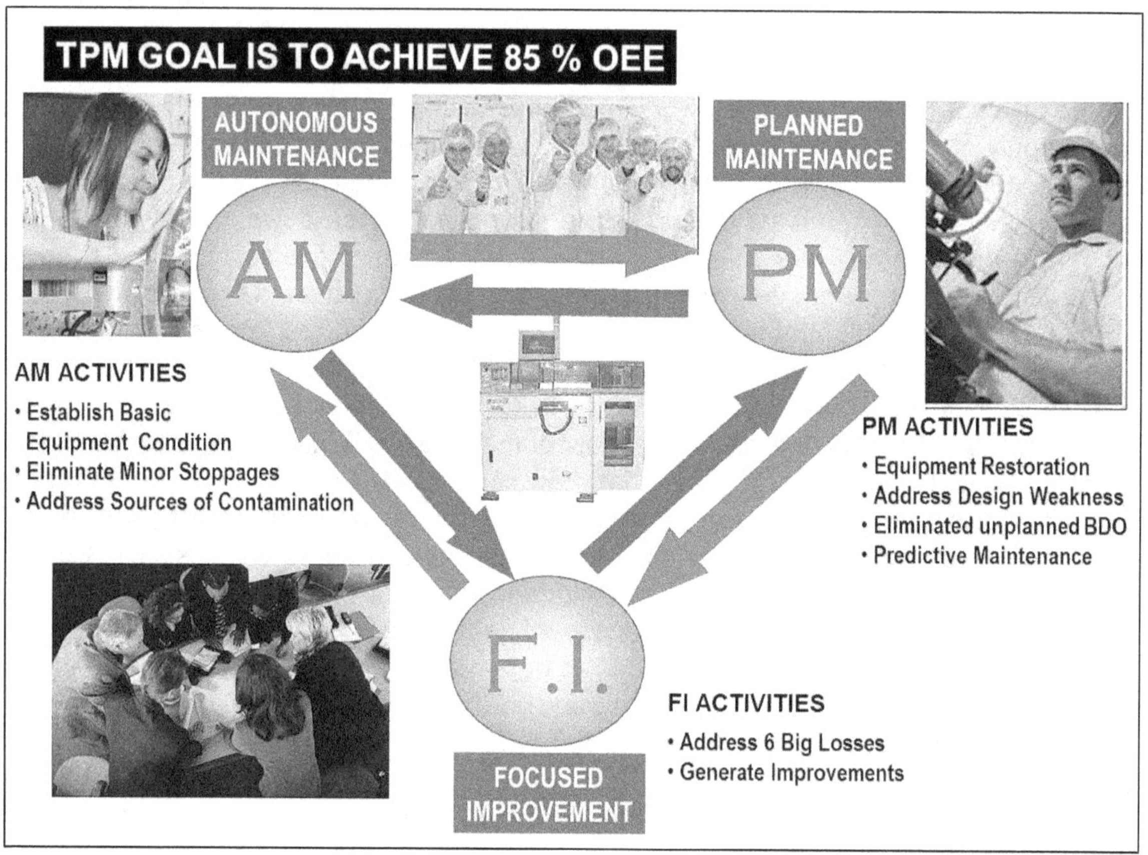

Figure 5.6: TPM Trinity

I hope that these tips on implementing TPM can provide you with some guidance and help you with your TPM journey. TPM is a long process and takes several years to complete depending on the pace you have set on it. Those assigned to the TPM Office working full-time expect resistance at the beginning of implementation. This is normal. What is not normal is if people do not show any resistance at all. Therefore one virtue you need to have is patience. Just remember one thing if other industries had benefited and achieved success in their TPM journey, so can your industry, too, as long as you do the right thing. Hope you enjoy reading this newsletter. Rock On!!!

5.9: September 2011: Survey on Top Problems on MRO Spare Parts

When a machine fails, it is mostly caused by a part that fails to fulfill its function. Mechanical parts wear out and need to be replaced and what is important is to keep the downtime to a minimum; hence a storeroom is a place to store parts that we need to keep our equipment running but managing spare parts simply means how fast we can respond in acquiring the right part during the time when maintenance and operations needed them most. Here are some of the top problems on spare parts most industries face in no particular order

1) Disorganized storeroom and time to get parts in the storeroom eat too much time: Most industries seem to develop their storeroom in a hit-and-miss fashion. If the equipment fails and the part is unavailable in the storeroom, they will usually purchase in excess quantities while some maintenance will be tempted to keep their own spares and create squirrel stores. In layman's terms, these are maintenance secret hiding places for their spare parts. Maintenance does this because they experience a lot of stock-out on their stores, meaning the system will reflect that the part is available. Still, it is not physically present in the storeroom. As a result, maintenance completely lost their trust and confidence in the storeroom and purchased the excess parts. At the same time, they keep the remainder with themselves.

One of the main causes of interrupted maintenance work in the industry is the lack of spare parts. In some cases, spares were not identified in advance or simply of stock when needed. The worst is that the storekeeper cannot locate the parts in the storeroom since everyone has access to the storeroom during peak hours when the storekeeper goes home since the storekeeper is from 9 am to 5 pm while their operation is 24 hrs. The company thought that having a one-shift person to handle the storeroom will save them money. If you want to improve your storeroom's housekeeping, just go to a supermarket, and you will have some idea of how to start with. Having a disorganized storeroom adds time in locating the parts and adds stress and fatigue to the stockroom people. It is much more comfortable to work in an organized environment.

2) Everyone access to the storeroom during peak hrs. Since their storeroom is from 9 to 5 pm: The saying that "Honesty is the Best Policy" does not simply apply to spare parts and storerooms. The storekeeper should be in control of what goes in and out of his storeroom. In short, in Spare Parts, you need to apply the golden rule, which is "**TRUST NO ONE**." Top management and decision-makers must understand that if their operations are 24 hours or 3 shifts, then the storeroom should also be 3 shifts. Trust and honesty systems do not apply in Spare Parts Management.

3) No Record of the Transaction: Since the part is needed ASAP, front-line employees and maintenance people will rush to acquire the item and not be inclined to fill up the paperwork. This usually happens when the storeroom keeper completed his day's work and leaves the storeroom key to the guard on duty. Two problems will be evident in this situation. First, expect your inventory to be lower than what is actually listed on the system. Second, there is a strong temptation for maintenance to store the part elsewhere in the plant, near their equipment, or even in their home for personal use, and theft happens. All items, parts that go out of the storeroom should always have a record as this will be the basis on what parts needs to be stored and reordered. If there is no control over the parts that go out of the storeroom, then expect parts to be unavailable when needed.

4) Physical Inventory and System Inventory Do Not Match: Suppose the actual inventory is lower than the system record; the risk is high that an out-of-stock condition can occur because parts will not be ordered on time. In comparison, if the actual inventory is higher than the system record, then parts will be flagged for re-ordering by the system even if not needed. It is critically important that the storeroom people should understand the importance of inventory accuracy and its users. If the inventory does not match the system, the users will lose their confidence and start stocking parts independently. One of the main reasons why system inventory and its physical inventory do not match is that everyone has access to the storeroom during peak hours, resulting in a stock-out of parts when the users needed them most.

5) Same Parts but Different Part Number: A part can be reordered because some departments may need them without knowing that some alternative parts can be used in the storeroom. Another reason is that the same parts can have different part numbers because they came from different vendors, or the part may have some very slight differences. The part number should remain consistent, or an alternative part number can be provided but should reflect on the CMMS when inquiring about a part.

6) Large Quantities of Obsolete Parts in the Storeroom: Often, many parts are held in storage that does not belong to any facility equipment since the equipment had already been decommissioned or retired. The facility's equipment may be retired and no longer on-site, yet the parts of that equipment are still in the storeroom. And worst, since the storekeeper does not know that the equipment is retired, purchasing or the storekeeper may still be reordering these parts that are deemed obsolete since they are fast-moving parts. Another reason why the part becomes obsolete is that the part's shelf life had already been reached. When equipment is retired, the spares associated with it must be identified so it can be free up from the storeroom. This will free up space in the storeroom. Remember that there is a cost of storing or holding an obsolete part in the storeroom. Storing obsolete items cost money and space in the storeroom. A plant can terminate plans to expand its storeroom by freeing up space consumed by obsolete parts.

7) Equipment Vendor Is No Longer Around: The machine you have been using had been around for more than 15 years, and the vendor may no longer be producing the parts you required, or worst, the vendor is no longer around and had shifted to another business. Another reason for the vendor being extinct is that most of the equipment's parts had already been modified by the vendor due to technology and advancement and may no longer use the old parts in your asset. Whether to continue running this equipment or retire can be done by conducting some economic and feasibility study appropriate for this situation. Suppose the decision is to allow the equipment to run; parts that frequently failed in this equipment should be identified and source other vendors that can fabricate the part.

8) Emergency Buying and Keeping Excess Parts In Maintenance in their Secret Hiding Places: When the machine fails, and the part is not in the storeroom and needs to be purchase immediately, there are many cases where a dozen of these parts are ordered for fear that the failure can happen again in the future. The excess parts will be kept not at the storeroom but at maintenance secret hiding places, and Joe is the only guy who knows this place. Since every maintenance has its own secret hiding place to store its parts. One day the same part fails that evening when Joe was no longer around, and since the storeroom has no record of the part, it was again purchased the following day, and now Steve kept the rest in his own secret hiding place. In starting up a Storeroom Improvement Strategy, the first thing to do is surrender all the parts that maintenance stores independently and let the Spare Parts Team handle them.

9) Low Quality of Parts, Purchasing Always Go for the Lowest Bidder: There is a tendency to reduce costs by buying parts with the lowest possible cost. Reducing costs and improving reliability are not the same; in fact, there are many cases where reducing costs will affect the equipment's reliability. If we want to save on cost, we should focus not on the initial cost but on the Life Cycle Cost of the component. This is where true and meaningful savings can be realized on maintenance. Every maintenance should focus on improving reliability and not on reducing cost. "Why?" Because if reliability starts to improve, then cost will definitely go down; it cannot be the other way around. Remember that there will be times that reducing costs will affect reliability. Having a low maintenance cost is always a consequence of good maintenance practice.

10) Maintenance do not know the code of the part and takes too long to retrieve the part in the storeroom: Numbering the part or spare should make life easier for the users to retrieve the part at the storeroom, but since the person who devises the part numbering is no longer around the plant or had been transferred to other plants and details on how he determines the part number is unknown, hence, idle time in acquiring the part is lengthened together with the machine downtime. In other plants, everyone has their own way of developing their own numbering system, which creates duplication of parts in the long run. Assigning part numbering or codification for every spare or part in the storeroom should be left to the storeroom people or to a centralized group and should not be done by anyone in the organization.

Survey On Top Problems on Spare Parts	As of June 2018		
		Take Survey	RANK
1) Disorganized storeroom and time to get the parts eat too much time for the maintenance	34		9th
2) Everyone access to the storeroom during peak hours because the storekeeper is from 8:00 am to 5:00 pm	28		11th
3) No record of the transaction was done. Maintenance just get parts and leave especially during peak hours	39		7th
4) Physical inventory and system inventory do not much	46		4rd & 5th
5) Same part with different part number	44		6th
6) Large quantities of obsolete parts in the storeroom	33		10th
7) Equipment vendor is no longer around. No idea were to get this spare part	36		8th
8) Emergency buying and keeping excess parts in maintenance secret hiding places	49		2nd
9) Low quality of spare because purchasing always go for the lowest bidder	46		4th & 5th
10) Maintenance do not know the part number of the part and takes too long to retrieve	29		10th
11) Large Amounts Of Non-Moving Parts in the Storeroom	71		1st
12) MRO Spare Parts is Not Manage By Maintenance	48		3rd

Figure 5.7: Survey on Top Problems on MRO Spare Parts and Storeroom (Updated)

11) Large Amount of Non-Moving Parts in the Storeroom: When the equipment was commissioned 5 years ago. The vendor told us to stock some critical parts of this asset, your warranty will be affected, and it scares you a lot. Today, these parts have not moved for the last 5 years. And when someone asks you what to do with them, your sentiments are to keep the part since we might use them someday. Stocking critical parts as recommended by OEM and vendors will always be a debatable issue, but the best approach to deciding whether to stock or not to stock this part can best be answered if maintenance and reliability people have an algorithm or decision diagram. A survey shows that an ABC Analysis shows that 70 to 80% of your inventories' total cost will come not from fast-moving items but from those non-moving items. These parts are big, as well as their cost.

12) MRO Spare Parts is Not Manage by Maintenance: The storeroom can contain many items from office supplies, janitorial services, raw materials, facility supplies, and MRO Spare Parts. The best people to manage the storeroom for MRO parts and all spares, items used in the storeroom should always be the users of these spares, and these are the maintenance people. Here are the sentiments of other departments.

- We at Finance and Accounting take care of the company's finances and funds, so we are in the best position to manage the storeroom.
- We at Purchasing are the ones in direct contact with the vendors, so I think that we are in the best position to manage the storeroom
- We at Maintenance are the users and know the parts very well; we should be given a chance to manage the MRO Storeroom and spare parts

To conclude this newsletter, 99.9999% of the storeroom and spare parts problem is simply man-made. The software doesn't think; they just do what you command them to do. Improving the stores does not only depend on the people managing them. Improving the storeroom is a collective effort of the users, purchasers, and other plant departments. Purchasing the top of the line software for your storeroom and spare parts will be useless if your plant is simply reactive, and I think that's all I have to say about that.

5.10: October 2011: Decision on Whether to Stock or Not To Stock Spare Parts

If I have 3 items with you and you need to decide whether to stock the part or not, most of you will perceive that the most critical and used part should be the one that needs to be stock. Therefore, let us analyze each of these three items and let us find out which is the part or item that really needs to be stock or not.

In Item 1: In figure 5.7, we have a gas tank that we used perhaps 3 times a day or more for cooking. Almost every kitchen has this one, but when I ask if they keep a stock or spare of this item, around 8 out of 10 people will most likely say, Rolly, we don't keep a stock of that one. The rest have some more money that they keep a stock of this item. Although this item is being used daily, the consequences of its failure when it ran out of gas would not be that critical or devastating as people have many options at hand as follows:

- Option 1: When the gas runs out, then we buy or order another one; downtime to cook will perhaps take 15 to 30 minutes delay, and since we use this daily, we can have an estimate on when it will run out so we can prepare for that situation.
- Option 2: Since the cost of the cylinder gas range is somewhat from 620 Philippine Pesos or roughly around 15 USD, if you don't have the money at the moment, there are many mini-restaurants in almost all subdivisions in the Philippines. In my native language, we call them "carinderia" that sell different varieties of cook food and rice which you can easily purchase. If you have more money, you can go to McDonald's, Pizzas, Kentucky Fried Chicken that you can call to order, and they will deliver as long as you don't live 500 miles away from them.
- Option 3: If you have the money, buy a spare gas cylinder.
- Option 4: You can use charcoal to cook or an electric stove if you have one until you have the money to purchase another gas tank for cooking. Therefore, the decision here is simply we don't need to stock this item.
- Option 5: If you are fat, don't eat and start fasting.
- Option 6: Call a friend and eat there.

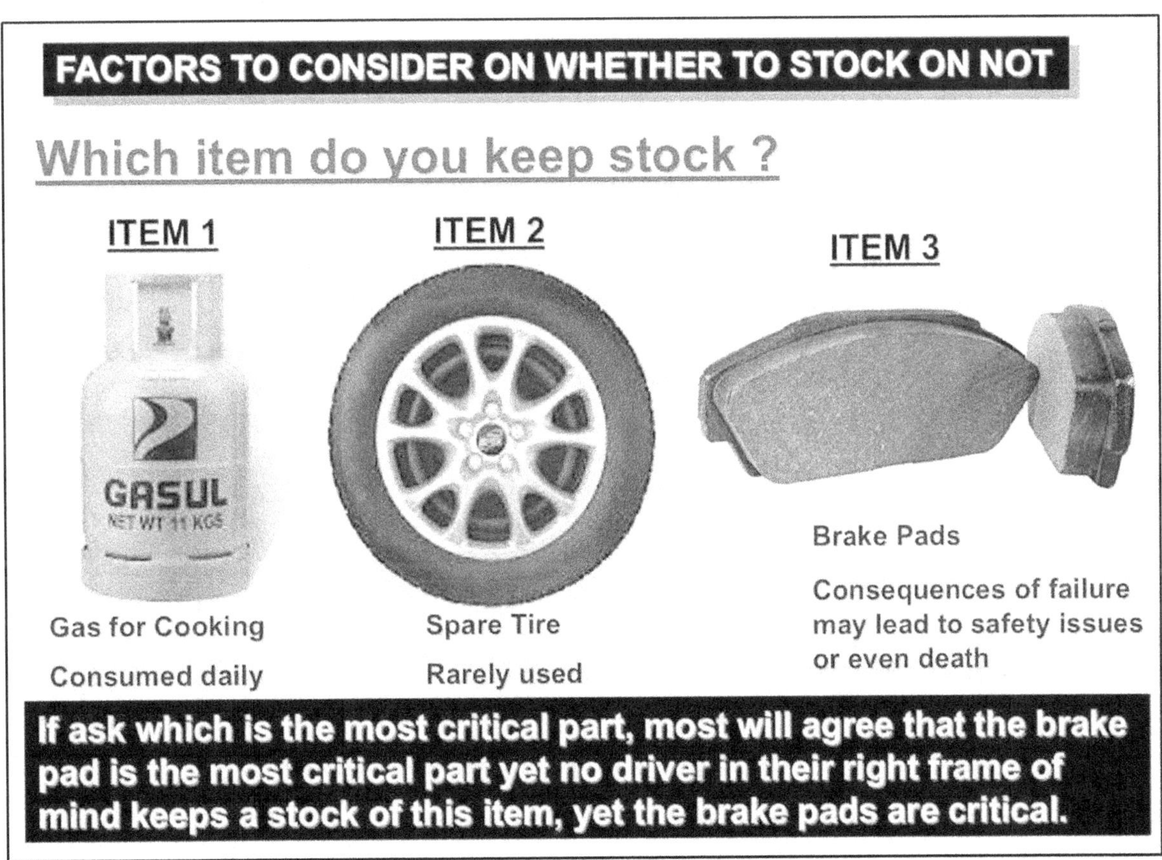

Figure 5.7: Decision on Do We Stock or Not

In Item 2: All cars keep a spare tire, and perhaps if you have been driving your car for 5 years now, you may have used the spare tire a couple of times or have never used them at all. Yet it seems logical and practical to keep a spare tire for emergency purposes since the consequences of having a flat tire will cause you to delay in your travel and money on your pocket since you will be towed away for good, and not to mention the traffic jam you will cause if this happens in a congested part of the city. Therefore, the decision here is to keep a spare tire with you every time you travel.

In Item 3: I think everyone would agree that the brake pad is the most critical part of all, yet I don't think any driver or motorist keep a stock of them. Although the consequences of failure are greater than the spare tire and may lead to harm or even death, we do not keep stock of this because not all drivers know how to replace it. The decision here is not to stock this part because reliability and maintenance come into place. Although it is unnecessary to keep a stock, we need a constant inspection of the brake fluid and inspect the brake pads from time to time for any traces of abrasion on the brake disk and brake pads' depth. These pads will gradually wear depending on the usage, so knowing how many km the car run would guide us when replacing these pads and the brake drums on the rear tires. Therefore, it is a good decision not to stock this part unless you disagree with me; then, by all means, keep a spare of these brake pads.

Frequently, the approval for adding a new item to stock is sometimes a complex situation, time-consuming, and likewise subjective, while sometimes it is a debatable process if people involved come from different areas of the organization. But there is a rational way of deciding on whether to stock the part or not. To perform this logically, a good background and understanding of reliability principles are required. The best people to make decisions on this will be our humble and down-to-earth maintenance people.

Your OEM may tell you to stock this because these are critical parts, and that was 5 years ago when he said that, and the part up to this time still remains untouched in the storeroom. As a result, visit your storeroom, and you can see spares that have not moved for two years or more in our storeroom. Do not believe everything your OEM tells you to do. Hence, here are some points and things to consider in making sound decisions on whether to stock or not to stock an item or part in your MRO storeroom.

Factors to Consider in Stocking a Part or Not In the Storeroom:

1) Type of Failure Modes: Some failure modes will warn that they are in the process or on the verge of failing. One of the key factors in the decision to stock is the warning of the components or parts they provide before the breakdown. The failure patterns can be classified as infant mortality, random, or wear-out failure. Many failure modes give some sort of warning that they are to occur. The P-F (Potential and Functional Failure) curve shows how a failure starts and deteriorates to a point where it can be detected and deteriorate at an accelerating rate if it is not detected. Hence, the P-F Interval is the interval between the occurrence of a potential failure and its decay into a functional failure. The failure development period can take as little as hours, days, weeks, months, and sometimes years. Determining the potential failure can only be possible if the plant's maintenance people are currently using Predictive Maintenance instruments. Hence, if the failure development period is longer than half the lead time to purchase the part, it is good not to stock it. For components with a very short warning or potential failure, such as fuse, shear pins, electronic parts, the decision will be to keep stock of the part. A normal failure mode for a truck will be the tires. For as long as the tires are aligned, properly camber, the tire will wear out gradually concerning the truck's running hours; hence, a decision not to keep a spare in the storeroom will sound be feasible. On the other hand, if the tire always fails because of sharp objects such as rocks, it is justified to keep a spare tire in the storeroom, as in underground mining industries.

2) Is the Spare part of The Equipment a Bottleneck or Not? Bottleneck equipment is equipment that constrains the flow of the process. In a manufacturing plant, if you have a series of equipment that produces a product, the machine with the lowest output per hour produced or UPH would be considered the bottleneck equipment since the product's inventory is expected to build on that piece of equipment. Therefore if the equipment is considered to be the bottleneck, then stocking some critical parts can be justified, but on the other hand, if the equipment is not considered as a bottleneck, or some redundancy or standby is in place, then some amount of downtime can be allowed to happen because it will not affect the overall plant output for the day.

3) Shelf Life of the Spare Part or Item: Some items have a shelf life or similar to an expiration date (Best if used before) and are better obtained fresh from the vendor than stocking them in large quantities. Examples include paint, chemicals, batteries, fire-extinguishers, lubricants, etc. For lubrication, oxidation occurs in all oils in contact with air, including stored lubricants. The base oil and additive combination affect the rate of oxidation. For grease, the presence of thickener in grease increases the degradation rate. Environment and storage conditions have the greatest effect on the rate at which the lubricant will degrade. Suppose we increase the temperature at which the lubricant is stored by 10°C (18°F), the oil will double the oxidation rate, which cuts the usable and useful life of the oil in half. The presence of water usually introduces temperature variations, and condensation increases its oxidation rate.

4) Does the Equipment Has a Back-Up or Redundancy: A run-to-fail situation is sometimes valid if some of the equipment has redundancy or back-up, reducing failure to non-operational consequences. When equipment has a backup or a standby, it makes the consequences less and can allow the plant or system to still run. The duty equipment that failed will then be switch on to the standby or redundant component. The duty equipment can be fixed, and parts that have been affected can be ordered during the time in which it actually failed. In this case, it is justified that the part affected need not be stocked in the storeroom but rather on a need basis. Critical in this setup is performing a functionality test on the standby equipment to ensure it will run when needed. There are many instances that your equipment also exists in other departments that are not beyond your control or responsibility. It is just a matter of communication with other people in that department to utilize their equipment. The only exemption of stocking parts is if the lead time will definitely be long.

5) The Use of Alternatives, Substitute, or To Fabricate On-Site: There will be parts where the lead time to acquire them may take quite some time. If a part or item can be substituted or an alternative part that will not compromise the quality of the equipment performance, it is advisable not to stock it. Purchasing people are good at finding alternative suppliers or substitutes for parts. What is important in this case is to communicate with the maintenance people to evaluate the parts for a go or no-go situation. Some plants have fabrication shops or machine shops, and some parts or spares can be fabricated locally by their own shops. Likewise, there are independent machine shops that perform this type of work. Whether to stock or not to stock this part will depend on the lead time to fabricate it.

6) Part is Age-Related Base on Preventive Maintenance Replacement: If the part is age-related and the wear-out process would be longer than the lead time to acquire the part, and the failure is consistent and will survive to that specific age dictated, it is justifiable and feasible not to stock the part in the storeroom. Usually, wear-out is a gradual process and does not happen overnight. Maintenance can experience parts that tend to deteriorate or wear out concerning a specific age. If the wear-out process is constant for the part, then maintenance can re-order the part on that fixed period when the part will start to deteriorate. What is important for the maintenance will be to identify what these parts wear out concerning the age. An example will be brake pads, clutch, liners, tires, batteries, impellers, crusher jaws, or parts in direct contact with the product such as blades, dies, etc.

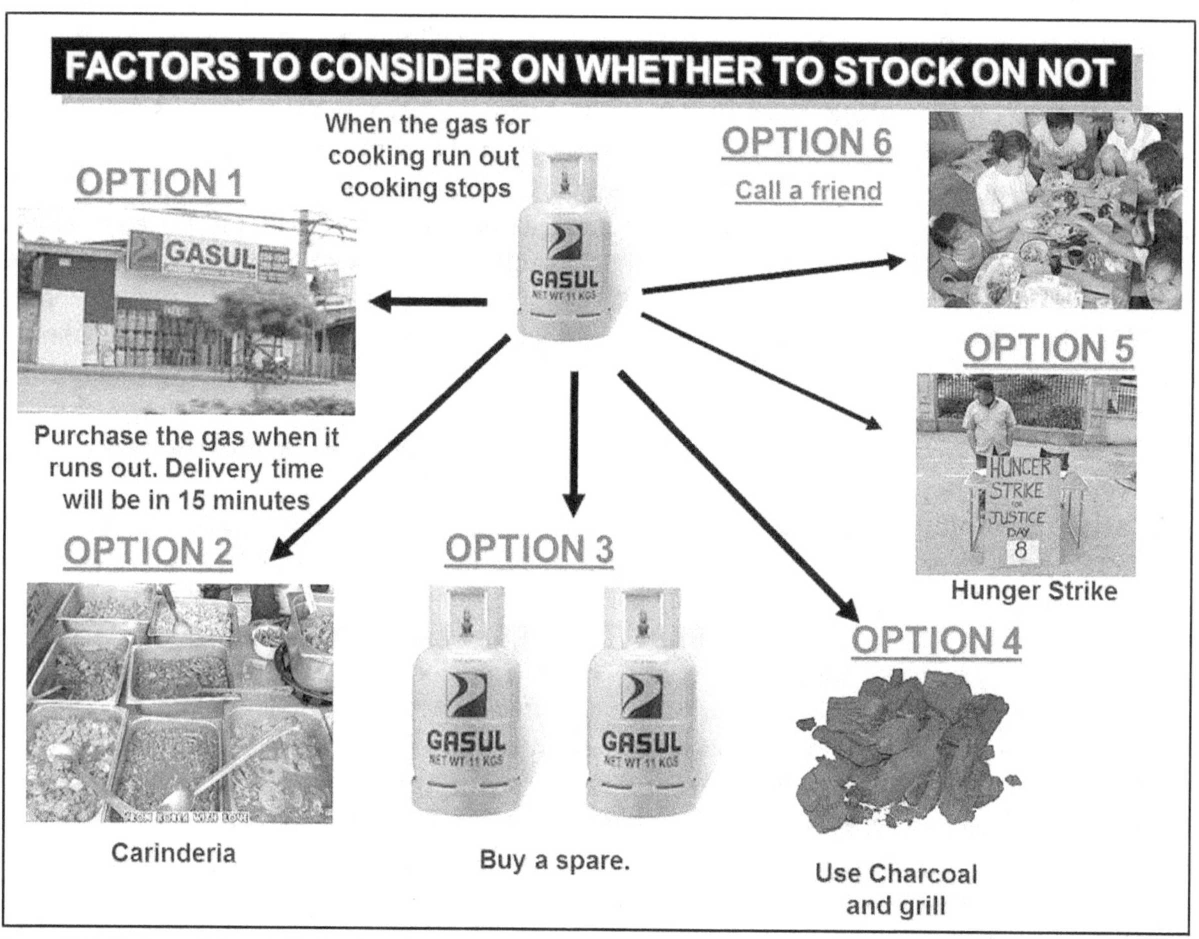

Figure 5.8: Options When Cylinder Gas for Cooking Runs Out

7) Parts Monitored by Condition-Based Maintenance: Although related to item 1, parts with a potential failure or failure development period will be under the watch list of a PdM or Condition-Based Maintenance program. Since the failure development period is consistent and can be known to the user of these instruments, these people can predict not only when the part will fail but also the best time on when to requisition the part or item that is on the verge of failing. With this, operations people can be notified when the equipment will be stopped for maintenance activity. Both Predictive and Preventive Maintenance groups can advise purchasing and storeroom to purchase the parts they need in this specific period. There will be no surprises of equipment failure since these failure modes can be predicted, and everybody is happy.

9) Market Demand is Low: In some manufacturing plants, there are times when the capacity or volume is high that you cannot afford a single operator or maintenance to go on vacation leave because of the demand for the product. For example, during summertime, the demand will be high for soft drinks, mineral water, canned juice, and ice cream. Machine uptime will play a vital role; hence some fast-moving spares need to be stuck in the storeroom. Likewise, in a semiconductor, there are periods where the demand is high and also where the demand is low where skeletal force is used; instead of 3 shifts, they reduce this into 2 shifts, and most people are forced to utilize their Vacation Leave and sometimes their Sick Leave as well even if they are not sick and around 50% of the equipment are utilized in this case.

Not all critical parts need to be stuck in the storeroom. When the machine was newly commissioned, there is a tendency for the OEM to convince us to stock parts in our storeroom since they are claimed as critical parts. Some or most of them right now had not been moving in the storeroom for a couple of years now or even more. The decision on whether to stock or not to stock will always be a debatable issue. The MRO logic tree diagram can provide us some basic guidelines and decision-making on whether to stock or not to stock the part in our storeroom. The more we can control the timing of failure, the better understanding we have on whether to stock or not to stock the part. The traditional thinking on most maintenance is to keep stock of the part just in case if we might need it someday. Suppose the machine fails and the stock was kept, maintenance says, "See, I told you we might need it someday," but, if after five years and the part is still around, maintenance again will say to still keep the part just in case we might need it someday. This is mostly when your industry is reactive because you do not know which machine will fail and what parts are needed. Therefore, the next time your OEM tells you to stock this part, consider the following factors for your team to make sound decisions

5.11: November 2011: Does Industries Know What Maintenance Is All About?

In most cases, I tend to say this over and over again in my training classes like a broken soundtrack, that most of the time, industries do not really understand the role maintenance play in their organization, and the problem is deeply rooted down in their organization. Suppose we speak about manufacturing plants, assembly, automotive, pharmaceutical, or semiconductor plants, for instance, when you visit their plant, almost every corner hangs a poster on what they vision to be in the future, and that is to be the best in terms of their product, quality, cycle time, and to be the fastest in terms of delivery by beating every single competitor in their field. On the other hand, if we speak about construction, mining, chemical, airline, shipping, and oil industries, quality is not their priority, but it will be safety. This will be their main priority, and all their employees should abide by all guidelines, procedures, policies, and protocols on safety. The bigger the plant and number of employees, the more people are deployed in the safety and quality department. Perhaps that is a good strategy for them. I even know of a plant that has more than 200 people in its quality department alone. I am not saying that it is a good or a bad thing, but I would like to stress out and emphasize that doing too much quality and safety is not the same as improving the equipment's reliability. And giving it some thought performing too much safety or quality can end up doing more police work in the plant and highlighting every non-conformance they can audit, which may have a detrimental effect on the equipment's reliability. So the question is, which industries focused on reliability, I can only think of one, and those are Power Plants.

Let me explain this matter with an example. A quality auditor in a manufacturing plant conducts an internal audit on their production floor and found out that only 20 out of 50 activities on PM had been performed on the month's equipment. The auditor slaps a non-conformance ticket to the maintenance crew. But what the quality people failed to realize is that their Preventive Maintenance group had been trained on the different patterns of failure and finally understands that placing too many activities on their Preventive Maintenance tasks ends up in a lot of Infant Mortality Failures, which is the right thing to do so they decided to skip some of their activities which are deemed irrelevant mostly on replacement and overhauling. Still, the quality people slapped a non-conformance ticket, thinking that he was doing his job. It is also the right thing to do for him.

A safety officer from an oil and gas plant again slaps a non-conformance ticket to the maintenance team because of a severe leak in one of their rotating equipment since the leak might induce some safety issues on the operator or from the maintenance themselves. From the operations point of view, this is not an issue since their capacity and production will not be affected by the leak of the equipment, while from the maintenance point of view, the seal will be replaced in their next week's PM schedule. In this case, we have three different groups, which are safety, operations, and maintenance. The problem with the three of them is that each one has their own opinion on what eventually constitutes a failure, and they have some different points of view. In this case, the question to raise is that if these three people have different views on failure, what are we trying to avoid, and which of these should we prioritize.

Every industry has its own procedures, specifications, rules, policies to follow on almost everything they do. Every industry wants also wants to adopt change so they can better themselves off in the future. Still, to change, we also need to revise some policies. It is not easy to change most of them, if you know what I mean, since some have been in place since the beginning of time and may take around 3 or more signatories who will ask you why you change this.

I have been teaching in a plant recently when an announcement was handed down to me that my training will be interrupted and delayed for an hour for a safety meeting with the training delegates, and I was lucky they did not kick me out of the room and allow me to listen to their safety meeting in which I practice my right to remain silent and allow the safety officer to speak. One of their main concerns and highlights by the safety team is that maintenance conduct work without the proper work permit and job order. Every time maintenance was caught working without a job order, the safety people penalize them by giving them a non-conformance ticket. During the break time, I ask some maintenance why is this the case and they told me that in most cases, it takes an hour or more to get or obtain a job order for a specific type of work in their plant, and if we comply on every single safety protocol, then we lose a lot of maintenance man-hours with our people in a week. My point here is simple, perhaps instead of thinking about very having a high compliance on safety issues and deviations, why don't we also try to solve how to lessen the time from hours to minutes on how to obtain a job order in the quickest possible time, perhaps by automating the process.

In a plant in Nigeria where I was lucky to be invited, they have a safety policy that when you climb on the stairs, you need to use the handrail for safety precautions, since I do not know the rules and the training were in the 3rd floor, I used the stairs and overtook as many people walking on the stairs. Everyone kept staring at me and asked me politely to hold the handrail of the stairs. Still, one guy holding the rail was coughing severely, that he places his hand to cover his mouth whenever he coughs. You can see his eyeballs almost plucking out of his face whenever he coughs. Hence, all the germs are in the rails, but maybe if there are two handrails on the left and right side and one for fast lane and the other for slow and at every floor, there is a liquid disinfectant that you can spread in your hands, then we can comply to that.

My point is that if we study human error, some are unintended, which are categorized as slips and lapses, and those which are intended, which are categorized as mistakes and violations, and under violations, they can be grouped into routine violations, acts of sabotage, and exceptional violation. And both quality and safety people should know what an exceptional violation is all about. A driver from the funeral parlor will drive at a very slow pace causing traffic to build up. In contrast, a driver from an ambulance will drive very fast since the driver knows whether since this is a matter of life or death situation. Hence, the driver of the ambulance took the opposite road. Have you heard of an instance where a traffic police officer gave a ticket to the ambulance driver for overspeeding? Hence, if one of your friends suffers from a stroke and you and your friend carry him out and drive the car in your plant, and a sign says the maximum speed at 20 kph and wears a seat belt always, we can disregard the 20 kph sign and travel beyond that speed because this is a case of emergency. We violated the safety rules, but we did it for a good reason to save a life. I hope I make some sense here, as well as your safety and quality people as well. We do not want any disputes or arguments with these quality and safety people since what we all want in life is for the good of the plant, but if there will be some problems and repercussions in the end, as indicated in my examples, then perhaps both maintenance and these groups can work out in harmony for our plant to run better and more effective.

Worst in some industries, every maintenance is handed down a sheet in which they need to update it on an hourly basis daily on what equipment they have repaired every single hour, and if operations people cannot find any maintenance in the line, then operations people yell and want maintenance to be visible all the time in the shift and tell them to look for problems. If I am the maintenance on sight, I will go in front of this person and stare at him, and when this person says, I told you to look for a problem, I will reply, "I am looking at the problem."

Take a look at what industries call their maintenance people. It will end in calling them technician 1, technician 2, technician 3, 4, mechanic 1, mechanic 2, mechanical 1, 2, 3, and so on. Their work is purely 90 percent troubleshooting, repair, and fixing breakdowns. The remaining 10 percent will be to do any other task assigned to them from time to time.

Sad to say that this is not maintenance work; maintenance and mechanics are two different people. A mechanic will perform repair until it comes a time that he becomes efficient and good at it, while maintenance, in the beginning, will likewise troubleshoot and repair the problem; but the difference is if the problem persists or recurs, then he now uses his brain and starts analyzing the problem. This is what the real and true role of maintenance in any organization.

It is really tough, challenging, and will take a lot of pressure involved to survive the maintenance department nowadays, but I will never steer you, wrong, good people, that the real role of our maintenance organization is to preserve and maintain the plant and its environment and this cannot happen if most of our people are deployed on the right side of failure which is the reactive side. You may visit this link to refresh yourself on the two sides of failure on my June Issue of our RSA Reliability and Maintenance Newsletter. Operations people want maintenance to be reactive because this is how we do things around here in the plant, and this type of culture has existed way before the second world war, so why in the world should I change things if this plant had been working this way since the beginning of time.

I do not know if there is actually a plant or industry that exists with a complete maintenance organization chart like this or but what I believe is that there are now industries and organizations that have been trained in different reliability and maintenance strategies and are now aiming slowly to have a holistic approach in their maintenance organization. Little by little, these industries are now aiming to have this type of organization because they believe that their plant can run better if most of their maintenance people are deployed on the proactive side of failure. The problem is, if the highest maintenance person in your plant reports directly to their operations manager and their operations manager is a non-technical person, then the vision of having a maintenance organization like this can never happen and can exist only in mind. But if the operations manager has some forehand knowledge on maintenance and reliability or if the highest person on your maintenance has the same position as the highest person on your operations, then this type of maintenance organization is possible.

I hope that industries finally realize what the role of maintenance plays in their organization and that people from different departments should work together hand in hand in and should not only focus on safety, quality, but they should also understand what reliability and maintenance are all about and create a balance on what they think is right or wrong. If people create a problem, then people should also be the ones to create a solution to their problem.

5.12: December 2011: Most Common Types of Mechanical Wear Explained

When a part or spare is replaced in equipment, and the boss asks for the reason why maintenance replaced it, maintenance most common response will be the part worn-out, wear and tear, or the part had deteriorated, but if the boss asks us a bit further such as what type of wear do you think it is, most maintenance will simply scratch their head.

Maintenance people should have a basic understanding of the most common type of wear and its causes. According to Dr. E. Rabinowicz, equipment may lose its usefulness because of three factors. First, the equipment becomes obsolete, where the equipment or the product it is producing will no longer be used, and the equipment will therefore be decommissioned. This consists of around 15% of why equipment can no longer be used. The second factor is accidents where the equipment encounters a major breakdown or failure. The cost of rebuilding or repairing the equipment will almost equal purchasing new equipment, which constitutes around 15% of why equipment loses its usefulness. But the bigger percentage of why equipment loses its usefulness is surface degradation, which constitutes around 70% of why equipment loses its usefulness. And under surface degradation, 20% will be due to corrosion, and about 50% will be caused by mechanical wear. In this case, mechanical wear will be caused by abrasion, adhesion, erosion, and fatigue. Dr. E. Rabinowicz states that surface degradation cannot be eliminated, and it will come and happen, but the good news is that the process of surface degradation can be controlled and can be prolonged. What is important for our dear maintenance function is distinguishing between the different types of wear, how they look on parts, their markings, and, most importantly, what causes them to occur on the mechanical part. By having this kind of knowledge and information, our maintenance can understand how the wear process happens and how to prolong its process, and what should be done to control them.

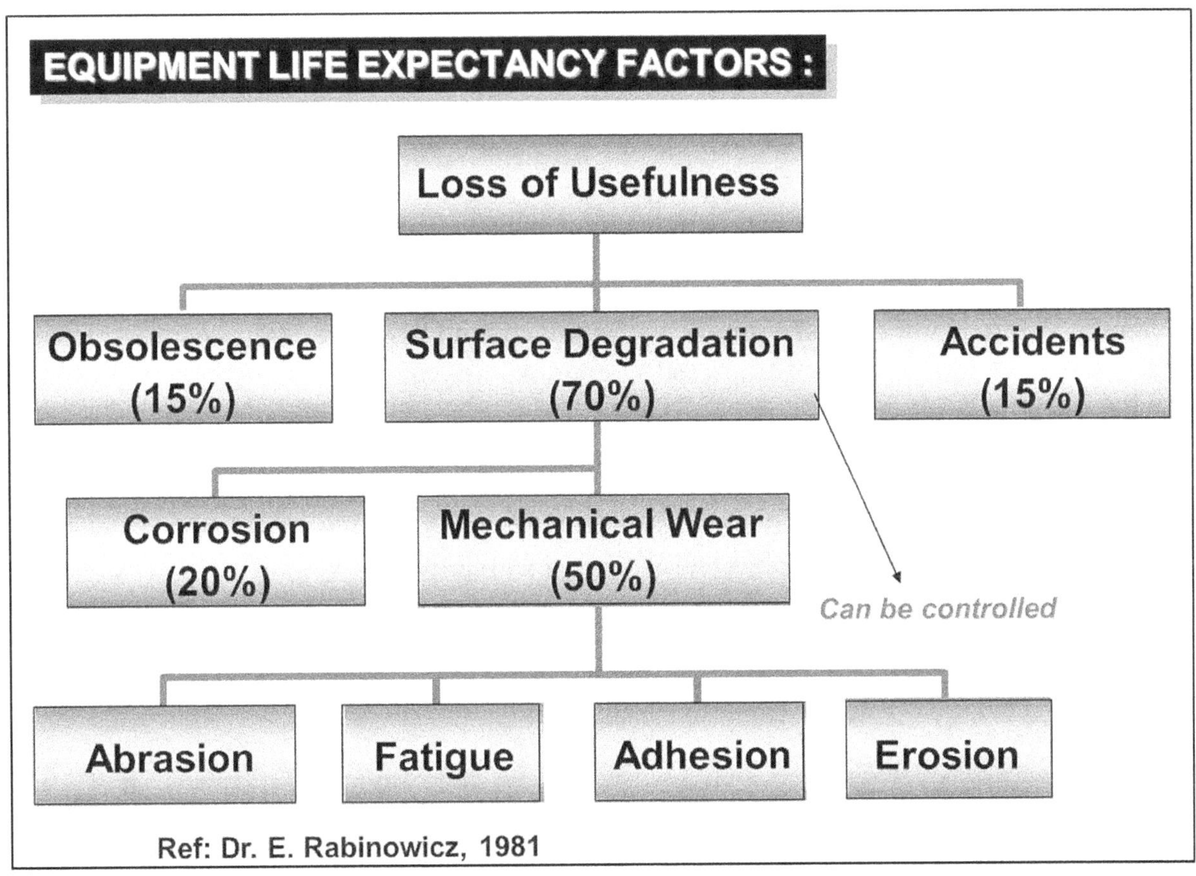

Figure 5.9: Why Equipment Can No Longer Be Used

By definition, wear may be defined as damage to a solid surface caused by removing or displacement of material using mechanical action of a contacting solid, liquid, or gas. It may cause significant surface damage, and the damage is usually thought of as gradual deterioration. This means that wear is a slow process. This is also a loss of material from a surface using some mechanical action. In any failure analysis involving metal fracture, the subject of both metal strength and the stress within the metal part must be considered, for they cannot be divorced nor separated from one another. Stress is defined as the force per unit area, often considered as a force acting through a small area within a plane. Strength in the metallurgical sense is the metal part's property, which resists the stress imposed upon the part. Therefore, all fractures will be caused by stress; a version of the weakest link theory will apply that a fracture will occur if the local stress will exceed its strength.

Our equipment is not like a television set or a plug-and-play instrument. When you plug it, then it will run continuously, or when you click on the remote control of your TV set and watch your favorite sitcom, you fall asleep, and when you wake up, the TV is still turn on, although the channels may have already run out and what you see is a snowy screen. Remember that your equipment that runs in your plant contains mechanical parts such as a gear meshing around with one another, a piston moving up and down, mechanical components that run either horizontally or vertically, separated by a very small clearance in mils or microns. Each movement, whether it is moving vertically, which is in an upward, then back to the centerline, then a downward motion, and then back again to the centerline, will be equivalent to one stress or one complete cycle. A 360-degree revolution of a bearing or rotating equipment will also be equivalent to one stress or one cycle. This means that after reaching N times the number of stress, the material's strength will deteriorate, and the stress in the end wins.

We have a given die punch that moves on an up and down movement and comes in direct contact with the product; the die punch may be made of a hard alloy. Each up and down movement will represent one stress, and so on. Hence, after reaching 1000 movements or stress, the strength of the material of the punch remains the same and unchallenged, but this small incremental stress or movement will take its toll once the punch reaches 100,000 movements. Stress will now take its toll on the punch, and whenever the punch contacts the product every time it moves, then it will start to wear, which means that small fragments of metal from the punch tear up. After perhaps 200,000 up and down movements of the punch, then it can no longer provide the correct measurement on the product and will now have an effect on the quality of the product hence; therefore, it is now time for the maintenance to stop the equipment and replace the punch because it already wears out but in the metallurgical and technical sense, the punch stress out after reaching its nth number of movements. The stress in the end, which is just merely an incremental movement, had finally won the battle against the punch's material's strength. I hope you can follow my explanation since I cannot explain it to you simpler than that. Let us discuss the most common types of wear.

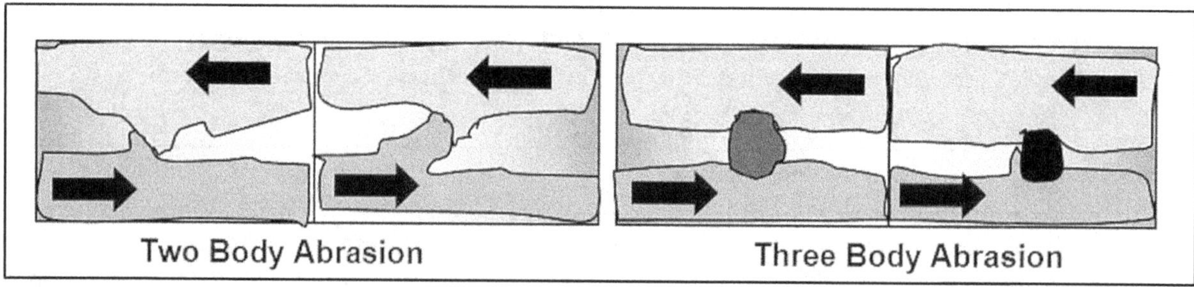

Figure 5.10: Two-Body and Three-Body Abrasive Wear

Abrasive Wear: This type of wear can be categorized by a single keyword, which is cutting.

This type of wear occurs when hard particles are suspended in a fluid or projections from one surface roll or slide under pressure against another surface, thereby cutting the other surface. This wear mechanism is basically the same as machining, grinding, polishing, or lapping that we use for shaping materials. When you go to a machine shop and a lathe is in operation, cutting a shaft produces long metal debris pieces as the cutting tool comes in contact with the metal; this is actually how abrasion occurs.

There are two types of abrasion: the two-body abrasion, which occurs when the lower portion of the mechanical component's peak or asperity comes in contact with the upper component's peak or asperity since they are moving in opposite directions. The weaker of the two asperities will break off, creating a third party, which later on can become the other type of abrasion, which is the three-body abrasion. In the three-body type of abrasion, a contaminant or particle usually as hard as or harder than the two mechanical components gets trapped in the two mechanical components' clearance. Suppose the contaminant gets trapped in the upper portion of the component; it will cut or rub the lower portion of the mechanical component. Signs or distinguishing marks from abrasion will be some hairline scratches on the component or part. There are two ways to control abrasion. The first is to harden the mechanical part since the principle for abrasion to occur is that the contaminant or particle trapped should be as hard as or harder than the mechanical part itself. But making the material harder will create a secondary problem, which will make the material more prone to a brittle fracture. A brittle fracture is usually a fracture with little or no deformation. Another way of controlling abrasion is to remove the contaminants themselves through the oil by applying high efficiency and beta rating type of oil filtration in which the filter can guarantee to remove more than 99% of the contaminants present in the oil. The higher the beta rating of filters in lubricating hydraulic or engine systems, the more contaminants it can remove. The cleaner the oil, the longer the wear process to occur in the mechanical part.

Adhesive Wear: This wear occurs when a peak or unevenness from one surface comes in contact with the other surface's peak or asperity. Adhesive wear has been commonly identified by the terms galling or seizing. There may be instantaneous micro-welding on the part taking place due to the friction or heat involved due to the loss of film on the lubricant. For adhesive wear to occur, there must be a metal-to-metal contact. When the metal comes in contact with each other, friction occurs, and heat is generated, and some form of discoloration occurs in the process. Micro-welding occurs, and the final stage will be a metal breakage or fracture on the mechanical component. Adhesive wear is produced by the formation and subsequent shearing of welded junctions between two sliding surfaces. For adhesive wear to occur, surfaces must be in intimate contact with each other. One way of controlling adhesive wear is by applying the oil's correct viscosity on the mechanical component. Surfaces held apart by very thin lubricating and oxide films will reduce the tendency for adhesion to occur. Adhesive wear starts out on a small scale but rapidly escalates as the two sides alternate weld and tear metal from each other's surfaces. That is what adhesive wear is all about, microscopic welding. The heat produced at the contact interface is very high near the two metals' melting points touching each other. Although it started as adhesive wear, the metal that breaks can also become abrasive wear.

Erosive Wear: This type of wear happens due to mechanical interaction between a surface and a fluid, a multi-component fluid, or impinging liquid or solid particles. Erosive wear occurs by the impact of particles upon a surface with a resultant loss of material due to fracture. This can occur both on metals and ceramics. The continuous bombardment of particles, either solid, fluid, or air, contacting the surface cuts a tiny particle from the component over time. This tiny particle that is removed is insignificant at the moment but, over some time, will have its impact. Erosive wear can be expected to occur in metal parts and assemblies where the above conditions are present. Common problem areas found are pumps and impellers, fans, steam lines, nozzles, inside of sharp bends in pipes and tubes, and sand shot blasting equipment. Sudden sharp curves or bends cause more erosion problems than gentle curves in tubes and pipes. One way to control erosion is to strengthen the material's hardness, which can prolong the erosion process, but again can subject the metal more prone to a brittle fracture.

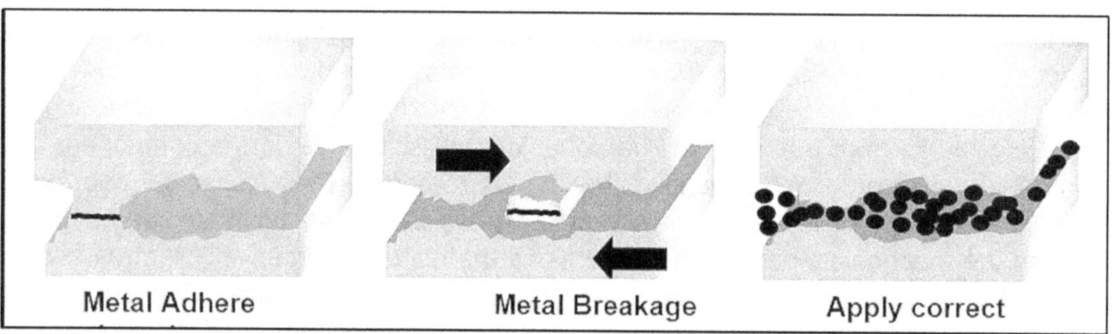

Figure 5.11: How Metal Adhesion Takes Place

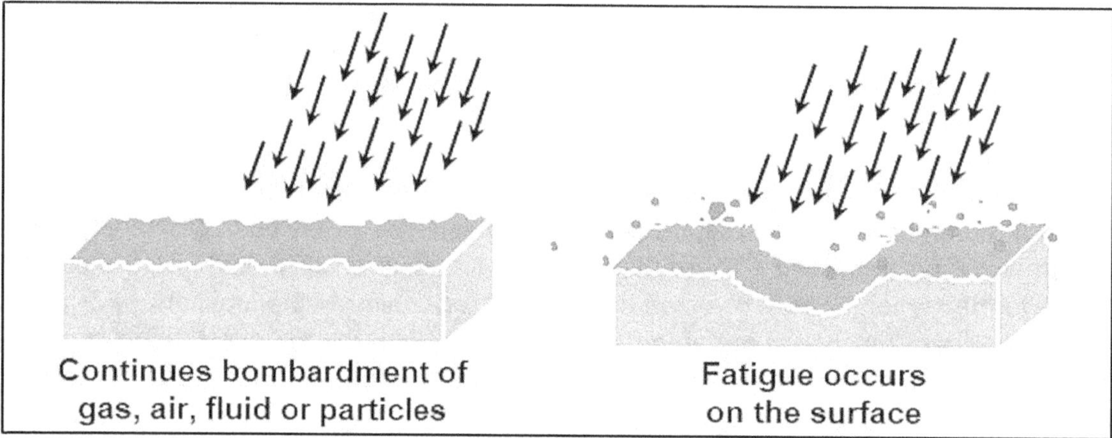

Figure 5.12: The Process of Erosion

Fatigue Wear: Fatigue wear results from continuous cyclic slip under repetitive load applications for many thousands or millions of load cycles. These removed fragments caused abrasive wear and other damage when carried by the lubricant to other parts of the mechanism. Some fractures will start as microscopic cavities and may stay microscopic if the equipment is no longer loaded. Still, it will start off as a microscopic fracture most of the time but gradually become larger after continuous load. The last stage is a large fracture or crack. Fatigue wear is also a phenomenon leading to fracture under repeated or fluctuating stress having a maximum value less than the material's tensile strength. Fatigue fractures are progressive, which can start as micro cracks and propagates and grow into larger ones that cause the destruction of mechanical components. For fatigue wear to happen, a contaminant or particle is caught between clearance, and the surfaces are dented, and crack will be initiated. After N number of cycles, the particle trapped is released, and the crack starts to propagate. The last stage will be a fracture of the component.

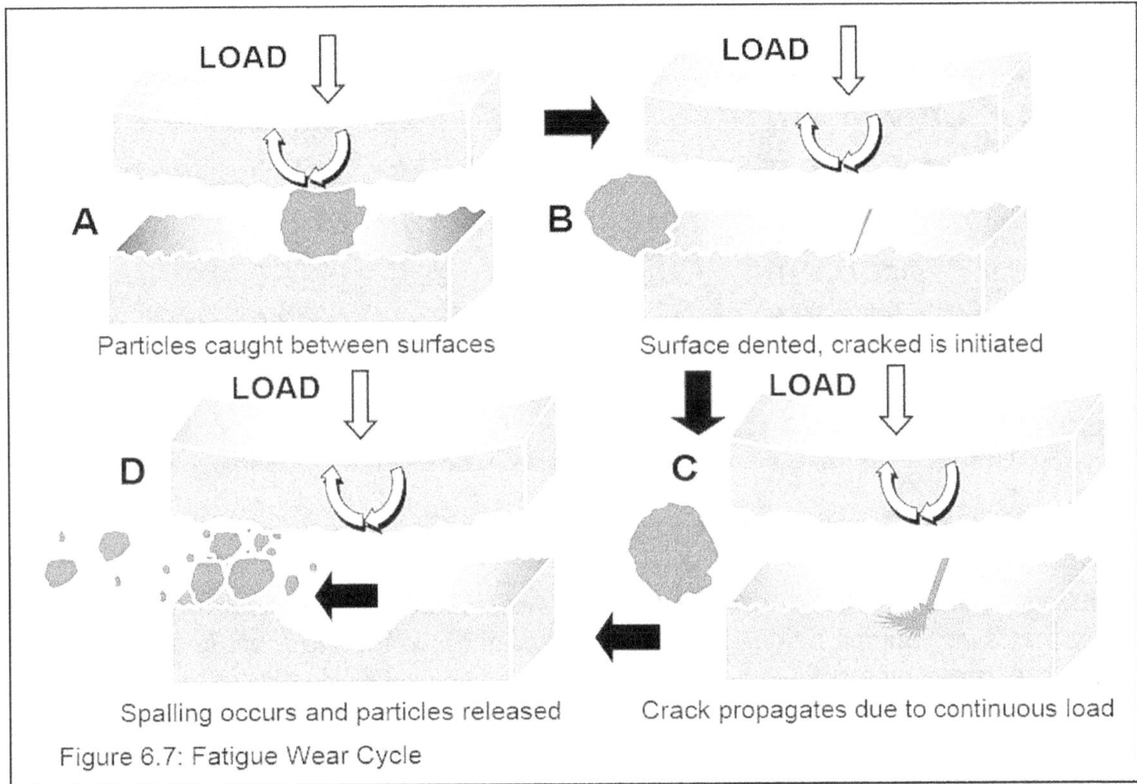

Figure 5.13: The Process of Fatigue Cycle

2012 RSA Reliability Newsletter Vault Archive

> *We can only learn from failures if we take the time to analyze them. It is an opportunity for us to begin more intelligently. Whether from the lessons of life or from the industry's problems, failures will tell us something. It tells us that to be fast, we need to take things slowly. Fixing the symptoms will just cause a recurrence or repeat of the same problem.*

6.1: January 2012: Condition-Based Maintenance Explained

A person is gifted with five senses: the sense of smell, touch, taste, hear, sight. While some may say that every human being has 6 senses and the sixth is the most important of all the senses, which is common sense, but nevertheless, we can use these senses to detect problems on their equipment. Similarly, Condition-Based Monitoring will check equipment condition through sophisticated non-destructive measuring instruments with high precision accuracy. Predictive Maintenance instruments are simply just a higher form of the human senses. For instance, if we talked about the sound spectrum, then what we can hear is simply those from the audible or sonic range, from 20 hertz to 20 kilohertz. Lower than these frequencies will be the infrasonic range, which some animals can hear, such as elephants. On the other hand, higher than 20 kilohertz will bring us to the ultrasonic level where some animals can hear these frequencies range such as dogs, bats. No human being can hear both the infrasonic and the ultrasonic range without some form of instrument.

On the other hand, if we speak about our sense of sight, we can only see a fraction of light in the entire electromagnetic spectrum, which is the visible light. These are the colors of the rainbow. Again, no human eye on the planet can see an x-ray, ultraviolet rays, radio waves, or infrared image without some form of instruments to use.

Condition-Based Maintenance (CBM) basic idea will answer the question, should maintenance or replacement be carried out on a piece of equipment or not? The answer here is that if the equipment is in good condition, then it should remain in the running and in service, while if it is in a slightly degraded condition, then maintenance or replacement must be carried out or planned before the failure itself to prevent an unplanned breakdown from happening. Condition-Based Maintenance is carried out because not all failures can be prevented. So when the failure is random, some of these failures will emit some form of warning that they are in the process of occurring. These failures will be good candidates for Predictive and not

Preventive Maintenance.

According to Miriam Webster Collegiate Dictionary, Predictive means to declare or indicate in advance, especially foretell based on observation, experience, or scientific reason. It is extracted from the Latin word pre (before) and diction from the Latin word dicare (to proclaim). Since our goal in maintenance is to keep our physical assets in an existing state of condition, prediction is a declaration in advance that something is going to happen, and from the dictionary, Predictive Maintenance is a proclamation or declaration in advance based on observation to preserve something from failure or sustain it against danger. Therefore, Predictive Maintenance is a maintenance activity that indicates where a piece of equipment is on the critical wear curve and predicts its useful life.

CBM combines diagnostics and prognostics to predict critical equipment failures long before they occur through sophisticated NDT or Non-Destructive Testing, carried out by highly trained and certified people from maintenance. While most maintenance practices have been Time-Based, with components being replaced based on the number of hours or elapsed time, which often results in inexpensive parts and labor costs due to unnecessary replacement of parts that are still working perfectly. Since Time-Based Maintenance or Preventive Maintenance adheres to the principle of JIC, which is just in case, it might fail, the decision is to replace to comply with their routine overhauling and replacement activities in their equipment.

Predictive Maintenance is a maintenance activity geared toward indicating where a piece of equipment is on the critical wear curve and predicting its useful life or is on the verge of failing. This is done with the aid of specialized diagnostic instruments. Although some industries termed or call this Predictive Maintenance. It is almost similar to Condition-Based Maintenance, but there is a slight difference between the two as Predictive maintenance tends to be equipment-oriented while CBM is system-oriented. CBM deals with the entire system as an entity. This means that when we consider the instrument and the activities involved, we speak about Predictive Maintenance.

On the other hand, if we speak about the entire system, the organization, the software involved, the monitoring, and all this stuff, Condition-Based Maintenance is most likely used. I thought you might know the difference just in case one day some of you might belong to this group so that you can decide what to place on your calling card. John Moubray, in his book RCMII termed this activity as On-Condition Tasks, which entails any activity about monitoring the condition of the equipment. He also indicated that human inspection and reading of gauges are also part of On-Condition tasks because they are used to detect the condition of the equipment.

The role of CBM is to predict intelligently when maintenance is required by checking the condition of the equipment. Besides, CBM applications do not only monitor the condition but can also be used to diagnose the problems and, most importantly, to identify the causes and suggest corrective actions. PdM inspectors are merely more than just data collectors. They should analyze the readings and recommend a solution to potential problems. For example, a vibration analysis indicating a low frequency can indicate misalignment, while a high-frequency reading can detect an early bearing failure.

Preventive Maintenance is when you change the oil in your car every 5000 miles, whether it needs it or not, while Predictive Maintenance is when you sample the oil from time to time and check for any changes in its characteristics. This is possible with Oil Analysis. You found out that you need to change the oil more often since the TAN is more than 2. Proven PdM Techniques increase reliability and decrease the costs of unexpected failures. Predictive Maintenance does not extend life, while the maintenance team's decision will be the contributing factor that will extend the life of the part or asset they are monitoring. For example, doing an infrared scan does not extend the life. The scan may show a hot connection; cleaning and retightening the hot electrical connection due to the diagnosis will extend the asset's life. Predictive Maintenance allows maintenance to decide by peeking inside your components and replacing them just before they actually fail. We are not basing our maintenance decision on the schedule time interval but rather on our components' actual condition. It will be the maintenance and reliability team's decision that extends the part's life being monitored. These non-destructive instruments will just tell us if there is something wrong with them or not. This also allows the maintenance to save on cost because, unlike Preventive Maintenance, only parts with problems will be replaced. Since we are now basing our maintenance on its condition rather than going on time-based or running hours.

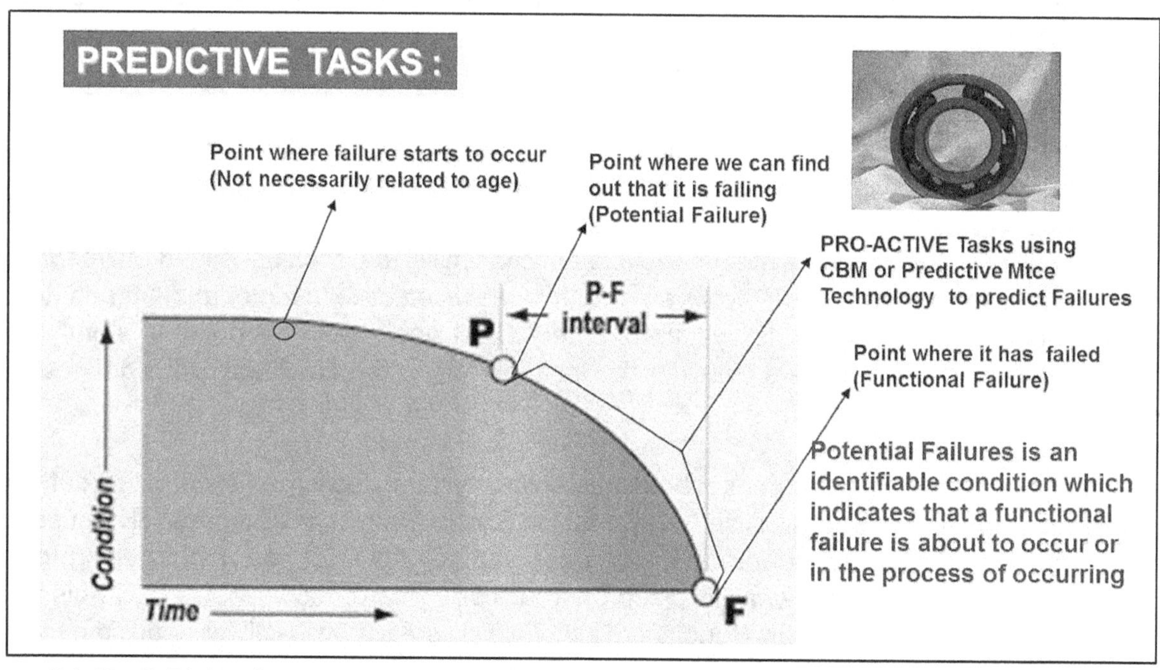

Figure 6.1: The P-F Interval

Condition-Based Maintenance tasks will entail checking the equipment for any signs of potential failures. Action can be taken to prevent a functional failure or avoid the consequences of a functional failure. What we are after in any Condition-Based Maintenance is the P-F Curve. The P-F curve is the interval between the emergence of the Potential Failure and its decay into a Functional Failure.

Potential Failure is defined as an identifiable physical condition that indicates that a functional failure is either about to occur or is in the process of occurring, while a functional failure is defined as the inability of an item to meet a specific performance standard. In short, if we speak about potential failure, some parts are providing symptoms that they are nearing their rupture point, while if we speak about functional failure, then the failure has actually has taken its toll. We can say from a standpoint that the P-F interval is the warning period or the lead time to failure, or simply the failure development period.

Many equipment failure modes will give some sort of warning that they are about to occur. The P-F curve will show how a failure starts, deteriorates to a point where it can be detected (Point "P"), and if it is not detected, it will deteriorate at an accelerating rate. The P-F Interval is actually the interval between the occurrence of a potential failure and its decay into a functional failure. Predictive Maintenance aids us in determining the potential failure or symptoms that equipment is in the process of failing. Changes or increases in the following include an increase in vibration, increase in heat, temperature, pressure, noise, flow rate, corrosion, oil contamination, or any changes in its characteristics such as resistance, conductivity, etc., denotes a potential failure is in place.

While the frequency of performing Preventive Maintenance or its interval in undergoing PM activities is the same and constant. The frequency of performing Predictive Maintenance is different; if the equipment is within the range, then the next frequency will be on a quarterly or monthly as indicated by the user, while if there is a slight increase in the reading which is being performed every quarter, then the frequency or interval is shortened from quarterly to monthly. If the reading still continues to increase, then the monthly frequency of monitoring will be changed weekly until the CBM team's recommendation takes place to stop the equipment for repair. The good thing here is that operations people are advised in advance, and the consequences of neglecting their recommendation will also to a failed equipment which will be certain to happen.

Predictive Maintenance is never a replacement for Preventive Maintenance. PM will always have its place in the overall improvement strategy of Maintenance Management. What we are saying here is that there are instances that checking the condition of the equipment is a much better strategy than overhauling the equipment itself. And there will be times that overhauling the equipment will induce infant mortality or start-up failures. A good CBM strategy is not about monitoring failures through Predictive Maintenance instruments, but rather information from Predictive Maintenance can also help determine the real cause of the problem.

Sometimes what I cannot comprehend in my thick skull is most maintenance know that Predictive Maintenance is a must-have instrument and not a nice to have in every industry where there are assets to maintain, but not every industry has one in place. Others simply do not understand the value of having these instruments because their thinking is that we have survived the plant for many generations, and we used to do our maintenance this way, so why bother to have these CBM or PdM instruments? And they also cost a lot of money. But on the other hand, every international airport on the planet has a form of infrared thermography, which they used to scan people walking on the aisle before entering immigration. I believe that it started during the outbreak of the SARS virus. As far as my small brain is concerned, infrared thermography is not a foolproof device for capturing the SARS virus. What it will capture is a person with fever since a temperature change will be read by the instrument, and the person with a fever will be isolated. Since the SARS virus can be transmitted through the air, then during the first few minutes or hours a person is transmitted, then infrared thermography cannot simply capture this one. Suppose you speak about its utilization of these instruments in international airports; in that case, it depends on how busy the airport is, and almost all international airports utilized it 90 to 100% of the time. But if these airport people really understand the true value or use of these instruments, which is not only to detect people with fevers but to check every single electrical connection, panels they have.

I sometimes wonder how come all airports have infrared thermography while most industries do not have one, which I know that industries will benefit more from this instrument rather than detecting people with a fever at the airport. I hope that one day, industries will find the value not only of infrared thermography but in the use of other Predictive Maintenance instruments and invest in it for their own benefit and transition them from a reactive to a more proactive mode of doing maintenance in their equipment and assets

6.2: February 2012: Is OEE A Perfect Measurement?

The primary measure of TPM is OEE or termed as Overall Equipment Effectiveness, not Efficiency. According to TPM books, OEE has three components: availability multiplied by the Performance Rate multiplied by the Quality Rate and expressed as a percentage. An availability of 95% multiplied by a Performance Rate of 90% and a Quality Rate of 99% (OEE will be 0.95 x .090 x 0.99 = 84.65%) will provide a value of 85%. According to TPM books and other consultants, OEE should be at least 85% or more considered World Class. OEE indicates the relative productivity of a piece of equipment compared to its theoretical performance. It identifies bottleneck equipment critical for the improvement of equipment productivity.

When utilization or availability is low, the equipment may be suffering from either the following equipment losses: a breakdown, set-up adjustment, conversion, cutting tool change, start-up losses, and shutdown losses. These downtime losses will definitely affect the uptime of the equipment. When the equipment's Performance Rate or Efficiency is low, the equipment may be suffering from either a lot of idling or minor stoppages or design speed loss. Lastly, for equipment that manufactures products, if your equipment suffers from many defects, quality issues, re-runs, reworks, and something of this sort, the equipment's quality rate will be low. What is important is trying to analyze your OEE is not only to provide the single number of OEE in percent but rather to break it down into the three components, which are availability, performance rate, and quality rate, so that we can analyze the weakness of your equipment and determine the corresponding equipment losses the equipment is suffering. This is how OEE is being analyzed and interpreted, and it is really quite challenging to improve the OEE of your equipment. So, let me start off with the simple ones first so that the reader can understand and grasp what my point is all about.

First question: If the equipment availability is 100%, is the equipment said to be at its peak and reliable? Answer: No, the equipment may be 100% available but may have a low utilization rate or may not have been utilized; after all, availability is not a measure of reliability, but it is a measure only when the machine is available. Suppose the equipment is running but is not being utilized; the availability is 100% while the utilization is 0%.

Second question: At this point, let us assume that if both the equipment availability and utilization are 100%, is the equipment said to be at its peak and reliable? The answer is still NO. The equipment may be 100% utilized and 100% available, but the machine may suffer from errors and speed loss. Example: The speed of a candy machine producing 1,000 units per hour is reduced by the maintenance to 800 units per hour since running at 1,000 units per hour provides many quality problems on the product. A pump have a rated capacity of 1000 gallons per minute, but over time, the impeller will start to erode as it is in direct contact with whatever it is discharging, so over a given period, the rated capacity drops to 920 gallons per minute. Still, the pump or equipment is utilized 100% of the time and is available for use since it is running. Therefore, when the machine speed is reduced, it still runs smoothly. Hence, if the machine is loaded, it will give us a high utilization rate, but the efficiency will be low, or the performance rate of the equipment will suffer.

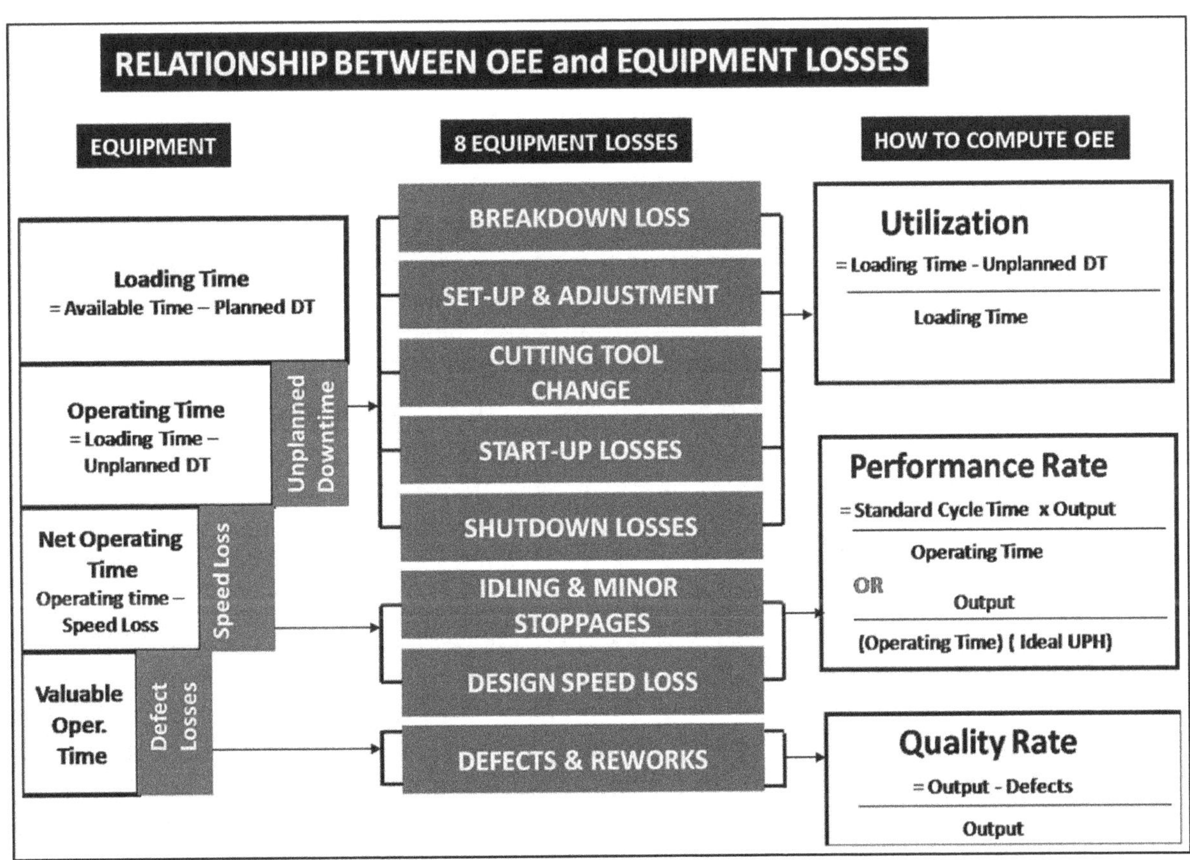

Figure 6.2: Relationship Between OEE and Equipment Losses

Last question: When you consider the efficiency and quality rate, then this is where OEE will come in. Now let us proceed with OEE. If the equipment has an OEE of, say, around 98%, is the equipment peak and reliable? Answer: Not exactly, for two reasons. First, the real objective of OEE is not about achieving 85% or more, but OEE is about satisfying the three components. In the case below, the equipment suffers from many defects, which is simply not acceptable. The equipment utilization here is perfect, so the Performance Rate has a high rate, but the Quality Rate indicates that the equipment is producing many defects in the product; hence, there is no need to celebrate, and we are just fooling around with the data.

OEE = 100% (Avail.) x 97% (Performance. Rate) x 94% (Quality Rate) = 91.2%

Every single piece of equipment has more than one function. For manufacturing industries, the primary function of the equipment will have something to do with producing products. This means that if a certain part of the equipment is not functioning, then we have some forms of function reduction breakdown in our equipment or simply some of the secondary functions of the equipment are not capable of working at all, and there are cases where the failure of secondary functions are much more important than the failure of the primary function. Let me provide you the following cases to illustrate my point:

Case 1: Molding Machine with an OEE of 96%: In a semiconductor industry that manufactures IC where I used to work many years ago, there is a Mold Station. Molding is the process of encapsulating the device in plastic material or molding compound that will be heated to cover the chip to protect it from moisture. The molding compound is heated until it melts and flows into the cavities, containing the lead frame for encapsulation. During my time in this semiconductor, we have lots of this equipment with different types. This molding equipment produced the highest OEE in the plant, 92%, 94%, some reaching up to 97%; since we were on the TPM Office, we were very happy with the results. One day, my boss told me to make a surprise audit on one of the molding equipment since he has some doubt on the integrity of the data because it was always high, and so I went to their station to observe if it gave us the true data on OEE. Obviously, the equipment performed very well with no failures and breakdowns, and there was not even an operator in the equipment since it is running smoothly. I observed for an hour until I got bored and returned back to my office. The following day, I went again and observed for an hour. Still, the molding machine was running perfectly until I asked the maintenance to open the molding machine's lower compartment. I was shocked by what I found out. The molding machine was flooded with oil. And there was a leak all over. They never addressed the leaks, but their OEE was great. This means that this machine with a high OEE produces output perfectly but fails in its secondary functions.

If you have a car, the primary function of a car is to take you from one place to another, but we do not just make our judgment on purchasing the car based on the primary function since we also need to consider its secondary function. This means that a car may be capable of traveling, but the car's headlight is busted, or the car which is supposed to have 4 doors have only 1 door, or the car's wiper is not functioning. These are called secondary functions. Like equipment and machines, they also contain many secondary functions and what is important is that both primary and secondary functions must be working on the equipment.

Case 2: Marking Machine: Also, in the semiconductor industry I previously worked, after molding the IC package for encapsulation, the next process would be either to go to the Deflash Station, plating, and then marking station. Anyway, if you have seen an IC, there is a mark with numbers, letters, and so on that, you can see on the package. This is the black thing with legs that you see on electronic boards. This process is done in the Marking Station. Previously, most of the marks used were white ink, but nowadays, most marks are done using a laser. We have several Marking Machines, and during those times, the top-of-the-line Marking Machine was a Dongyang Marking Machine. Although the marking machine never obtained a high OEE than the Molding machine, let me explain the importance of the equipment's secondary functions. The IC package we used to produce was a PLCC package. It was square in shape and of different sizes depending upon the number of leads it has. The more leads, the larger the package. This Dongyang machine marks different packages such as 28 leads, 32 leads, 44 leads, etc. The IC package is still in a strip form, which means that it is still not cut. It is grouped in sets of 10 or more depending on the package; marking is done per stroke for every strip or group of 10 IC's. Since the package was square, the mark's orientation is critical; hence, before the operator can place the strips on the Dongyang machine, the operator must check the strips randomly in the tray. One day, the production department received a love letter from one of their customers. It was a customer complaint. We call it CC, which stands for customer complaints, and actions are taken to correct this issue are called CCART or Customer Complain Action Response Team.

A cross-selection of engineers and maintenance was formed to address the issue. The operators checked the strips 100% for our containment plan until a more suitable action was derived. The equipment manuals were checked, and the engineers and maintenance noted that this equipment is designed to have a sensor that can detect misoriented strips. If the IC package orientation was not correct during loading, the equipment was supposed to stop automatically, but when they tested it by loading some dummy strips and deliberately placed some misoriented strips on the equipment, the equipment continued running. The sensor was functioning, but it was not detecting the misoriented strips because it was clogged with dirt. So they placed a routine cleaning and inspection for this sensor, but it was very late since the customer walked away and placed his business elsewhere. It was only a small piece of sensor, a failure of a secondary function of the equipment, yet the impact of this failure was a complete loss of business for the company. There are several cases where the criticality of the secondary function poses a much bigger threat than the failure of the primary function of the equipment, and OEE will not capture those failures of the secondary function since OEE is only designed to measure the primary function of the equipment. This means that for as long as the equipment is utilized with high efficiency and few quality rates, then the OEE is high even if your equipment suffers from many leaks and half of its sensors and protective devices and gauges are not working. My message here is that both the primary and secondary functions of the equipment must be satisfied. A pump can provide a discharge rate based on its rated capacity, but the pipes are terribly corroded in such a way that leaks are present or the tank itself is leaking.

From the book of John Moubray RCMII, he quotes: There is often a tendency to focus too heavily on primary functions when assessing maintenance effectiveness. This is a mistake because trivial secondary functions often embody bigger threats to the organization if they fail rather than the primary functions of the equipment. As a result, every function must be considered when setting up the maintenance effectiveness measures and targets from page 302 of the book Reliability-Centered Maintenance by John Moubray.

The OEE, as defined above, only relates to the primary function of any asset. This is misleading because, as in the case of the gasoline storage system, every asset, machine tools included have many more functions than the primary function, and each of these functions has their own unique performance standards; consequently, OEE is not a measure of the overall effectiveness at all but only a measure of the effectiveness with w/c the primary function of the asset is being fulfilled. From page 304, Reliability-Centered Maintenance by John Moubray. So think again, if you have equipment with a high OEE yet failed in many secondary functions, then you are in a dangerous situation as there might be implications or consequences and are just waiting for the right time to happen.

6.3: March 2012: Who Should Manage Your MRO Spare Parts Storeroom?

In connection with my reliability newsletter issue for September 2011 regarding the Top Problems on MRO Spare Parts and Storeroom, out of 57 respondents who had participated in the survey, 23 voted that MRO Storeroom is not managed by maintenance. I strongly believe that there is a strong connection between this problem and most of the Spare Parts Storeroom's problems. For the reader's benefit, the storeroom can contain many items so let us classify the items inside your storeroom.

Raw Materials Inventory: Raw materials are inventory items used in the manufacturer's conversion process to produce components, subassemblies, and finished products.

Work-in-process (WIP) Inventory: These are made up of all the materials, parts, components, assemblies, and subassemblies that are being processed or are waiting to be processed within the system.

Cleaning Materials Inventory: These are needed for sustaining and maintaining the cleanliness of the plant and its facilities, comfort rooms, toilets, office spaces. These are mostly used by janitorial services, which are most often contracted by the plant. Facilities supplies for utilities/facilities such as lighting bulbs, fuse, and other supplies used by the utilities/facilities department to maintain the plant's facilities are sometimes joined here.

Office Supplies Inventory: These are supplies ranging from printer ink, pencil, ball pen, bond paper, folders, envelopes, scissors, stapler, etc., used by different departments in an organization, mostly in offices.

Finish Goods Inventory: This reflects the number of manufactured products in stock available for customer purchases. On an income statement, the finished goods inventory is considered an asset to the company. These products will either be shipped to their clients, picked up, or ready to sell in the market.

Safety Supplies Inventory: These are regularly being supplied to employees, such as gloves, hard hats, goggles, safety shoes, overhaul dress, hearing aids used as part of the plant protocol in conducting their regular work routine at the plant, mostly provided to operators and maintenance people working in industries.

MRO Spare Parts Inventory: Maintenance, repair, and operating supplies, or MRO goods and spares, are used to support and maintain the equipment used to produce the production. In the survey provided in my September 2011 reliability newsletter, 23 out of 57 respondents stated that their MRO Storeroom is managed by other departments in their plants such as Purchasing, Accounting, Administrative, Warehouse, or other departments not by the maintenance department themselves. This means that either the storeroom people report directly to them or are staffed by people from these departments. So my question in mind is, how can we manage something that we do not control? The answer is you simply cannot.

When the storeroom people report directly to Purchasing, they believe that these are the people who are directly involved with the vendors and suppliers and talk to them directly and make the calls and place orders on the parts needed. While these people can maintain the integrity of supplier and customer relationships with no favoritism among vendors. They have knowledge of the holding cost and purchasing cost of the parts, and they know when the best time to place an order on the part and the correct quantity as based on some EOQ calculation (EOQ stands for Economic Order Quantity). Purchasers establish company-wide transactions and award contracts to vendors and suppliers, and to ensure that each facility gets the most from the contracts, a purchasing manager may finally decide that they are the best people to handle the MRO spare parts storeroom.

The downside here is that purchasing people always go for the lowest bidder and sometimes sacrifice the spare part's quality. The original spare may cost a higher than a replacement part, but it will last longer when used on the equipment. For example, a 3-micron bypass oil filter may cost around 300 US dollars with a beta rating of 1000 and an efficiency of 99.9% guarantee in removing contaminants 3 microns and above, but since they might think that this filter is expensive and might go for a nominal rating of beta 2 or efficiency of around 50% since the cost is 50% lower than the high-efficiency rating filter Since purchasing have no knowledge about the beta rating and efficiency of the filters, then they go for the lowest bidder. Because of this, maintenance is always on the battlefield troubleshooting some hydraulic problems on their equipment. True and meaningful savings can only be realized if the part's life cycle cost had been considered. Also, since there are many instances that a part is supplied by different vendors, then most likely, these same parts will have different part numbers in the storeroom, so purchasing can have some control over the different vendors, and the part might be located on different shelves. Storeroom people staff might not have knowledge that these parts are identical, but their part number is different. Suppose the new part number had not been communicated with the storeroom and maintenance people; a zero inventory might reflect on the system on the original part number, not knowing that the same part exists but with a different part number. Purchasing people may actually think that they have been saving the company some cost, but the truth is that they add more problems to the maintenance.

In other industries, the storeroom people report directly to their Accounting or Finance Department. And these people are the ones managing their storeroom. They note that proper store inventory control is important. Since they manage money matters and manage their company's finances, an audit might occur from time to time to reduce corruption's tendency to occur during the transaction with vendors. Since most of these people are accountants and non-technical people, they may be good at some money matters but might lack technical knowledge on what part or spare work best for the equipment. I firmly believe that they are not the best people to manage the storeroom. Their goals and objectives might not be the same as the maintenance people. Both this department might think that the best way to lower the cost is to go for a cost-cutting scheme, while for the maintenance people, the best way to reduce cost is to improve the life cycle of the part and focus on the reliability of the equipment because if the reliability increase then costs will go down, and it cannot be the other way around. Buying the cheapest oil, for example, saves these departments a few dollars, but maintenance keeps on adding oil to their equipment because the flashpoint of the oil is low, and the volatility rate is high.

Remember that both these departments work on a 9 to 5 basis and may have their day off on a Saturday and Sunday. Assuming a failure happens at 10:00 pm in the evening or on the weekend, maintenance waits for the storekeeper to arrive, prolonging the downtime duration. What I believe is that most of the problems in the MRO spare parts storeroom are, in a way is connected to who is controlling or managing the storeroom. Once the MRO storeroom users lose their confidence in the storeroom people, expect maintenance to source the part themselves, buy them in excess, and keep the remainder in their secret hiding places and for squirrel stores to exists.

Some industries attempt to operate with no MRO spares management inventory control; parts are ordered on a need base. Alliances and partnerships with their vendors, suppliers, and consignment are some of the strategies deployed by these departments to reduced inventory in their storeroom. However, since the parts do not arrive fast enough for maintenance to perform their routine work and repairs, maintenance is often pressured by operations people to fix the equipment in the shortest possible time and get the parts in any way possible. Maintenance thinking is that if the part needed will not be there in the storeroom, I need to get hold of the part and keep it myself. On the other side, the purchasing and accounting people are now satisfied that their inventory had been reduced dramatically, but the real truth is that they just do not know that most of the spare parts needed by maintenance are kept in the maintenance bay or within themselves and is no longer accounted in the books because maintenance has already lost their trust and confidence in their storeroom. A well-managed storeroom and spare parts mean that all parts that go in and out of the storeroom should be known and accounted for.

Because most material transactions involved maintenance parts, materials management, and storeroom, people should report to the Maintenance Department. If the storeroom people and materials management report to other functions such as Purchasing or Finance, there is a greater possibility that the objectives will not be aligned with those of both operations and maintenance. Purchasing and inventory management functions should be separated because the skills required to be a good purchaser or negotiator are different from those required to be good inventory management. In a good storeroom, people should not only have knowledge of the inventory of the part but of the spare part itself. Both purchasing and accounting goal is to reduce the cost of operations of the plant while maintenance and the goal for maintenance people is to improve the reliability of their equipment and many cases reducing cost mostly on spare parts will have a harmful and detrimental effect on their assets and equipment. But that is fine with these departments rents because they have saved a few bucks by sourcing the cheapest bidder or vendor, while on the other side, maintenance suffers as a result.

Maintenance people are the MRO Storeroom; hence they should be the ones to manage the spare parts. I know I am biased when I say this, but the largest and the most common de-motivator for maintenance are waiting for the materials, and worst if the parts are not in the storeroom or are not around when they needed it most. And since the part was not around when maintenance needed them, then expect maintenance to make a McGuyver move, which they know is not right for the job. They do this because there is pressure from the operations people to keep the equipment running at all costs even if the part is not in the storeroom. In short maintenance, people try to cannibalize some parts from idle equipment to make the equipment run, which had been part of their day-to-day lives. Maintenance people must finally convince their Top Management and decision-makers that they are the best people to manage and control the storeroom. Savings can be generated from the MRO storeroom if the right people manage the storeroom. Some of the parts inside the storeroom are termed as obsolete parts since the equipment was already decommissioned, and if these spare parts had not been communicated with the storeroom people and a system for reordering is in place, then expect a high chance that these obsolete parts will still be reordered and being held in the storeroom even if their usage is already extinct.

Spare parts management is one of the strategies needed to improve the reliability of the equipment. When the plant finally decides to improve its assets' reliability, the spare parts storeroom should likewise be improved. If a certain part or spare's lifespan had been increased, the reordering quantity would definitely be decreased, and the interval of placing the call to make an order will likewise increase. When predictive maintenance is in place, then the part can only be reordered once a potential failure can be detected on the equipment since they have some knowledge that the delivery time of the part is shorter than the failure development period. Obsolete parts are finally being taken care of and decided on the best course of action. ABC analysis of their non-moving parts can be analyzed. Their physical and systems inventory is 95% or more accurate all the time. Parts are properly cataloged. Spare parts with different part numbers are now being consolidated and made known to every maintenance and controlled storeroom. Their MRO spare parts are now operated 24 hours because their operations also run 24 hours. Spare parts inventory is low, and this time it is not because maintenance is keeping the parts to themselves, and the most important part is that the maintenance organization have regained their faith, trust, and confidence in the storeroom people because they are now united with the goals of the maintenance side.

Remember that maintenance people are the humble users of the MRO storeroom. They know the parts better than anyone else on the plant and what parts work best for their equipment. They know which part will keep on failing as well as the lifecycle of parts they replace frequently. Spare Parts Management can only be managed if the right people are the ones managing the storeroom. And so I think that is all I have to say about that.

Every time I conduct my class on MRO Spare Parts and Storeroom Management, this survey will be answered by the delegates, and at the end, I would sum it up and update my materials for the next class. Although the figures are not yet conclusive, I would like to give it some time to arrive at a more meaningful figure.

- Warehouse/Supply Chain: 24 votes
- Maintenance Department: 55 votes
- I Don't Know: 1 vote
- Number of respondents 80

6.4: April 2012: The Importance of Conducting Oil Analysis

When we are sick, we schedule an appointment with our doctor, and they will try to analyze what is wrong with us. The doctor may recommend drawing a blood sample from us and send it to a medical laboratory to have it checked and analyzed. Upon receiving the blood test result from the medical lab, the doctor renders his medical opinion or advisory that we are free to accept or reject. Similarly, the oil inside our equipment is like our blood. It can dictate its condition through what we call an "Oil Analysis Program." Therefore like blood, the oil contains a great deal of information about the equipment itself. Much useful information can be learned by analyzing the oil inside our equipment. While many industries are still changing their oil on a time-dominated pattern, some industries shift from running hours to a condition-based approach by sending oil samples to an independent oil analysis laboratory. Other industries tend to have their own laboratory inside their plant, which gives them an added advantage. By conducting a regular Oil Analysis sample, it will answer the following questions.

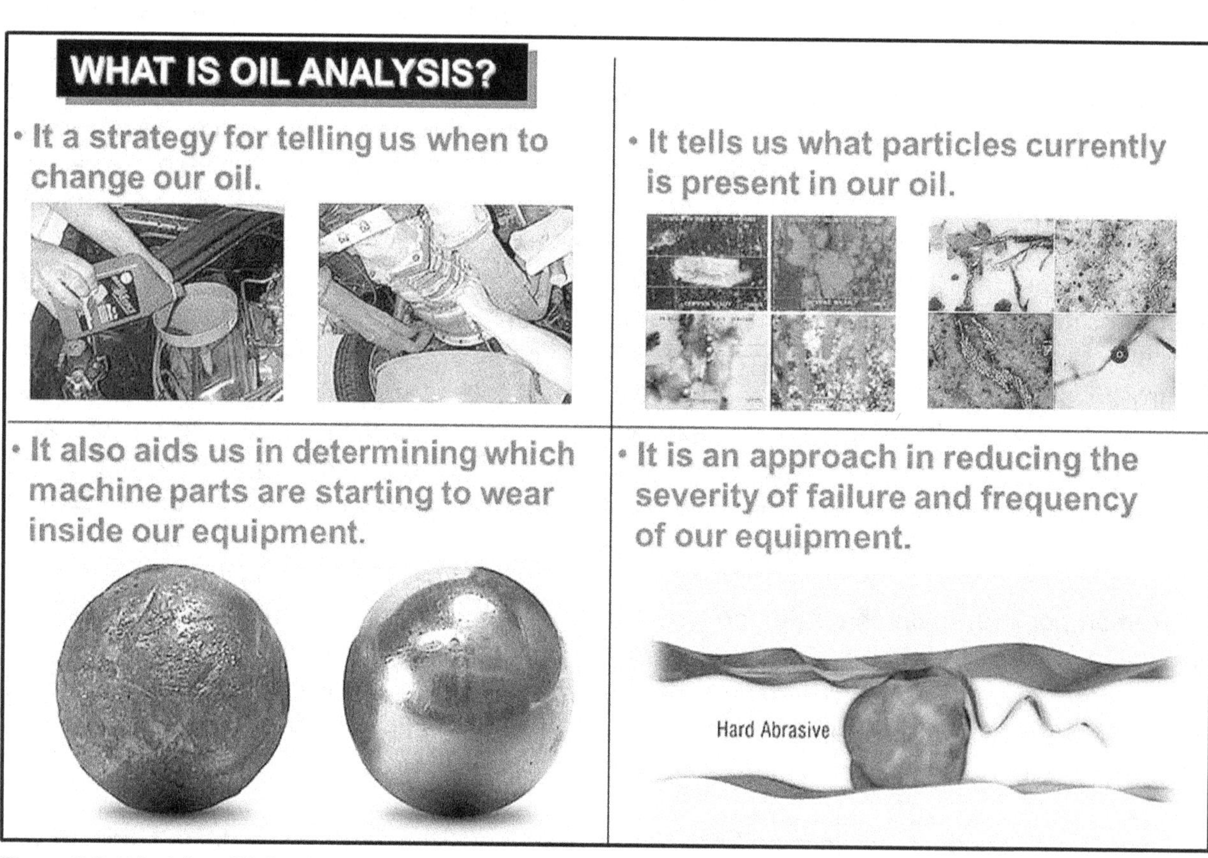

Figure 6.3: What Can Oil Analysis Do?

Is it a strategy for telling us when to change our oil? Yes, by knowing how to interpret the report provided by the oil analysis laboratory, maintenance, and reliability, people can make a sound decision on whether to continue running the equipment or stop it and change the oil.

Does oil analysis tell us when and where machine parts fail? Yes, the wear metal debris analysis test tells us what metallic and non-metallic elements keep increasing over time. By trending and monitoring these elements every time a sample is sent and understanding each element's limits, maintenance can decide what part is starting to wear inside their equipment. By knowing this information, maintenance can advise operations, order the part, and schedule the equipment for repair instead of waiting for it to fail and perform emergency work. In short, oil analysis, like any other condition monitoring instrument, is a strategy that allows the user and maintainer to look inside the equipment without actually dismantling it. The report will provide a crucial and critical information on what is going on inside our equipment.

Does oil analysis tell us what particles currently present in our oil? For example, if we send an engine oil and the report indicates large traces of silicon present and is increasing over time, there is probably dirt ingression happening in the equipment. Many different types of tests can be conducted through oil analysis, and each test will try to look at a different aspect of the oil. Therefore, what is important is knowing the test that best suits your application because this is where the industry can derive the most benefit from its application. Oil Analysis makes it possible to look inside an engine, transmission, gearboxes, or hydraulic system by monitoring the oil condition and determining contaminates' presence inside. It can tell us if the additives in the oil are still present or had already been depleted, or why is this oil been burning so fast? There is a test that can detect how much moisture is present in the oil. Moisture also promotes oxidation of the oil's base stock as well as inhibits rust and corrosion.

An oil analysis program can save us in lubricant and spare parts costs, energy costs, labor costs, and fuel costs. In fleet applications, oil analysis has been applied to determine when additives had been depleted so we can decide to change the oil or not. There are also test that can determine if the wrong oil had been added to the system. There are many benefits to having an oil analysis program for maintenance functions, such as the following.

Prolong the change interval of oil: Oil contains two things, the base, and additives. When the additives deplete, then the oil is no longer fit to be used and needs to be changed. Hence, the best decision on when to change oil is based not on the number of hours or miles the equipment has run but on the number of contaminants present in the oil. An oil analysis program will also suggest methods to reduce accelerated wear and contamination on mechanical parts. Thus, oil now becomes a working history of the equipment itself.

Reduction in maintenance cost: Oil analysis plays a major role in reducing maintenance costs and increasing its life. By improving oil contamination, engines improve its' combustion efficiency by monitoring and optimizing fuel system efficiency, thereby decreasing harmful emissions into the atmosphere. The most traditional way of maintaining our equipment is going on a time-based schedule mode or Preventive Maintenance in which assets, systems, components, and equipment are overhauled on a time-dominated pattern, and parts are replaced even when there is nothing wrong with them just to comply with the scheduled PM. With Oil Analysis, there are actually two ways to save on cost: the amount of lubricant consumed and the amount of reactive maintenance or breakdown maintenance that needs to be attended to. My experience tells me that the cost of doing repair work is doubled or tripled compared to the plant's number of lubricants. Let us say that if your lubricant's cost every month in a mining industry is around $100,000.00, multiply this by a minimum of 2, and that is the rough estimate cost of your reactive maintenance work. The cost of parts that fail, labor, downtime cost is always higher than the lubricant consumed.

Early Detection of Problems: An oil analysis program can provide early detection of problems on mechanical parts. It can anticipate problems in advance and provide maintenance adequate time to inform operations people to schedule repair work to avoid long downtime during its critical time of use. Studies show that failures can be predicted 18 months in advance. By establishing trends from oil analysis, tests can indicate a change in engine or lubricant condition, and the information they provide may be used to correct abnormal conditions before they cause damage or failure.

Evaluating used equipment: A complete record of oil analyses performed can prove to be a great tool when selling or buying a piece of used equipment. It shows potential buyers how you have maintained the equipment and any adjustments you made during its entire life. Also, it can allow the buyer to make sound decisions in buying used equipment.

Determine the frequency of changing oil: Through oil analysis, it can allow maintenance to decide on when to change their oil and determine the correct interval needed for an oil change to determine if it is necessary or not. Since we are now shifting from a running hours mode to a condition-based mode, deciding whether to change the oil will depend on the oil analysis report and not its running hours. If additives are still present and within the target limits, the default decision will allow the oil to run.

So how does oil analysis works? The oil that has been inside any moving equipment or asset for some time reflects the assembly's exact condition. Oil is in contact with mechanical components and parts as wear metallic particles mixed with the oil. These particles are so small that they remain in suspension. Many products of the combustion process also become trapped in the circulating oil. The oil becomes a working history of the equipment. Since particles will mix with the oil, established trends from oil analysis tests indicate a change in engine or lubricant condition, and the information they provide may be used to correct abnormal conditions before they cause more serious damage or failure. These impurities and contamination present in the oil are measured using different tests; hence maintenance has an indication of the rate of wear and contamination present inside their engine or equipment. An oil analysis will recommend methods to reduce accelerated wear and contamination.

I am sure that there are many more benefits derived from implementing an oil analysis program in your industry. Still, the question most maintenance managers have in mind, perhaps if you are reading this newsletter, is how come I still have a lot of crisis. Still, many emergency repairs work in my operations, even if I send oil samples outside an independent oil analysis lab.

There may be more than a hundred reasons why industries do not benefit from their oil analysis program, and first, it will all boil down to training. Before implementing or initiating an oil analysis program in your plant, whether you will be sending samples outside on an independent oil analysis lab or planning to build your own oil analysis "ALL," I mean "ALL" maintenance should be trained and must have knowledge on what it is all about. It is not only the lube man or lab people that need to be knowledgeable about this.

If the oil samples are to be sent outside, kindly note their response time. If the report is provided to you in a couple of weeks, look for another oil analysis lab. There are cases where the failure already took place before the report was provided. This is one benefit of having an oil analysis in-house rather than sending them outside since you can have the results in minutes. You see, when you send your samples outside, your bottle needs to queue or fall in line with the other bottles since most outside laboratories follow a strict pattern on a first-come, first-serve basis. Remember that you are not only the industry on the planet that they are serving, and the oil analysis lab also needs to earn a profit. In oil analysis, there are sensitive tests and less sensitive tests, meaning there is a test that needs to be performed while the oil is still warm, and in my experience, the most sensitive test is the particle counter. Suppose the test is conducted a couple of days after the sample had been sent; the data might not be conclusive since it no longer represents the actual sample since some of the contaminants had already settled on the bottom part of the bottle. Hence, when you have your own oil analysis laboratory inside your plant, you have a tier of importance on which type of test needs to be prioritized.

Another main reason I see for the oil analysis program's failure is setting up the sampling points. The golden rule to follow is that the sampling points should be consistent and done on the same spot or location. A no-no on oil analysis sampling is trying to get the sample from used oil or acquiring a sample from the drain plug. Remember that this is where all the dirt and contaminant are settled, so the oil's actual sample can be declared failed. Kindly note that the lab technician only recommends what to do with your oil and might not know how your equipment actually operates, but we, on the maintenance side, are in day-to-day contact with our equipment. Always use the same sample ports since the oil variation can be seen at different reservoir locations. It is best to place a minimess faucet-like on every sampling port if necessary so the sample's extraction point is consistent. Suppose you have a lubrication system with several components; in that case, the sampling points should be located after the component or part and not before the part to monitor the part itself.

When I visited an independent oil analysis laboratory here in my country, I observed many bottles used. Some industries used a bottle of a mineral bottle; others used a bottle of catsup, mayonnaise, soft drink, a bottle of vinegar, and some even a bottle of beer or whiskey. Kindly note that bottles that are being sold for oil analysis samples should be certified, and most of them are disposable. But the important part is that these bottles have their certification level, clean, super-clean, and ultra-clean. So purchase oil analysis sample bottles that are certified. Bottles used must be free from impurities, dirt, and other foreign materials.

The oil sample must be drawn while the unit is in operation. If that is not possible, the sample must be drawn right after shutdown, so that separation of any particles or water in the reservoir does not occur yet, and contaminants in the oil are still suspended. Suppose you have around 2000 pieces of equipment; it is unlikely that the laboratory technician will be the one to extract all the samples here. Once again, training will play a key role in properly extracting samples from your equipment.

Some industries might be hesitant to have their own oil analysis laboratory, but my advice is that if your lubrication cost is very high, I think you need to make some thinking and reconsider having one. Remember that oil analysis instruments require two investments. These are the instruments, and second is training and justifying the cost; you need to provide a feasibility study, present that to your decision-makers, and answer the most important question, which is the payback or rate of return so that you can convince your top management to have them approved onsite. Make a feasibility study and include the following cost of failures due to lubrication for the past 2 to 3 yrs., your plant's oil consumption costs, which you can easily get from your purchasing department. Explain in layman's terms what oil analysis is and its benefits. What tests and instruments are required and why you need them, the estimated cost of each instrument, where the lab will be located inside your plant, rate of return, estimated savings, success stories from those that have adopted onsite or in-house oil analysis. By doing this, then there is a greater possibility that management can rethink their decision on the importance of having an oil analysis inside their own plant, and I think that's all I have to say about that.

6.5: May 2012: Wisdom on Maintenance Part 2

Let me share this wisdom on maintenance with you, which are the guiding principles of all my training and courses on reliability and maintenance. These reflections are what I believe to be what maintenance and reliability people in industries should be focusing on and thinking. It is never too late to improve the reliability of your plant. Improving reliability is a consolidated effort not only for maintenance but for operations people as well. It is important that people from both operations and maintenance be educated and must have the correct maintenance paradigm.

Figure 6.4: Wisdom on Maintenance Part 1 on Youtube.com
Watch it on youtube.com. Here is the link to Wisdom on Maintenance Part 1,

https://www.youtube.com/watch?v=97BA0kQcrCo

1st Wisdom on Maintenance

- Every maintenance should focus on improving reliability and not on reducing costs; why? Because if reliability starts to improve, then cost will definitely go down; it cannot be the other way around. Remember that there will be times that focusing on reducing costs will hurt reliability, a lesson we should reflect upon. Having a low maintenance cost is always a consequence of good maintenance practice.

2nd Wisdom on Maintenance

- Troubleshooting is no longer an effective strategy. Remember that when your people become really good at repairing failures, then something is definitely wrong with your maintenance organization; why? Because they are doing it much too often, but when we expect a different result from the same things that we are doing, it's just not possible; the Chinese called this "Insanity."

3rd Wisdom on Maintenance

- Big problems start with little ones. The best time to address a problem is when it is small. It is very hard to advance to any specialized maintenance activities and improvement efforts if Basic Equipment Condition had not been well established. Always remember that your equipment remains a shared responsibility for both operators and maintenance people. A lesson we must all learn from the Japanese.

4rth Wisdom on Maintenance

- There is no silver bullet solution or rocket science program and strategy that can transform a plant's reliability overnight, all will start with its basic foundation, and that is through "Education." Remember that reliability is not a program with an end but a culture that never ends; it is the same as any other continuous improvement philosophy.

5th Wisdom on Maintenance

- The distinction between a maintenance and a mechanic is that a maintenance uses his brain more often rather than his hands in analyzing problems, while a mechanic uses his hands more in fixing failures and later on becomes good at it. Let us treat them as maintenance people and not as mere mechanics.

6th Wisdom on Maintenance

- Why industries remain reactive may lead to a thousand reasons or more, and those who feared that improving reliability is achieved by reducing or cutting manpower do not really know what maintenance is all about. Improving reliability means slowly getting out of the repair business so that new doors will open for the maintenance function.

7th Wisdom on Maintenance

- The real challenge in any equipment reliability initiative is improving in a reactive world with the same resources and time. Remember that world-class industries started reactive themselves.

8th Wisdom on Maintenance

• Our equipment is always a shared responsibility for both operators and maintenance working together, so instead of blaming each other whenever a failure occurs, why not work together, not against each other.

9th Wisdom on Maintenance

• Operations can never be maintenance customers. Why? Simple because they are maintenance partners, and they can never be divorced.

10th Wisdom on Maintenance

• The traditional belief leaves maintenance people to think that all parts have a life and that replacing or overhauling them before they fail on a specific period will eventually restore the parts back to their original condition. If you think that this is right, then you are so wrong! This thinking is the main reason why most Preventive Maintenance activities result in being reactive. Itself.

11th Wisdom on Maintenance

• TPM and RCM are not designed to contradict each other, but rather, they complement one another, and they can work together. It is the independent RCM and TPM consultants that do the contradiction and not the methodologies

12th Wisdom on Maintenance

• Maintenance is not a department. It is not an organization nor a function, but rather maintenance are humble and down-to-earth human beings; hence, let us treat them with the respect they truly deserve because that is all they ask for.

6.6: June 2012: The Problem with Predictive Maintenance

Although there are many vendors on Predictive Maintenance (you can see them on LinkedIn) promoting their instruments, software, wireless technologies 24 hours a day. A seldom talked about topic on the subject of Predictive Maintenance is the user's attrition rate. By definition, the attrition rate is a calculation of the number of individuals or items that vacate or move out of a larger, collective group over a specified time interval. Here is the irony behind this, while it is true that Predictive Maintenance will capture more failures than Preventive Maintenance, it is unlikely that a Predictive Maintenance user will stay long in a company just to collect data on the equipment for the rest of their employment days. This means that industries must strengthen their Predictive Maintenance users as well. The question is, how can we do this if the Predictive Maintenance user always leaves your plant after getting certified.

I can still remember vividly during my employment days, we organized a group of 10 people we called the Predictive Maintenance group; 2 people will be handling vibration monitoring, 2 people for thermography, 2 people for ultrasonic monitoring, 2 people for an in-house oil analysis laboratory, 1 female secretary and 1 Predictive Maintenance manager. According to rumors, one of our Predictive Maintenance users, a level 2 certified thermographer, was eating in a mall when all of a sudden, someone approached him and introduced himself. He told the user that he recognized him from a recent convention where he wrote a paper regarding savings generated from the use of infrared thermography. The person handed the PdM user his calling card, which was a senior Vice-President from a big firm here in the Philippines, and after 6 months on the Predictive Maintenance group, he disappeared for good. At the end of just one year, only three people remain the secretary, the manager, and one person from oil analysis who is nearing his retirement. The question here is, how do you select a Predictive Maintenance user who will stay with your company for a long time or perhaps until they retire. My advice is do not get young people, get the Rolling Stones, or those that will die in your plant, meaning they will reach their retirement in the plant. The problem is that if you place a new or younger generation in this case, even if you have a good package of benefits for employees, they are still inclined to leave, especially if they will be offered a much higher salary. These users will definitely leave their plant and look for a greener pasture, most especially if these users are certified on higher levels, say level 2 or level 3. Most users I personally knew are pirated. Even if you provide them a contract to pay for their certification if they leave, if an industry wants them, then the industry that wants them will also pay the contract just to have the user working in their plant. They know that many industries are looking for them.

Turnover for these Predictive Maintenance users, who monitor the equipment, acquire data, and readings on vibration, infrared, and so on, is always high. It is a dirty hot job and gets old very quickly. Industries can do two things to retain them, but still, it is not a foolproof guarantee that they will remain and stay. First, allow them to grow as they increase their skill level. This will keep some of them as they become analysts, but this is not foolproof, and it does not completely solve the data collection problem. The second thing you must do is accept the turnover as normal and create a process to find, hire, and train a replacement. Perhaps one in a million people would want to collect data for the rest of their life. The rest will either want to move up in your organization or get a job elsewhere as an analyst. Perhaps I can also recommend that if a user is certified, he needs to train and pass what he knows to a couple of people in the plant so you can have a backup. Having this as a requirement before signing a contract can provide a continuity of Predictive Maintenance activities in the plant. Remember that if the user resigns, then the equipment will remain idle.

With the introduction of smart sensors that can be permanently be fixed on the equipment, these wireless sensors that are now available and sold in the market today, there is a possibility that we can eliminate the need for manual data collection usually done by these Predictive Maintenance users. These wireless sensors can be linked to a central software automatically without going to the actual location of the equipment. When you compare the cost of hardwiring to a remote data collector and then wireless transmitting of data to that of a Predictive Maintenance user, these wireless sensors will be less costly. This way, we can still predict the failure with precision accuracy, which is far more accurate than the human senses. The wireless sensors that are hooked up in the equipment will provide a 24-hour inline or online monitoring directly to a central control room where data will be directed automatically. This can be possible for vibration, some oil analysis tests such as an online particle counter, but for thermography, you still need a user to get the thermal image of the asset's condition. In some cases, we can install thermocouples and hardwire them through the same data logger being used to acquire these measurements.

6.7: July 2012: Why Maintenance is a Complete Unknown?

In many industries and organizations, the maintenance department seldom exists. Instead, they rephrase or change their organization name to "Equipment Engineering," which is a cool name, I recall being a regular trainer in one of the training firms in the Philippines, and they change my courses from maintenance and reliability training to equipment engineering training because industries prefer to use that term. They simply hate the word "Maintenance" because the word maintenance connotes a negative and evil word to them. They would prefer to use the term "Equipment Engineering," even if not all their people are engineers themselves. But the funny part is that they call their people names such as technicians, mechanics, sustaining group, electrical, mechanical, artisan. I really do not understand up to this point in time if industries understand the role maintenance play in their organization because if they do, then I won't be calling them technicians or mechanics since the function of these people is meant to fix, repair, and troubleshoot equipment. This simply means that a breakdown or failure needs to happen first before they can perform their function, and this is what is happening in many industries. Being reactive creates more overtime for these craftsmen, which I think they like, but on the other hand, they also experience a lot of pressure from the production people, which I think they hate. They aren't proud, after all, to be called maintenance because most industries perceive maintenance as often a neglected word and are mostly considered a cost center since the emphasis is given on output and productivity. As a result, Basic Equipment Condition is often neglected on their equipment.   Today industries suffer from high maintenance costs due to a lack of an effective Planned Maintenance structure.   From the book of John Campbell "Uptime" page 2, the ratio of direct maintenance cost to the total value-added cost is as follows:

- Mining   20 – 50 %
- Primary Metal   15 – 25 %
- Manufacturing   5 – 15 %
- Processing   3 - 15 %
- Fabrication & Assembly   3 -   5 %

Suppose it takes 100 dollars to manufacture a product; in that case, around 20 to 50 dollars will be the cost of maintenance. Recent surveys of maintenance management in different industries indicate that one-third or 33 cents out of every dollar of all maintenance costs is wasted due to unnecessary or improperly carried out maintenance. This results in ineffective maintenance management, representing a loss of more than 60 billion dollars per year or more.

Most industries insist that all breakdowns can either be prevented or eliminated; hence, too much focus is given to doing Preventive Maintenance on their assets. Most maintenance managers admit that even with a sound PM program in their plant, they seem to realize that failures still exist. They often strike without warning or when they are least expected to occur. When this happens, most maintenance managers and operations people discuss countermeasures to prevent its recurrence, and the default is to add an activity in their current or existing PM activities. Hence, the list of PM activities never stops increasing.

And so, if I tell a little story about how they function, it will start from a technician's point of view that all parts will eventually wear out, so a need to schedule equipment will be done for every asset in their plant. But when the schedule comes for a Preventive Maintenance then, PM will be missed out because operations will simply not give the equipment for a PM since their thinking is output is all that matters. Their thinking is that the equipment should be like a TV that when you plug it, then it plays without any further need for intervention. Since PM is missed, then breakdowns increase and strike us at the time we least expect them to happen, and so all maintenance is busy recuperating and troubleshooting their equipment. If a part or spare is affected, most often the part is unavailable at the storeroom, and maintenance is left just 2 options, to buy the part themselves outside and buy in bulk and keep the remainder in their secret hiding places or simply cannibalize other parts on equipment that are not used since operations are now pressuring them to get the line going at all cost. Now once again, the equipment will be running with some Mc Guyver fix. These craftspeople will attempt once again to get the asset for a sound Preventive Maintenance. Still, operations will again deny them of that chance since they need to cope with their backlog. These humble craftspeople will work for long hours, and more pressure will be given to them to keep the asset running like a TV set without any chance of failure. As a result of the massive amount of breakdowns and failures suffered from their assets, maintenance costs increased due to repair, and fixes continue to increase. The funny thing is everyone knows the reason for the high cost of doing maintenance; instead, your boss will question you why your maintenance cost is high, and they want you to provide them a detailed answer for that. Management is now looking for a scapegoat or a fall guy for an escape, and the best candidate are these humble and down-to-earth maintenance people themselves. Morale among maintenance declines, and life goes on. Well, I think that's about it.

In a reactive environment, most maintenance people's work occurs right after a failure or breakdown happens. Repairs and fixes should be done at the quickest possible time. Some planning and scheduling are being done and set in place, but even with the planning, the machine fails without warning; repairs and troubleshooting initiatives are set. This will most likely always be the accepted norm of their culture, and maintenance is praised and gets a pat on the back for being reactive, which is being promoted by following this course. This is what the incumbent maintenance taught to their new ones. They have been doing this since the beginning of mankind; training is a waste of time and money for these industries since they have been existing with this kind of thinking, and this is how they do things in the plant on a day-to-day basis. They do not believe in using any Condition-based maintenance or any other form of non-destructive diagnostic instruments since they think of them as expensive items and some nice to have features for the maintenance function. These industries will not invest in these instruments, but if the part continues to fail, again and again, they have the money to invest in the part or spare itself even if it fails 1000 times or more. This is what I cannot comprehend in my thick skull. The truth of the matter is not all failures can be prevented. Some failures will provide some signs and symptoms that it is on the verge of failing, and they can be predicted with precision accuracy with the aid of these Predictive Maintenance instruments.

The longer you can tolerate the pressure from operations people, the better chance you can stay longer in this industry type. There are many instances where your boss will call you and ask you the root cause of the problem? You just simply cannot state that boss you are the root cause of the problem if you still want to be employed even if your boss is really the root cause of the problem, you need to think of other causes, and the easiest excuse to say is "wear and tear" and for the corrective action, "replace," and the boss will tell you to make sure it does not happen again. The saying goes, if you cannot change them, then just join them.

On the other hand, being Proactive simply means being ahead of the failure itself. When equipment reliability improves, then there will be lesser people working in a reactive or firefighting mode. It does not mean that these people will be retired nor terminated. Top management must understand that maintenance is a diversified and noble profession. These people should understand that there are many positions on Maintenance that can be filed in. When maintenance and reliability improve, then new doors on the maintenance function can finally be open.

The word maintenance is not synonymous with repair or troubleshooting. Suppose you look at the meaning of the word "maintain" in Oxford Dictionary; it simply means "cause to continue." At the same time, Webster Dictionary will tell you that "maintain is to keep in the existing state." Maintain is simply an activity that is meant to preserve something.

On the other hand, maintenance ensures that physical assets continue to do what the users want them to do. With that said in mind, we maintain only the things that the users use and are important to them, nothing less. I have an Erickson old-model cellphone, and my kids always tell me to buy a modern touch screen mobile phone. I used my phone to SMS to call and check the time; that is all I need on my mobile phone; it also has a built-in camera that is no longer functioning. I have never used that function since the day I purchased it, and now it is no longer working, so it really does not matter to me much if that function is not working.

When we set out to maintain something, the questions to raise will be, what is the existing state that we wish to preserve? And what is it that we wish to cause to continue? Therefore, when we maintain an asset or equipment, someone wants it to do something. These users expect their equipment to fulfill a specific set of functions because it is what the users want it to do. And for industries, the question will always be asked why do we need to maintain our assets? From an Operations point of view, we maintain so that operations can push through with production. From a maintenance standpoint, we maintain preserving and taking care of our equipment. But basically, the most important reason why we need to maintain our equipment and assets is that we maintain because there will be times when the consequences of failure are more important than the failure themselves or will far exceed the failure itself, and I hope industries realize this reason for their own benefit and not when it is already too late.

If we can just provide our people with the right amount of knowledge and education on how to do and maintain our assets and equipment the right way, then perhaps we can sleep better at night without any interruptions from the plant asking you to report to work in the middle of the night for a failure or breakdown they cannot fix themselves. I always believe in the thought that all breakdowns are man-made, either from how the equipment was design since there will always be some weak points in the design, the way it was commissioned, the way it was being operated, and mostly the way we maintain our equipment. If man created the problem, then let man provide the solution to the problem itself.

To conclude, changing your organization's name to other names such as equipment engineering or other names to eliminate the word maintenance makes it sound really nice but technically does not really make your maintenance organization more effective or proactive. I will still stick to the word maintenance because this is who we really are. Maintenance should not be a cost center, but it can be transformed into a profit center if both operations and maintenance people truly understand what maintenance is all about and the role they play in it. Maintenance is one area where cost can definitely be saved, and I hope that one day, industries realize it.

6.8: August 2012: Why Operations and Maintenance Went Their Separate Ways?

In almost all or many industries, the saying goes that when production for the day is met, then all credit goes to operations, and if it's the other way around or productivity for the day is missed, then all fingers point to the maintenance, and the maintenance asks, why can't you include us in your celebration even for ones. There is always a feud or friction between these two, and operations' most lame excuse when productivity is not met will be maintenance. They look at maintenance as their scapegoat for not hitting their output. On the other hand, maintenance will retaliate by saying that this won't happen if you only release the equipment for a Preventive Maintenance schedule, and the war goes on with maintenance always on the losing end. Sounds familiar? Believe me when I tell you that I've been there. I have met the most horrible operations managers in my lifetime, and they stink!

- **Operations Sentiment:** If we cannot deliver, then I'll definitely blame the maintenance guys.
- **Maintenance Sentiment:** The problem with you is that you always flock the equipment to death, and you don't allow us to perform our regular Preventive Maintenance on the machine.

Even with the slightest problem with their equipment, these operators would leave everything to the maintenance for fear of making the problem worse, and besides, operators do not take on these maintenance responsibilities because it is not their job. On the other hand, traditional mechanics would love the smell of a breakdown. They know that they have become indispensable specialists in the trade. Maintenance thinks that they are assured of a stable job every time a fix is needed on their equipment. Maintenance loves to work overtime because of the added pay they get even if they miss some important events in their life. And as the vicious cycle goes on and on, the aftermath of which is an immense amount of waste in man-hours, production time, loss of opportunity, and the ballooning of maintenance expenses. And when the maintenance expense is high, even if everyone knows why, they want maintenance to explain, provide some corrective actions and deadlines for making it happen.

It is sad to note that many operators have not been allowed to learn more about their machines in most industries rather than being a switch flickers. In power plants, operators do some walk-around checks without realizing abnormal situations and irregularities on their systems and assets. It creates situations in which major accidents and disasters are much more likely to occur. Much worst is the lack of communication between operators and maintenance because their roles are separated. When a breakdown occurs, operators tend to disappear after calling the maintenance. And when the concept of Operator and Maintenance partnership is bought forth to the table, some operation managers would simply kill the idea immediately and tells maintenance, "Hey Boy, That idea simply won't work here. In implementing Autonomous Maintenance, expect a lot of resistance from annoying people initially, from their operations managers. These will be the number one resisting force in this initiative. In the beginning, they will do everything at their disposal to stop its implementation.

Their way of thinking is that operators are here to operate and not to maintain your function, so don't pass your burden to us. Empowering operators requires change, and this cannot be done overnight; it takes years to build an empowered workforce since you are changing the person and the culture. Implementing any kind of improvement initiative is difficult but not impossible why because it is not only about improving the ways things that being done here in the plant, but the most difficult part is about changing the people's culture as well since this is what they have been doing since the beginning of time.

I do blame neither operations nor maintenance for this situation; both have their own excuses, and if you ask them individually why they cannot work together, it always ends up in a merry-go-round that always makes me dizzy. Both departments have some sort of sacrifice to make these things happen. It will be the maintenance people who will make the first step; like in a chess game, maintenance will take on the white pieces. Operators must understand that maintenance is a shared responsibility for both operations and maintenance. There are maintenance activities that must be done by the operators themselves and not by maintenance. Maintenance must understand that Autonomous Maintenance or having an operator maintenance partnership can only work if they teach operators about their equipment. Maintenance should be the ones to make the first move. But most importantly, for Autonomous Maintenance to work, the industry needs to change from the traditional I operate you fix syndrome to "We operators and maintenance are both responsible for taking care of our equipment." Blaming and finger-pointing have to be stopped, and culture needs to be changed; otherwise, simply forget implementing Autonomous Maintenance because this can't be forced down to operators.

Operators must understand that maintenance is a shared responsibility for both operators and maintenance people. There are maintenance activities that must be done by operators themselves and not by maintenance people. On the other hand, maintenance should not develop the checklists for operators, but their role is to teach operators about their equipment. Once the operators know their equipment well, then they know what is important to check, and the chances of them doing it consistently and not faking it will become a reality because now they know not only how to operate but why it is important to do these things on their equipment. They must provide continuous communication with the maintenance people because it is important to understand what eventually happens before the failure or breakdown. If there is some sort of noise, smell, or anything that the operator senses, maintenance should know about it. Remember that it will be the operator who mostly witnesses the breakdown in their equipment and not maintenance. One problem with traditional industries is that when the equipment breakdown or fails, operators call the maintenance, and when maintenance arrives on the equipment, the operators disappear or take a break; in short, communication is lost between the two.

Autonomous Maintenance's message is simple; instead of operations fighting, blaming, and finger-pointing on maintenance on the fault, both parties should work together to address small problems before they become big. Big problems are simply an accumulation of small things that are always left unattended on the equipment. Coupled with the right tools and training, operators take on these necessary skills to address mechanical or equipment-related issues. Calling the engineers and mechanics is no longer necessary since operators are already prepared and confident in dealing with minor problems.

Many years ago, machine operators were limited to operating their equipment and handling their respective posts for those industries that have matured in their Autonomous Maintenance. Whenever there is a mechanical trouble or breakdown, operators stop working and call in the maintenance to fix the problem. So going back to the basic question on why did operations and maintenance go their separate ways? According to Fredrick Taylor, the best way to manage an organization was to standardize the activity into simple repetitive tasks and then closely supervise them into doing it

In effect, management people do all the thinking and decisions, while the supervisors act as the watchers making sure the decision is followed to the letter. Workers are focused on doing what they are told to do and just follow instructions until they get bored to death. While Western countries focused more on producing big volumes, capacity, and production, the Japanese people learned that the best way to run an organization is to focus more on their people's voices by making decisions to perform their work better. Western countries slowly realized that they are being beaten badly by their Japanese competitors, but many are still stubborn and remain trapped in the old Taylor paradigm. Thus, most American and western style factory management clearly separated the production and maintenance departments' roles. Managers were convinced that this style was the most effective way to utilize human resources. Operators concentrate on production with little or no knowledge of the structure and function of their equipment. Concurrently, maintenance received work orders and perform repairs on the equipment. As a result, both operations and maintenance went their own ways instead of following the path to mutual cooperation and shared responsibility. That is why today's feud is pretty much alive on both sides.

6.9: September 2012: Why Establishing Basic Equipment Condition Matters Most?

I do not know if you are familiar with a Smoky mountain or if other countries have it also? It is where all the garbage in a particular city is being dumped by the garbage collector trucks. Slums are living in that area. Their work is to go to this dumpsite daily and collect things like plastics, metals, newspapers that can be recycled and sell them to junk shops to buy food and survive. People are used to the smell of garbage all their lives. Children get sick all the time, and sad to say, sometimes they die. Imagine eating your lunch or sandwich in this place; where would you dump your trash after eating?

The answer is obvious since this is a dumpsite; you throw your trash anywhere you, please. On the other side, around 100 km from this area is a country club, a cozy restaurant, a park with a garden of flowers and a golf course, where company CEO's, executives, politicians and those sort of people go and take time off with their work, or perhaps a meeting place to discuss business with their clients while playing golf. Let us forget in the meantime about the type of people who go here and just focus on these two places. Imagine you are in a country club, eating your sandwich. You are in a nice garden. Where will you dump your trash after eating? Is it anywhere such as the case if you were eating in a Smoky Mountain? I doubt if you will throw your trash anywhere in the garden. What is the difference, or what is my point in this? A country club is a well-maintained place, while the Smoky Mountain gets worse and worse over time. There is no difference at all in how we treat our equipment. So the question is, how do we treat our equipment? Is it like a Smoky mountain or a Country club?

Figure 6.5: Do We Treat Our Equipment and Assets Like a Smokey Mountain?

World Class industries understand the essence of establishing basic equipment conditions in their equipment, while others often overlook and ignore them completely, yet no one can deny that almost everyone understands its importance. The major difference is that world-class industries implement what others only dream of and talk about. They try to walk the talk. They understand the importance of these basic things. Just imagine dust in our equipment. We see it all the time, even if it's very tiny. Some can ingress inside the equipment, and when inside the equipment, this can get trapped in moving surfaces and later on cause abrasion in some of the equipment's mechanical parts. You see, establishing basic equipment conditions in our equipment is not just about doing it once in a lifetime. It must not only be a habit but it must also be considered a discipline itself. When we truly understand what these small things can do to our equipment, we realize that we need to get started and involve ourselves with the basics first. World Class Industries started reactive and understood the importance of establishing these basics before moving on with the more advanced maintenance management strategies.

Several plants acquire vibration analysis and other non-destructive and diagnostic instruments to monitor potential failures in their equipment, but they forget one simple thing, which is to address the basics first. I am not against these instruments. I am driving at that before wishing for these instruments in your plant; why not address these simple basics first. I see many conflicts even on the best improvement strategies and methodologies, especially with TPM and RCM. RCM will start by identifying all sorts of failure modes in the equipment, and through an algorithm or decision diagram, the team will decide what the most feasible task for maintenance to perform for each failure mode uncovered. But my question, if we start with TPM, TPM will address the basics first, and when the basics are addressed and well established, we can expect fewer failure modes that are possible to occur in our equipment. Am I right? I hope the readers can understand how many failure modes can be addressed by doing the basic things of cleaning, lubrication, and having complete bolts. A couple of years ago, we recently performed a Root Cause Failure Analysis investigation regarding a gear driving a shaft. The shaft is screwed to the gear but is not rotating together with the screw. The set screws that used to tighten the shaft had easily worn out. Why? Because maintenance keeps on using the wrong tools to tighten the screw, the head of the screw gets worn out easily, and no one can remove that screw since the head is damaged severely. Big failures are just the accumulation of small things that are left ignored all the time. When you dig up the causes, you can see many things that should have been done, or simply there are many cases where we need to drill a hole in our skull and say hello, don't forget the basic equipment condition first.

Establishing basic equipment conditions means eliminating the causes of accelerated deterioration. This is when a certain part of its component in our equipment does not reach its useful life. It means cleaning to remove dirt and sources of contamination, proper lubrication to prevent early wear, and understanding that bolts need to be complete and secure. Parts do not achieve their desired lifespan due to dirt and contamination. The wrong lubricant was poured into the equipment because the guy is new. The machine fails simply because several bolts are missing. The maintenance uses the wrong tools that later on wreck the equipment and so on.

Equipment is subject to stress simply because there are parts movements inside. These mechanical parts that move are subject to stress and wear. In hindsight, dirt, dust, foreign material, contamination, inadequate lubrication, and excessive vibration cause parts to fail prematurely. This is called accelerated deterioration. Contaminants and other foreign objects can cause abrasion and come in direct contact with rotating parts, hydraulic systems, and other critical parts of the equipment. Establishing basic equipment conditions exposes these abnormalities and flaws and allows these parts to reach their natural deterioration and useful life. Many failure modes can be addressed if the basics have well been established in our equipment.

When operators establish these basic equipment conditions, equipment reliability starts to improve, and maintenance can focus on specialized activities and start performing improvements in their equipment. This is the importance of having Autonomous Maintenance in place because our equipment's basic things are finally being taken care of. When Autonomous Maintenance is missing, feud between operations and maintenance occurs, and maintenance will always be blamed if equipment lags behind productivity. The mindset of "I operate you repair" must be changed. I am inclined to think that industries have a way of hypnotizing these operators and telling them that their job is just to operate and remove everything from the mind and think of just one thing, "output," and since you are the maintenance, your frame of mind should always be on how quickly you can repair the equipment. Operator, output, maintenance repair. This is what is happening to most industries. It is sad to say that this thinking now becomes a part of their culture and both operators and maintenance become the gorilla in the cage. World-class industries understand the importance and value of addressing the basic equipment condition, and this includes:

• Keeping the equipment clean

• Performing proper lubrication
• Completing the bolts and screws in the equipment
• Addressing leaks
• Detecting problems using operators' senses

Cleaning: This is the process of removing any form of unwanted object from the equipment. It removes dirt, contaminants, grime, excess oil, grease, and other foreign objects that can affect equipment parts and components. Cleaning is performed not just to satisfy the equipment's cosmetic looks but because dirt and contaminants shorten the life of certain parts and components of the equipment. Detailed cleaning of components and equipment is often a no man's island because while everybody agrees that it is important, nobody wants to do it actually. World-Class Companies perform these activities. Components and equipment are kept clean in every detail. Such an organization realizes that inspections cannot be done without this level of cleaning. Cleaning definitely extends the life of parts and components. We can expose problems on the equipment that is left ignored for a very long time through cleaning. When operators clean their equipment, they touch parts, and by touching parts, they are actually inspecting them. Leaks, cracks, and fractures can be detected and exposed when operators touch and clean their equipment after a long time. Dirty and untidy equipment will tend to deteriorate more rapidly than the equipment that is well maintained and cleaned at all times.

Lubrication: All machines contain lubricants in one form or another. The purpose of lubrication is to reduce friction for mechanical parts moving inside our equipment. It is important to lubricate the equipment using the correct quantity and viscosity. Equipment that lacks lubrication or over lubricated can likewise induce problems. Operators should be taught about the importance of checking the lubrication in their equipment and maintaining the lubricant's correct amount. Many failures can be attributed to lubrication that can be avoided if only these basic equipment conditions can be established. Maintenance must teach the operator regarding important points in their equipment to lubricate. Just like humans, lubrication is the blood of the equipment, and it should be maintained clean and adequate all the time. Many failures are attributed due to lack or inadequate lubrication. These things can be avoided and controlled in our equipment if we understand what lubrication plays.

Bolting: Machinery contains bolts, nuts, and fasteners as part of their construction, and they serve a particular purpose. The equipment functions properly only if fasteners are securely tightened. All equipment vibrates, but excessive vibration can be destructive as this can induce secondary damages to other parts and components affected by the vibration. When we touch the equipment, we can feel its vibration. What we feel is simply the sum of all vibrating forces moving inside the equipment. Bolts, screws, and fasteners are placed to minimize the vibration in the equipment. When vibration increases, it can result in cracks and fractures, and these fractures can propagate to the point of rupture due to excessive vibration. Remember that it only takes one loose bolt to create a chain reaction of destruction and havoc in our equipment. As a result, other bolts become loose, and vibration increases. Match marks are placed on critical bolts and nuts to easily detect if bolts have been loosening due to excessive vibrations for some time. Imagine driving your car, and each of your wheels has 3 instead of 5 stud bolts on each of your tires. Would this be all right with you? I guess not. Then why don't we treat our equipment in the same way? We can see many bolts missing or loose as a result of many activities performed on the equipment. For a start, why don't we complete them?

Establishing Basic Equipment Conditions is a shared responsibility by both operations and maintenance and should regularly be performed on the equipment. Catastrophic breakdowns can greatly be reduced if Basic Equipment conditions are in place. Lacking and lose bolts often lead to excessive vibration, which produces secondary damages on parts affected by the vibration itself.

One of the major reasons why maintenance remains reactive and often trapped in the repair business is that basic equipment conditions were not established. If asked why it had not been established, they really have no time for it since maintenance is always undermanned. When we asked why they are undermanned, operators simply don't take part in their responsibility. Most maintenance finds it hard to advance to any continuous improvement, some had advanced to Predictive Maintenance stage, yet they find it hard to control the failures, and parts lifespan had not been reached; why? Because the Basic Equipment Condition had not been well established. Remember that ignoring the Basics Simply Leads to Failures

6.10: October 2012: Frequently Asked Questions on Oil

1) Does the oil really wear out and need to be changed?
• Oil does not wear out; many times, the question should be answered in the negative. *Reference Lubrication Engineering Volume 17*
• Oil does not wear out; it only gets dirty. *Reference U.S. Standard Bulletin No. 86*
• Oil is like any mineral and cannot wear out. Oil can become dirty and contaminated, but like copper, iron, or silver, they are as good as new when reprocessed. *Reference, Theory and Practice Lubrication for Engineers, 2nd Edition P590-591*
• Oil does not wear out, break down, or otherwise deteriorate to such an extent that it needs to be changed; it becomes contaminated with water, acids, carbon particles, and sludge. Average Oil Filter can remove solid particles above a certain size. It cannot remove water, acids, carbon particles, all of which pass through the oil filter just as readily as the oil. *Reference Mobile Oil Technical Bulletin # 863*
• Oil does not wear out; it simply becomes contaminated due to inadequate filtration, which to the point that it needs to be drained. The actual base stock never wears out; only the oil additives become depleted due to contamination. *Reference Proactive Maintenance Solution Newsletter*
• Oil doesn't wear out; it just gets dirty. It can be cleaned, re-refined, and used again and again. *Australian Government – Department of Environment and Heritage*

2) How do I know if I need to change my oil?
2) How do I know if I need to change my oil? Many people believe that you have to change the oil regularly. One of the indicators that would tell you that you need to change your oil is the TAN (Total Acid Number). If your TAN is greater than 2, you had been shot, and your oil needs to be changed. If TAN is under one, then the oil is in excellent condition unless you have an unusual amount of water in the oil. If your oil has a high amount of particulates, don't change it; if the TAN number is good, filter it.

3) What is the main reason why oil breaks down? Heat and contamination are the main cause of oil degradation or why oil breaks down. Both heat and contamination can be controlled; therefore, there is no reason to change the oil itself if it can be controlled. Contamination, this is anything that should not be present in the oil, solid, liquid, or gas. Depletion of Additives, The lifespan and protective ability of your oil's additives package is primarily dependent upon the amount of contamination present in your oil and not by the number of miles or hours the oil has been in service.

4) At what temperature ordinary mineral oil can withstand? According to most engine-related studies, oils are designed for optimal protection at 185 degrees Fahrenheit. Above or below that temperature, you give up some protection. Most petroleum oils start breaking down at around (163 to 177 ° C or 325 to 350 ° F, and no amount of fancy additives will directly protect the base oil from degrading.

5) Why do we need by-pass filters aside from our current OEM full-flow filters? The factory filtration usually installed on capital equipment is placed as protective filtration; however, most full-flow filters are designed to remove particulates in the 30 to 40 microns. This means that particles smaller than 30 microns remain in the oil and are causing wear in our system, which can be removed only with a good bypass filtration system.

6) Is there a relationship between contamination and the engine's life or system it is lubricating? According to Cummins Technical Center, wear can be reduced to 91% using a bypass filter combined with a full-flow filter Cummins Technical Center. More than 70% of hydraulic failures can be attributed to contamination based on filter study Contamination in the lubricant of engines, transmissions, and hydraulic systems cause up to 70% of equipment failure by SAE. General Motors Corporation states that a study on contamination confirms that 80% of internal wear is caused by particles in 40 microns. According to Caterpillar, dirt and contamination are the no. 1 cause of hydraulic system failures.   The case states that with regards to hydraulic systems, they must be kept clean.

7) How is the acid form in the engine? Soot is formed by incomplete combustion of the fuel. Fuels, especially diesel fuel, contain sulfur. The sulfur in soot, when mixed with water and heat, forms sulfuric acid. This acid causes corrosive wear in the engine. While some water will naturally evaporate when oil gets hot, commuters and other short-trip drivers do not travel far enough to heat oil to a level that will prevent acid formation. The only reliable method of removing water is to trap it. Full-flow filters made from paper or synthetic media cannot trap water or moisture. It will just pass through the filter just as readily as the oil.

8) Will additives "As Seen on TV" helps eliminate friction? The Federal Trade Center (FTC) conducted a similar test on these additives performed on TV and found that increased performance and reduce engine wear were unsubstantiated. Consumers Report attempted to reproduce the no oil test where oil was drained out of an engine and treated with Prolong. The engine runs for 13 seconds, then it seized. The same test was conducted on Duralube, and the engine lasted for a staggering 11 seconds and seized. Blue Corral, the manufacturer of Slick 50, lost to FTC and had agreed to pay upwards of $20M in damages to affected customers.

Motor-up also faces charges of the lawsuit on false claims on their products.

9) Can moisture be removed from oil? Several absolute filters can remove moisture, these filters are made from cellulose, and some may even contain cotton to contain the moisture. The use of desiccant breathers can help reduce moisture in our equipment, especially in hot and humid places.

10) How do you compare synthetic and mineral oil? Concerning the molecular structure of mineral-based oil, this composes of small and large molecules, while synthetic oil is composed of uniformly shaped molecules when small heated molecules present in mineral oil boils off. Synthetic oil can withstand a much higher temperature than mineral-based oil. Synthetic Oils have a higher resistance to heat than mineral-based oil. Synthetic oil vaporize at a much higher temperature at 316 °C, or 600 ° F compared to petroleum oil at 177 ° C, or 350 ° F. Synthetic oil reduce friction and provide higher film strength (mineral-based has a film strength of 400 psi while synthetics usually exceed 3000 psi)

11) What is the amount of moisture that can damage the oil? The table shows the limit of moisture for a 100ml sample of the oil

General Water Concentration Limits		
• 100 - 300	• Alert	Check seals, breathers, coolers, etc., for ingression sources, watch the trend
• 300 - 800	• Danger	Aggressively investigate and correct the source of ingression and implement an effective water removal activities
• 800 above	• Extremely Danger	• Immediate action is required to eliminate ingression and effect removal of water to minimize damage to machine and lubricant

Figure 6.6: Limits of Moisture in Lubricating Oil

12) How much do you think is lost due to the oil leak? This table tells how much is consumed due to oil leaks assuming the cost of oil is at $ 4.00/gallons.

Leakage Rate	Monthly Rate	Yearly Rate
1 drop/5 sec	6.6 gal = $ 26.40	80 gal = $ 320.00
1 drop/sec	34 gal = $ 136.00	409 gal = $ 1637.00
3 drop/sec	113 gal = $ 452.00	1243 gal = $ 4972.00
Steady Flow	720 gal = $2880.00	8640 gal = $34,560.00

Figure 6.7: Estimated Cost of Oil Leak on Equipment and Assets

13) What is the best brand of oil to be used in my car? To answer this question is first to understand some basic properties of the oil. Flash Point is when the oil gives off vapors ignited with a flame over the oil. The lower the Flash Point, the greater the oil's tendency to suffer vaporization loss at higher temperatures. Therefore, the higher the flash point's temperature, the better the oil. The oil property describes the degree and rate at which oil will vaporize under given pressure and temperature at 250 ° C. Oils have chemical ingredients that tend to evaporate, boil when it is hot. Petroleum base stocks also boil off in the heat, which changes the viscosity of the remaining oil. It determines the evaporation loss of lubricants in high-temperature service. NOACK is the amount of oil lost (light molecules) over time at a given temperature and pressure. When oil suffers from high volatility, lighter molecules evaporate, making the oil thick and heavier, making the oil circulate poorly, reduce fuel economy, increase oil consumption, and excessive wear and emission. Lubricants with low Noack scores are preferred. Low Noack scores indicate an oil that is less prone to evaporation at high operating temperatures. Low Noack scores indicate less oil consumption, less deposit formation, longer oil life, stable viscosity, and fewer emissions. The engine oil with the lowest Noack Volatility rate will be the best type of oil to be used.

6.11: November 2012: Why Planned Maintenance Should Be the Strongest Pillar in TPM

Although an industry trying to initiate a TPM strategy in their plant will focus their priorities first on Autonomous Maintenance. First, because it is dictated in the TPM books but given myself a chance to initiate TPM in a plant, I would be starting off by setting up and strengthening the Planned Maintenance pillar of TPM ahead of Autonomous Maintenance. Although Autonomous Maintenance will be the pillar with the highest population, most especially in a manufacturing site, Planned Maintenance must be considered the strongest pillar in any TPM implementation. Why? Simple because if Planned Maintenance is weak, then Autonomous Maintenance will collapse. Planned Maintenance is the TPM pillar that will provide all the needed support for any Autonomous Maintenance activities.

One of the main key activities that Planned Maintenance will be conducting is starting off with a preparatory stage before moving on with their 1st Phase of activities on Planned Maintenance, equipment restoration. The first step in the Planned Maintenance journey will be for maintenance people to undergo basic training on Planned Maintenance. If your plant is active in TPM, a facilitator should be the one to conduct the training on Planned Maintenance, or other third-party consultants can be hired for the training. It must clearly be stated that the maintenance goal is not on how much equipment they had completed in their PM schedule. It is not how many repairs a maintenance technician had done in a day's work, or how many parts they've repaired during their shift, but rather a more indebt understanding of the Basic Equipment Condition and the role they will undertake most especially with their active partners which are the Autonomous Maintenance. One of the highlights of Planned Maintenance is to educate operators so that they can do minor repairs and set up while the maintenance team can focus more on restoring the equipment and reducing chronic problems and breakdowns. The machine often needs a surgical operation, but due to cost-cutting budget constraints, maintenance performed what they called the "Band-Aid Therapy." Believe me, this is not the way to cut corners and reduced costs. This approach can be more costly in the long run. Having a system and investing in restoring equipment will be your main responsibility. Likewise, you need to have a heart-to-heart talk with your management team. Just do not take this message very lightly.

Phase 0 details the requirements needed to set up an effective Planned Maintenance System in your company. These are the basic essentials needed for a strong foundation. The preparatory stage usually takes from 1 to 3 months, depending on the plant's size and the number of equipment you have. It is very important not to ignore this stage; otherwise, building a Planned Maintenance System without establishing a basic foundation builds a house on sand. It is best for a company starting on Planned Maintenance to hire persons to act as Facilitators. (Mostly, this will be a full-time person from the TPM Office) In this regard, I would prefer hiring since there are actually a few people who give consultancy. The important thing is that maintenance people must own Planned Maintenance, and it must not be treated as a separate program from their day-to-day activities.

Building Your Planned Maintenance Organizational Structure

The whole Maintenance Department should comprise the Planned Maintenance Organizational Structure. Some companies have a separate organization for line maintenance or sustaining group and Preventive Maintenance group. Both must work together towards achieving a common goal. Top Maintenance Manager must spearhead the Planned Maintenance implementation, with close communication from the Planned Maintenance Facilitator from the TPM Office, whose function is to develop the legwork needed, conduct training, audits, develop the roadmaps, and recognition on Planned Maintenance, set-up the teams, and anything related to it. Suppose your company is big with different divisions; in that case, it is best to set up a Planned Maintenance Committee. They shall represent their division; a maintenance section manager usually is best suited for this position.

There are several points to be undertaken on your Planned Maintenance Implementation. During the preparatory stage, all divisions in the Planned Maintenance must be involved and should participate. All Top Management must not only support but provide commitment and ownership on Planned Maintenance Activities. Remember that TPM and Planned Maintenance is a long-term approach. It will be a regular part of your day-to-day activities on maintenance. Not committing and owning the program besides announcing it will negatively impact all maintenance people to lose their motivation and enthusiasm.

Another important part of your Planned Maintenance activities will be to include your Facilities/Utility people in your Planned Maintenance Strategy since they will play a vital part in its implementation. Remember that these are the people handling the most critical equipment in the plant. If their sub-station fails, then all production equipment fails. I cannot state it any simpler than that. Doing TPM activities involve investment in time and money; ensure that your management team is aware of that. This will hurt your implementation as you control your maintenance cost, but this will pay you off in the long run. Imagine a team conducting restoration and found several parts neglected and prepared a Purchase Order to be denied and rejected by your management group. Give priorities and approve the ones that really need to be restored. Below will be the different responsibilities of different people in Planned Maintenance.

Roles and Responsibilities of Top Maintenance Manager

• Launches the Planned Maintenance implementation in the company
• Conduct review with the PM Committee selected (weekly or bi-weekly)
• Track progress of completion on each activity involved in Planned Maintenance
• Approves logistical support and requirements needed by the teams
• Must own at least one Model Team
• Provide simple recognition to teams that completed each Phase with good results
• Review overall indices set forth by the Planned Maintenance

Roles and Responsibilities of Planned Maintenance Facilitator (TPM Office)

• Define roles and responsibilities for each maintenance involved
• Conduct training on Planned Maintenance Steps
• Guide the team in their implementation
• Conduct Initial Audit together with the Maintenance Section Managers for each Phase completion
• Planned Maintenance facilitator must act as the internal consultant on Planned Maintenance

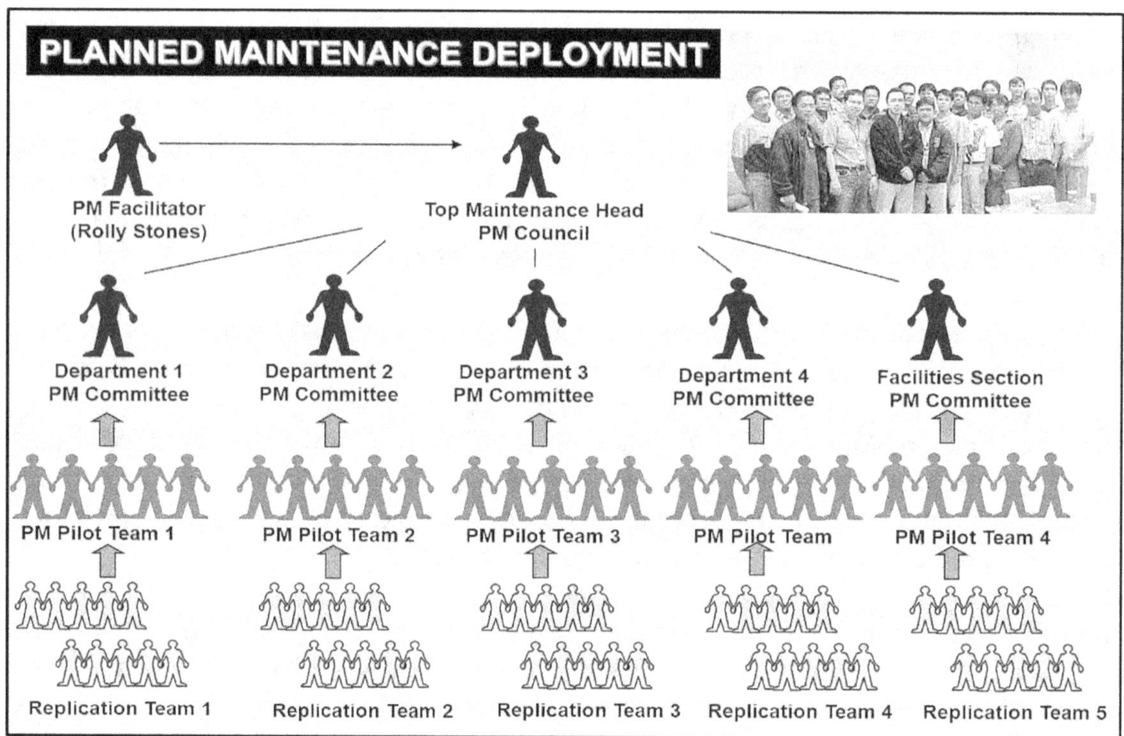

Figure 6.8: TPM Planned Maintenance Pillar Deployment

Roles and Responsibilities of Maintenance Managers

• Must own at least one Model Team

- Review logistical support needed by their team and elevate them to Top Management
- Conduct Audit on their Team's completion and Machine Certification
- Review the teams on their Planned Maintenance Journey
- Summarize results obtained from implementing Planned Maintenance
- Set-up goals on maintenance indices, completion stages
- Attend training on Planned Maintenance

Roles and Responsibilities of Maintenance Pilot Teams

- Complete the Planned Maintenance Activities
- Provide OPL (one point lessons) to Operators and Fan-Out Teams
- Attend training on Planned Maintenance
- Monitor equipment performance and results as they complete each Phase activity
- Set up a bulletin board and update them to show their activities performance
- Serve as a model to other teams and provide guidance and education on other teams
- Generate standards on maintenance and revised existing standards or specs
- Train operators on Basic Equipment Operation and Condition

Phase 0 is the foundation of building a solid Planned Maintenance structure. Try to aim at completing this initial stage and determine a timeframe of completion. These are the basic activities needed for you to start a strategic maintenance program. Not all equipment will undergo each step/phase of Planned Maintenance but only those classified and ranked as A and B. The aim of Phase 0 is to plan your activities ahead on Planned Maintenance, create a Timeframe of completion. It will take 1 to 3 months to complete this stage. Once the planning had been achieved, your next step is to proceed to the detailed step/phase activities of Planned Maintenance.

6.12: December 2012: Planned Maintenance Phase 0 Preparatory Stage

Phase 0, which is the preparatory stage in Planned Maintenance, details the requirements needed to set up an effective Planned Maintenance System in your industry. These are the basic essentials needed for a strong foundation. Depending on your industry's size and the number of equipment to undergo Phase 0 Preparatory stage, it usually takes 3 to 6 months to complete this step. It is very important not to ignore this phase; otherwise, building a Planned Maintenance System without establishing a basic foundation is like building a house on sand. It is best for a company starting on Planned Maintenance to hire a person to act as the Planned Maintenance Facilitator specializing in this regard; I would prefer hiring since few people give consultancy. The important thing is that maintenance people must own Planned Maintenance, and Planned Maintenance activities must not be treated as a separate program from their day-to-day activities. Here are the basic guidelines in staging Phase 0 of Planned Maintenance.

1) Build Your Planned Maintenance Organizational Structure: The Maintenance Department should comprise the Planned Maintenance Organizational Structure; some companies have a separate organization for line maintenance or sustaining/repair group and Preventive Maintenance group. Both must work together towards achieving a common goal. The maintenance Manager or head should be the one to spearhead the Planned Maintenance

Implementation, with close communication from the Planned Maintenance Facilitator who usually is assigned as full-time staff on TPM office, whose function is to develop the details and legwork needed, conducting pieces of training on Planned Maintenance, setting-up the maintenance teams, and anything related to it. Suppose your company is big with different divisions; in that case, it is best to set up a working Planned Maintenance Committee. They shall represent their division. A maintenance section manager usually is best suited for this position.

2) Planned Maintenance Kick-Off: This can be done by simply having a general meeting for all maintenance personnel in which the maintenance head announces the launching of the Planned Maintenance pillar of TPM in their organization. Some companies provide food, slide shows, invite guest speakers specializing in the subject matter, and spending 2 to 4 hours for this launching. Highlights of this event usually explain to the maintenance people the need for launching this program in your plant, the vision on what the maintenance department would want for the next 3 to 5 years, and most especially the support they will be providing for the Autonomous Maintenance pillar of TPM.

3) Define Roles and Responsibilities: The Planned Maintenance Facilitator will be responsible for defining the roles and responsibilities of each member of the maintenance organization from the head down to the maintenance crew. Each member involved must participate in the Planned Maintenance implementation from the top down to the maintenance shop floor people. There are several key points to be undertaken in initiating Planned Maintenance. During the preparatory stage, all Planned Maintenance divisions must be involved and should participate in this pillar. Most importantly, Top Management must be committed to this initiative and be committed to completing the Planned Maintenance Activities. Remember that TPM is a long-term solution to your day-to-day problem with maintenance. Not owning a program besides announcing it will negatively impact all maintenance people to lose their motivation and enthusiasm. Another important function of Planned Maintenance is to include your Facilities/Utilities in the Planned Maintenance activities to play a vital part in its implementation. Remember that they are the ones responsible for the most critical equipment in the plant. Doing TPM activities involved investment in time and money; be sure that your management understands it. As Autonomous Maintenance will focus its activities on detecting and correcting abnormalities, Planned Maintenance will be responsible for restoring the equipment as they approach Phase 1 of its implementation. Expect that doing Phase 1 will increase your maintenance cost depending on how the equipment is neglected since we are restoring the equipment and bringing it as close to its original basic condition. Maintenance costs will increase during the initial phases of implementation, but rest assure that it will pay you off in the long run. Imagine a team conducting restoration and found several parts that had been neglected and prepared a Purchase or Requisition Order only to be rejected or denied by management. Give priorities and approve the ones that really need to be restored. Below are some guidelines to help you and the facilitator with their respective responsibilities.

4) Set Up Vision and Mission for the Planned Maintenance: This will be your philosophy that every maintenance personnel will carry out and be guided by. It shall be your vision or where your team is headed forth in the future. Be sure that this is observed and that all activities that you perform on your Planned Maintenance activities are aligned to your vision and mission. Set a time frame for achieving the vision.

5) Generate the Planned Maintenance Master Plan: A Planned Maintenance Master Plan is where you put the activities with their completion timeframe. It is a piece of cake for the overall TPM Master Plan; below is an example of a PM Master Plan.

6) Conduct Machine Inventory and Rank Equipment: List all your equipment by preparing a Machine Inventory Lists and conducting machine ranking for all your plant's equipment and assets. Rank A will be the worst, and Rank C is the good machine. Only machines that are rank A and B should undergo the Planned Maintenance activities. Prepare a category for each piece of equipment and rank them accordingly. Perform this on all your equipment lists. The definition of breakdown must be clear to our Maintenance people, and it must not be confused with Minor Stoppages. Planned Breakdowns or Schedules Time Based Maintenance must not be considered as breakdowns since they are Planned downtime. Once completed, a summary of figures on machine ranking will give the maintenance team adequate data for selecting their 1st pilot machine that will undergo the four phases of Planned Maintenance. This also gives the maintenance people which equipment to focus on improving them; remember, the maintenance goal is to improve the equipment that frequently fails. These machines will be the focus of the Planned Maintenance Category on Machine Ranking. Equipment may differ from one industry to another, and the severity of failure varies accordingly. Try to weigh heavily if there are failure modes experienced on the equipment that have severe consequences on safety and environment that happened in the past or may happen in the future. Equipment that will undergo Planned Maintenance activities must belong to either Rank A or B only.

• Rank A Machines: Failure occurrences of more than 2x per month
• Rank B Machines: Failure occurrences of less than 2x a month
• Rank C Machines: Good Machines

7) Selection of Pilot Machine and Backtrack Breakdown Data: A good index for tracking Planned Maintenance activities will be BDO (Breakdown Occurrences) and MTBF or Mean Time Between Failure.   That pilot machine should have similar equipment types belonging to Rank A or B for Horizontal Replication purposes or fan-out activities. After deciding which equipment will be piloted, backtrack data on maintenance costs, MTBF, breakdown by checking history logs and records, backtrack BDO (Breakdown Occurrences) for the past year as this will be your benchmark and reference and the measure of performance on your activities. It is important that the team monitor the data on Breakdown and should be updated regularly.

8) Determine Basic Training Requirements for Pilot Team: Planned Maintenance Facilitator in your plant must see that all team members are trained on the basic requirements needed to undergo each step or phase. This training should be conducted in-house by hiring a Planned Maintenance facilitator who specialized in this field. This training does not include specialization training on their equipment as training may differ depending on the type of plant and equipment. Training is an essential factor that is required for the success of Planned Maintenance. As each team completes each step or phase in the Planned Maintenance implementation, their skills are enhanced, and they gain more knowledge and have an indebted understanding of the daily needs of their equipment. One factor often overlooked in the maintenance process is putting everything on time-based inspection and Preventive Maintenance tasks. Many industries do not use condition-based maintenance, and many maintenance people have never heard of this concept since they are too busy dealing with repairs and failures. Invest the time needed for this training. These are investments that your company needs, and when each had been applied, whatever time and money being spent on them will be returned to you 10 to 100x fold.

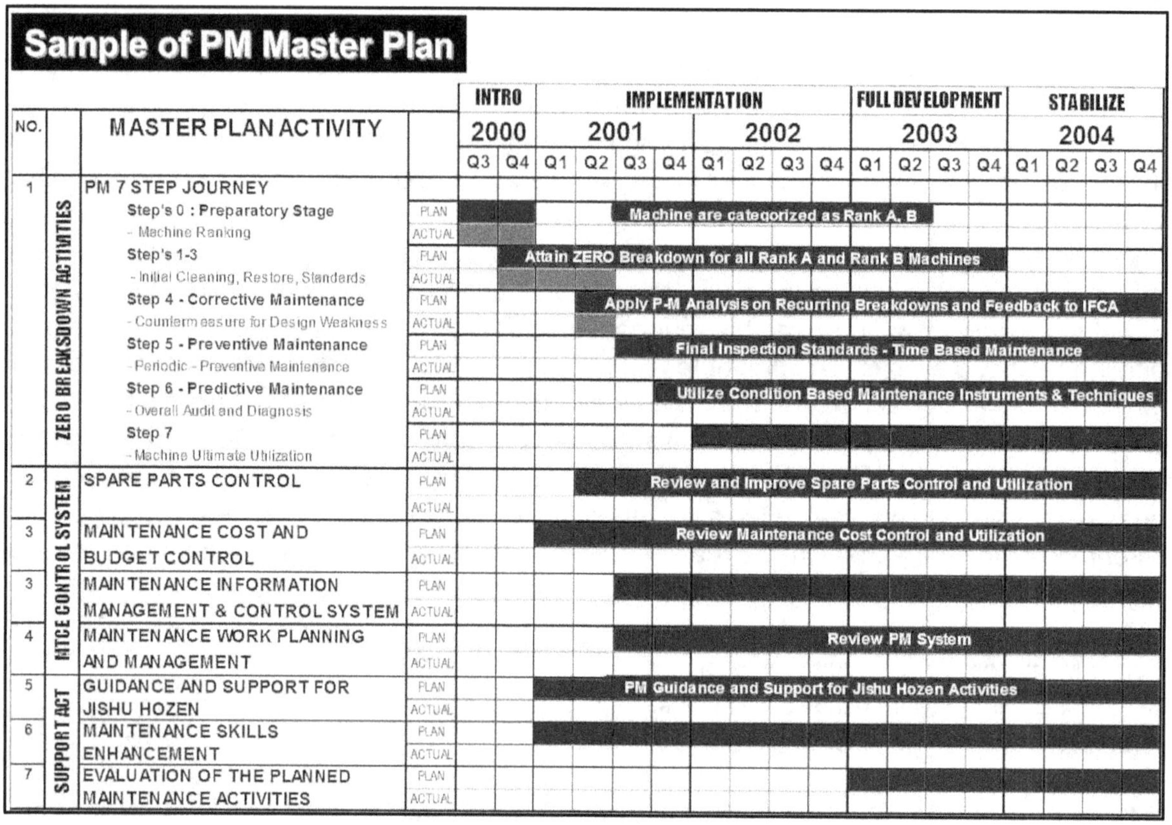

Sample of PM Master Plan

NO.	Category	MASTER PLAN ACTIVITY		INTRO 2000 (Q3–Q4)	IMPLEMENTATION 2001–2002 (Q1–Q4)	FULL DEVELOPMENT 2003 (Q1–Q4)	STABILIZE 2004 (Q1–Q4)
1	ZERO BREAKSDOWN ACTIVITIES	PM 7 STEP JOURNEY					
		Step's 0 : Preparatory Stage – Machine Ranking	PLAN / ACTUAL	■	Machine are categorized as Rank A, B		
		Step's 1-3 – Initial Cleaning, Restore, Standards	PLAN / ACTUAL		Attain ZERO Breakdown for all Rank A and Rank B Machines		
		Step 4 - Corrective Maintenance – Countermeasure for Design Weakness	PLAN / ACTUAL		Apply P-M Analysis on Recurring Breakdowns and Feedback to IFCA		
		Step 5 - Preventive Maintenance – Periodic - Preventive Maintenance	PLAN / ACTUAL		Final Inspection Standards - Time Based Maintenance		
		Step 6 - Predictive Maintenance – Overall Audit and Diagnosis	PLAN / ACTUAL		Utilize Condition Based Maintenance Instruments & Techniques		
		Step 7 – Machine Ultimate Utilization	PLAN / ACTUAL				
2	MTCE CONTROL SYSTEM	SPARE PARTS CONTROL	PLAN / ACTUAL		Review and Improve Spare Parts Control and Utilization		
3		MAINTENANCE COST AND BUDGET CONTROL	PLAN / ACTUAL		Review Maintenance Cost Control and Utilization		
3		MAINTENANCE INFORMATION MANAGEMENT & CONTROL SYSTEM	PLAN / ACTUAL				
4		MAINTENANCE WORK PLANNING AND MANAGEMENT	PLAN / ACTUAL		Review PM System		
5	SUPPORT ACT	GUIDANCE AND SUPPORT FOR JISHU HOZEN	PLAN / ACTUAL		PM Guidance and Support for Jishu Hozen Activities		
6		MAINTENANCE SKILLS ENHANCEMENT	PLAN / ACTUAL				
7		EVALUATION OF THE PLANNED MAINTENANCE ACTIVITIES	PLAN / ACTUAL				

Figure 6.9: Sample of TPM Planned Maintenance Master Plan

Year 7

2013 RSA Reliability Newsletter Vault Archive

> *Many maintenance people believe that all failures can be addressed by doing Preventive or Scheduled maintenance. If you think that this is right, then you are so wrong. This thinking is the main cause why most industries are still trapped in the reactive world of doing maintenance.*

7.1: January 2013: Definition and Function of Lubricating Grease

According to ASTM or the American Society for Testing and Materials, the lubricating grease is defined as a solid to a semi-fluid product of dispersion of a thickening agent in a lubricant. It comes from the Latin word Crassus, which means fat. Grease is a thick, oily lubricant containing inedible lard, the fat of waste animal parts, mineral or petroleum-derived or synthetic oil, and a thickening agent. In layman's terms, grease can be referred to as oil that is soaked in a sponge. While a lubricating oil contains two basic ingredients: the base and the additives, the difference is that grease contains a third ingredient called a thickener, which holds the lubricant in place. A lubricating grease's function is to remain in contact with lubricated moving surfaces without leaking out under the force of gravity, centrifugal action, or being squeeze out under pressure. Grease must retain its properties under shear forces at all temperatures it experiences during its use, depending on the grease's dropping point.

Grease is most often used instead of oil, where a lubricant is required to maintain its original position in a mechanism, especially where the chances for frequent lubrication may be limited or not economically feasible to sustain. The requirement may be due to the physical structure of the mechanism, its motion type, sealing, or the lubricant's need to perform part of any sealing function in preventing the loss of the lubricant or the ingression of solid contaminants. Since grease can either be solid to liquid, unlike lubricating oil, grease does not dissipate the heat, nor can it be used for cooling and cleaning functions associated with a fluid lubricant. With these exceptions, greases are expected to accomplish other functions, different from lubricating oil. Several functions of grease include:

- Reduce friction and wear of mechanical components.
- Protect mechanical components against rust and corrosion.
- Grease can also act as a seal to prevent dust, dirt, and moisture ingression.
- Grease can resist leakage or drip from the lubricated surfaces.

• Grease retains its viscosity, shear, and temperature throughout the grease's useful life in a mechanical component that can subject the grease to shear forces.

• Grease does not harden excessively to cause undue resistance to motion during cold environments.

• Grease should be compatible with elastomer seals and other construction materials in the lubricated portion of the mechanism and must tolerate some amount of contamination, such as moisture, without any loss of its significant properties. Thus, it can quarantine contaminants instead of being able to ingress into the system.

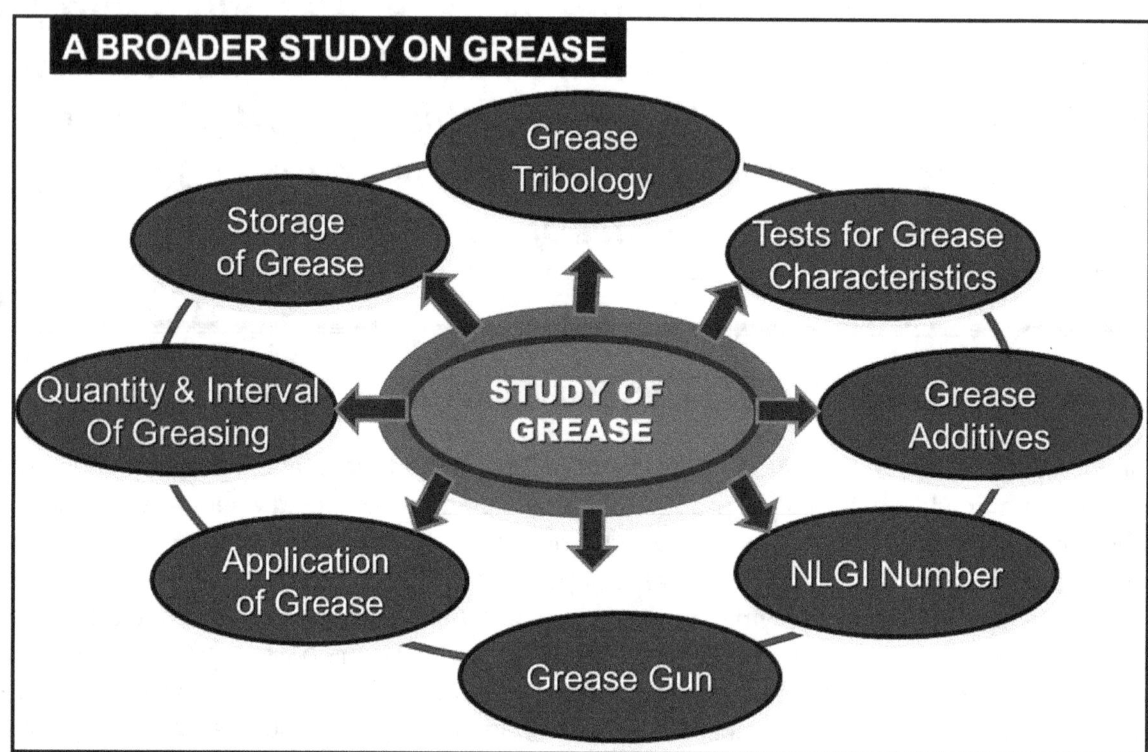

Figure 7.1: Broader Study on Grease

While the Society of Automotive Engineers (SAE) states that the grease's function is mostly for automotive and vehicles, grease can also have a wide range of stationery and industrial equipment applications. The grease is composed of the following:

Base: The base oil of the grease reduces friction between two moving surfaces. It has exactly the same role that the base oil plays in lubricating oil. The base oil, just like in lubricating oil, can be made of petroleum, semi-synthetic, or fully synthetic for grease. In extremely high and low temperatures, synthetic base oil for grease will provide better stability and protection than mineral or petroleum-based grease.

Additives: Like lubricating oil, the additives play several lubricating roles to enhance its different properties and functions. The additives supplement the base oil's lubricating capability, improving its wear protection and rust prevention characteristics. The most common additive packages for grease include oxidation inhibitors, anti-wear agents, extreme pressure (EP) additives, rust, corrosion inhibitors, pour point depressants, friction modifiers, dyes, adhesive agents, etc. Molybdenum disulfide can also be added in certain grease types where applications include heavy loads, low speeds, restricted or oscillating motion. Most of the additives used in lubricating oil are also the same as the additives used in grease. However, grease has a higher concentration of additives compared to lubricating oil. Likewise, the color of the grease has nothing to do with its overall performance.

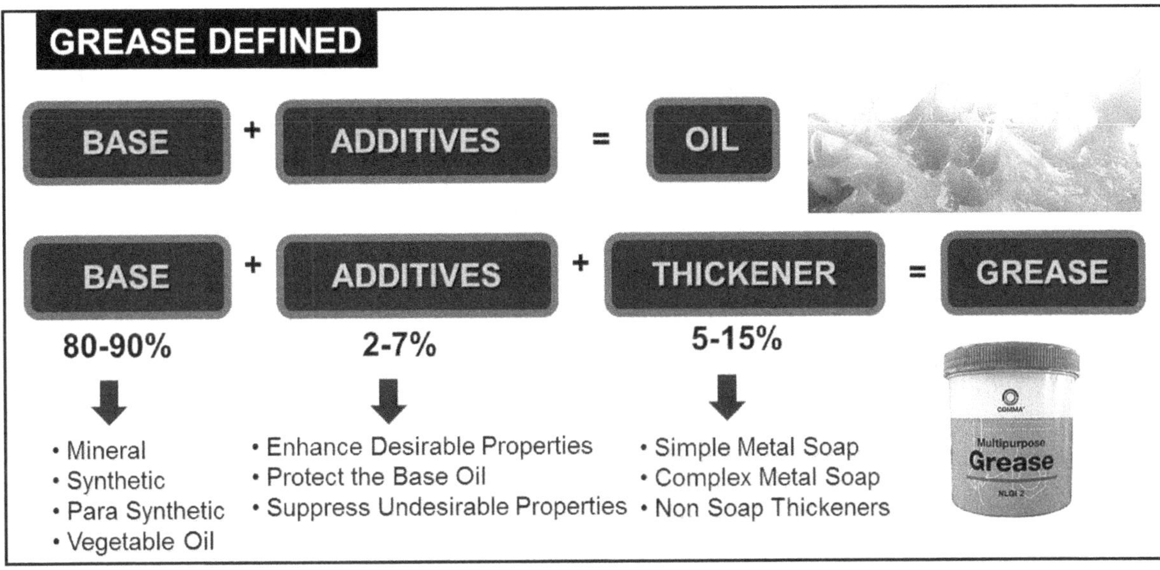

Figure 7.2: Definition of Grease

Thickener: Thickener is a material that, combined with the selected base and additives, will produce a solid to the semi-solid structure. The most common type of thickener is metallic soap. The soap can either be lithium, aluminum, clay, polyuria, sodium, silicon, and calcium. The thickener acts like a sponge and holds the grease in reserve until it is lubricated. The grease can also function as a sealant to minimize leakage and to keep out contaminants by trapping them. It also prevents the entry of corrosive contaminants and foreign materials. The grease holds the lubricant in total suspension. Molybdenum disulfide and graphite are mixed with grease at a temperature of 157 °C or 315 °F in an extremely high-pressure application. Every type of thickener is unique as each of them have their own dropping point and individual properties. The majority of these thickeners can have good water resistance. Different grease types have their own range of minimum and maximum temperatures to be used and applied. Lubricating oil is often classified by machinery types such as engine oil, gear oil, compressor oil, hydraulic oil, turbine oil, transformer oil, etc., while lubricating grease is classified by thickeners used such as metal soap, complex metal soap, or non-soap thickeners. Saponification is the process used to develop the grease thickeners where fatty acids are used to react with an alkali to form a chemical soap.

7.2: February 2013: Benefits of Condition-Based Maintenance for Industries

The dictionary defines prognostics as "Foretelling or Prediction." It is possible to predict machinery and equipment failures by running them to destruction, but this is costly and will probably take some time. Condition-Based Monitoring combines diagnostics and prognostics to predict critical equipment failures long before they occur (NDT or Non-Destructive Testing). In a traditional industry, most maintenance practices have been Time-Based, with components being replaced based on the number of hours, resulting in inexpensive parts and labor costs due to unnecessary replacement of parts. Some of the risks involved in performing a Preventive Maintenance replacement and overhauling are that the parts scheduled to be replaced may still be in good working condition, and human errors do happen when attempting to put the equipment back in one piece again.

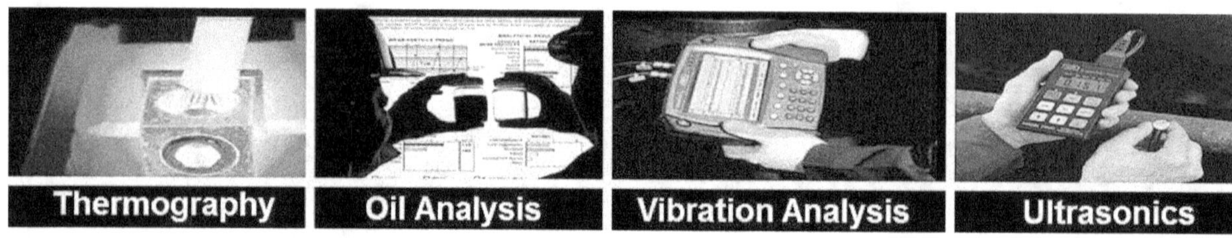

Figure 7.3: Different Predictive Maintenance Analysis

Predictive Maintenance is a maintenance activity geared toward indicating where a piece of equipment is on the critical wear curve and predicting its useful life or is on the verge of failing. This is done with the aid of specialized diagnostic instruments. Although John Moubray RCMII author includes any activities that entail checking the equipment's condition and will fall under "On-Condition Tasks," including human senses, reading pressure gauges, charts, and so on. The role of CBM is to intelligently predict when maintenance is required by checking the condition of the equipment. The CBM application also monitors the condition, diagnoses the problems, identifies the causes, and suggests corrective actions. The users of these instruments are merely more than just data collectors; they should analyze the readings and recommend a solution to potential problems. For example, a vibration analysis indicating a low frequency can indicate misalignment, while a high-frequency reading can detect an early bearing failure.

A human being is gifted with five senses: smell, touch, taste, hearing, and sight. He can use these senses to detect early problems with the equipment. Similarly, Condition-Based Maintenance checks the condition of equipment by using sophisticated measuring instruments with precision accuracy. The only difference is that Predictive Maintenance instruments are just a higher form of the human senses. Our human senses are limited. These instruments can sense problems in the equipment far more than our human senses can detect. Let me provide a clear example. If we speak about sound, there are three spectrums. We have the infrasonic range, sonic range, and ultrasonic range. A human ear can only hear the audible range or call the sonic range, around 20 hertz to 20 kilohertz. No amount of human ear can hear an infrasonic or ultrasonic range. Frequencies below the sonic range are called infrasonic, usually below 20 hertz. Few animals in nature can generate or respond to these sound ranges. For example, elephants can hear infrasonic sounds, and they use this ability to communicate with other elephants at very far distances. Elephants have big ears since they need to hear the infrasonic sound, which has large wavelengths. They can hear up to 15 hertz. On the other hand, the frequencies above the sonic range are called ultrasonic. Bats use a very high-frequency burst of sound to locate insects in the dark since insects emit some form of ultrasonic sound. High-frequency sound waves do not bend around objects and reflect in straight lines. This allows the bat to navigate and locate objects without the benefit of using its sight. Dogs can also hear ultrasonic frequencies. They can hear up to 40 kilohertz and can respond to dog whistles. In hindsight, an ailing bearing can emit an ultrasonic range of around 31 kilohertz.

This can be detected using an instrument called ultrasonic detection. This instrument converts the ultrasonic range to an audible or sonic range through the process called heterodyning. Heterodyning translates these frequencies down into the audible range. Heterodyning is the mixing of two waves, which produce both the sum and the difference of their original waves, which allows the shifting of a high-frequency sound to the audible or sonic range. Assume that we have a bearing that is generating a sound signal of 31 to 33 kilohertz. A human ear can hear up to 20 kilohertz. Mixing a constant frequency of 30 kHz sound wave with this signal, we will get the following:

• Adding up the frequency: 61 to 63 kHz, which cannot be heard by the human ear.
• Subtracting up the frequency: 1 to 3 kHz, which can be heard by the human ear.

While ultrasonic allows us to hear the problem, infrared thermography allows us to see the problem. This time, if we speak about the spectrum of light, a person can only see a small fraction of light in the electromagnetic spectrum. God designed man to see only the visible light rays, which is the color of the rainbow. There are many forms of light that we cannot see. In fact, we can only see a very small fraction of the entire range of light in the electromagnetic spectrum. The electromagnetic spectrum includes gamma rays, x-rays, ultraviolet, visible, infrared, microwaves, and radio waves.

Although Predictive and Condition-Based Maintenance are almost similar, there is a slight difference between these two. The difference between Condition-Based Maintenance (CBM) and Predictive Maintenance (PdM) is that PdM tends to be equipment-oriented while CBM is system-oriented. This means that if we speak about using the instruments to monitor incipient failures in the equipment, we speak about Predictive Maintenance. On the other hand, if we speak about the organization involved in doing this type of maintenance, the software used, the monitoring using both online through the use of probes and monitoring them in a control room, the scheduling or in short, if we speak about the entire system on how this is being done, then we speak about Condition-Based Maintenance. Although some books refer to this type of activity as Predictive Maintenance since this rhymes with other strategies such as Reactive, Preventive, and Proactive Maintenance. Here are some of the most common benefits of having a Condition-Based Maintenance Strategy in your industry.

1) CBM/PdM Is More Proactive than Reactive: Condition-Based Maintenance is simply just-in-time maintenance. It involves no fire-fighting, no unscheduled or unplanned repairing of equipment, and specifies when the equipment is sick by monitoring its vital signs and condition. This means that the equipment will be overhauled on a need basis only. Since maintenance replacement and overhauls are based on need only, parts are not replaced based on a scheduled calendar frequency. If correct decisions will be made, the life span of critical parts is maximized. CBM will not replace parts when they can still be used, unlike in Preventive Maintenance, where parts will be replaced whether they are still functioning. CBM alerts maintenance personnel of the problem, enabling both operations and maintenance to plan their equipment activities.

2) CBM is performed while the machine is running: We cannot perform vibration monitorring or infra-red monitoring if the equipment is not running. These instruments allow us to peek inside the machine without dismantling the equipment itself. It is an accurate way of establishing the equipment's condition during normal operations since no shutdown is required. Potential failures can be spotted at an early stage before catastrophic failure occurs. Since CBM can predict that a failure is in the process of occurring, both operations and maintenance can definitely plan for effective and precise scheduling of their equipment for overhaul or replacement purposes.

3) CBM/PdM can help determine the cause of the problem: Predictive Maintenance does not extend the life of parts like bearing service life. However, the maintenance of the decision-making resulting from Predictive Maintenance data allows the users to make sound decisions on performing just-in-time maintenance, detecting and identifying the cause of the problem. Both Infrared Thermography and Oil Analysis can accurately pinpoint the precise location of the failure. Predictive Maintenance is one step towards failure analysis. Once a symptom is detected, the cause can be identified and corrected, extending the parts' service life. When problems are detected early, cost savings can be realized by repairing rather than replacing a part that has failed, like a bearing, which is impacted by misalignment, lubrication, or other damaging conditions. In rolling element bearings, vibration results from the impact generated by a ball rolling over a fractured raceway. Vibration measurements are taken and analyzed over time, which can indicate possible bearing damage. In fact, vibration is the best operating parameter to judge rotating machinery for balance, alignment, bearing stability, stress, and fatigue factors.

4) Reduce Environmental and Safety Consequences: The worst failure modes are those that have environmental and safety consequences when it occurs on its own. CBM plays an important role if the failure has environmental as well as safety consequences. The tragedy in Bhopal, India, can be avoided if ultrasonic leak detection was present during those times. CBM can reduce the chance of dangerous malfunctions of plants that can endanger the lives of its employees and the environment.

5) Reduce Maintenance Costs: As the equipment is only repaired when needed instead of time-based maintenance and routine disassembly, the cost of maintaining the equipment is reduced as resources, labor, equipment, and parts are only used on a need basis. Predictive Maintenance reduces inventory costs because, as a substantial warning of impending failures is provided, parts can be ordered as required, rather than keeping a large inventory in the storeroom.

6) Extend Bearing Service Life: CBM is an essential element in detecting and identifying bearing defects. Once a symptom is detected, the cause can be identified and corrected, thereby extending service life. When found early enough, cost savings can be realized by repairing rather than replacing a bearing impacted by misalignment, lubrication, or other damaging conditions. In rolling element bearings, vibration results from the impact generated by a ball rolling over a defect. Vibration measurements taken and analyzed over time can be a good indication of bearing damage. In fact, vibration is the best operating parameter to judge rotating machinery for balance, alignment bearing stability, and stress factors

7) Peace of mind for maintenance: As failures can be predicted and planned, maintenance can expect less call in the middle of the night for emergency repair work at the plant when they are at home with their family. This means less stress and pressure on maintenance. Although overtime pay will be less, more time can be devoted to having precious and quality time with their family and loved ones. Remember that when Reactive Maintenance is reduced, new doors will open on the maintenance function. The best position there is for industries always belongs to the maintenance function.

Chris Staller believes that Condition Monitoring Program cannot become a successful CBM practice unless the operational market links are made, and the financial metrics are determined. He also explains that the Condition Monitoring market has been relatively low as 2 - 4 % per annum. He adds that maintenance programs that were once responsible for cost reductions and productivity improvements are now subject to the same cost-cutting

All parts have a reason for failing, while some will reach their life most parts will provide symptoms that it is in the process of failing, organizations which adopt CBM strategy have accepted the fact that not all equipment failures are based on a Time-Dominated Frequency. CBM is not an isolated group or an independent entity in the maintenance department; it is not here to replace the PM department; lines of communication between CBM and the other maintenance departments should always remain open.

7.3: March 2013: Step by Step Activities in Implementing RCM

Initiating a Reliability-Centered Maintenance strategy in your plant can be one of the most challenging things to do as it will involve determining the correct maintenance tasks on the asset and should likewise be challenged by the culture itself on maintenance. The analysis can be completed once they have undergone the training, but the challenge lies in initiating the derived RCM tasks. I have laid down these step-by-step activities as a basic guideline for starting your plant's RCM strategy.

Step 1: RCM Prototype Team Selection and Register, the Team: The RCM team should purely be composed of a cross selection of experience maintenance with at least one incumbent operator. On page 266 from the book of John Moubray on Reliability-Centered Maintenance, on who will compose the RCM team, he quotes that these people with the most extensive knowledge and experience of the asset of the process which it forms part. To ensure that all the different viewpoints are taken into account, this group should include a cross-selection of users and maintainers, a cross-selection of the people who do the tasks, and those who manage them. The team should have a regular and fixed meeting date, either once or twice a week, depending on the members' availability. The rule of thumb is that the member's composition should be a minimum of five but not more than eight team members.

Step 2: RCM Team to Undergo 3 Days RCM Training: As a minimum requirement for the RCM process to commence, it is important to note that all the team members should undergo the RCM three days course to provide them the basic understanding and principles on what RCM is all about and the importance of implementing RCM in their plant. The three-day training will provide the members with the basic knowledge they need to conduct the RCM Analysis.

Step 3: Determine the Asset or Equipment Being Analyzed: Industries vary from one another, and performing RCM on the asset can be a little bit tricky. Most consultants will recommend performing RCM on a system based. But if you belong to a manufacturing or process plant, then it's a bit different. Performing RCM can either be done on the most constrained or bottleneck equipment or process. If the equipment or asset is a complex one, RCM can be performed on the worst sub-assembly. It is not recommended to perform RCM on new equipment for two reasons. First, new equipment is under warranty, and the RCM derived tasks might differ from those recommended by conservative OEM and vendors; second, the impact will not be felt since this is a new equipment. We highly recommend RCM to be done on the most problematic equipment where maintenance costs are high.

Assets, components, sub-systems, or systems to be analyzed will depend on the type of industry and equipment. For oil and gas powerplants where their assets are interconnected, candidates for RCM will be critical assets, bottlenecks, and constraint areas. Experience has rather clearly shown that the most efficient and meaningful function lists for RCM analysis are derived at the system level. Having established that a system-based level is the best practical level of assembly to conduct the Classical RCM analysis, we can now focus on which system to address and in what order. In selecting systems, use the Pareto Principle or 80/20 rule. As a rule of thumb, systems should be represented by no more than five functions. Limiting the number of functional systems to be used is to avoid overlapping on the system. RCM is not recommended to be done on every system but only on the plant's most critical system. There should be precise knowledge of what has or has not been included in the system so that accurate lists of components can be identified or will not overlap components in an adjacent system. Backtrack KPI and maintenance indices on your asset for the past 6 months to one year to provide a clear benchmark for downtime, breakdown, cost of repair, mean time between failure, and maintenance costs. If done correctly cost of maintaining the asset should dramatically improve, and breakdowns should likewise be reduced. For the team to realize its benefits, RCM should be done on its most critical assets/systems and not on new machines.

Step 4: Have a Fixed Regular Meeting for the RCM Analysis: Have a fixed and regular meeting weekly, once or twice a week. The meeting will be from 3 to 4 hrs. Appoint a secretary and have all relevant documents of the system or asset analyzed in the room for convenience, such as schematic drawings, OEM, PM Manuals, history records, PdM and PM records, etc. Provide house rules during the meeting and have some refreshments and snacks as this will be a lengthily meeting. The purpose of the analysis is to capture the experience of the members on their assets. Members should be committed to the completion of the RCM Analysis. Management should refrain from calling their people while conducting the RCM analysis; doing this will de-motivate the team's morale in completing the RCM Analysis. There might also be cases where the team needs to go to the actual sight from time to time.

Step 5: Determine the RCM Operating Context Statement: Draw a schematic diagram of the asset and determine its operating context. This is a critical point in the RCM Analysis because the maintenance tasks of the asset to be analyzed will depend on how it is being currently operated. Indicate how the asset is running and write down the most common failures and problems encountered. The criticality of this system on the plant and why this had been selected by the team. Members should be committed to the completion of the RCM Analysis. Management should refrain from calling their people while they are in the process of conducting the RCM analysis. In writing down the operating context statement, state whether the asset to be maintained has redundancy or a stand-alone, affecting the whole plant. Will the asset be running for 24 hours? per day, 7 days a week, 5 days a week? Specify how much time the asset will be running. Specify which parts and subsystems frequently fail during operation. List down the most critical parts of your asset. What months are most critical in operation? Are there specific months that the equipment runs at its peak and optimum performance? Detail any environmental breach as well as well as safety incidents that have happened in the past. What is the average repair time being done on the asset and estimated the total loss of production hours? Draw a schematic diagram of the asset to be analyzed

Step 6: Derive the RCM Information Worksheet (FMEA): The RCM information worksheet is simply an FMEA where you need to derive the asset's functions, functional failures, failure mode, and failure's effects. Determine the asset's primary and secondary functions together with their functional failures, failure mode, and failure effect. The RCM team should be in agreement on when the asset is performing below performance. The team should also be in agreement as to what precisely will constitute a failure/breakdown or not. Next is to group the failure modes into mechanical failures, electrical failures, electronic failures, etc. Detail the failure effect as much as possible. Assume that the team that is doing the RCM analysis will not be exactly the group that will be performing the actual maintenance tasks itself. Remember that the best source of failure modes will always be the asset operators since these are the people that come face to face with the breakdown.

Step 7: Determine the RCM Decision Worksheet: Using the RCM Decision Diagram or algorithm, the RCM team should derive the most feasible tasks for every failure mode. There will be cases where more than one maintenance task will be required to address a certain failure mode. Upon completing the RCM Decision worksheet, sort the spreadsheet by tasks and remove all No Scheduled Maintenance or run to fail tasks. Sort the spreadsheet for the second time by interval and combine all similar tasks. List all protective devices on the system or asset being analyzed and determine its Failure Finding Interval. Include all Failure Finding Activities on the final tasks of the RCM Decision Worksheet. Be sure to combine all redundant tasks.

Step 8: Complete the RCM Documentation: Once the RCM documentation had been completed, review the document for duplicated tasks and combine them. This RCM living document will serve as the correct maintenance task on the asset analyzed through the RCM process.

Step 9: Present To Top Management and Defend the RCM Analysis: Management should understand why the RCM tasks had been derived; hence, management should also know and background regarding RCM. It will be highly recommended that someone from the team should provide an hour or a couple of hour's management overview on what RCM is all about so that when the team and management finally meet then, they are of the same language. The team's purpose is to defend the RCM analysis and that this will supersede their current Preventive Maintenance tasks that they are currently doing on their asset being analyzed. Another reason for the management presentation is to seek their approval for some logistical support that may be needed for the RCM analysis to be done on the equipment. Suppose there are broken gauges to be replaced or Predictive Maintenance instruments needed for the RCM to push through; it should be noted in the management presentation.

Step 10: Final Review and Encode on the CMMS: Note that once the RCM is encoded on the CMMS, the current Preventive Maintenance tasks being undertaken should no longer be performed and must be superseded by the newly derived RCM tasks. Make a timeframe for initiating the RCM Analysis.

Step 11: Monitor Key Performance Indicators: The success or failure in the RCM analysis will depend on the Key Performance Measurements being monitored, such as the number of breakdowns, Mean Time Between Failure, downtime, and the cost of maintaining the asset. These indices should improve as the RCM analysis is being done on the asset or equipment. The RCM analysis should be reviewed by the team every six months for any addition or deletion of failure modes and any changes in their corresponding RCM tasks and interval.

<table>
<tr><td colspan="2" align="center">Gearing Towards A
Pro-Active Maintenance System</td><td colspan="14" align="center">RCM DETAILED ROADMAP
For Pilot Equipment</td><td>RCM</td></tr>
<tr><td colspan="2">AREA ____________
LEADER : ____________</td><td colspan="14">EQUIPMENT TYPE : ____________   Teamname : ____________
EQUIPMENT No. : ____________</td><td></td></tr>
<tr><td colspan="17">OBJECTIVE : To determine's shell asset and equipment to undergo the RCM Analysis and provide it's maintenance requirements</td></tr>
<tr><td rowspan="2">NO.</td><td rowspan="2">ACTIVITY DESCRIPTION</td><td rowspan="2">FORMS USED</td><td></td><td colspan="3" align="center">TRAINING</td><td colspan="10" align="center">RCM ANALYSIS</td><td rowspan="2">REMARKS</td></tr>
<tr><td></td><td>1</td><td>2</td><td>3</td><td>1</td><td>2</td><td>3</td><td>4</td><td>5</td><td>6</td><td>7</td><td>8</td><td>9</td><td>10</td></tr>
<tr><td>1</td><td>Selection of the RCM Elite Prototype Team</td><td>RCM Pilot Regitration Form</td><td>P
A</td><td></td><td></td><td></td><td></td><td></td><td></td><td></td><td></td><td></td><td></td><td></td><td></td><td></td><td></td></tr>
<tr><td>2</td><td>Fan-out team to get initial KPI on each fan-out team</td><td>MTBF Form</td><td>P
A</td><td></td><td></td><td></td><td></td><td></td><td></td><td></td><td></td><td></td><td></td><td></td><td></td><td></td><td></td></tr>
<tr><td>3</td><td>Undergo RCM 3 days training</td><td>n.a.</td><td>P
A</td><td></td><td></td><td></td><td></td><td></td><td></td><td></td><td></td><td></td><td></td><td></td><td></td><td></td><td></td></tr>
<tr><td>4</td><td>RCM Team to have a fix and regular meeting</td><td>n.a.</td><td>P
A</td><td></td><td></td><td></td><td></td><td></td><td></td><td></td><td></td><td></td><td></td><td></td><td></td><td></td><td></td></tr>
<tr><td>5</td><td>Prepare the operating context of your pilot equipment
- Detailed The explanation per sub-assembly</td><td>Operating Context Form</td><td>P
A</td><td></td><td></td><td></td><td></td><td></td><td></td><td></td><td></td><td></td><td></td><td></td><td></td><td></td><td></td></tr>
<tr><td rowspan="4">6</td><td>Define Primary and Secondary Functions of the specified equipment</td><td>RCM Information Worksheet</td><td>P
A</td><td></td><td></td><td></td><td></td><td></td><td></td><td></td><td></td><td></td><td></td><td></td><td></td><td></td><td></td></tr>
<tr><td>6.1 Define Functional Failure of identified Primary and Secondary functions</td><td>RCM Information Worksheet</td><td>P
A</td><td></td><td></td><td></td><td></td><td></td><td></td><td></td><td></td><td></td><td></td><td></td><td></td><td></td><td></td></tr>
<tr><td>6.2 Define Failure Mode of each Functional Failure identified (probable causes of failure)</td><td>RCM Information Worksheet</td><td>P
A</td><td></td><td></td><td></td><td></td><td></td><td></td><td></td><td></td><td></td><td></td><td></td><td></td><td></td><td></td></tr>
<tr><td>6.3 Define Failure Effect of each Failure Mode (Determine what happens when it fails)</td><td>RCM Information Worksheet</td><td>P
A</td><td></td><td></td><td></td><td></td><td></td><td></td><td></td><td></td><td></td><td></td><td></td><td></td><td></td><td></td></tr>
<tr><td>7</td><td>Proceed with the Decision Worksheet to determine if maintenance will be NSM, PM, PdM, FFT, RED</td><td>RCM Decision Worksheet</td><td>P
A</td><td></td><td></td><td></td><td></td><td></td><td></td><td></td><td></td><td></td><td></td><td></td><td></td><td></td><td></td></tr>
<tr><td>8</td><td>Complete the RCM Documentation Process</td><td>n.a.</td><td>P
A</td><td></td><td></td><td></td><td></td><td></td><td></td><td></td><td></td><td></td><td></td><td></td><td></td><td></td><td></td></tr>
<tr><td>9</td><td>Team to perform RCM Self-audit and proceed with the audit</td><td>RCM Certification Request</td><td>P
A</td><td></td><td></td><td></td><td></td><td></td><td></td><td></td><td></td><td></td><td></td><td></td><td></td><td></td><td></td></tr>
<tr><td>10</td><td>Present RCM Case study to management for approval</td><td>n.a.</td><td>P
A</td><td></td><td></td><td></td><td></td><td></td><td></td><td></td><td></td><td></td><td></td><td></td><td></td><td></td><td></td></tr>
<tr><td>11</td><td>Final review and encode into CMMS and revise PM tasklists</td><td></td><td>P
A</td><td></td><td></td><td></td><td></td><td></td><td></td><td></td><td></td><td></td><td></td><td></td><td></td><td></td><td></td></tr>
<tr><td>12</td><td>Monitor results on new maintenance checklist using RCM</td><td>MTBF Form</td><td>P
A</td><td></td><td></td><td></td><td></td><td></td><td></td><td></td><td></td><td></td><td></td><td></td><td></td><td></td><td></td></tr>
<tr><td>13</td><td>Recognize the RCM team</td><td></td><td>P
A</td><td></td><td></td><td></td><td></td><td></td><td></td><td></td><td></td><td></td><td></td><td></td><td></td><td></td><td></td></tr>
</table>

LEGEND : P = PLAN   A = ACTUAL   Revision 1 : August 30, 2016

Figure 7.4: RCM Details of Roadmap

The message of RCM is that a holistic maintenance strategy should be to develop a maintenance strategy that will address both Infant Mortality, Random, and Age-Related Failures. A common misconception among industries is that if the failure is random in nature, this will be added to the never-ending growing lists of Preventive Maintenance checklists giving rise to the equipment's infant mortality failures. If the failure is random in nature, then Preventive Maintenance is not the tasks to be undertaken, but rather maintenance should look on other feasible tasks such as run to fail if the failure has a low impact and consequences, predictive maintenance, if the failure will emit signs and symptoms that it is on the verge of failing or modification/redesign if both runs to fail or predictive maintenance is not a feasible option.

7.4: April 2013: Planned Maintenance Phase 1 Restoration Stage Part 1

Phase 0: Planned Maintenance Preparatory Stage
Phase 1: Stabilize MTBF through Restoration
Phase 2: Lengthen Equipment Lifetime by Addressing Design Weaknesses
Phase 3: Periodically Restore Deterioration
Phase 4: Predict Equipment Lifetime

There are three main activities in Phase 1 of Planned Maintenance, including performing restoration, recurrence prevention of deteriorations uncovered, and standardization. All these activities will establish Basic Equipment Condition for the Planned Maintenance team, where equipment deteriorations are exposed and corrected. In Phase 1, the goal is to bring the equipment as close as possible to its original state. These TPM activities will be done in parallel with Autonomous Maintenance Step 1 Initial Cleaning Activities, performed by the operators themselves. Before conducting Phase 1 activities, equipment is run continuously, and its basic condition is often neglected. Hence equipment is subject to accelerated deterioration, and failures occur randomly and frequently. Deteriorations are left unchecked even if maintenance is very well aware of them. These deteriorations are perceived as normal, and breakdowns are difficult to control since most of the time, the boss does not approve any request for the purchase of spare parts which needed replacement, and to add to the burden, operations waive the equipment for a sound Preventive Maintenance activity to cope with their day to day output demand. This attitude frequently spells disaster since one single breakdown may lead to another and another, causing more downtime while the cost of doing maintenance and repair increases over time. Planned Maintenance Phase 1 aims to change the traditional reactive and firefighting approach by doing the basics and restoring equipment to its original condition.

The goal of Phase 1 is to prevent accelerated deterioration of the equipment. This task involves extending the equipment lifespan by prolonging MTBF or its Mean Time Between Failure. The longer the equipment's failure on the equipment, the lesser the downtime and breakdowns realized. This activity will begin by tagging the equipment for any traces of deterioration and abnormalities uncovered and correcting each of them. Once most of these deteriorations had been corrected, the team can expect a dramatic reduction in the number of breakdowns they experience on the equipment. The team also performs a simple why-why analysis on recurring deteriorations that occurred, and they perform simple corrective measures to prevent the recurrence of accelerated deteriorations.

A machine that had been left unattended for quite some time suffers from accelerated deterioration; it is both the responsibility of operators and maintenance to establish basic equipment conditions on their equipment. Whatever improvement strategy or initiatives being undertaken, one's basic equipment condition had not been established, then the efforts will be deemed meaningless and hopeless. Since most industries are reactive by nature, the maintenance department simply has no time to perform Planned Maintenance because they are too busy dealing with repairs daily, not to mention the pressure operations, people, and maintenance. It is impossible to established Planned Maintenance in this situation. Frequently due to the equipment running on continuous non-stop operations for 24 hours a day, 7 days a week, these simple basic conditions were neglected.

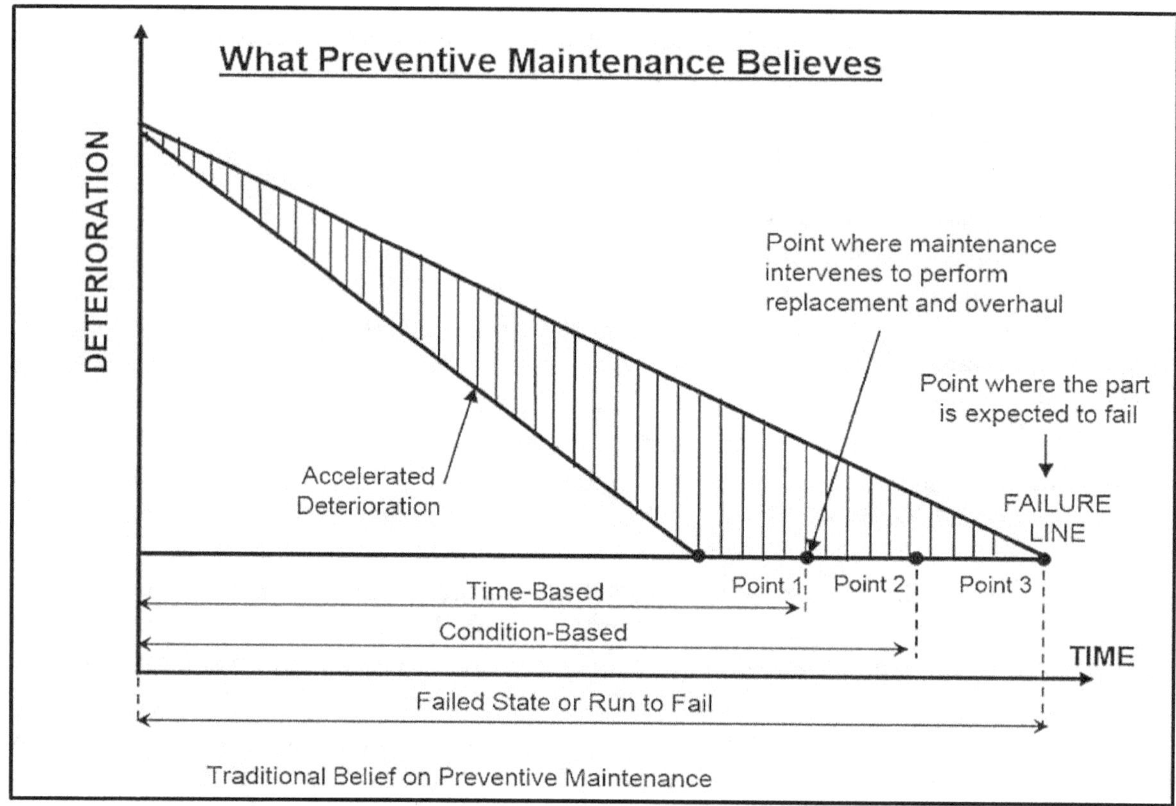

Figure 7.5: Traditional Belief on Preventive Maintenance

Deteriorations on the Equipment:

• Dust, dirt, and abrasion
• Poor working environmental condition
• Severe oil leaks and contamination
• Corroded pipes and internal part of machines
• Non-calibrated and non-working gauges
• Non-functional parts
• Busted sensors and protective devices
• Corrosion of metal parts and cracks
• Missing bolts and nuts due to assembly/disassembly

In this figure, a given part or spare is expected to wear out at a specific point in time, referred to as natural deterioration. Since basic equipment conditions had been neglected, the part or spare fails even sooner before reaching its lifespan. This is called accelerated deterioration. The difference between points 1 to 3 is the amount of time you could have saved, which is the objective of performing restoration on the equipment. Our challenge is to bring these parts to their natural deterioration and avoid failures caused by accelerated deterioration. Once the equipment had been restored, measures must be undertaken to prevent the same problem's recurrence.

The Difference Between Restoration and Improvement

Our basic goal in Phase 1 of Planned Maintenance is to eliminate parts with accelerated deterioration by bringing the equipment back to its original condition. Restoration is simply replacing the parts and not modifying them; we shall leave the redesign and modification subject in later phases/steps of Planned Maintenance. What is important in this restoration phase is that the parts with deterioration are just being replaced. Therefore when we speak about restoration, we simply mean establishing basic equipment conditions by eliminating accelerated deterioration.

To perform this phase on Planned Maintenance, divide the equipment or asset into different sub-assemblies, and an audit is performed on the actual equipment itself. On the other hand, for operators performing Autonomous Maintenance activities, their equipment's basic equipment condition simply means that the machine is cleaned, correctly lubricated, and all bolts are secured and tightened. While Autonomous Maintenance will focus more on correcting the exterior part of the equipment, the Planned Maintenance will focus on the machine's interior part. The goal of the two TPM pillars is to bring back the equipment to its original condition.

On the other hand, modification or improvement means changing something from the original part's specification. This is usually done when the machine suffers from design limitations or weaknesses. These are improvements, modifications, redesign, or countermeasures after analyzing its root cause to ensure that deterioration had been identified and eliminated. An example is that equipment may have an oil leak because the seal was damaged; restoration simply means replacing the seal, while Phase 2 of Planned Maintenance will determine the cause of why the seal was damaged by determining its failure analysis and perhaps recommending for a change in the strength of the material for the seal.

While the operator's Autonomous Maintenance performs initial cleaning and tags abnormalities on their equipment, the Planned Maintenance team is not exempted from this step. The main difference between the initial cleaning performed by operators and maintenance is that Autonomous Maintenance will focus on correcting the exterior part of the machine, while the Planned Maintenance team performs their restoration activities on the interior part of the machine. Addressing these accumulated dirt, dust, rust, corrosion, leaks will be the focus of the PM Team's restoration activities. The PM teams must list all deteriorations uncovered during their initial cleaning stage and correct each one of them.

Note for Autonomous Maintenance Team: While most of the machine's deterioration includes peeled-off paint, do not spend the time repainting the machine; but rather, they should understand what caused the paint to peel off. Once the cause had been determined, only then can they repaint their equipment and assets. Thorough cleaning of the equipment means taking equipment apart to clean internal parts that have never been exposed to before. This kind of cleaning leads the team to discover deteriorations on the equipment. Over time you will learn the correct way to inspect equipment for deteriorations and distinguish them from normal conditions. By performing initial cleaning, the teams address the following on the equipment,

• Dirt and dust on equipment
• Equipment leaks on hydraulic fluid and lubricants
• Overflowing oil pans
• Cutting debris scattered beneath the equipment
• Layers of oil mist on motors
• Strange noise on motors, overheated motors
• Too much vibration on some equipment parts
• Clogged drains and pipes
• Unharness wires and octopus wiring connections
• Oil leaks on the floor and machine compartments
• Dirty and disorganized dies and fixtures
• Broken and missing gauges
• Rust and corrosion on equipment parts and piping
• Worn out or overused parts and spares
• Malfunctioning parts in which equipment still runs
• Dirty and dead sensors

For The Maintenance Bay Area

• Disorganized maintenance bays
• Slippery floors and cab webs on the ceilings
• Improper storage of spares
• No fixed location of tools
• Disorganized files and bulletin boards not updated
• Disorganized spares cabinet, no proper labeling

Many equipment breakdowns occur because of the lack of care on the equipment since most maintenance is too busy dealing with the failures themselves. Performing restoration on the equipment can dramatically reduce breakdowns by 80% or even more if properly executed. Most industries make it a habit to clean their equipment and workplace or perform thorough housekeeping when they know that there will be an audit by one of their major customers or perhaps a visit by a high-ranking person from the corporate level plans to visit the shop floor. Planned Maintenance must not be done this way to satisfy others. If this basic equipment condition is well established, then both operations and maintenance will be satisfied with the outcome. We simply perform maintenance so that we can satisfy the requirements of the user. Often, negligence on these basic factors happens because people perceive breakdowns as normal and part of their routine job. They are custom to perform repair and fixes on equipment as part of their basic day-to-day function. Performing Planned Maintenance requires a change in direction, a change in the way people think, a change in the way people act, and a culture change. Unless these changes are accepted by all levels of the organization in the maintenance department and are willing to perform these necessary steps required, it can determine the turning point or spell the difference of success or failure in its implementation.

Restoration plays a crucial factor in Planned Maintenance, and Top Management must support this activity and be committed. There are cost factors involved in restoring the equipment. To be honest, expect the cost of doing maintenance to increase ones' Planned Maintenance Phase 1 is initiated since we are simply restoring the equipment to its original basic condition depending on how neglected the machine or asset is. Imagine you have been approved of a loan, and you purchase a second-hand car; the moment you acquire the car, you will try to inspect and perform initial cleaning on your second-hand car and note down what needs to be restored and perform restoration on these parts that need to be replaced so that you feel more confident and safe in driving your car. The same thing goes with your equipment, which had been running non-stop to cope up with production demands. There are functional reduction breakdowns that need to be analyzed and restored. Often, this does not affect the equipment; however, they serve a particular function. Spend the time to correct and restore these basic functions through Phase 1 restoration activities of Planned Maintenance.

Detailed Drawing of Equipment Sub-Assembly

It is highly recommended that a drawing on the different major sub-assemblies of the equipment should be done by the Planned Maintenance Team. There are two purposes for doing this, first so that we can place a mark on where each breakdown occurs for easy reference for all maintenance involved, and second, these drawings will serve as training materials for the operator's Autonomous Maintenance activities since one of the major functions of maintenance is to educate and teach operators on their equipment and assets. These training materials will be used later on for teaching operators teams once they moved to the higher steps of Autonomous Maintenance

Planned Maintenance should be the strongest pillar in any TPM implementation because if Planned Maintenance is weak, then Autonomous Maintenance will collapse. After all, one of this pillar's maintenance functions is teaching and educating operators on their equipment. These people will serve as the mentors of Autonomous Maintenance. Rolly Angeles

7.5: May 2013: Planned Maintenance Phase 1 Restoration Stage Part 2

Phase 0: Planned Maintenance Preparatory Stage
Phase 1: Stabilize MTBF through Restoration
Phase 2: Lengthen Equipment Lifetime by Addressing Design Weaknesses
Phase 3: Periodically Restore Deterioration
Phase 4: Predict Equipment Lifetime

Imagine the level of knowledge an operator can have once maintenance teaches them about their equipment and assets. This is not an overnight event, but the point is that maintenance will take the time and initiative to teach these operators about their equipment from time to time.

Communication between the two is the key since operators also know something about the equipment because they are the first person who witnesses and experience the failure first before the maintenance people. These can aid as communication in pinpointing the exact location of the failure, which occurs frequently.

When operators know their equipment, it can facilitate easy communication on what problems or failures occurred in their equipment. Later on, other operators can be taught to perform some very basic minor repairs and troubleshooting. When operators reach this level, then maintenance can improve their equipment and address more severe failures. These will be the main activity highlights of Phase 1 of Planned Maintenance.

Analyze Deteriorations and Prevent Recurrence Of The Failure: After completing the deterioration lists, the Planned Maintenance team meets up and discusses every deterioration uncovered, prioritizes them, performs simple why-why analysis on recurring failures, and single out corrective measures to prevent the problem's recurrence on the equipment. Most of these deteriorations had been accumulated over time because the equipment had been neglected. As the team journey on their roadmap, the basic equipment condition of the equipment is taking place. Maintenance finally understands why these deteriorations had occurred and what effect does it have on their equipment. Although some parts restored may have no relation to breakdown and are considered as functional failure reduction, they may still contribute to some future problems and failures on the equipment.

Document Evidence Before and After: It is highly recommended that a before and after photo of the deterioration be taken. These will serve as their living document on completing each restoration phase activity. Later on, they can also serve as a teaching aid to other maintenance teams and operators performing autonomous maintenance. A digital camera or a mobile phone will be useful in taking these photos in today's world. The team's goal is to correct all deteriorations for Planned Maintenance and abnormalities for Autonomous Maintenance uncovered during their restoration and initial cleaning stage.

Standardization of Documents: Standardization of documentation is an activity for Planned Maintenance Phase 1, where each deterioration uncovered is being crossed checked in their existing Preventive Maintenance Checklists on whether a suitable maintenance task should be performed or not and why such tasks seemed to be overlooked. If no standards had been made, a temporary inspection standard is performed, and a tentative revision of their Preventive Maintenance checklist will be done. This activity aims to include all uncovered deteriorations, with timely inspection and frequency, which must be developed, especially when the failure's consequences would be of high impact. The standards listed here will just be considered tentative standards and must be performed on the specified date provided. These tentative checklists will be finalized when the Planned Maintenance team moves on to the higher phases of Planned Maintenance. Also indicated in this phase activity is to select who will be the most suitable person to perform this kind of inspection, whether this will be done by the maintenance or the operators themselves.

Suppose the responsibility is to be performed by the operators of the equipment; it is important to educate and train these operators through some form of visual aids such as OPL (One-Point Lesson) or GTF (Good to Find) so that operators will understand why they need to perform such a task. Suppose Autonomous Maintenance is not performed regularly or no activity exists; operators will just think that this is an added burden to them and will just pass these tasks back to the maintenance. These must be explained to operators carefully since they are the people closest to the asset whenever a failure or breakdown strikes their equipment. TPM believes that operators are the first line of defense on any equipment improvement strategy.

Planned Maintenance Activity Board: Although JIPM Consultants require Autonomous and Planned Maintenance to have their own activity board in place, it serves as a landmark for operators and maintenance, where they can highlight their progress. Some important things to place on their activity boards will be their KPIs that they monitor on their equipment, such as breakdown trend, downtime, Mean Time Between Failure, maintenance costs, availability, and other measurements. Maintenance should ensure that these indices should be improved dramatically as they complete their equipment restoration activities. Not only does this activity board serve to highlight the team's progress, but it also serves as a reference to all maintenance and operators that, by seriously performing the basics of Planned Maintenance, breakdowns will be reduced dramatically. The goal of Planned Maintenance is to minimize unplanned Breakdowns on their assets and equipment. Teams must update their activity boards regularly and monitor the trend of their graphs. These will serve as evidence of their accomplishment in successfully completing their Phase 1 activities on their Pilot Equipment.

Performing The Planned Maintenance Audit: Audits are done as the Planned Maintenance completes each phase to check that the team complies with the activities performed on Planned Maintenance.   This is initially done by a TPM Facilitator together with the Maintenance Managers. Maintenance managers should own the audit since they will also benefit from these activities. They are also responsible for their equipment. During the initial audit, the Planned Maintenance Facilitator briefs the Maintenance Manager on what to look for during the audit process. The Planned Maintenance Audit will compose of two parts. They will audit both the team and the equipment.

Team Audit - Room Audit or Activity Board Audit: The first part of the audit will be a team presentation, which can be done on the activity board and the auditors. The team briefs the auditors on their respective activities on Planned Maintenance. The presentation usually takes from 30 minutes to 1 hour. The team will highlight their activities. To make sure the team is ready to take on the audit, the Planned Maintenance team performs a self-audit on themselves, and once they are confident about the results, they will finally request an actual schedule for the audit to take place. The audit's objective is to see that the team takes pride in improving their equipment's breakdown rate.

Equipment Audit: After the team presentation, both the PM auditor and the Planned Maintenance team proceed to the equipment. The auditor checks the validity of the documents concerning the equipment's actual condition. They will also look for other unwanted items. If something is found, it will be included in their pending requirements.   The maintenance managers performing the audit will ask the teams regarding the benefits of doing Planned Maintenance and how they will sustain these activities, especially on restoration, so the equipment will not deteriorate and be left unattended again. Previously, maintenance is focused more on repairs on their equipment, battling the day-to-day pressures to keep equipment up and running, but doing Planned Maintenance will minimize breakdowns, and maintenance can have more time to do more relevant work on their equipment. Standards being set must be performed since this will serve as a means to sustain the equipment, otherwise, if these standards will not be carried out, then everything will just go back to the way it was before, and nothing will ever change. Sustaining the equipment should be done at all measures.

After some question and answer portion between the team and the Planned Maintenance auditors, the audit is concluded. The Planned Maintenance team is graded for their performance; if the team had attained the standards and requirements expected from them, they would deserve a passing mark, usually a rating of 85% and above. Suppose many items should be addressed by the team found during the audit; the PM team may receive a remark of unconditionally passed unless otherwise all pending requirements are settled on the date both agreed by the auditor and the team. Phase 1 audit should compose of the following: Completion of the Preparatory Stage, how the pilot equipment had been selected, understanding the current condition of the equipment, restoration activities and its completion, simple why-why analysis on recurring deteriorations, document evidence of before and after restoration photos, KPI indices such as breakdown trend, downtime, MTBF, and maintenance costs and document on how to sustain their restoration activities on their equipment.

Horizontal Replication or Fan-Out Activities: Depending on the number of equipment to be replicated or fan-out, the original Phase 1 team members break themselves into several groups and perform Phase 1 activities to similar equipment with new members from the maintenance joining in. Since the original Planned Maintenance team had already gained some experience in performing Phase 1 activities, they will be responsible for coaching new members on the activities to be performed on Planned Maintenance.

Planned Maintenance Phase 1 Completion: Depending on the number of deteriorations unfolded, Planned Maintenance Phase 1 activity usually takes around 4 to 6 months to complete, and once this is done according to the step activities provided, unplanned breakdowns should dramatically be reduced by as much 80 % or even more, and we have just been doing the basics on the equipment. The pilot or model equipment chosen should belong to Rank A (worst equipment category) and fan-out; replicating restoration activities should be prioritized on similar equipment, which belongs either to Rank A and B (close to worst) category.

Once breakdowns had been reduced on the 1st Pilot Equipment and their corresponding replication activities had already been carried out and completed to other Rank A equipment. The team will select their next pilot equipment and perform the same activities on their 1st model or pilot equipment. The team can compose the existing Planned Maintenance, or new members can join. As each piece of equipment had completely undergone Phase 1 activities, a reduction in the breakdown is achieved. Breakdowns that can stop and paralyze operations can be controlled, and the most important experience is that a change in paradigm from a reactive mode of doing things will be change. Both operators and maintenance change the way they treat and maintain their equipment and assets. Both operator and maintenance skills had been elevated and upgraded since they have fully understood the value of doing both Autonomous and Planned Maintenance. Recurring breakdowns and deteriorations had been analyzed and corrected. As the teams mature in their activities, it now becomes a habit. The team that had undergone these activities realized that the most important part of their job is not how many failures or breakdowns had been repaired for the day but how these breakdowns were dramatically reduced and controlled.

Place Company Logo	Phase 1 ACTIVITY ROADMAP	QUALITY
Gearing for Pro-Active Maintenance	For Pilot Machine and Fan-Out	NO EXCUSE

AREA _____________________ Machine (Pilot) : _____________________ Teamname _________

LEADER : _____________________ Machines (Fan-Out) _____________________

Step 1 : Perform PM Initial Cleaning on Pilot Machine
Step 2 : Restore Deterioration Uncovered and Prevent Recurrence
Step 3 : Preparation of Standard Documents

Step	Act	ACTIVITY DESCRIPTION	Code	FORMS USED	WW
1	1	Track down BDO of machine ensure team understand category when to list Breakdowns	S1-01 / S1-01a	Equipment's MTBF/BDO Analysis Chart / BDO Graph and Trend	Plan / Actual
	2	Perform Initial Cleaning on PM Pilot Machine	------	n.a.	Plan / Actual
	3	Lists all deteriorations seen during PM Cleaning and correct all deteriorations uncovered	S1-03	Deterioration Lists and Countermeasure	Plan / Actual
2	4	Sectionalize the machine per sub-assembly, draw & describe phenomenon and update BDO per section	S2-04	Equipment Section & Phenomenon Form	Plan / Actual
	5	Analyze the equipment for deterioration and prepare why-why analysis on major deteriorations	S1-03 / S2-05	Deterioration Lists and Countermeasure / Why-Why Analysis	Plan / Actual
	6	Set Permanent countermeasure for recurrence prevention of the problem	----	(Put permanent countermeasure on S1-03)	Plan / Actual
	7	Prioritize the worst section and apply restoration	----	n.a.	Plan / Actual
	8	Repeat activity 5 to 7 on other sections until BDO is eliminated, write documentation on main section	----	(Include documentation on main section)	Plan / Actual
	9	Document Evidence Improvement (Before / After) for major restoration	S2-09 / S4-08	Restoration & Countermeasure Documentation (Before / After)/MP Design for Improvements	Plan / Actual
	10	Classify Top Deterioration & prepare GTF or OPL (To be done on major deteriorations uncovered)	S2-10	Good to Find form / or One Point Lesson	Plan / Actual
	11	Update Total machine MTBF / BDO Trend Chart for Phase 1 restoration	S1-01 / S1-01a	Equipment's MTBF/BDO Analysis Chart / BDO Graph and Trend	Plan / Actual
	12	Summarize the lists of deteriorations restored on the different sections or sub-assemblies	S2-12	Horizontal Replication Form for Deterioration	Plan / Actual
3	13	Prepare Inspection Standards on Deteriorations Uncovered Countercheck existence on PM Checklists	S3-13	PM Inspection Standards	Plan / Actual
	14	Post all Step 0, Step 1 to 3 documents on the PM Activity Board, Follow PM Activity Board Guide	----	PM Activity Board Guide	Plan / Actual
	15	Documentation and posting of Activity Board	S3-15a / S3-16	Request for PM Certification / Assessment / Step's 1-3 Certification Sheet	Plan / Actual
	16	PM Step 1 to 3 Audit Certification	S3-16	Step's 1-3 Certification Sheet	Plan / Actual

STEP 1 to 3 ACTIVITIES FOR THE FAN-OUT EQUIPMENT's

Step	Act	ACTIVITY DESCRIPTION	Code	FORMS USED	WW
1	1	Group the machine by module or batch (if many) and back-track BDO for 2 months	S1-01F	BDO Fan-Out Analysis Chart	Plan / Actual
	2	Divide the Original members into 2 and add new members (1 to 2) for Fan-Out and register		Step 1 to 3 Registration Form	Plan / Actual
	3 NEW	Perform initial cleaning on Fan-Out and lists deteriorations uncovered	S1-03	Deterioration Lists and Countermeasure (Use Separate for Fan-Out)	Plan / Actual
2	4	Check Horizontal Replication of deterioration of Pilot's Fan-Out and update for new deteriorations	S1-03 / S2-05	Deterioration Lists and Countermeasure / Why-Why Analysis	Plan / Actual
	5	Perform restoration as required on Fan-out (Same Procedure for Pilot)	-----	n.a.	Plan / Actual
	6	Team to assess PM Activities on Fan-out and Update results of BDO on Fan-Out Module	S3-14	BDO Fan-Out Analysis Chart	Plan / Actual
3	7 NEW	Perform Inspection Standards derived on Pilot on Fan-Out Module	S3-13	Note : Use Inspection Checklists for Pilot	Plan / Actual
	8	Documentation and posting of Activity Board	S3-15a / S3-15	Request for PM Certification / Assessment	Plan / Actual
	9 NEW	Check completeness and compliance on Step's 1 to 3	S3-08 / S3-15a	Compliance Checklists / Request for PM Certification / Assessment	Plan / Actual

PM Phase 1 Roadmap for Pilot and Fan-Out

Figure 7.6: Planned Maintenance Phase 1 Roadmap

Phase 1 is not only about improving the equipment, but it is also about improving the way operators and maintenance think about their assets and equipment. Just as each person needs love and care, the same thing goes with their equipment. When the equipment and assets of your plant are operated and being maintained properly, in that case, they will deliver to perform the task required of them.

It is a good gesture for the maintenance manager to provide simple and inexpensive recognition to the teams that have undergone these activities, as it will boost their morale and be motivated to move into the higher phases of Planned Maintenance. The team should be proud of their accomplishment. A certificate of achievement can be provided by their Human Resource Department with your recommendation, or a simple lunch outside with the team may do. This is important in letting them know that the management team acknowledges their efforts. It will also improve the relationship they have with their people.

The team that comprises the 1st pilot equipment will serve as a success model for every new team doing the same activities; they will be benchmarked and consulted by other teams in their activities on Planned Maintenance. As more and more teams are created, more and more breakdowns are reduced, and maintenance skills are upgraded.

7.6: June 2013: Honesty is the Best Policy" does not apply to MRO Spare Parts

All industries have problems and issues in their storeroom and MRO spare parts, yet no one will admit that your spare parts inventory is in worse shape than you can ever think, and your team does not really know how to fix the problem. We usually stock parts that we really do not want to use, just like a spare tire in the car. We just simply love to stock them; if you ask why, the answer is simple, just in case.

Most industries seem to develop their storeroom in a hit-and-miss fashion. How many times have you went to your storeroom and the part you are looking for is unavailable or not around. If this happens to you frequently, then maintenance will purchase the part themselves and will buy not just one but in excess quantities and making him the keeper of the rest, why, because they have already lost their trust in the storeroom and storekeeper. One of the main causes of interrupted maintenance work in the industry is the lack of spare parts. In some cases, spares were not identified in advance or ran out of stock when needed for a repair. The worst is that the storekeeper simply cannot locate the parts in the storeroom since everyone has access to the storeroom during midnight and peak hours.

In a reactive industry, the spare part is defined as those available in plenty when maintenance doesn't need them and is not available when maintenance needs them most. A recent survey indicates that 60 to 80% of maintenance expenditure is accounted for by spare parts consumption in industries. Many plants and organizations admit that spare parts and storeroom functions are the most neglected functions in any industry. In fact, it may be the missing link on any reliability and maintenance strategy.

If a machine fails and the part is not available in the storeroom, maintenance has two options. The first option is that they will cannibalize idle equipment and get the part that they needed; the second option is that maintenance will result in emergency buying, buy in excess, and keeping the part near them on the assumption just in case it might fail again, now I have the part. Almost all the storeroom problems are man-made, or in layman's terms, the problem was created by man. One of the problems a storeroom experience is that if their operation is 24 hours, the storeroom is just from 8 to 5 pm. In this case, everyone can have access during peak or late night hours when a breakdown finally struck their equipment, and a part is badly needed. The saying that "Honesty is the Best Policy" does not apply to Spare Parts and Storerooms. The storekeeper should control what parts and spares come in and go out of the storeroom. In short, in Spare Parts, you need to apply the golden rule, which is "**TRUST NO ONE**."

If everyone can access the storeroom, always expect that the physical inventory and system inventory will never match. Management and decision-makers must understand that if their operations are 24 hours or 3 shifts, the storeroom should also be 3 shifts. Trust and honesty systems do not apply in Spare Parts Management. Since the part is needed ASAP, front-line employees and maintenance people will be in a rush to acquire the item and will not be inclined to fill up the paper works and documents. This usually happens when the storeroom keeper completed his day's work and leaves the storeroom key to the guard on duty. Two problems will be evident in this situation. First, expect your physical inventory to be lower than what is actually listed on the system. Second, there is a strong temptation for maintenance to store the part elsewhere in the plant, near their equipment, or even in their home for personal use. Perhaps not all.

All spare parts that go out of the storeroom should always have a record, as this will be the basis on what parts needs to be stored and reordered. If there is no control over the parts that go out of the storeroom, then expect parts to be unavailable when needed. When the actual inventory is lower than the system recorded, the risk is high that an out-of-stock condition can occur because parts will not be ordered on time. When the actual inventory is higher than the system recorded, in that case, parts will be flagged for re-ordering by the system even if needed. It is critically important that the storeroom people and the users should understand the importance of inventory accuracy. If inventory does not match the system, the users will lose their confidence and start stocking the parts independently. When the machine fails and the part is not in the storeroom, maintenance will cash advance and purchase it excessively. These excess parts will be kept by the maintenance in their secret hiding places or what they called "Squirrel Stores." They will be the only one who knows where the part is hidden. If the same part failed in the evening when the storekeeper was no longer around and since you are the only one who knows where the part is, it will be reordered.

To start a storeroom improvement strategy is to surrender all the parts maintenance people store independently and let the storeroom handle them. These stocks out log should be kept by the storeroom people to determine the number of times an item is requested but are not currently held in stock or has run out of stock. A good target value for stockout items would be around 0.5% each month, but the ultimate goal is to have a zero stock-out. Remember that keeping or hoarding the part will only satisfy you and not other maintenance in the plant, and if everyone has been doing these for quite some time, then we actually have no idea how much inventory there is outside the storeroom.

Stock-Out Explained: A stock-out condition occurs when the inventory reaches zero before the replenishment stock arrives. A safety stock level and its lead time should always be known to avoid stock out. Frequent stock-out can make maintenance lose their confidence in the storeroom and start to keep the parts themselves. Although it is not a recommended practice to keep every single part in the storeroom, neither it reduces inventory by creating large stock-outs. Storekeepers should also be aware of the consequences of a prolonged downtime because the part is not available when needed. There are three causes of stock-outs: First, the holding parameters may have been incorrectly set for the expected quantity or incorrect demand for the current situation. Second, there were some delays in the demand or supply chain, including the internal processes associated with purchasing, which may include reordering and its lead time, and third, the part is not stocked.

The disadvantage of Having Squirrel Stores: There are many disadvantages of having squirrel stores in the plant as it will only satisfy the parts' keeper. By its simplest definition, squirrel stores are unofficial storage of spare parts. They are often called maintenance secret hiding places for spare parts.

• Parts kept on oneself are not controlled by the storeroom
• Other users of the part do not know you have the part they needed and may reorder the same parts on their own
• Parts are not controlled and maintained and may cause the parts to rust and accumulate dirt
• Overall costs of inventory on spare parts can never be known

Storeroom Security: Many companies are aware of the importance of maintaining security around their storeroom and inventories. They secure their spares in lock-up areas, ensure that staff is on hand to manage spares requests during their shift period, but one thing they forget is that they left the back door open. Storeroom staff cannot police all store entries, but this is rarely the case. Diverted attention while making deliveries, managing inquiries, or attending to a request could mean that attention to demand is not there when needed. And to make things worse, there is often a door left open through which anyone can access. Storeroom security ensures parts required for maintenance are available when needed. Ensuring that parts are on hand and available when maintenance needs them requires control.

MRO storerooms must be walled or fenced off from surrounding areas. The walls should be high enough to prevent entry using ladders or other means. Storerooms should have lockable gates and doors and restrict entry to only authorized staff. Door locks should incorporate key-card technology that many companies use to limit building access. When possible, managers should provide 24/7 coverage or have dayshift coverage by storeroom people. No parts should leave the storeroom, warehouse, or satellite without being accounted for by the Storeroom people. If anyone can have access to the parts, both actual and systems inventory will never match. Removing items from the storeroom without proper record, no matter how honest the intent is, will not trigger reordering when required.

Storeroom control is more than just having neat shelves and good labeling. Storeroom control requires efficiency in processing items, both in and out of the store, and control over access so that only quality parts are stored and all withdrawals are recorded. When your plant's operation is three shifts, your storeroom should also be 3 shifts. Reordering, purchasing, receiving, and restocking functions should be accurate, efficient, timely, cost-effective, and a stock-out chance of less than 0.5%. Our main objective in securing the storeroom is that we want to maintain a high level of inventory accuracy of around 95% or more, and we want to make certain that whatever goes in and out of the storeroom will be documented and put on the record

Surrendering All Items On Your Squirrel Store: Maintenance team members take parts and put them away in their own secret stores, and when really needed, the part cannot be found. Maintenance does this because they think it is convenient or saves time for him but not for the rest. Operations and maintenance are two different people. Everyone knows that when things go wrong, maintenance gets the blame, but when things go right, production gets the credit. As a result, maintenance will keep the parts for themselves just in case. That's why these unofficial stores are often referred to as squirrel stores. Squirrel store exists because maintenance has lost its trust in the Storeroom people due to frequent stock-out. If your storeroom is unreliable, then I better keep the part myself. Everyone has been squirreling away parts, so I might better do the same, just in case. If the equipment fails, it's your job to fix it, and if there is no spare, you get the blame from production, even though you did not cause the asset to fail.   The maintenance mindset is to reduce the response time whenever a breakdown happens. This thinking drives them to keep the parts themselves. Hoarding parts benefit you and not the industry. When you order the part in the storeroom in excess to keep in your squirrel store, it will show higher demand than what is actually being consumed.

Squirrel stores are just one of the many problems experienced in the storeroom.   The problem with squirreling is that we just satisfy ourselves but not the others. Squirreling costs money, since if one is keeping a spare which is not available in the storeroom, then this will trigger a reordering. All spares should be controlled under one roof, and that is your MRO storeroom. Squirrel stores exist for two reasons. First, the stock-out part is mostly experienced, and you purchase the part, or second, you order excess parts in the storeroom and try to remain the keeper of the parts. When your storeroom is a distance away from where you operate, it is best to decentralize your storeroom. Still, the inventory will reflect the same as in the central hub of the main storeroom. Physical location will be reflected upon in the system. Remember that there are also other users of the parts that you intend to keep, and if they don't know that you keep this part, then the storeroom will replenish or reorder this part in their system. So with this said, I think it is time to give it some serious thoughts on whether we will continue with squirreling parts or surrender them back to the storeroom.

7.7: July 2013: Equipment 6 Big Losses Part 1

Industries vary from one another and the type of equipment they used. Equipment losses also vary from one another depending on the type of industry. Below is the TPM list of losses equipment can suffer.   This does not include interruptions or downtimes caused by outside factors such as loss of power, facilities requirements, operator skills, etc. It is important to note what type of losses your equipment encounters most of the time and who should be responsible for addressing them.   It is recommended that each loss be treated separately, as this type of loss needs different types of people involved.   Each of the losses specified below must not be solely addressed by the maintenance people alone but by a cross selection of expertise from the plant.

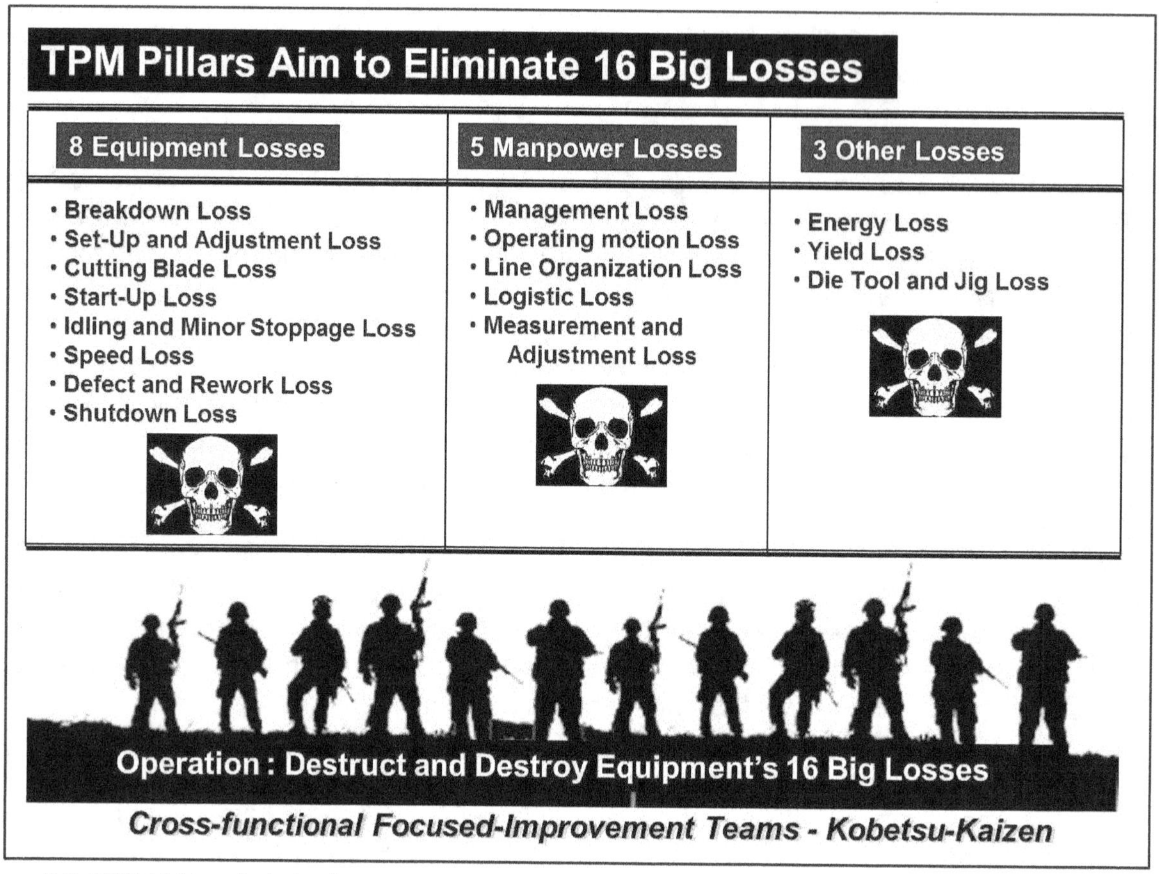

Figure 7.7: TPM 16 Manufacturing Losses

1st Equipment Loss - Breakdown Loss: Sometimes called equipment failure, it is a loss when a machine stops due to loss in its specified functions. There are two types of breakdowns, Function-reduction breakdown, which is when deterioration of equipment causes other losses in function even when the equipment can still operate. Imagine your car running and stating that a car's primary function is to travel from distance A to distance B. It can comply with the function, but one of the car's headlights is busted, the side mirror is missing, the hub of the wheel is missing, the car has excessive oil leak due to damage to the seal, car's air-conditioning does not work, the wiper is missing, the seat belt is not functioning and so on, but the car can deliver its primary function which is to travel from distance A to distance B. This is what function reduction breakdown is, and most equipment suffers from this type of loss. It seems that most of the time, this type of loss is frequently being neglected. These are breakdowns that account for the largest proportion of overall equipment losses. The second type of breakdown is a function-loss breakdown, a failure in which the equipment stops completely. These are losses in which production is stopped. Another term is unplanned downtime or simply termed as breakdown maintenance.

Breakdowns halt production, deliveries are delayed, quality problems arise, maintenance cost is expected to increase, and a single breakdown can create havoc throughout the plant. Analysis in the breakdown failure can be attributed mostly to human problems, basic neglect of design problems, lack of operator skills, lack of repair skills, and failure to addressed equipment basic condition. That is why a basic understanding of the failure must be address thoroughly.

Breakdowns are caused by many factors, and most of the time, slight deteriorations are overlooked, which contributes highly to equipment breakdown. Improvement in equipment performance can be done by simply addressing minor problems such as loose bolts and missing screws; abrasion, debris, and contaminant are addressed. Zero unplanned breakdowns can be established by addressing the following:

• Prevent accelerated deterioration
• Maintaining Basic Equipment Condition
• Maintaining Operating Condition
• Improvement of Maintenance Quality
• Addressing root causes of breakdowns
• Guidance for Autonomous Maintenance
• Correcting Design weaknesses

Once a breakdown occurred, be certain to learn everything you can and study the causes. Conditions, maintenance tasks involved, repair methods, much can be learned to prevent it from happening. The basic responsibility of each maintenance is not to repair breakdowns but to analyze what had caused the breakdown problem and draw measures to prevent the recurrence of the problem. If this is not being performed, you will just be wasting your precious time on doing repairs and quick fixes on the same problem over and over again.

2nd Equipment Loss - Conversion and Set-Up Loss: Conversion or set-up loss is the time required to remove dies, jigs for one product, clean-up, prepare dies and jigs for the next product, reassemble the equipment, adjust the equipment, perform trial runs and make further adjustments until the product of acceptable quality is finally obtained. There is equipment mostly on manufacturing plants that are not dedicated, and it is used to process more than one kind of product. This usually begins when the production of one product is completed and ends when standard quality is attained in producing the next type of product being processed. Shigeo Shingo's (SMED) Single Minute Exchange of Dies deals in techniques in reducing set-up time and adjustment time without reducing its accuracy. According to him, a good set-up time in manufacturing must fall between 10 minutes and below. Findings of Shigeo Shingo, why set-up time is prolonged is due to the following factors:

• Preparation of materials, jigs, tools, and fittings   20 %
• Removal and attachment of jigs, tools, and dies   20 %
• Centering and Dimensioning   10 %
• Trial Processing and Adjustments   50 %

The first step in improving set-up is distinguishing the activities performed while the equipment runs from those performed when it is shut down. Differentiate external from internal setup. External set-up are those activities that can be performed while the machine is running, while an internal set-up are those activities that can be performed when the machine had been shut down for conversion. The goal of set-up is to minimize the time to perform it, and one of the techniques is to write down all the steps performed in doing your set-up, then converting internal to external set-up. This can be accomplished using a standard one-touch jig, comparing the shapes of tools and jigs for different products, and considering preparing a standard or universal jig that can be shared by all. Eliminate adjustments during internal set-up time by using intermediate jigs.

Figure 7.8: Shigeo Shingo's Single Minute Exchange of Dies (SMED)

Here are a couple of steps that can be taken to eliminate the need for adjustment. First, in many cases, adjustments can be scaled down simply by improving the precision of the equipment, jigs, and tools. The accumulation of imprecise settings creates the need for many avoidable adjustments. Second, standardize procedures. Lack of consistency in the standards for measurement, quantification, and other operation and maintenance procedures is another cause of unnecessary adjustments. Set-up losses cannot be eliminated, but they can be reduced dramatically. Poor set-up and conversion can cause other losses such as breakdown, quality problems to occur. It is best that procedures that the conversion of each product must be standardized. One maintenance can set up one piece of equipment differently and take 15 steps, while one may take as much as 20 to 25 steps. It is important that whoever is performing the set-up procedure has a set of detailed step-by-step standards to follow. Much can be learned by knowing the difference between what are those that are considered as internal and external set-up practices, and most of them only need a modicum of common sense.

<u>Some Tips on Shortening Internal Set-Up Time</u>

• Simplify clamping mechanism by using quick fitting jigs
• Adapt parallel operations; two people working together can perform a set-up faster
• Optimize the number of workers and division of labor, especially for large and complicated set-ups; doing these simple basic steps will greatly reduce your set-up time

Eliminating Small Losses in Set-Up

- What type of preparations needs to be made in advance?
- What tools must be on hand?
- Are the jigs and tools to be installed in good condition?
- What type of workbench is needed?
- Where should jigs and dies be placed after removal?
- How will they be transported?
- What types of parts are necessary?
- How many maintenance people are needed to perform the set-up?

Traditional Setup Approaches

Processing set-up and conversions are the main emphases in most plants. First, Shigeo Shingo (author of SMED, Single Minute Exchange of Dies) addresses the traditional strategies for improving setup in operations. In traditional manufacturing operations, efficient set-up changes require knowledge and skill. As a result, efficient set-ups require highly skilled workers or "set-up engineers." While a set-up engineer performs the set-up, the machine operator is idle or performing other insignificant or miscellaneous tasks. Shigeo Shingo points out that this approach on set-up and conversion is a common misconception and is very inefficient. Many companies have set up policies to raise the skill level of workers, while few have tried to implement strategies that lower the skill level required by the set-up itself. Traditionally, manufacturing companies have also increased lot size to hide the effects of longer setup times. Large-scale production seems the easiest and most effective way to minimize set-up time's undesirable effects because the set-up time is minimal compared to each unit's operating time. This large lot size also increases inventory. The inventory itself does not produce added value, so the space it ties up is wasted, and the inventory costs increase. Inventory stock is at the risk of becoming outdated or even damaged. This approach assumes that reductions in set-up time are impossible. The SMED system makes these reduced set-up times a reality and a better alternative to increasing lot size.

The SMED System (Single Minute Exchange of Dies)

Shingo found that set-up operations are composed of two different types, internal set-ups, and external set-ups. Internal set-ups are set-ups that can be performed only when a machine is stopped. External set-ups are set-ups that can be conducted while the machine is in operation. This distinction between internal and external set-ups is one of the driving factors behind SMED; reduce set-up times by converting internal set-ups to external set-ups. SMED was developed over nineteen years due to closely investigating the theoretical and practical aspects of set-up improvement. It is a scientific approach to set-up time and reduction applied in any factory to any machine. It was implemented into the Toyota Production System and has helped them to become the leading production system. In traditional setup operations, internal and external setups are mixed up. Some set-ups that could be done externally are performed as internal set-ups, causing machines to remain idle for extended periods. The first and most important stage in implementing SMED is to identify which set-ups are internal and external.

The next stage is to convert the internal set-ups to external set-ups. The third and final stage is streamlining all aspects of the set-up operations. Many different kinds of waste occur in traditional set-up operations because internal and external operations are not distinguished. For example, in some manufacturing facilities, the machine will be turned off when the finished goods are transported to the warehouse, wasting valuable time.

The first stage of SMED is to separate internal and external set-ups. This task sounds easier than it actually is. Therefore Shingo suggests using various techniques to accomplish this task. Checklists, charts, check tables, functionality checks, and identification and transportation improvement to and from the machines are all suggested to help facilitate this stage. The next stage is to convert the internal set-up to external set-up operations. This stage is different from each process, and Shingo provides various examples. Shingo introduces function standardization, where standardizing only those parts whose functions are necessary from set-up operations. Efficient function standardization requires analyzing each piece of apparatus's functions, element by element, and replacing the fewest parts possible. When converting internal set-ups to external, standardizing the process as much as possible will reduce the set-up time. Standardizing is accomplished through different jigs and clamps, making everything the same, so set-up changes are minimal. Standardization reduces set-up time substantially, simplifies the organization, eliminates the need to search for appropriate tools, and eliminates the need for adjustments. The final stage in SMED is streamlining all aspects of the set-up operation. Improvements in transportation and storage of all parts, products, and tools can assist in streamlining operations. This stage doesn't necessarily reduce set-up time by itself, but it does aid SMED in creating a continuous flow.

Techniques to Implement SMED

Some work in a manufacturing facility involves work at both the front and the machine's back. A single person working at one of these machines wastes time and movement when they continually walk around the machine. Shigeo Shingo suggests parallel operations involving more than one worker in this situation. The most important issue in parallel operations is safety. Some sort of signal system has to be worked out so that workers in the process know when to do their respective jobs. The concept of parallel operations is also a technique that can be used to streamline operations. A functional clamp is an attachment device that is used to hold objects in place with minimal effort. Shigeo Shingo points out that a nut and bolt are used to fasten or tighten a clamp most of the time. If the bolt has fifteen threads on it, it can't really be tightened until the last turn, loosened in the first turn. The other fourteen turns are wasted. It is with this observation that Shingo spent time developing one-turn attachments and one-motion methods.

There are also several techniques used to eliminate wasted adjustments. Adjustments and test run normally account for as much as 50% of the set-up time. Therefore, if there is a decrease in the adjustments, there is a reduction in set-up time. Inaccurate centering, dimensioning, and other procedures in the internal setup necessitate test runs and adjustments. To eliminate these adjustments, Shigeo Shingo says that we must improve the stages of internal setup. Minimizing these adjustments can be accomplished by fixing numerical settings, setting centers on machines, using gauges, and using reference planes. Shingo also introduces the Least Common Multiple systems as a technique for eliminating adjustments. Finally, after every attempt has been made to improve set-ups, mechanization could reduce set-up times. Mechanization should be considered last to reducing set-ups because it is an inefficient setup operation that will achieve time reductions, but it will do little to remedy the basic faults of a poorly designed setup process. Mechanization can also cost a lot of money to implement. It is more efficient to mechanize set-ups that have already been streamlined. Shigeo Shingo strongly believes that SMED success involves knowing why the system works rather than just knowing how to implement it. Therefore, he provides a plethora of actual examples of the SMED system in operation so that the reader can gain further insight into the concepts and their principles.

Effects of SMED (Single Minute Exchange of Dies)

The goal of SMED is to reduce set-up times and conversion of equipment that produce different products, but there are also other effects that SMED has on a production system. One effect that SMED has on a production system is that inventory is minimized. An inventory reduction can lead to more efficient use of plant space. The unusable stock due to model changeovers or mistaken estimates is eliminated. Production is increasing because stock handling operations are eliminated. Goods are no longer lost through deterioration or damage. The ability to mix the production of various types of goods leads to further inventory reductions. Some other beneficial effects of SMED are that machine work rates and productive capacity are increased since downtime is reduced due to set-up and conversion. There is an elimination of human errors on set-up and conversion, and the elimination of trial runs lowers the occurrence of quality defects on products. Quality and safety are both improved. Standardization reduces the number of tools required, and those that are still needed are organized more functionally for easy retrieval and access. Tool changes are quick and simple. Eliminating the need for highly skilled workers to perform the set-up, therefore lowering the skill level requirements for a process. SMED increases the manufacturing flexibility because of faster time in doing changeovers. Therefore a company can increase its production flexibility because it will respond rapidly to changes in demand. SMED can also have an effect on the attitude of people in the company. SMED welcomes employee involvement towards continuous improvement. Its effects consist of more than just reduced set-up times, conversion, and improved work rates. Companies can gain more from Shigeo Shingo's SMED technique than they realize. To be continued.

7.8: August 2013: Equipment 6 Big Losses Part 2 (Continued)

3rd Equipment Loss - Idling and Minor Stoppages: As industries' equipment becomes more complex and automated, more losses are attributed to Idling and Minor Stoppages (some companies term this as assist or errors). First, let us define this type of loss. A minor stoppage occurs when a failure or an error in automatic handling, processing, assembly of parts and workpieces, or an equipment stoppage due to quality-related abnormality. This type of loss arises mostly in automated processes and includes the following:

• Workpiece flow stops

• Operator resets workpieces correctly
• Operator reactivates process and machine runs

The problem with this type of loss is that the number of occurrences far exceeds the breakdown, and fixing this type of failure requires a short time. This type of error is often left unrecorded since the time to fix it can be done by just resetting the buttons, and the operator will spend more time recording and writing down the error. Idling and Minor stoppages sometimes are mistaken for breakdowns, which must not be the case. These types of losses must be separated from breakdowns. These losses can be addressed by Autonomous Maintenance through initial cleaning. JIPM experts believe that having equipment free from dirt and dust can reduce minor stoppages from 20 to 60%. While more complicated minor stoppages that cannot be reduced through cleaning must be addressed by a cross-functional Kobetsu-Kaizen or Focused Improvement teams composing engineers, quality, production, and maintenance people. Some examples of Idling and Minor Stoppages include a workpiece jamming, resetting of sensors, an error reading on the monitor of computerized equipment where the machine is stopped temporarily, operator resets, and the machine starts runs again. These problems occur more frequently in automated equipment. In addressing minor stoppages, ensure the following, observe what is happening. These can be done by carefully observing the equipment until a minor stoppage occurred and planned corrective measures. Correcting slight or minor defects, a small dent in the chute may cause minor stoppages, and frequently, we overlook this factor; maintaining the cleanliness of the equipment can reduce minor stoppages.

4th Equipment Loss - Design Speed Loss: Design Speed Loss is the loss of production or output caused by the difference between the design speed or theoretical speed and the equipment's actual operating speed. Lack of care at the design stage of the equipment may result in speed reduction. Equipment is operated beyond its operating speed limit, quality defects and breakdowns are encountered. Although Japanese and European maintenance systems had conflicts on this type of loss, Japanese TPM experts still consider this a type of loss. According to them, to reduce speed loss, we must address the following:

• Level 1: Achieve Standard Operating Speed for each product
• Level 2: Increase Standard Operating Speed for each product
• Level 3: Achieve Design Speed
• Level 4: Surpass Design Speed

Equipment may be run at less than the design or ideal speed for various reasons, contributing to mechanical problems, quality, defect problems, history of past problems, and sometimes not knowing the optimal speed. On the other hand, deliberately increasing the operating speed actually contributes to problem-solving by revealing latent defects in equipment conditions. The goal is to eliminate the gap between the actual speed and the design speed of the asset. Design speed is reduced due to problems with quality or mechanical trouble, past trouble resulting in short lifespan, equipment being operated with its specification unknown, or increasing the speed will reveal concealed problems either on the product or equipment itself.

Running the equipment on its rated or design speed will produce problems with the quality of the product. Thus, maintenance will reduce the machine's running speed; however, in most cases, production will increase the equipment's speed to deal with productivity issues and provide more serious downtime. The maintenance needs to know what parts are affected when the speed of the equipment is increased. Study the part concerning its design, shape, the strength of materials, and so on. TPM Planned Maintenance Phase 2 will focus on lengthening the equipment lifespan by addressing inherent design weaknesses.

5th Equipment Loss - Start-Up Loss: Start-Up loss is a type of loss that occurs during an equipment start-up or run-in. Problems arise during starting the equipment. Start-Up Loss means that the material loss caused at the initial stage of product launching, namely the loss caused during the start-up of production to the stabilized production stage. Its frequency depends on several factors, such as unstable machining conditions, poor maintenance and operator skills, human errors, and infant mortality failures.

Start-Up loss can also occur after a poor set-up or conversion, after overhauling equipment subject to a Preventive Maintenance Schedule. It takes some time for a machine to stabilize. Start-up loss is difficult to identify. Their scope includes the stability of processing conditions, workers' skills, training, the loss incurred by test operations, and other factors. A thorough understanding of why this type of loss occurs can be addressed through RCM or Reliability-Centered Maintenance. RCM term this type of failure as infant mortality failure.

6th Equipment Loss - Defect and Rework Loss: This is the loss caused when defects are found and the product has to be reworked. In general, these defects are likely to be considered a waste that should be disposed of, but even the reworked products need wasted manpower to repair them.   Included in this type of loss are products being shipped back due to customer complaints? Products are returned and reworked in the production area.

Quality Defects and rework losses are caused by malfunctioning production equipment. In general, sporadic defects are easily corrected. Chronic defects require a thorough investigation and innovative remedial action. The conditions surrounding and causing the defect must be determined and then effectively controlled. Tools such as P-M Analysis (P stands for Physical, Phenomenon, while M stands for Mechanism and the 4M Man, Machine, Method, and Materials) are suitable in dealing with chronic problems. This tool will aim at reducing the defects to zero. However, P-M Analysis must only be used after extensive use of conventional tools had been used, and around 1 to 5 % of the problems still exist; this is where P-M Analysis must be used ideally. Sporadic defects are easy to solve, and it is more difficult to solve chronic defects.

Although other TPM books include 8 Major Equipment Losses, the other two losses mentioned are shutdown loss, which means stopping the equipment for a periodical or Time-Based Maintenance, and Cutting blade change, which is loss caused by line stoppage for replacing the grinding wheel, cutter, bits which might be broken or worn out after excessive usage.

In implementing TPM, expect a lot of resistance in the beginning. This is normal; what is not normal is that if no one shows resistance at all. TPM is a long journey with its ups and downs, but once the industry reaps its benefits, your efforts will be well rewarded. Remember, we are not only improving the equipment, but we are also improving the people behind the equipment.

7.9: September 2013: Lessons on TPM (Total Productive Maintenance)

Once carried out according to JIPM 12 Developmental Steps, TPM would yield significant results, and lessons can be shared and learned. Results from TPM will come after a couple of years or more of implementation. Remember that TPM is never an overnight affair and will take time to yield results. Also, TPM is all about people, a lesson that we all must realize and that only the right people will improve their equipment, and it is not the other way around. All people from the plant will have their roles in TPM from the top-down to the shop floor. Results motivate people and provide pride in their workplace. Rewards and recognition are essential to sustain enthusiasm among the teams implementing TPM, especially Autonomous Maintenance, Planned Maintenance, and Focused Improvement. These pillars will work together as teams to realize an improvement in their OEE indices. Here are some basic lessons on TPM that we should reflect on for a while:

TPM Lesson 1: Most failures start from small things left unattended and neglected; TPM emphasizes establishing Basic Equipment Conditions on their machines. Basic Equipment Condition includes cleaning, lubrication, and completing bolts needed for the machine. The message is straightforward, and this what gives TPM its simplicity, whatever improvement strategy being initiated by the plant on their assets and equipment to improve its reliability, whether it is TPM, RCM, FMEA if these simple and small problems will be neglected, then the outcome may not be as what you want it to be. Most equipment failures and breakdowns are simply an accumulation of very small problems neglected over time, which we constantly perceive as normal in our day-to-day operations. For example, if 10 bolts need to be secure on the equipment and only 8 bolts are fastened, it increases the equipment's overall vibration.

TPM Lesson 2: Implementing TPM is a Top-Down Approach and cannot never be a Down-Up Approach. Both the company and TPM goals must be aligned for TPM to be successful. The company's President or CEO must make a formal announcement that TPM will be adopted.

This makes TPM a bit different from the rest of the improvement strategies since its implementation is a TOP-DOWN Approach. This means that the CEO or President of the company should be the one to make the formal announcement that TPM will be in place in the organization. Once the President/CEO makes the announcement, the rest of the TPM details and pillar implementation will begin.

TPM Lesson 3: Performing TPM requires a change in direction, a change in the way people think about their assets, a change in the way people act, and most importantly, a change in the culture of the organization and unless these changes are accepted by all levels of the organization and are willing to perform the necessary steps required for such a change to take place then only can we realize the difference between success and failure in any TPM undertaking. TPM wants to create a change in thinking on the people working within the organization. Training will play a key role and will serve as the foundation for any TPM initiative or start-up. TPM requires a change in the way things are being done in a day-to-day fashion. TPM dreams of an ideal factory with zero breakdowns, zero accidents, and zero defects, and slowly TPM does these incremental, step-by-step improvements on each pillar to bring the plant closer and closer to this ideal factory.

TPM Lesson 4: Companies that achieve results in their TPM journey usually challenge the JIPM Awards to attain certification. There are around 100 criteria that must be satisfied, and mostly the reason for attaining JIPM certification is to provide an edge in their business competition. What makes TPM unique from other improvement strategies is that those that have benefitted from TPM activities and have reaped its fruit can apply for the most distinguished award on TPM, which is the TPM Excellence Awards Category 1 and 2, which is provided by the JIPM or Japan Institute of Plant Maintenance. Industries that achieve the JIPM Excellence Awards provide an edge in their competition.

TPM Lesson 5: Although all pillars are important in TPM, the strongest pillar must be Planned Maintenance as these are the people who will educate and coach Autonomous Maintenance. Although AM is the larger force, PM must be equipped with the right training and tools to coach operators. The Planned Maintenance pillar usually compose of a cross-selection of maintenance people in the plant. Planned Maintenance's goal is two-fold, to reduce the number of unplanned breakdowns in their asset and equipment and educate operators about their equipment. Planned Maintenance will partner with Autonomous Maintenance in working hand in hand to establish the basic equipment conditions on their assets and bring the equipment back to its original condition. TPM believes that maintenance is not solely for the maintenance function alone, but it is responsible for both the operators and maintenance working together. From the traditional thinking that I am the operator, so I just operate; when it fails, then I'll call you, and you fix it to a much more powerful paradigm that both operators and maintenance takes care of their equipment. TPM believes that maintenance is a shared responsibility for both operators and maintenance working together.

TPM Lesson 6: The pillar that will bring the most impact on OEE (Overall Equipment Effectiveness) will be the Focused Improvement pillar. Both Planned Maintenance and Autonomous Maintenance and the rest of the pillars will also be important and will likewise impact on improving OEE. Planned Maintenance will be focusing more on reducing unplanned breakdowns on the equipment, while Autonomous Maintenance will be inclined to reduce idling and minor stoppages of the equipment by keeping their machines clean and establishing basic equipment conditions.

TPM Lesson 7: TPM will always be considered a separate program when the company and TPM's goals are not aligned. The best way to internalize TPM and make it a way of life for the company is if all company improvement initiatives must be under the umbrella of TPM and not TPM under the umbrella of any said program. Focus on one continuous improvement strategy and get the most out of it rather than having so many improvement strategies in your plant such as lean, six sigma, RCM, which will only confuse their people about which strategy should be prioritized. Just imagine a band, and each of the members sings a different song and different beat and when you mixed them up together, what you have is noise. The message here of TPM is simple, all people and teams should just sing one song.

TPM Lesson 8: TPM is a continuous journey; in the book "TPM World Congress," around 3,000 companies have started their TPM journey, and only 10% have achieved the results, although there are a variety of reasons for the failure. Management commitment is the number one reason why TPM fails, which must be learned at the beginning of any TPM initiative. For TPM to be successful, it is only about soliciting management support from management but what we need to start is management commitment. Managers can support but never commit to any TPM initiative and once the management team finally accepts TPM, then what we need is much more than management commitment but management ownership.

TPM Lesson 9: Do not imitate the Japanese; as Deming quotes, companies expect miracles and are looking forward to a silver bullet that will make things happen; it simply does not exist, the American Management thinks they just can copy from the Japanese, but the problem is they do not know what to copy. Adopt TPM in a way that best fits the culture of your company. Adopt TPM how your culture will accept it and do not perform TPM how the Japanese do them. Each industry is unique and has its own fingerprint, and no two people have the same fingerprint. In industries, that fingerprint is your culture, and no two industries have the same culture. Most of the TPM pillars can be applied to different industries but what is best is to adopt TPM the way you want them to be done and not by 100% copying the Japanese way of doing TPM.

TPM Lesson 10: TPM is not a maintenance program alone but a plant-wide process involving everyone from top to bottom. With TPM, maintenance takes the mindset that preventing failures is more important than repairing them. TPM wants to transition an industry from a reactive to a proactive stage where maintenance will be ahead of its failure. Although improving the equipment will prioritize TPM, other pillars will also improve how they facilitate things around, such as office TPM, Quality Maintenance, Early Equipment Management, Environmental, Health, and Safety.

TPM Lesson 11: TPM and RCM are not designed to contradict each other, but rather they complement one another, and they can work together. It is the independent RCM and TPM consultants that do the contradiction and not the methodologies. Both TPM and RCM came from different origins; TPM from the east and RCM started from the west or the United Airlines to be precise, which was first used for the airline industries. The two methodologies never contradict each other; in fact, RCM can exactly fit in the later Phases of Planned Maintenance when both restoration and improving design weaknesses on the equipment had been completed.

TPM Lesson 12: Operators can never be maintenance customers. Why? Simple because they are maintenance partners, and they can never be divorced. Maintenance should treat operators not as their customers but as their partners. The customer is the one that pays your salaries and why your organization is still in business. What TPM wants to achieve is whenever a failure occurs on the equipment, instead of blaming maintenance or operations for the failure, why not work together to finally resolve the problem instead of placing our fingers on them. It contradicts Fredrick Taylor's belief that there is only one best way to do everything and standardize it. TPM believes only in the word better because it knows that the best can still be made better.

7.10: October 2013: Total Productive Maintenance Eight Pillars - Part 1

History of TPM: The term "Total Productive Maintenance" was first used in the late 1960s by Nipponese, an electrical parts supplier from Toyota. It was a slogan for their plant improvement theme, "Productivity Maintenance with total employee participation. In 1971, Nipponese received the Distinguished Plant Award (The PM Prize) from the Japan Institute of Plant Maintenance (JIPM). Nipponese was the first to receive the award due to implementing TPM, which marked the beginning of JIPM association with the improvement methodology. Eventually, Seiichi Nakajima, a vice chairman of JIPM, became known as the father of TPM since he provided implementation support to hundreds of plants, mostly in Japan. Nakajima describes TPM as "Productive Maintenance" carried out by all employees through small group activities. He considers it an equal partner to Total Quality Management in the attainment of world-class manufacturing. According to TPM principles, optimizing equipment lies not just with the maintenance department but with all plant personnel. Although many definitions for TPM had been gathered, for the purpose of this book, we shall define TPM as follows:

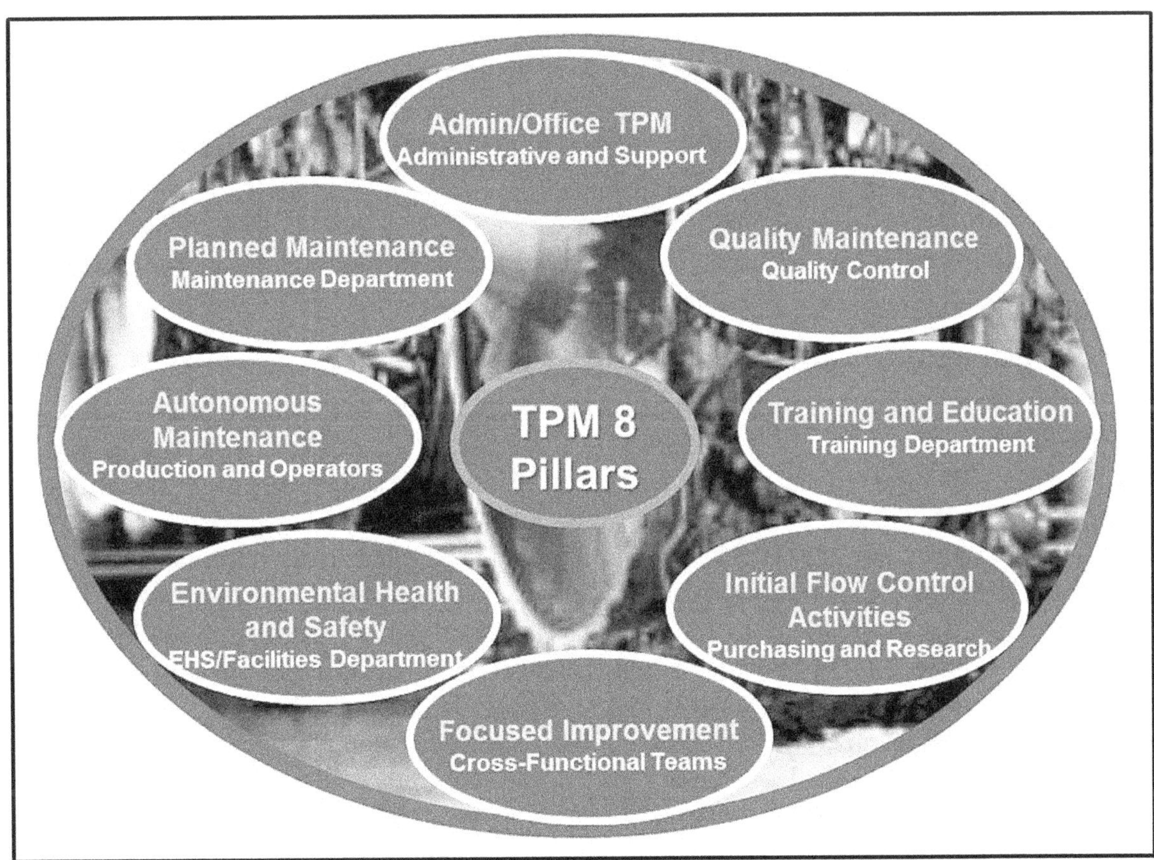

Figure 7.9: TPM Eight Major Pillars

TPM is a method for bringing about change. It is a set of structured activities that can improve plant assets management when properly performed by individuals and teams. The culture of a plant does not evolve solely from TPM but may also reflect other improvement processes underway, such as TQM, Six Sigma, Lean, Kaizen, Root Cause Analysis, etc. A critical aspect of TPM is that improvements should be rapid as well as continuous. Today's marketplace requires new paradigms. The story between the race of the hare and tortoise had to be modified. The industry's current and future winners will combine the rabbit's quickness and speed with the tortoise's perseverance. To attain or maintain leadership at a rate that is much faster than their competition.

TPM is a plant improvement methodology that enables continuous and rapid improvement of the manufacturing process through employee involvement, employee empowerment, and closed-looped measurement of results.

Performance targets must always be dynamic and not static. Suppose a company sets goals and measures to reach their best-in-class competitor's performance levels in two years; they will lag behind since their competition will have improved over that same time. To be the best in class, a company must leapfrog its competition by setting goals beyond where its competition is projected to be.

Likewise, in TPM, employee involvement is a necessary part of the TPM process. The goal is to tap into the entire plant or facility's expertise and creative capabilities through small group activities. The total involvement of plant personnel generates pride and job satisfaction, and financial gains for the organization. Despite the advent of self-managing teams, employee involvement is still new and just starting in most western countries. TPM requires employees to take a more active role in decision-making and to accept responsibility for the plant and its physical condition. They have a heightened role in defining their job content, along with work systems and procedures. The intent of TPM is that each employee must take pride in their equipment, and all efforts must be directed to the plant's objectives. For example, JIPM recommends that management adopt the theme of "My Plant" to increase autonomous maintenance.

Western plants typically emphasize performance measures that are related to production and financial results. Numbers are tracked, reported by accountants, and made available to selected members of the organization. There are two problems associated with classical results measurements. First, the results are not reported to all involved parties, and secondly, results that are reported do not effectively measure performance. In TPM, the plant establishes the key performance indicators that measure performance relative to plant goals and objectives. These key performance indicators measure results in areas over which the plant has control. Typically, they include availability, quality, productivity, and cost-efficiency, and measures of the improvement process's effectiveness. The indicators are reported in a closed-loop manner back to the individuals who have the power to impact them. Hence, information is passed on to everyone, including the shop floor people.

There are three main reasons why TPM has spread so rapidly throughout the Japanese industry and why companies outside Japan are becoming interested in it. It guarantees dramatic results, visibly transforms the workplace, and raises knowledge and skill in production and maintenance workers. Companies practicing TPM invariably achieve startling results, particularly in reducing equipment break downs, minimizing idling and minor stoppages (chokotei in Japanese), and lessening quality defects and claims boost productivity, trimming labor costs, shrinking inventory, cutting accidents, and promoting employees morale, as shown by the increase in improvement suggestions. TPM involves everyone from the organization and is structured through the eight pillars, explained in detail.

TPM Pillar 1: Training and Education: A company's workforce is a priceless asset, and all companies must train and educate their employees systematically for them to do their work correctly. Industry workers are becoming scarcer, increasingly elite, and more multi-skilled, so training must be an integral part of every employee's career development path. Imagine the type of people you want your training program to produce; in other words, identify the specific knowledge, skills, and management abilities you want them to have and then design a training program that suits them and achieves this vision.

Training must also be tailored fit to serve their individual needs. Assess every person's training needs and measure the gap of required knowledge, skills and pinpoint things they can improve. Use these results to make the training more effective. Workers and their supervisors should discuss these assessment results and use them to set the next year's targets and plan the next phase. Also, set firm schedules for achieving program targets. Decide the kind of people you want to have in a year, two years from now, and so on, then draw up a comprehensive plan for the job and off-the-job training design to achieve this, including seminars and workshops.

Training also inevitably plays a major role in any improvement process. The continual investment in an employee by upgrading their skills and capabilities is as critical as investing in plant equipment. Remember that it is the people that will improve their assets and equipment. People are the company's greatest asset no longer holds true because the right people are the company's greatest asset, and the wrong people are called liability. And we can only have the right people if they are equipped with the right knowledge to perform their jobs correctly. Their importance is recognized and promoted by the TPM process.

The traditional roles of the production operator, maintenance craftsperson are now being reinvented. Operators accept greater responsibility for the health and performance of their equipment and assets as they take on certain maintenance responsibilities that traditionally were solely performed by maintenance craftspeople. These maintenance people, in turn, are relinquishing many so-called routine maintenance tasks such as checking, adjusting, and lubricating the equipment, which is now done by the operators. Their efforts are increasingly allocated to higher value-added activities such as Predictive Maintenance and analyzing equipment failures. Rather than simply being repairmen, they now are problem solvers performing these highly skilled analytical tasks of Root Cause Failure Analysis, Reliability-Centered Maintenance, and redesign. These changes in responsibilities for operators and Maintenance craftspeople have required new emphasis on basic and advanced technical training. Besides the additional technical skills development, behavior, modification, and process training facilitate historical work practices. This training type usually focuses on the change process and covers group dynamics, communication workshops, one-point lessons, and disciplinary systems and procedures. Education and Training are being established to elevate the skills of operations and maintenance. It is limited to classroom training but also aid in the use of visual controls and one-point lessons. The Training group identifies the knowledge needed, prepares the training curriculum, and finally assesses their people's skills.

Training and education play a very important role in TPM. A training needs analysis should be performed for the different people working in the plant. Each pillar, such as Planned Maintenance, should be trained on maintenance-related courses, while the Focused Improvement group should be well-versed in different analytical tools and techniques. Likewise, operators should be trained with proper operation and safety on their equipment. Education and training is a continuous process and should be provided to all employees.

Four Levels of Skills:

• Level 1: Lack of both theoretical and practical ability (needs to be taught)
• Level 2: Knows theory but not in practice
• Level 3: Has mastered practice but not theory
• Level 4: Mastered both practice and theory

TPM Pillar 2: Autonomous Maintenance: TPM improves the corporate business results and creates pleasant and productive workplaces by changing how people think about their work with equipment through their company. Autonomous Maintenance, which operators perform, is one of the most important basic building blocks of any TPM program. Two keys to developing a successful autonomous maintenance program are thoroughness and continuity. A further decisive factor is a smooth integration with two other TPM pillars: Planned Maintenance, Training, and Education Pillar of TPM. The production department's mission is to produce good products as cheaply as possible. One of its most important roles is detecting and dealing with equipment abnormalities promptly, which is the goal of a good maintenance program. Autonomous maintenance includes any activity performed by the production department operators with a maintenance function and is intended to keep the plant operating efficiently and stable to meet production plans. The goals of an autonomous maintenance program are:

• Prevent equipment deterioration through correct operation and daily checks
• Bring equipment to its ideal state through restoration and proper management
• Establish the basic conditions needed to keep equipment well maintained

Another important goal is to use the equipment to teach people new ways of thinking and working. In the past, plant operators in process industries were expected to keep their equipment working by checking it regularly and performing minor services. Although different companies had different practices, many expected operators performed simple repairs and routine inspections on their equipment and assets. In general, plants practiced a high degree of Autonomous Maintenance.

During the high-growth era of the 1950s and 1960s, equipment was simple, and not much emphasis is provided on maintenance; however, as time passed by and equipment became more automated and sophisticated plants grew larger and production technology becomes more advanced, the need for a maintenance strategy is critical as some failures will not only lead to downtime but may have safety and environmental consequences as well. With the introduction of Preventive Maintenance, equipment became increasingly specialized and more sophisticated. At the same time, many companies were making significant technical progress in automation. Faced with two oil price explosions in succession, western companies reduced the number of plant operators to reduce costs. For many years now, production departments have played an exclusively supervisory role, concentrating on production and leaving maintenance to a specialist, which traditionally is known as the I Operate, You Fix Syndrome. However, the future is uncertain; many companies hope to survive by cutting costs to boost their competitiveness. As a result, Autonomous Maintenance has become an indispensable program in the drive to eliminate losses and waste from the production floor. It requires a change in thinking and paradigm that both operators and maintenance are responsible for their equipment because maintenance is a shared responsibility.

The participation of operators in maintenance activities is one of the features of TPM, which is Autonomous Maintenance. Although Autonomous Maintenance will be where the largest population on TPM since production people will be involved in it, Planned Maintenance should be the strongest TPM pillar because if Planned Maintenance structure is weak, then so goes with Autonomous Maintenance since will be the maintenance people who will guide and educate these operators to partner with them in helping them establish the basic equipment condition of their assets. The importance of maintenance activities is recognized by the corporation to survive in a fiercely competitive environment. That is why both operator and maintenance need to create a strong bond and partnership, a necessary part of the TPM strategy.

Under these circumstances, the Quality Control department and ZD (Zero Defects) campaigns have been gaining wide popularity in manufacturing industries, and the concept that one's work should be voluntarily maintained by oneself has taken root and developed into the Jishu Hozen or Autonomous Maintenance concept of owning the equipment itself. Autonomous Maintenance is the activities in which each worker performs daily inspections, lubrication, parts replacement, minor repair, troubleshooting, accuracy check, and so forth on their own equipment, aiming at achieving the goal of keeping one's own equipment in good tip-top condition.

With the advancement of technologies, the equipment has become more sophisticated, complex, and with the expansion in the operation scale in industries, maintenance functions have been divided into specific areas. The so-called "I operate you fix" syndrome in which operators only engages in production while when the machine breaks, then maintenance fix it had been a wide tradition in most industry's culture. As a result, many people came to think that the people who were engaged in production should only handle the work and check their quality, and such activities as the maintenance of equipment, lubrication, and other care of machines and equipment should be left to the maintenance people. Today many industries are now transitioning from this state of thinking.

Critical attitudes, such as poor practice employed by maintenance and improper introduction of equipment by the division, should be blamed for the trouble, and we don't have any responsibility for the problem to be discarded. A little attention to additional tightening, lubrication, and cleaning can often prevent future trouble in advance. A little touch or care on the equipment would often help find any slight problems and abnormalities to prevent the problem from further escalating. In TPM, the point is that operators are trained to be proficient in equipment mechanisms and understand the importance of doing the basics on their equipment and assets. To satisfactorily perform Autonomous Maintenance, operators should be knowledgeable about their equipment. Operators are expected not only to be merely just an operator, but they should also know how their equipment works and the care they should provide to their equipment; after all, this is their bread and butter. In designating what cleaning tasks are of most value, the experiences of a qualified TPM trainer can be invaluable and can be disseminated to new team members. Although individual tasks are assigned, these small group activities (SGA) retain total control over this overall improvement strategy. The production operator who is regularly responsible for the machine operator must also be part of the cleaning team if the group is to achieve and act as a partner for the Planned Maintenance team. The more the equipment is automated, the more the operator should be equipped to perform this basic equipment condition on their assets.

One of Autonomous Maintenance's requirements is to acquire the ability to find an abnormality on their equipment. Another requirement is to sense early problems and abnormalities on their equipment or products by feeling suspicious behavior. To acquire this type of attitude, the operator should have the following basic abilities:

• Ability to tell normality from abnormality precisely
• Accustomed to strictly keeping the rules of basic equipment condition on their assets
• Ability to take quick and proper action against these abnormalities

Those who can perform this task can be useful in predicting signs of defect or failure and take necessary steps to prevent such problems from developing into serious ones. Autonomous Maintenance development is performed by a team under the leadership of an operator based on the process, primarily established to increase the level of equipment and workers' performance efficiency through these seven steps of Autonomous Maintenance. The ultimate goal is to empower operators to make decisions on their equipment.

A detailed step-by-step development should be implemented in 7 steps to maintain higher productivity and cultivate proficient operators. A preliminary step is needed from the top management side to explain why TPM is necessary to survive the business competition, while the technical people should teach operators the adverse effects of forced or accelerated deterioration in their equipment. The source of motivation can be found in the process of action understood; the action is a prerequisite. The preliminary preparation is made by considering why forced or accelerated deterioration occurs and understands such activities' importance.

Hence, before getting laid down to the actual developmental steps on Autonomous Maintenance, operators should be given basic education about machine safety and initiate safety measures before commencing Step 1 Initial Cleaning on Autonomous Maintenance. Remember that when a serious accident occurs during initial cleaning, that would be the end of your TPM activities. The team doing Step 1 should list all of the predictable unsafe actions, unsafe conditions, and countermeasures for each predictable accident that should be completed before doing the initial cleaning activities in Step 1. Step 0 will include activities where maintenance teaches operators about their equipment and some safety precautions to be undertaken before doing Step 1 initial cleaning activities. To be continued.

7.11: November 2013: Total Productive Maintenance Eight Pillars - Part 2

Here are the 7 Steps of Autonomous Maintenance

Step 1: Perform Initial Cleaning: An all-around detailed cleaning of dust, dirt, tagging abnormalities, pinpointing lubrication points on the equipment will be the main activity for Step 1. This is the main purpose of conducting this step. First, small group activities can team up together in accomplishing a common goal, the cleaning of a particular machine or process. Second, it promotes a better understanding and familiarity with their assets and equipment, and third, the actual Step 1 activities uncover hidden defects that, when corrected, will have a positive effect on equipment performance. The activities associated with initial cleaning are typically performed by members of these small group activities as part of their full benefits. The team must acquire safety training before commencing on initial cleaning.

Training before doing Step 1 on Initial Cleaning activities is critical. The initial cleaning is not intended to be an overhaul or turnaround of the equipment. The focus is to increase the team's understanding of their equipment through the cleaning process; in cleaning, they touch their equipment, and by touching them, it reveals problems that were long concealed from them. If one were to perform initial cleaning on an automobile engine, key tasks would include steam cleaning the engine exterior, checking the head gasket bolts for looseness, and possibly replacing the fan belt and doing a compression check. But one would not remove the head, clean up the cylinder walls, or replace the bearings. Autonomous Maintenance will focus mostly on the exterior part of the equipment.

The time required to complete the above steps may exceed the available time for a given shutdown period. It may be necessary to schedule the above steps over several shutdown periods to accommodate operating requirements. Dozens of equipment abnormalities or improvement suggestions may result from the initial cleaning activities on a single machine. Many of the equipment issues will be fixed or improved immediately during the initial cleaning activities. Others that will take time to restore will be scheduled and prioritized and passed on to skilled trades, maintenance, or external contractors.

Depending on existing equipment conditions, initial cleaning activities may require major time, money, and resources. Most companies underestimate the size of this commitment. Once they recognize the extent of resources required, they either cut back on the amount of equipment to be included or simply perform superficial cleaning. Some companies invest as much as 160 hours per plant employee on initial cleaning activities. This will inevitably depend on the size of the equipment.

Step 2: Eliminate Sources of Contamination and Hard to Reach Areas: Initial cleaning activities identify contaminant that resides in the machine or area. Each contaminant type should be identified and documented. Typical contaminants include leaking process fluids, leaking lubricants, dust, corrosion, process scrap, material handling scrap, worker-generated trash, and other external pollutants. Contaminant identification cannot always be accomplished during the initial cleaning activities due to the volume of the contaminant build-up through the years. The team should identify after cleaning the sources where this contaminant came from. In a manufacturing plant, let's say that management was concerned that the plant was not able to match the productivity numbers of its sister plants in Europe, even though the European plants were at least twenty years older. The plant had hundreds of grinding machines. Each machine was equipped with a catch basin loaded with metal shavings, oil, cigarette butt, and candy wrappers. It was impossible to identify whether the oil was cutting oil, hydraulic oil, or another type of lubricant. When asking why the catch basin was not clean to increase machine reliability, the tour guide's reply was they would just get dirty again. People in this plant failed to understand that clean machines lead to improve equipment performance.

Step 3: Establish Tentative Equipment Standards: These standards will include three main areas: developing a tentative standard for cleaning, inspection, and lubrication. Once a clean work environment is established, and steps are taken to prevent deterioration, new, higher standards can be set and documented; the goal is to combine inspections for cleanliness with lubrication checks so that both activities can be performed together efficiently as possible. Standards for lubrication and cleanliness also should be developed concurrently. These small groups will be responsible for developing standards for their equipment. By developing these standards, operators become more responsible by taking care of them since cleaning and lubrication are both inspection forms. Many potential equipment problems can be spotted visually or enhance their senses, causing a major malfunction or breakdown of their equipment. People with cars establish a standard for cleanliness and lubrication on their automobiles, and they periodically open the hood of their car to check fluid levels or change the oil in their car. They also may inspect the engine block for cleanliness to ensure it is free from spilled oil and road grime. While the hood is up, a quick glance identifies whether the head gasket is leaking and any loose belts or hoses. Those are simple activities that anyone with minor training can perform. The car owner is motivated to meet these cleanliness standards so that he can avoid future repair bills. An individual who follows through with this type of inspection is more likely to exercise greater caution in their vehicle's operation. They can recognize that it is an acceptable risk to drive a car with a leaking head gasket to the repair shop, but he would not start a car with a broken oil pump in the process as this can cause overheating of the engine. The cleaning and lubrication standards should be similar to a sound Preventive Maintenance on the equipment, but the only slight difference is that they are performed by operators and not by maintenance. What makes their inspection standards unique is by adding some visual controls to make their inspection much simpler.

Step 4: Develop a General Equipment Procedures and Training: Cleaning and lubrication comprise the bulk of routine inspections needed for most equipment. Additional inspection and adjustment can be grouped into an all-inclusive category called general inspection. These activities include bolt torque or tightening, minor calibrations, adjustment, replacement of wear part, and other process-related visual, temperature, pressure, or flow checks. The general inspection also includes more detailed inspections on sub-systems such as hydraulic, pneumatic, and electrical subsystems. Small group teams should accept responsibility for the development of a general inspection standards. The plant maintenance mechanics possess' ideal skills for this process, but production and engineering should also educate the operators from time to time.

One purpose of the general inspection is to raise operator understanding of their equipment and its maintenance requirements. By the operator's participation in developing these general inspection standards, valuable equipment knowledge is gained for future use in diagnosing equipment abnormalities. This method of developing inspection procedures is consistent with the TPM objective of empowering small group activities to address equipment-related issues. Although small groups are empowered to determine what procedures are performed, they must act within the plant's policies or regulations. Their actions are also shaped by the plant's overall strategy, thus ensuring data and formats' commonality. Commitment and ownership over the process are heightened because operators are the primary decision-makers on what is best or not for their equipment and asset. Every plant has a standard or checklist being provided on their equipment and assets but seldom do they accomplish them. The difference with Autonomous Maintenance is that the operator will be the ones to develop their own set of checklists on their machine with the guidance of the Planned Maintenance teams. One of the highlights of this step is that operators are continuously trained by maintenance on what they should know about their assets and equipment. This may be the longest step in Autonomous Maintenance since operators need to digest the learning slowly but surely because once they know their equipment intimately, they will fully understand why they need to do these inspections on their equipment.

Step 5: Conduct General Equipment Inspection Autonomously: The cleaning and lubrication standards set in Step 3 and the tentative inspection Standards prepared for each given category in Step 4 are now combined in Step 5 into a unified Autonomous Maintenance Standard. Cleaning, inspection, and lubrication standards should be reexamined to accomplish a higher check efficiency and eliminate the possibility of human errors. If problems occur, operators work with maintenance to develop inspection points that can prevent the problem from recurring. At this step, operators continue to think of ways to shorten the time to perform cleaning, lubrication, and inspection on their assets. It is important to consolidate the activities performed by both maintenance and operators on the equipment in this step to avoid duplication and redundancy on the maintenance side. If such activities as cleaning, lubrication, and tightening activities are included in the current PM lists, CMMS is revised, and the responsible person is changed from maintenance to operator since this will be carried out by the AM teams. Inspection finalized and generated on this stage is very different from the checklist provided by maintenance in a traditional industry since the operator now understands the value of doing it. An autonomous Maintenance inspection is the actual transfer of responsibility for equipment inspection performed by the operators themselves. For this to become a reality, two specific activities must be accomplished. First, the previously developed lubrication, cleaning, and general inspection standards must be combined into one checklist and institutionalized into the system. Second, production operators must be trained in that system in the technical aspect of the inspection.

It is important to develop a checklist for operators and organize them by performance interval. In other words, develop a check lists for all daily activities, weekly and monthly, quarterly inspection checks. The checklists should be distributed logically to operators on the different shifts so that the workload is balanced and considers different shift operating conditions. The check sheets should use visual controls and photos or drawings for operators to easily grasp and understand the tasks to be performed. The illustrations may be an actual part, making it easy for the operators to understand what tasks will perform a particular inspection.

Step 6: Manage Systematic Autonomous Maintenance in the Workplace: Teams that have completed the first five steps of Autonomous Maintenance achieve optimal equipment conditions and establish standards to sustain these conditions. The good point about Autonomous Maintenance is that as the operators' level of knowledge improves, their equipment reliability also improves. Operators now fully understand their equipment and can detect and prevent abnormalities in advance through proper checking and operation. One of the goals of this step is to allow operators to perform a sound, comprehensive autonomous maintenance of their entire process and extend their activities into a realm of quality maintenance. Activities that promote this include standardization of various control items, preparing process flow diagrams and quality maintenance manuals, and deepening operators understanding of the relationship between equipment and quality. Operators expose quality defect sources by performing general quality maintenance inspection, notes on these process flow diagrams and simple equipment structural diagrams, and gradually builds a system that enables them to detect and promptly rectify abnormalities affecting quality. Also, in this step, the teams improve their work areas as well. The concept is somewhat like this. In steps 1 to 5, we want to improve the inside of your house; in step 6, we move on further on improving the outside of your house like your garage, surroundings, gardens, and so on.

Step 7: Achieve Empowered Autonomous Maintenance Workforce: An empowered Autonomous Maintenance operator is the point in which the operator's process becomes self-sustaining. External forces are no longer required to drive the pursuit of the concept of zero breakdown and defects. Empowered small groups can interpret company goals and policies and self-manage their continuous improvement initiatives and activities. They now become mentors to new members. Minimal outside assistance and management direction are necessary since operators can decide what is best for their equipment. This state of being in the natural progression of the Autonomous Maintenance process. Although Step 1 to 6 has accomplished results in concentrating all activities, in changing equipment and changing workplaces, in Step 7, one's own ability is recognized and the emotion to participate is high, solidarity, creation, and emotion are enjoyed by indefinitely challenging problems and issues by performing kaizen and incremental improvements. This last step also aims to build operators who act as electric trains with self-energy and motivation, completely changing their attitude towards their equipment and making their work no longer a routine. To be continued.

7.12: December 2013: Total Productive Maintenance Eight Pillars - Part 3

TPM Pillar 3: Planned Maintenance: There seems to be a misunderstanding and feud between operations and maintenance people in most industries. The never-ending saga of feuds between the production and maintenance people is evident and rampant where maintenance accuse the operations of flocking their equipment to death and when the machine is due for a Scheduled-Maintenance Overhaul, production will waive the equipment for a sound Preventive Maintenance activity to cope with their day to day production demands, and when the machine fails or breaks, then it is production's turn to strike heavy on their enemy. If this feud exists in your company, don't worry, you are not alone. Does the maintenance lack skills in performing their duties, or the system in place need to be replaced?

Doing maintenance is a serious business since it can constitute 3 to 50 % of production costs. Without a good system in place, not only can maintenance be costly, but it will also affect the company's survival. Some may advert to continuous improvement programs, modifications, purchasing the best maintenance software in town, benchmarking other companies, cost-cutting programs, but mind you, each of them have their own sets of limitations. Let me give you an example; in the company I previously work, there seems to be a battle of programs; more than a dozen continuous programs exist, TPM was in place, and modifications and improvements were done, but the team missed out on something, a very important point, the equipment is capable of producing was going to be phase out in a couple of months meaning the machine was also going to be decommissioned as well. And after a couple of months, the equipment was pulled out and removed. Thousands of dollars in improvement, modification costs were gone.

Developing a strategic maintenance system must be done two-fold, both for the short term and for a long-term plan. The long term is where most maintenance managers and people lack. It is not sufficient to be just doing the band-aid therapy or the "Stop the Bleeding Syndrome," neglecting the fact of the possibility that other parts might be affected as well, which can take a month or more to take effect. A good maintenance leader always sees this aspect in two ways: the short-term and the long-term plan. The sad fact is when we compared equipment 50 years ago, and today, we can see a more complex, automated equipment; the change had been rapid, but if we speak about how we perform maintenance on the equipment, we still use the same technique the caveman used to do, and that is firefighting (fix when it broke).

When we speak of a holistic and strategic maintenance system in place, we must speak about lower cost and higher availability, but a maintenance perspective on the following must seriously be considered.

• Higher Plant Availability and Reliability
• Greater Safety
• Better product quality
• No damage to the environment
• Longer equipment life
• Greater cost-effectiveness

Therefore, in a nutshell, if we speak of a structured maintenance system, the ideal dream for every maintenance manager will be is to select the most appropriate strategy to deal with each type of failure process to fulfill all the expectations of the owners and users or of the equipment, in the most cost-effective manner maximizing it's total Life Cycle Cost with the active support and participation of both operations, maintenance, and other departments together. Therefore, the concept and basic understanding of Planned Maintenance are truly important; it is not a cure-all strategy for all our equipment problems; however, when you applied the Steps/Phases involved, not only can we have some peace of mind at night where maintenance can sleep without being bothered for late-night calls where your boss tells you to come back to the plant to fix some problems the shift cannot handle. A structured and effective Planned Maintenance system can guarantee that the machine is expected to perform its intended function at its optimum condition. Repairs on equipment should be minimum. Planned Maintenance is a long-term approach for your daily day-to-day maintenance problems. Maintenance should rethink their strategy and transcend from being reactive and fixing failures since an accumulation of such practice will not only hurt your plant financially but can cause to shorten the life of the equipment, delayed deliveries to customers, long cycle time, and high costs on spares. A company that wants to survive with its competition must focus on a rigid framework on its maintenance strategy and adopt a thoroughly structured Planned Maintenance System.

Planned Maintenance Defined

TPM comprises of 8 major pillars Autonomous Maintenance, Planned Maintenance, Office TPM, Focused Improvement, Training and Education, Quality Maintenance, Initial Flow Control Activities, and Environmental Health and Safety. Chapters on JIPM (Japan Institute of Plant Maintenance) books cover the Planned Maintenance, and its definition varies from one author to another.

- According to Tokutaro Suzuki, Planned Maintenance is the deliberate activity of building and continuously improving such a maintenance System.
- Planned Maintenance is defined as maintenance activities performed on a pre-determined schedule of activities. By Charles Robinson and Andrew Ginder

Although there may be more definitions of Planned Maintenance, Suzuki's definition is the simplest way of defining it. Every company had its own maintenance in place; Planned Maintenance's job is to continuously improve such a maintenance system. Planned Maintenance is a pillar of TPM that aims to achieve high equipment reliability while minimizing maintenance costs. These are achieved by properly applying all the maintenance tasks such as Breakdown Maintenance, Predictive Maintenance, Preventive Maintenance, and Corrective Maintenance, which we shall define later in this newsletter. Planned Maintenance also aims at targeting a complete zero reduction in unplanned breakdowns.

What Planned Maintenance Pillar Includes

A complete strategy on Planned Maintenance includes a Master Plan for the 8 major activities involved, although the major focus would be implementing the step-by-step activities on Planned Maintenance in 4 Phases. The best way to carry out Planned Maintenance activities is to set a time frame for each phase and have a Master Plan. Planned Maintenance activities include.

• Guidance and support for Autonomous Maintenance Activities
• Planned Maintenance 4 Phase Activities
• Lubrication Management Activities
• Setting up the Planned Maintenance Structure
• Spare Parts Management
• Reduction in Maintenance Cost Activities
• Enhancement and Upgrading of Maintenance Skills
• Success in Using Predictive Maintenance Instruments and techniques

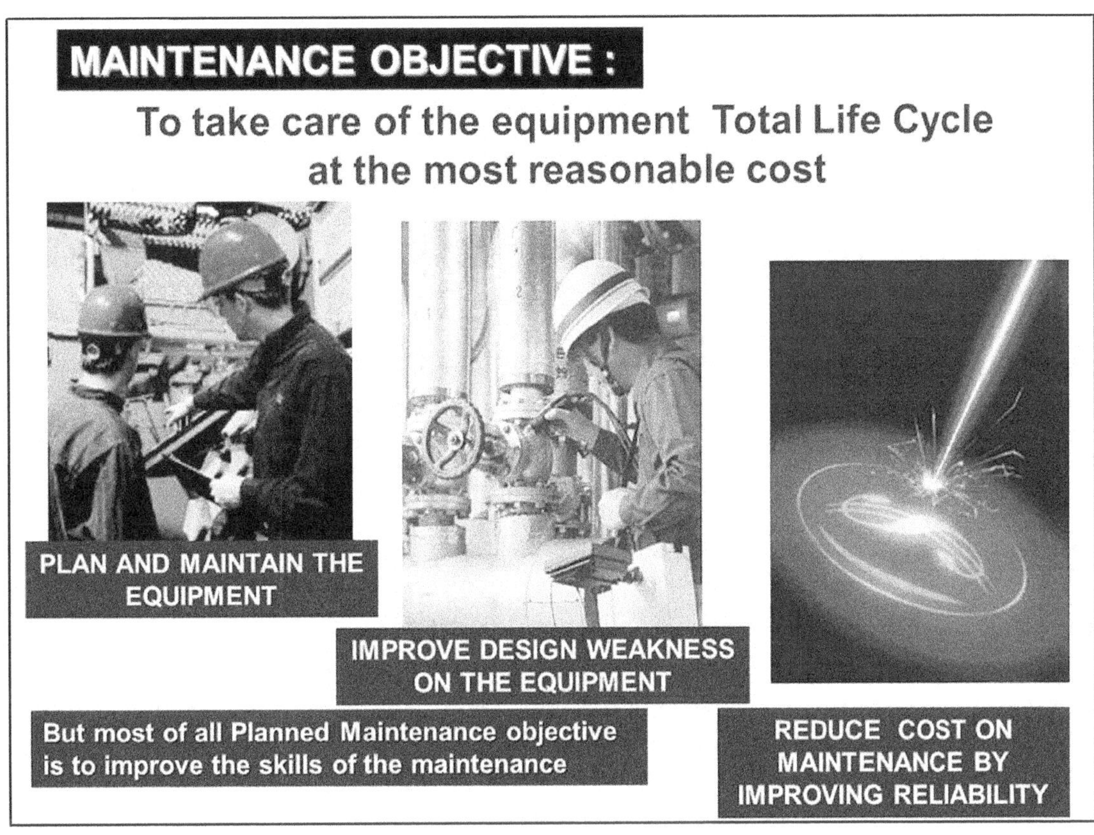

Figure 7.10: Planned Maintenance Objective

<u>What Planned Maintenance Wants To Achieve</u>

Doing Planned Maintenance is no different from coaching a basketball team. Imagine if I am the coach and I put my first 5 starting with the smallest person to be 6' 8", and the rest are 7 feet tall. Your team will be strong on the rebound but definitely, weak in the offensive, or what if you sent out 5 guards? Then statistics show you will be weak in the defense. In setting up your first five, it should be balanced, just like these maintenance tasks. You need to have 2 guards, 2 forwards, and 1 center to complete your team.

Similarly, in maintenance, you need to know what parts will undergo breakdown maintenance, what parts will be on scheduled preventive maintenance, and what parts are under predictive maintenance; then, you have the best maintenance system. Traditionally, most industries rely so heavily on just one strategy, which is Preventive Maintenance. It is like having five center positions in basketball, and if this is the case in your industry, then always expect your maintenance to be very costly and reactive in nature.

Implementing a long-term Planned Maintenance system will take years to accomplish, but the results are what every maintenance department can only dream of. Its benefits will have an effect on the following:

• Reduction in Maintenance Costs
• Higher MTBF and Reduced Planned Breakdowns
• Higher Equipment's Reliability and Availability
• Upgraded and Higher Maintenance Skills
• Making Maintenance Pro-Active and Not Reactive

Planned Maintenance activities are essential for any manufacturing or industry with equipment since maintenance must not be taken lightly.   Most of the time, maintenance is neglected. While a company can hire the best maintenance manager truly capable of knowing every detail of the equipment he manages, he can lack the most important thing about maintenance, having a clear, structured approach to performing maintenance. I strongly emphasize in this newsletter content that it is 100 times better to have a maintenance system than having the best maintenance software in town.   Prioritize first to have a system of Planned Maintenance structure built in your organization. If done properly and correctly, this investment can impact you on saving cost, which can amount to hundreds of thousands to millions of dollars in waste that can be avoided on doing maintenance.

Planned Maintenance in 4 Phases

Although there are many variations in how Planned Maintenance can be implemented, its goal is to achieve a Predictive Maintenance stage on their equipment by utilizing non-destructive equipment diagnostic techniques.   These instruments are useful in predicting equipment failures by determining failure modes with potential failures. Likewise, operators for Autonomous Maintenance are trained on enhancing the use of their senses to spot possible equipment failures.   Predictive Maintenance aids in helping maintenance understand that a component or part is on the verge of failing and can likewise plan for this event before much more damage can be realized.   Predictive Maintenance simply tells us whether to continue running the equipment or shut it for sound maintenance activity.

Phase 1: Stabilize MTBF through Restoration: There are three main activities in Phase 1: performing restoration, recurrence prevention of deteriorations uncovered, and standardization. All these activities aim at establishing basic equipment conditions while deteriorations are exposed and corrected. Before conducting Phase 1 activities, equipment is mostly subjected to accelerated deterioration, and failures occur frequently. Deteriorations are left unchecked even if maintenance is aware of them.   These are perceived as normal day-to-day norms, and breakdowns cannot be eliminated since the boss does not easily approve our request to purchase spare parts that need replacement.   This attitude often spells disaster since one breakdown may lead to another, prolonging downtime and increasing maintenance costs. Activities in Phase 1 aim to change the traditional approach to firefighting or reactive stage by doing the basics and restoring equipment to its original condition.

Prevent accelerated deterioration tasks involves extending the equipment lifespan by prolonging MTBF or its Mean Time between Failures. The longer the interval for the time to repair the equipment, the more the equipment is utilized. Planned Maintenance will begin by tagging the equipment for deterioration and abnormalities uncovered by the team and correcting each of them. This will be done in parallel with Autonomous Maintenance initial cleaning. Expect a dramatic reduction in breakdowns after Phase 1 had been completed. The team also analyzes why deteriorations keep on recurring and perform countermeasures to prevent the recurrence of accelerated deteriorations. A simple why-why analysis must be performed at this level.

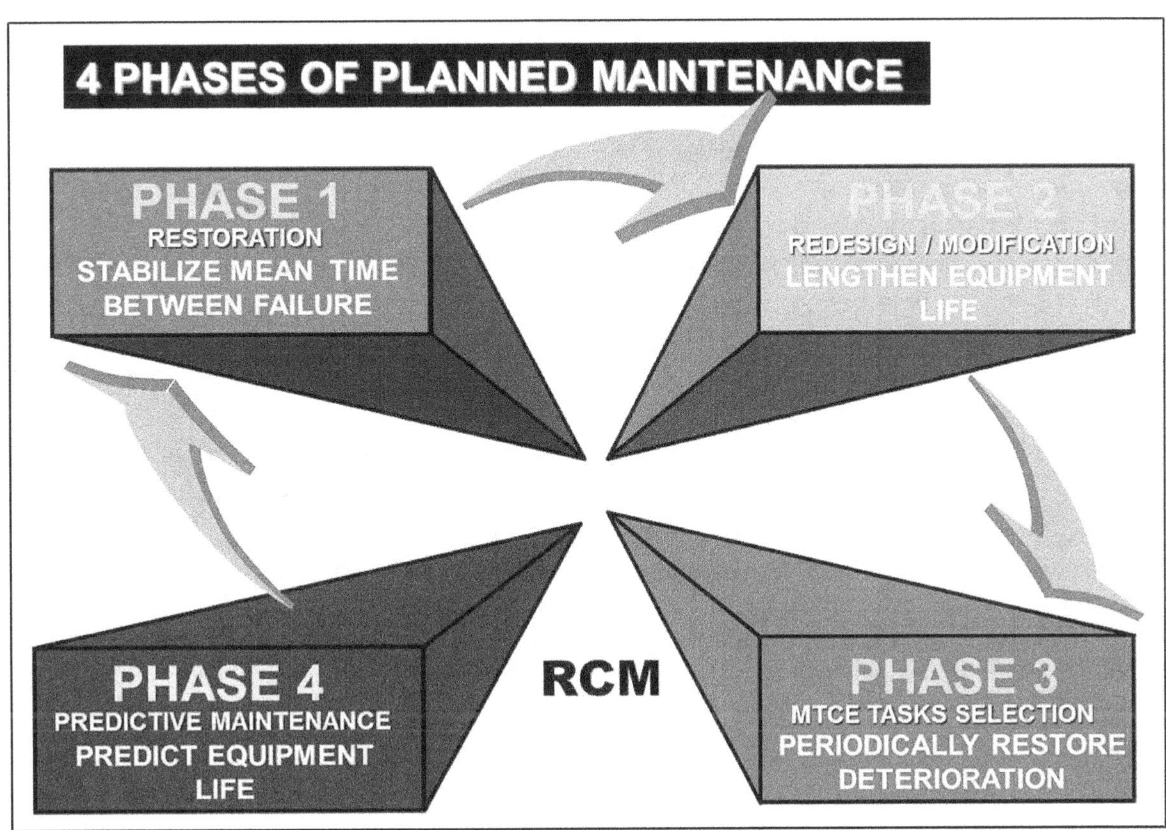

Figure 7.11: Planned Maintenance Generic 4 Phases

Phase 2: Lengthen Equipment Lifetime: Once accelerated deterioration had been eliminated, equipment will suffer only from natural deterioration. There are several spare parts of equipment that will deteriorate naturally. The team exposed themselves to study several parts of equipment with inherent short natural lifespan and correct design weaknesses by improving the part's dimension, the strength of materials, construction, etc. An MP or Maintenance Prevention Design Form is usually used for this activity and later on feedback to the IFCA (Initial Flow Control Activity Pillar) so that when the company decides to purchase equipment in the future, these improvements and modifications are discussed with the designers to be included in the new equipment purchase. A cycle must be established where MP Design improvements must be looped back to the Early Equipment Management or Initial Flow Control Activities pillar of TPM.

Correcting design weaknesses can prevent major breakdowns from recurring unexpectedly. Teams are trained on analytical and problem-solving tools such as P-M Analysis, FMEA, and Root Cause Failure Analysis for a more detailed approach to dealing with Chronic Breakdowns. Likewise, most breakdowns are caused by human errors. Hence, both operators and maintenance must upgrade their analytical skills to eliminate human errors; applying Poka-Yoke solution may help address human errors but not necessarily improve the level of understanding of the mistake caused by the person involved. Equipment failed parts with design weakness are modified; this must run in parallel in improving the skills on how the operator operates equipment and how maintenance performs repairs.

Phase 3: Periodically Restore Deterioration: A thorough study of the maintenance tasks must be done to establish the correct maintenance to be performed on the equipment. RCM (Reliability-Centered Maintenance) is a perfect methodology for this concept. Not all tasks need to go through a Time-Based or Preventive Maintenance System. This can be done by having a thorough understanding of how such tasks can be performed correctly. Each part has its own failure characteristic pattern. The key in this activity is to know the 6 failure pattern to derive the correct tasks for each function through an Algorithm or Decision Diagram. In this activity, we are not dealing with improving the equipment and maintaining maintenance on the asset. What parts must undergo time-based, what parts can be predicted, what parts need an inspection, and what parts do not have any consequences when a failure occurs and can be left to a run-to-fail task.

Phase 4: Predict Equipment Life: In the last Phase of Planned Maintenance, we introduce performing Predictive Maintenance or Condition Based Maintenance on the equipment. Although this is similar to using our human senses, the only difference is that it can sense problems much more than our human senses through the use of these non-destructive instruments. Several equipment parts can be predicted through these techniques; this is done by checking the equipment's condition. The key to using this technique is to know the P-F interval, which means that if a part shows potential failure symptoms, they are a good candidate for Predictive Maintenance. For example, a bearing may produce noise, an increase in temperature, an increase in vibration; these are symptoms that a failure is likely to occur. Therefore, maintenance can be prepared for this failure in advance. Predictive Maintenance's advantage is that maintenance can utilize the part to its maximum life without the need for a breakdown to occur. Parts can be replaced on a need basis and not based on time or running hours. Both Predictive and Preventive are Proactive maintenance tasks; for as long as you know how to adapt them properly. The key is to understand the 6 failure pattern. To be continued.

<u>2014 RSA Reliability Newsletter Vault Archive</u>

> *For c-level and decision-makers who do not understand maintenance and reliability, expect them to play a dangerous game called cost-cutting. These are merely small pebbles, a tiny fraction of temporary savings that often lead to repercussions. If industries are dead serious in saving money <u>big time</u>, then they should be the drivers, since reliability is everybody's responsibility in the plant.*

<u>8.1: January 2014: Total Productive Maintenance Eight Pillars - Part 4</u>

<u>TPM Pillar 4: Focused Improvement</u>

Focused Improvement or Kobetsu-Kaizen in Japanese includes all activities that maximize the overall effectiveness of equipment, processes, and plants through the elimination of losses and improved performance. Many people ask about the difference between focused improvement activities that they may already be practicing. The basic point to remember about TPM Focused Improvement is that if a company is already making all possible improvements in routine work and small group activities, then Focused Improvement activities are already in place.

Focused improvement aims to eliminate all kinds of losses. Identifying and quantifying these losses are, therefore, important issues. The traditional method of identifying losses, analyzing results statistically to identify problems, then searches back to find their causes. The method adopted in TPM emphasized a hands-on, practical approach and examined the production process, equipment. Materials, people, and methods and treat any deficiencies in this input as losses. Achieving a profitable TPM in industries can be difficult if improvement teams limit their fabrication and assembly industries' approaches. Proper mental and physical preparation is essential before starting any focused improvement project. Improvement teams should be prepared in the following ways:

• Understand the philosophy of focused improvement fully
• Understand the significance of losses and the rationale behind improving OEE fully
• Understand the production process well, including its basic theoretical principles
• Gather data on failures, breakdowns, and losses and plotting this overtime

• Clarify the basic conditions necessary to assure the proper function of equipment and define clearly what factors contribute to an optimal state
• Understand the necessary techniques for analyzing and reducing failure and losses
• Observe the work closely to discover what is actually happening

Focused improvement often requires a high level of engineering technology. In addition to improving the level of proprietary technology relating to a company's products, services, and equipment, it is also necessary to raise standards. It is easier and more effective to conduct incremental improvement activities step by step by documenting progress visually as you proceed. This approach has the following advantage:

• Everyone can see what is happening and take an active interest in the focused improvement pillar
• Plans for individual topics and teams are developed separately but integrated with plant-wide goals to maximize results
• The focused improvement committee can easily monitor the program and control the schedule
• Holding presentations and audits on completion of each step make it easier to consolidate the gains and sustain enthusiasm. Provide rewards and recognition to successful teams that achieve lasting results with their improvements since people take pride in what they do.

Step by Step Procedure for Conducting Focused Improvement

Step 0: Select an improvement topic
• Select and register the topic
• Form the project team and plan the activities

Step 1: Understand the situation
• Measure failure, defects, and other losses. Identify bottleneck areas
• Use baseline to set targets

Step 2: Expose and eliminate abnormalities
• Painstakingly expose all abnormalities uncovered
• Establish basic equipment conditions by restoring deterioration and correcting minor flaws

Step 3: Analyze Causes
• Stratify and analyze causes by applying analytical tools, such as RCFA, P-M Analysis, etc.
• Employ specific technology, fabricate prototypes, and conduct experiment

Step 4: Plan Improvement
• Draft improvement proposal and prepare drawings
• Compare cost-effectiveness of the alternative proposal and prepare the budget
• Consider possible harmful effects and disadvantages of the countermeasure

Step 5: Implement Improvement
• Carry out the improvement plan and perform test operation and formal acceptance
• Provide instruction on improved equipment operating methods

Step 6: Check the Results
• Evaluate results with time as improvement project proceeds

• Check whether targets had been achieved

Step 7: Consolidate the gains
• Draw up control standards to sustain results
• Formulate work standards, manuals, and instructions and feed information to maintenance prevention or early equipment management team

TPM Pillar 5: Administrative/Office TPM

Office TPM involvement will include all departments with offices. Companies must map out a clear strategy to respond to this essence of change and dramatically shorten their production time to market. At the same time, they must distinguish themselves from their competitors on both quality and costs. These are the most important challenges facing managers today.

Eighty percent of a product's quality and the cost are already determined at the development, design, and production stages. Development, design, and other staff departments must cooperate willingly to ensure that the production department does not produce wasteful products. Meanwhile, companies must set up manufacturing plants to enable the production department to fill orders on time at the quality and cost that the development and engineering departments prescribe. This is not the production department's responsibility alone; it requires a TPM program that embraces the entire company, including the plant's administrative and support departments.

TPM activities in administrative and support departments do not involve production equipment. Rather, this department increases its productivity by documenting administrative systems, reducing waste and loss. They can help raise production system effectiveness by improving every type of organized activity that supports their production. Their contribution to the smooth running of business should be measured.

The Role of Administrative and Support Departments

Unlike administrative and support departments, planning, developing, engineering, and administration do not add value directly. As experts in their particular area, their primary role and responsibility are to process information, advice, and assist with the production department's activities and other departments to reduce costs.

Their second task is to enable the company to respond rapidly to changes in the social and business environment and outperform their competition. This means improving their own productivity, reduce cost the right way, and helping the company accomplish the strategic developments that senior management envisions. Their third task is to win the customer's confidence and create an outstanding corporate image. To pursue these goals through TPM, administrative and support departments must define their mission by answering the following questions:

• How do we support the TPM activities of the production and other functions?
• What issues must we address to maximize our own efficiency?

The function of administrative and support departments can be improved by improving efficiency so each department can perform its particular function satisfactorily, developing people that can sustain and continuously improve a new, more efficient system. Improving efficiency means boosting output while reducing input. Boosting output means eliminating anything that reduces production system efficiency to enhance work functions and raise their effectiveness. Reducing input means eliminating administrative losses associated with the work and creating a cost-effective administrative system to provide high-quality, timely, and reliable information.

These are the goals of improving the organization and management of administrative and support departments. An even more fundamental goal is to use these activities to develop administrators who are extremely effective at processing information. Each company must tackle this challenge differently; however, a firm's present and future problems need to depend on its type, scale, management, history, present circumstance, and business environment. Each company must work out the best approach for its own situation.

TPM Pillar 6: Quality Maintenance

As equipment takes over the work of production, quality depends increasingly on the conditions of the equipment. Quality maintenance evolved as a major TPM activity in certain fabrication and assembly industries, becoming increasingly automated. In an environment where human intervention decreases, quality maintenance aims to maintain and constantly improve quality through effective equipment maintenance.

For manufacturing industries, quality has always been built into the product through the process. However, new product development is accelerating, and the greater diversity of raw materials and products currently necessitates more frequent changeovers ever. To cope with this, production departments must review their quality assurance systems to tackle quality through equipment management. Quality is built into the product through a process that provides the conditions needed for transformations such as reaction, separation, purification of materials as they become the final product. It is necessary to set appropriate process conditions such as temperature, pressure, flow rate, etc., for the particular properties, composition, and volumes of raw materials, reagents, and other substances being handled to produce close to perfect products. The equipment units are made up of the plant, and their component modules and parts must be installed to achieve this, so they function optimally to create no quality defects on the products.

Industries always aim for this, but the results often leave much to be desired. Quality defect losses and reprocessing losses still occur. Substandard products often have to be recycled and salvaged by mixing them with good products or downgraded. Customer complaints and dissatisfaction are perennial problems for industries. Meanwhile, in plants where chemical reactions occur, poor control of conditions affects quality and is also dangerous to human health. To create safe plants that produce only flawless products, a company must analyze processes and equipment rigorously to identify and maintain conditions that do not lead to defects. This is the role of quality maintenance. A quality defect is something that falls short or outside the specified range.

Quality Maintenance consists of activities that establish equipment conditions that do not produce quality defects to maintain perfect working conditions in producing perfect products. Quality defects are prevented by checking and measuring equipment conditions periodically and verifying that the measured values lie within the specified range. Potentially, quality defects are predicated by examining trends in the measured values and preventing corrective measures in advance.

Rather than controlling results by inspecting products and acting against defects that have already occurred, quality maintenance in TPM aims to prevent quality defects from occurring altogether. This is accomplished by identifying checkpoints for the process and equipment conditions that affect quality, measuring these periodically, and taking appropriate actions. The approach focus on the production inputs of equipment, materials, people, and methods as sources of quality defects. Establishing conditions means setting the range for material, equipment, or operating conditions that must be maintained to produce flawless and defect-free products. Once set, these conditions are maintained and controlled by competent operators extensively trained in production technology as part of their autonomous maintenance activities.

A quality maintenance program builds gains by doing fundamental TPM activities such as autonomous maintenance, planned maintenance, focused improvement, and maintenance skill training. Several preconditions for a successful quality maintenance program to abolish accelerated deterioration, eliminate process problems, and develop competent operators.

Who is Responsible for Quality Maintenance?

The Quality Control Department should be responsible for promoting the TPM Quality Maintenance pillar throughout their industry. Quality Maintenance projects vary considerably in difficulty. However, projects spanning a wide range of processors requiring advanced technology should be tackled by project teams headed by their managers. Easier projects can be addressed by small groups in the workplace. After teams established the conditions for zero defects, operators should maintain and control most of these conditions as part of their Autonomous Maintenance activities. More difficult problems should be addressed by project teams from the production department, such as product design, product engineering, maintenance and equipment engineering, and quality assurance control. To be continued.

8.2: February 2014: Total Productive Maintenance Eight Pillars - Part 5

TPM Pillar 7: Early Equipment Management

As products diversify and their life cycle becomes shorter, TPM teams find ways to make new product development and equipment investment more efficient as it grows in importance. The goal in TPM is to reduce the time dramatically from initial development to full-scale production and to achieve a vertical startup that is fast, free of bugs, and right the first time. In other words, the goal of Early Equipment, or others called them Initial Flow Control Activities, is to shorten the lead time or commissioning time of new equipment and products, and this is what early equipment management is all about.

It is vital to develop products of readily assured quality that anticipate the user's needs, competitive, easy to sell, easy to produce, and easy to do efficiently. Simultaneously, however, the transition from development to full-scale production must be rapid and problem-free. To accomplish this initiative, the teams must identify the production inputs required to bring the products to the market, eliminate the losses associated with the equipment that produces them, and maximize the return on investment. In other words, teams must ensure that production equipment is easy to use, easy to maintain, highly reliable, and well-engineered. With such equipment free of bugs, assuring product quality is simple.

Major equipment items are often customized to individual specifications; they are often designed, fabricated, and installed in a rush condition. Without strict early equipment, such equipment enters the test operation phase with many hidden defects. The truth of this is borne out by the frequency with which maintenance and production personnel discover defects generated by fabrication and installation during shutdown maintenance and start-up.

Early Equipment Management is particularly important in industries because large amounts of money are invested in processing units, and top management expects the plant to operate for a considerable number of years. After each shutdown, the restart operations must be managed by the same procedure followed when the plant was first commissioned. To accomplish this, all departments must cooperate closely not only from the research and development, design, engineering, production, and maintenance but also from planning, marketing, finance, storeroom/spare parts, and quality maintenance.

TPM gives equal importance to early product management, early equipment management, and other TPM activities. Early equipment management aims to gain economic performance evaluation optimizing life cycle costs and maintenance prevention or MP design improvements.

Maintenance Prevention (MP) Design

Maintenance Prevention design activity minimizes future maintenance costs and deterioration losses of new equipment by considering maintenance data on current equipment and new technology and designing high reliability, maintenance, economy, operability, and safety. Ideally, MP design equipment must not break down or produce non-conforming products. The machine should be easy and safe to operate and maintain. The MP design process improves equipment reliability by investigating weaknesses in existing equipment and feeding the information back to the equipment designers. Improvements done by Autonomous Maintenance Planned Maintenance and Focused Improvement are being consolidated, prioritized, and discussed with OEM and vendors of the equipment that these modifications should be imbedded when purchasing for future equipment.

Even when the design, fabrication, and installation of new plant and equipment appear to have gone smoothly, problems often emerge at the test operation and commissioning phases. Production and maintenance engineers struggle to get the plant working properly, and they achieve normal operation only after repeated modifications. After the plant has begun operating normally, checking, lubricating, and cleaning to prevent deterioration and failure may be difficult to carry out. It may also be for set-up and adjustment, and repair. When equipment is not designed for easy operations and maintenance, operators and maintenance personnel tend to neglect routine housekeeping, set-up and adjustment take too long, and even the simplest repairs necessitate shutting equipment down for unconsciously long periods.

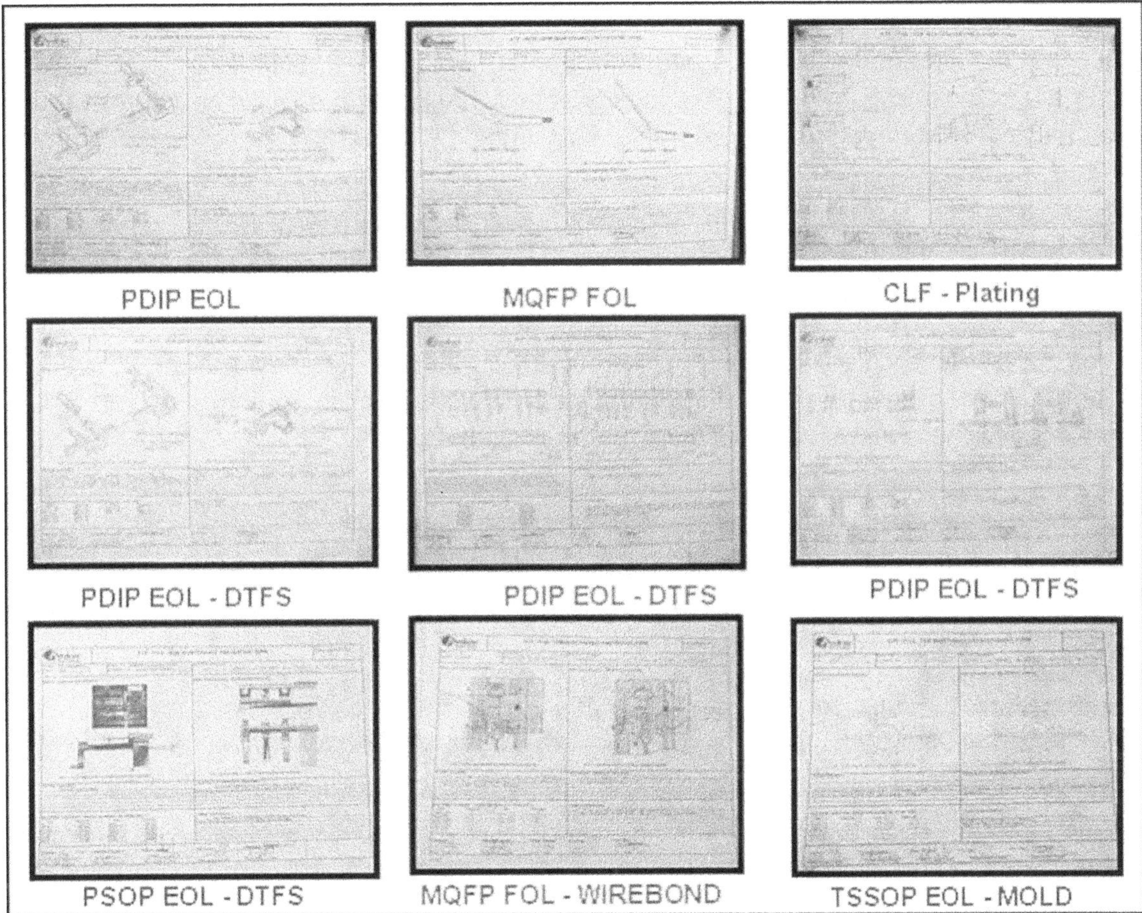

Figure 8.2: MP Design Improvements

Some people claim that numerous problems at the initial operation stage are inevitable because of the rapid advancement of technology, increased speed, size, and equipment automation. Never try to justify the problem like this. Equipment engineers must incorporate new processes and operating conditions into the equipment design conditions. To ensure that equipment is highly reliable, maintainable, operable, and safe to use, avoid outside purchasing. Make full use of in-house technology that your own production, design, and maintenance engineers have accumulated from problems they have overcome in the past. The thoroughness of the investigations performed at the design stage largely determines the maintenance a plant requires after installation.

TPM Pillar 8: Environmental, Health, and Safety

Eliminating accidents and pollution is a mandatory requirement for winning the most coveted

TPM PM prize in Japan. The safety records of prize-winning operations are, in fact, significantly better than before they introduce TPM. Review your own safety and environmental management system at the end of the TPM implementation phase and establish an environment that permanently maintains the improved safety record. The goal must be to have zero accidents and pollution.

Safety and Environmental Management

Ensuring equipment reliability, preventing human errors, and eliminating accidents and pollution are basic gorals of TPM. This is why safety, health, and environmental management are key for any TPM development program. Fully implementing TPM improves safety in many ways, such as:

- Faulty equipment is a common danger source, so zero failure, defect, and safety.
- Through applying 5s principles, the team eliminates leaks/spills and makes the workplace clean and tidy.
- Autonomous Maintenance and Focused Improvement should eliminate unsafe areas.
- TPM operators are better able to detect abnormalities and deals with them promptly.
- Operators takes on the responsibility for their own health and safety

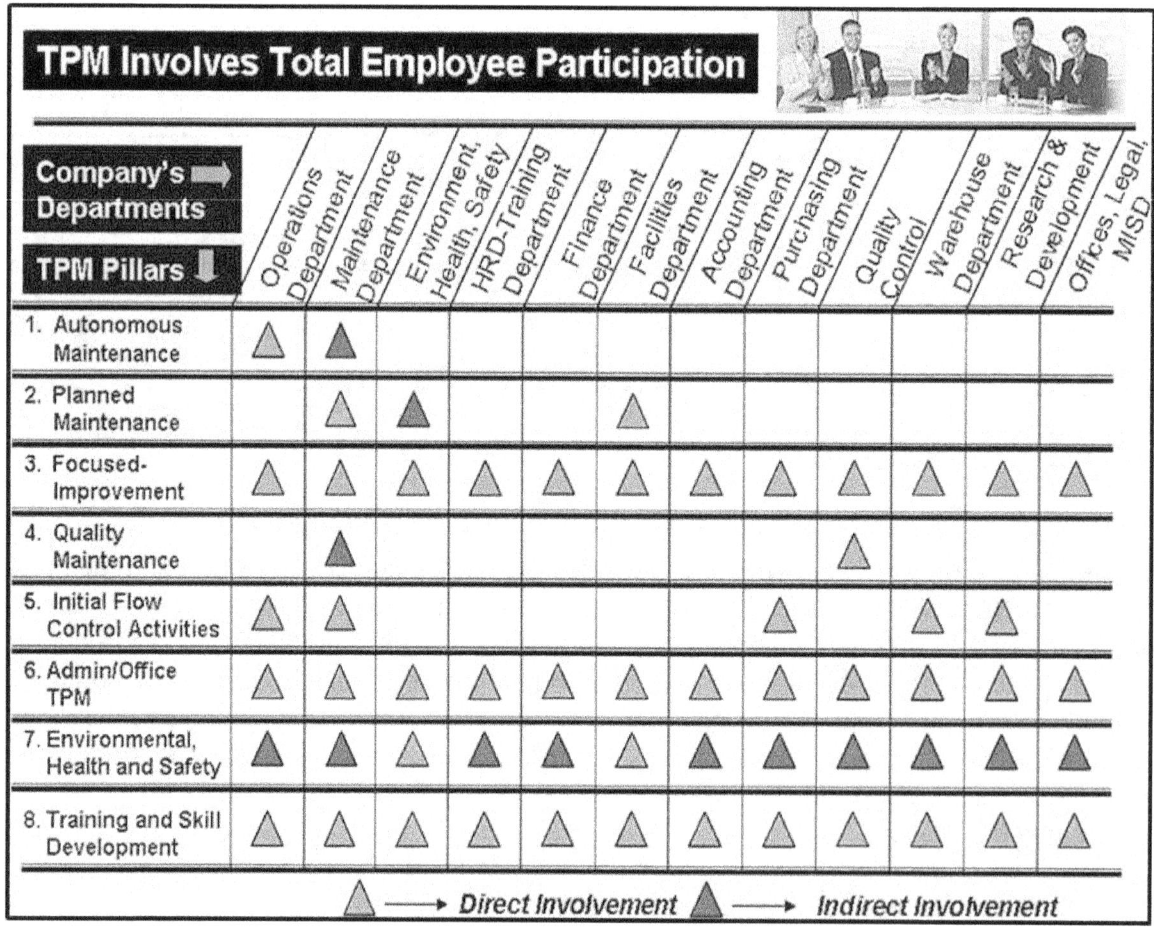

TPM Pillars \ Company's Departments	Operations Department	Maintenance Department	Environment, Health, Safety	HRD-Training Department	Finance Department	Facilities Department	Accounting Department	Purchasing Department	Quality Control	Warehouse Department	Research & Development	Offices, Legal, MISD
1. Autonomous Maintenance	△	▲										
2. Planned Maintenance		△	▲			△						
3. Focused-Improvement	△	△	△	△	△	△	△	△	△	△	△	△
4. Quality Maintenance		▲							△			
5. Initial Flow Control Activities	△	△						△		△	△	
6. Admin/Office TPM	△	△	△	△	△	△	△	△	△	△	△	△
7. Environmental, Health and Safety	▲	▲	▲	▲	▲	▲	▲	▲	▲	▲	▲	▲
8. Training and Skill Development	△	△	△	△	△	△	△	△	△	△	△	△

△ ⟶ Direct Involvement ▲ ⟶ Indirect Involvement

Figure 8.1: TPM Involvement Will Include Everyone in the Organization

Practicing TPM builds safety in the workplace. It also contributes greatly to a healthy and hospitable working environment. Perfect safety, health, and environmental cleanliness are basic manufacturing requirements. In practice, however, there is always a possibility of plants or equipment causing accidents and pollutions. The potential for disaster is always present, even in a plant with a perfect safety record.

Plants that handle large quantities of flammable, explosive, or toxic materials, use high-pressure gases, consume large amounts of energy or operate under extreme conditions are, particularly at risk. The danger of fire and explosion is very evident, and accidents may affect not only the plant but also the surroundings and environment. Pollution due to plant accidents or process problems is also highly undesirable since it can harm and affect the local environment and communities. Eliminate these risks by taking the same approach that you take to improve safety. Since major accidents and disasters are rare, zero accident momentum is easily dissipated. Stay alert for blind spots and remember that a tiny defect or problem can develop into a serious accident or pollution incident, so focus daily on zero activities toward zero accidents and pollution. Once all pillars of TPM have been deployed, involvement will include everyone in the organization. All regular employees will have their own respective pillars they will belong.

8.3: March 2014: A Passion for Maintenance and Reliability

Last February 2005, I started to venture out a career on my passion and specialization. I started offering courses in my specialized field, equipment maintenance, and reliability. These courses sum up my previous experiences with industries, such as my first work as an oiler on board MV Sea Raider until my last work as a training specialist at Lepanto Consolidated Mining Industry, where I started to develop most of my training materials on maintenance. I have also worked for 8 years at Amkor Technology, where I worked as a TPM practitioner for 7 years handling the pillars of Planned Maintenance, Focused Improvement, and Initial Flow Control Activities, where I have actually implemented and reaped its benefits. It also includes my knowledge and continuous research on best practices from the eastern and western parts of the world, their conflicts, and how they can be integrated and complement one another. I've conducted plant visits and training so far with various manufacturing, process, oil, and gas, and power plants where I have come to realize where the best standards in maintenance truly exist. My continuous communication with my circle of friends on reliability and maintenance experts from around the world and the case studies my counterparts have shared with me through the years, the previous training on reliability I've attended, and how we have initiated these strategies such as Lubrication Management, Reliability-Centered Maintenance, Condition-Based Management, Total Productive Maintenance, and other maintenance and reliability strategies and finally my passion for maintenance which is the main reason why I remain independent so that I can share this little knowledge I know with other industries.

These reliability and maintenance courses have gathered much attention for industries seeking ways to improve their equipment's reliability and upgrade their human maintenance resource's technical knowledge and competency to a level of world-class excellence by capturing Industries Maintenance Best Practices. These powerful and proven courses provide indebted details and a wealth of information on Reliability and Maintenance's true meaning. I am truly confident that these strategies are what every company needs to improve its equipment reliability as they lower down their operating cost.

In today's trend in industries and with the consistent focus on productivity and secondary to reliability and maintenance, our equipment often results in frequent breakdown and unexpected failures, costly unscheduled repairs, and much more our maintenance resources are often taken down by these untimely breakdowns, inevitable resulting to the high cost of maintenance. What we need is the adaptation of a more robust and effective maintenance strategy that is truly world-class. For maintenance, the question is often raised, is it really possible to manage maintenance, or the pressure over maintenance is the once managing us for a very long time? What I believe is that reliability is a culture that never ends. Like any other continuous improvement initiative strategy, we need to continuously improve the way we maintain our assets and equipment. Still, everything will start by cultivating the right mindset for all our maintenance people in our plant, and that is by providing them with the right knowledge and training on reliability and maintenance.

Finally, it is also undisputed that every maintenance manager's challenge is maximizing their equipment reliability through a traditionally and often self-centered approach called Preventive Maintenance system. Perhaps this practice is the very reason why the approach in maintenance management remains to be costly and reactive rather than proactive. Truly, these courses are designed for every maintenance engineer and professionals whose mandate is to optimize their equipment's capacity and reliability at the lowest possible cost.

If I am to write down in one paragraph the most important learning's from my courses, it would be summarized below, and I would like to share them with you. These are the seeds that we must plant in our people's minds before driving any improvement initiative.

Learnings' from Root Cause Failure Analysis: Never accept failures in your plant. Troubleshooting and repair is no longer an effective strategy. In today's competitive world of manufacturing, the analyst finds real solutions to equipment problems. Remember that when our people become really good at fixing failures, then something is definitely wrong since they are doing it much too often, but when we expect a different result from the same things that we are doing, it just ain't possible, the Chinese called this insanity. The reason why we perform Root Cause Failure Analysis is to learn from the things that go wrong with our equipment and assets, but we can only learn from the things that go wrong if we are brave enough or have the courage that we, too, are also part of the problem. The distinction between a maintenance and a mechanic is that maintenance uses more of his brain than his hand to deal with problems, while a mechanic uses it to deal with problems. Let us treat our people as maintenance and not as mere mechanics.

Learnings from Meaningful Measures of Equipment Performance: As maintenance professionals, we know the importance of these KPIs to our organizational goals and objectives. The difficulty is in translating the overall company strategies to a meaningful measure of performance. The old saying that "if you can't measure it, you can't manage it" is as true for performance as for anything else. Much of what is meant by performance often appears to be simply not measurable. It's fine at looking at how things happened after the event, and traditional measures can do this. What a manager need is a   much more dynamic, real-time view of performance as it happens. People are central to this, and qualities like motivation, confidence, leadership & perception need to be understood. Getting quantifiable results and dependable measurements are crucial if resources and time are used to good effect. As the saying goes, we measure what is important. And if what we measure is important, then our goals will definitely be rewarded. The best way to change the culture is to focus on results; remember, what gets measure gets done; if we don't measure our performance, then we are just another person with an opinion, and in the real world of manufacturing, opinions don't last forever.

Learnings from Lubrication Strategy and Oil Contamination Control: For equipment that fails due to lubrication, there is only one secret, just keep the oil clean. If the oil is kept clean, then there is no reason for it to oxidize and, moreover, no need to change the oil. Oil should be changed not based on the number of hours it has run but by the number of contaminants the lubricating oil has. When the oil is maintained clean, the additives can last longer.

Learnings from Total Productive Maintenance: More and more companies seek ways to improve their performance by applying continuous improvement tools such as TPM in today's manufacturing industries. Although it was ignited in Japan, TPM is not culture-bound, but its principles can be applied to any plant as long as the people accept TPM as a way of life. TPM improves the manufacturing process by utilizing employee involvement, employee empowerment, and closed-loop measurement of results. The best time to address a problem is when it is small. It is very hard to advance to any specialized maintenance activities and improvement efforts if basic equipment condition had not been established, always remember our types of equipment is a shared responsibility for both operators and maintenance people, a lesson we must all learn from the Japanese

Learnings from Reliability-Centered Maintenance: The best maintenance strategy to adopt is knowing when to use the different maintenance tasks simultaneously with the aid of a of an algorithm or Decision Diagram and that the degree of maintenance requirements should always be based upon the consequences of failure itself. The belief led to the idea that the more often an item is overhauled, the less likely it is to fail. Schedule Overhauls / Preventive Maintenance increases Overall failures by introducing Infant Mortality into otherwise stable systems. The resulting schedules are used for all similar assets without considering that different consequences apply in different operating contexts. This results in many wasted schedules, not because they are wrong in the technical sense, but in reality, they achieve nothing. Two discoveries evolved, which created a change in the evolution and thinking of the maintenance system worldwide. First, scheduled maintenance has little or no effect on a complex item's reliability unless it has a dominant failure mode. Second, there are many items for which there is no effective form of scheduled maintenance.

Learnings' from TPM Planned Maintenance 4 Phases to Zero Unplanned Breakdown: The real challenge in any equipment reliability initiative is improving a reactive environment with the same resources and time. Remember that best-in-class companies started from being reactive themselves.

Learnings from World Class Maintenance Management, The 12 Disciplines: The focus of every maintenance must be on reliability and not on reducing cost because if reliability starts to improve, the cost will definitely go down. This cannot be the other way around. There will be times that focusing on reducing cost will hurt reliability, a lesson we should reflect upon. There is no silver bullet program or strategy that can transform a plant's reliability overnight. Everything will start with its basic foundation, and that is through education. This knowledge must be used to change the mindset of our maintenance resources.

8.4: April 2014: Right Reasons Why We Need To Measure Performance

Measurement plays a vital role not only in industries but also in our day-to-day lives. Our kids' performance is reflected in the grades they obtain from school. Their grades indicate what subjects they excel in and those that definitely need to be improved. When we buy food in the marketplace, we always speak of weight in kilo depending on the amount of money and budget we have. If we want to go on a diet or simply gain weight, we measure our weight through a weighing scale in kilograms or pounds. When we travel to other places or countries, the amount of fare we pay depends upon the distance we simply want to travel and the comfort we would like to have. Likewise, in maintenance, we measure our equipment to indicate if we have been performing well or falling below our goals and targets. Measurement reflects how we do things around the plant, which takes some form of Key Performance Indicators. These measurements provide us with some benchmark or starting point in our quest to improve our asset performance. As the saying goes, if we can measure it, then we can manage it. Before thinking about measuring our performance and assets, let us first understand what we want to manage. What is important is to understand what it is that we want to measure so that we can determine where we are currently heading.

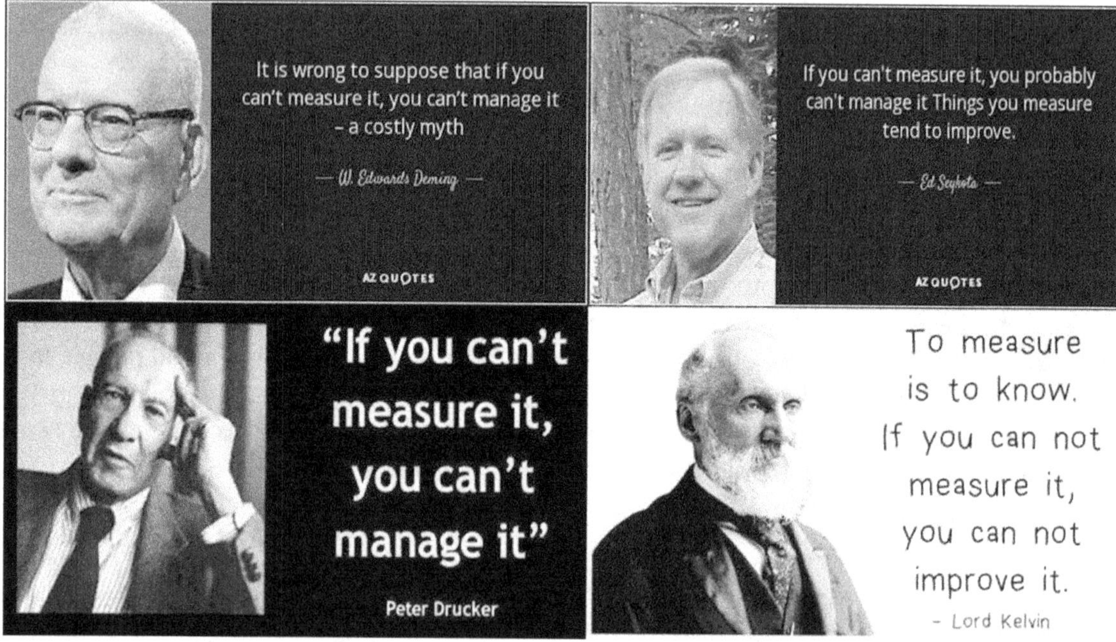

Figure 8.2: Famous Quotes on Measurements

Measurements and KPI (Key Performance Indicators) allow us to determine if our equipment is performing well or not. Data gathered can tell us that our equipment is functioning based on our targets and goals. We measure our equipment so that we can evaluate its performance. The purpose of the evaluation is to determine what strategies we need to adopt and those that we need to drop. Measurements tell us many things, most especially if we are in control of the situation or not. It tells us a lot about the way we do things in our plant. If we fall short of our goals, then we need to do something about it. When we are well on target, we need to sustain our efforts or further increase our goals. By measuring performance, we can determine which strategies, programs, activities deserve additional focus and priority. Measuring performance can tell us which of our equipment is within the budget. By measuring performance, we can allocate budget wisely on the things that need it most. Developing performance measurements allows the industry to set goals and identify targets for measuring how well the results are achieved. These measurements can aid our top management in making a sound decision based on the performance outcome. Performance measurement contains valued information that we all can learn. Learning can help with the budgeting of both money and resources, and through learning, we can therefore make sound decisions on improving our assets.

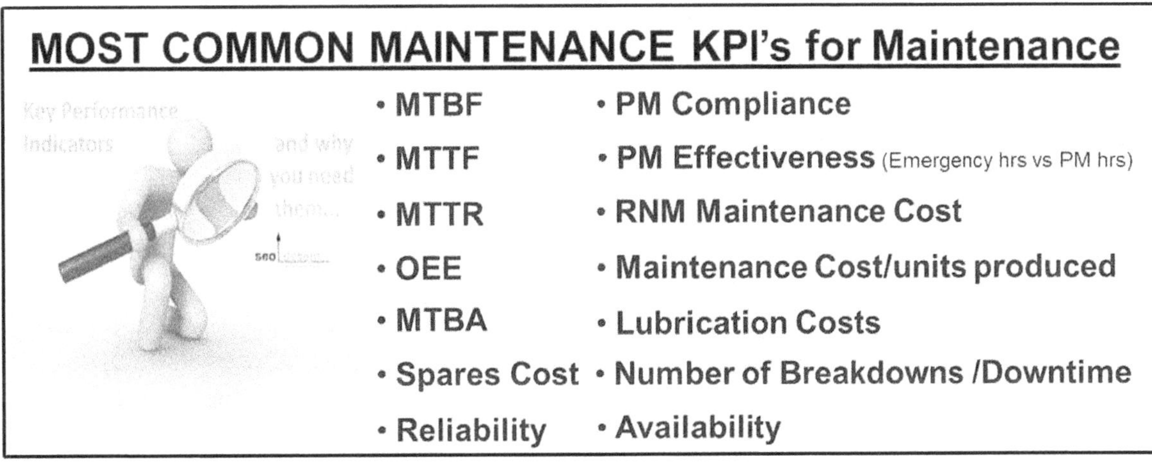

Figure 8.3: Common Maintenance KPIs and Indicators

However, the most important reason for measuring these KPIs is to improve our industry's performance. By knowing our performance, we ask ourselves, what exactly should we do differently to improve? Industries today challenge themselves by setting up goals and doing their best to improve them. Their goals cannot remain the same all the time, nor can they have the same goals repeatedly if they want to remain competitive. Moreover, they should be improving faster than their competition. To do this, they should challenge the goals and targets that they set from time to time and continuously improve them.

Selecting the Right KPI for Maintenance

It is quite difficult to recommend the right KPI and indices suited for your plant because industries differ in how they operate and do business. Certain key performance indicators and indices may apply to one plant but not exactly to the other plants' rest. Experience tells me that there is no single and perfect KPI to measure a plant's performance; you need various indicators. You need to have at least a minimum of 4 KPIs in your maintenance function. Your team needs to sit down and decide what KPI and measurements matter most that best suit your industry and your plant needs and requirements.

There is a wide selection of measurements for maintenance. We can measure the rate or number of breakdowns, downtime, availability, meantime indicators, OEE, set-up time, repair time, maintenance costs, spare parts costs, and so on. Therefore, the industry must understand what measurements and KPIs are needed that best suit their plant needs and requirements. It is unlikely for an oil and gas plant to measure OEE since it is difficult to quantify the quality rate of oil unless we can determine the exact number of bad molecules or the exact amount of contaminants in the oil. Each industry should focus on measurements and KPIs that are likely applicable to them. Indices from manufacturing industries will greatly differ in other industries, such as mining, oil and gas, and other assembly plants. For plants initiating TPM, the primary measure of their performance would be OEE or Overall Equipment Effectiveness.

During my TPM days in the semiconductor industry, I was assigned to handle Planned Maintenance's pillar. We organized a group of people from every single department to be part of this company-wide initiative. We also included the facilities and utility section of the plant in our Planned Maintenance journey. Our initial requirements were to agree on what KPI and maintenance indicators we should track. The breakdown was one of them, so we started to measure the number of breakdowns on our first pilot equipment. It was a simple task to ask, but the truth is that it is very confusing. There were eleven departments, to be exact, and each department was composed of two members. One from the front of the line, and the other represented the end of the line. We had around 22 members during that time, and our meeting was every week.

Since all of these departments produced IC packages except the Facilities and Central Lead Finish or Plating section, it was very common to have similar equipment types in each station of a semiconductor plant. One of the stations in front of line was the Wirebond Station. This station could have a minimum of around 100 to 200 wire bond equipment, and we had around 10 departments in the plant with a wire bond station. A wire bond is a machine responsible for adhering or welding a thin wire to a pad on a bare semiconductor die-chip and the other end of the wire to a conductive pad on a substrate. The wire is usually made of 99% gold and is usually very thin, often .001 to .0013 of an inch. Anyway, so much for the introductory and back to the breakdown part of the story. All members of the Planned Maintenance were instructed to select initial equipment to undergo our Planned Maintenance activities, and one of the instructions was to provide the number of breakdowns on their pilot machine that would undergo Planned Maintenance for the past 3 months so we could have some benchmarking data. When they submitted the data, I was shocked by the results. A couple of wirebond from different departments recorded breakdowns of 3 to 8 times in a month, but what was shocking was that some departments recorded breakdowns of around 30 to 50 times a month. This was where I sensed that something was wrong with the data. So what was wrong? They included the short or minor stoppages and interruptions as breakdowns.

The frequency of breakdowns and failures are easy to identify for oil and gas plants, mining, metal plants, airline industry, and so on but relatively difficult in some cases for manufacturing plants. My advice for some manufacturing plants was that before measuring the number of breakdowns you have, maintenance people should sit down and agree on what would constitute and what would not constitute a breakdown. Therefore, for the succeeding weeks, we deliberated and discussed this matter thoroughly with the members. There were many debates and arguments, just like passing a bill in congress and the senate, but finally, we resolved the matter and arrived at a consensus among the Planned Maintenance members regarding what to include and exclude as breakdowns, and here was what we agreed upon.

What We Included as Breakdown

- Breakdowns caused by poor set-up and conversion where the maintenance stops the machine replaced parts which are affected as a result of poor set-up and conversion
- When technician stops the machine due to breakdowns caused by defects and reworks such as worn-out punch, thereby producing product defects
- Function loss breakdown due to no availability of spare parts
- Unscheduled repair and overhauling of equipment
- Unscheduled replacement of parts
- Failure and breakage of tooling
- Run to fail items providing little or no consequences at all, like replacing tower lamp

What We Excluded as Breakdown

- Downtime caused by a material related problem
- Actual conversion and set-up or the process of changing from one product to another
- Machine stoppage caused by interruption, assists, and minor stoppages resulting from jamming of products due to dirt and so on. (Japanese call this chokotei)
- Machine downtime caused by PM Schedule such as overhauling of parts
- Machine DT caused by the replacement of parts as reflected from the tool algo schedule
- Repairs attributed by Predictive Maintenance findings
- Machine stoppage caused by quality audits
- Machine stoppages caused by Facilities stoppages such as no air, no power, cooling water, etc.
- PM Schedule, the shutdown of equipment, all planned downtime
- No inventory and all non-machine related downtime

Although some may not agree on the lists we provide on what to exclude and include on the breakdown, what is important is that the list generated is a result of a consensus among the members of our Planned Maintenance team. I think the message I would like to emphasize here is simple. Each of us may have our own understanding and definition of failure or breakdown. A failure may have a different meaning from a safety point of view. Suppose the safety person finds something that can sense danger or accident, such as an oil leak in the equipment; he can declare the equipment in a failed condition even if the equipment is still capable of functioning. Maintenance can declare equipment to be failed if there is some form of functional reduction failure in the equipment, while from an operation's point of view, the equipment has encountered a failure if it stops running. What is important is that before measuring breakdowns and failures, maintenance and other people from the organization must agree on what to include and exclude in breakdowns.

The frequency of breakdowns and failures should not only be the sole measurement or indicator used in the maintenance department. In fact, having a low breakdown frequency is not actually an indication of having a good performance.

Remember that not all breakdowns are created equal. What is important is not to eliminate breakdown but to understand the consequences of every breakdown. Measuring our maintenance performance indices and KPI clarifies the need to focus on long-term goals and strategies by comparing our actual performance against the goals we set forth. We can only control the situation if we measure what is meaningful and important to us because measurements allow us to make the right decisions in our organizations.

Determine what is important to measure for your industry. Remember that no one KPI will tell you everything. There should be a series of measurement maintenance is monitoring. Various measurements can be selected, and each meaningful measure will represent the different losses your equipment is suffering. Measure what you think is important with your maintenance organization. Different industries will have different measures and KPIs. KPI's true value comes from knowing how the number is calculated and not by manipulating them to look good to the management team. As the late Peter Drucker quote, the problem with management is they are measuring the wrong things. The first metric you used must drive the right behavior positively.

8.5: May 2014: What Is Machine Breakdown

The word breakdown may mean different things to different people, and it is not as easy as you can imagine. In my training, I give some specific cases and ask the delegates if they would consider this is a breakdown or not, and what do you think? Some will definitely say "YES," and some will say "NO," so hear me out.

• There was a compressor in the facilities, and one day it overheats. Its primary function is to supply compressed air to 10 pneumatic equipment on the production floor. When the compressor overheats up, all the ten pneumatic equipment stopped. Now, if we speak about the compressor, definitely, there was a breakdown. My question is, if we speak about the 10 pneumatic equipment, was there a breakdown? *Answer for the 10 pneumatic equipment we have no breakdown, but this is considered as a downtime.*

• There was a pump with a standby unit next to it, and it is pumping water. This pump is running 24 hours a day, 7 days a week continuously. One day the motor of the duty pump burned up, and the pump stopped its discharge. The flow then transferred to the standby pump, and it started automatically; hence, this time, it is now the standby pump running and not the duty pump. So my question, was there a breakdown that took place? *The answer is yes; there is a breakdown on the component even if there was no downtime on the system.*

• In a semi-conductor, most types of equipment are highly automated and are run where the operator type some keys on the keyboard. A virus prevented the software from starting the machine. My question, was there a breakdown on the equipment or not? *The answer is yes;*

this is considered a breakdown.

- A minor stoppage, by definition, mostly occur on automated equipment and is relatively easy to fix in the first place. The maintenance or operator will just have to reset the equipment, and then it will run again. These are usually errors in automated processes where the workpiece flow stops, operator reset, and the machine runs again. One piece of equipment stopped working, but Joe was the only technician on the shift and was still troubleshooting equipment that had failed earlier. It took Joe 1 and a half hour to complete the repair, then went to the other equipment that stopped. He resets the equipment, and then it runs again. It took him just 40 seconds to do this, but downtime was 1 and a half hours. Would you consider this a breakdown or not? *The answer is no, as this is a minor stoppage and not a breakdown.*

- There was hydraulic equipment in the production line that is leaking oil severely due to a worn-out gasket, but the equipment was still capable of its primary function of providing the necessary output. The safety manager decides to stop the equipment for safety purposes. The machine's gasket was replaced, and the downtime lasted for 4 hours. My question, would you consider this as a breakdown or not? *The answer here is no. A breakdown is only declared when it happened first and the maintenance repair it, but in this case, maintenance is ahead of the failure.*

- One of the plant substations failed, and all equipment stopped due to a power failure. The downtime was about 2 hours. If we speak about all the equipment that stopped production because there was no power, do we consider this a breakdown for the production equipment? *The answer is no, there is no breakdown, but this will be included as downtime for the production equipment affected.*

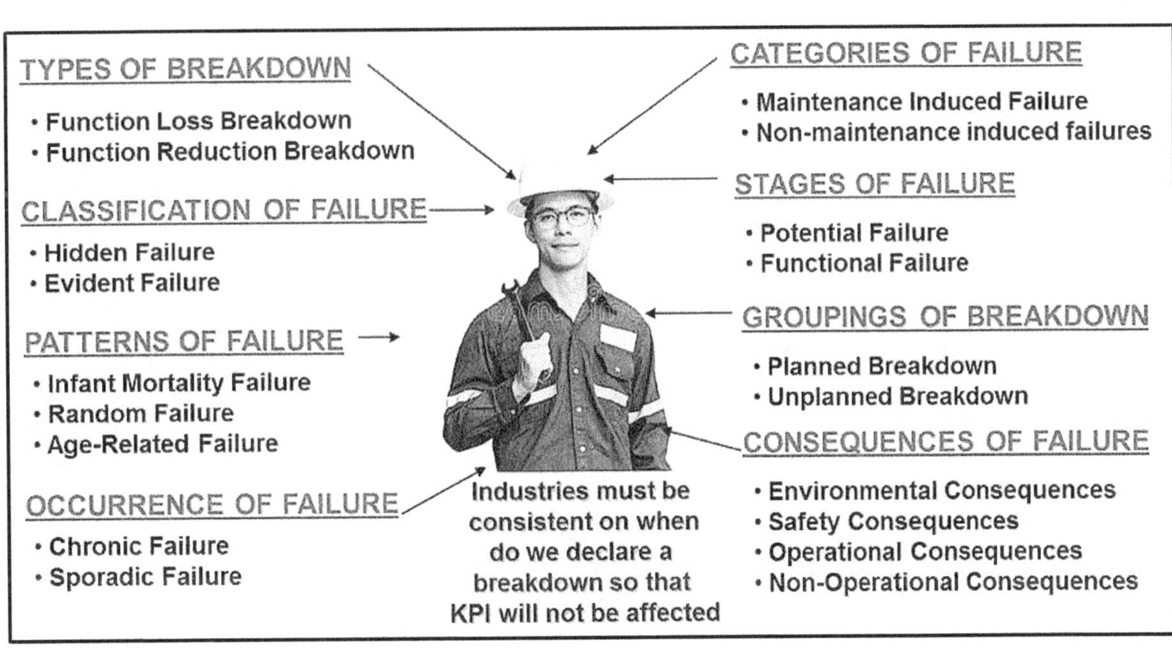

Figure 8.4: A Broad Subject on Failures and Breakdowns

Answers to be revealed in my next newsletter. So think about it, the word failure/breakdown is not exactly that simple as you think it is; hence, for those industries aiming for a Zero

Breakdown on the equipment, my advice is for your maintenance people to understand what eventually will constitute a breakdown or not. Otherwise, people will have their own perceived concept and definition of when to lists a breakdown or not.

If we go to Miriam Dictionary Online, a breakdown is defined as a machine's failure to function. An occurrence in which a machine (such as a car) stops working. The failure of a relationship or of an effort to discuss something. A failure that prevents a system from working properly. While Oxford Dictionary defines breakdown as simply a mechanical failure. Failure simply means the inability of equipment to perform its required function. The failure of a component is viewed as terminating its life. The word failure is technically broad in its very sense; therefore, before measuring the number of breakdowns occurring in your equipment, try first to discuss with your maintenance team what will eventually constitute a failure, breakdown, or not.

My point here is not to confuse the reader on what breakdown is or not. The type, kind, classification of breakdowns may vary from one industry to another. It is important to have a crystal clear definition of breakdown, which will be acceptable to both operations and maintenance. Suppose each maintenance has its own meaning of what breakdown is; in this case, some KPIs or Key Performance Indicators will be affected.

As stated in my previous newsletters, what is important is not to zero out breakdowns but to understand the consequences of every single breakdown. A consequence of a breakdown or failure of one sub-station is different from a breakdown of a production equipment in a production line. The message I would like to leave here is that in its very essence, the breakdown is a tricky word and may mean different things to different maintenance people or even to different people. A safety officer may declare equipment in a failed state if it poses a danger to its employees, while from a maintenance point of view, a breakdown is when the equipment had totally stopped because something fails or break. What is important is a cross-selection of people from different departments sit down and discuss what will eventually constitute a breakdown. Otherwise, they will have their own definition of breakdown or failure if this will not be done.

So think about it, the word failure/breakdown is not exactly that simple as you think it is; hence, for those industries aiming for Zero Breakdown on the equipment, my advice is for your maintenance people to understand what eventually will constitute a breakdown or not; otherwise, people will have their own perceived concept and definition on when to lists a breakdown or not. If we go to Miriam Dictionary Online, a breakdown is defined as a machine's failure to function. An occurrence in which a machine (such as a car) stops working. The failure of a relationship or of an effort to discuss something. A failure that prevents a system from working properly. While Oxford Dictionary defines breakdown as simply a mechanical failure. Failure simply means the inability of equipment to perform its required function. The failure of a component is viewed as terminating its life. The word failure is technically broad in its very sense; therefore, before measuring the number of breakdowns occurring in your equipment, try first to discuss with your maintenance team what will eventually constitute a breakdown and not.

8.6: June 2014: Step by Step Activities to Improve MRO Storeroom

It is not only important to place people in the storeroom just for the sake of withdrawing parts. It is important to place people who can control and manage the storeroom and spare parts list because MRO Spare Parts Management is about having the right part at the right time when maintenance needs it most. I believe that the best people to manage spare parts are maintenance. Spare Parts Management plays an important role in any maintenance improvement strategy. Much can be saved if we can just manage our spare parts intelligently. Every industry has its own storeroom for keeping parts needed for maintenance work, but not every industry has control over its spare parts. Here are lists of step-by-step activities to follow if you want to improve your storeroom and spare parts.

Step 1: Determine Your MRO Storeroom KPI: Success or failure depends on the KPI's that are important in our storeroom operations. There are wide lists of measurements available for your MRO Storeroom that you can use to determine your starting point in the overall MRO spare parts and storeroom strategy. For a start, recommended storeroom KPI's can include total line inventory and cost, inventory accuracy, number of stock-outs, overall costs of slow and non-moving, costs of obsolete parts, costs of emergency purchases, and so on. KPIs play a major role in gauging the success or failure of MRO and Stores. They provide a simple and concise method for comparing the actual performance from one period to another. They can even compare performance from one industry to another, plants within a corporation, and other MRO storerooms.

Step 2: Manned Your Storeroom: Industries should realize that if their operation is 24 hours, the storeroom should also be open 24 hours a day. Management and decision-makers should understand that they are not saving money on the storeroom if they only run for a single shift while their operations run for 3 shifts a day. We have to understand that the longer we retrieve the parts from the storeroom, the longer the downtime and repair time on the equipment. We need the storeroom people to provide us the parts that we need and analyze which parts need to be in stock in the storeroom and not how many we should stock. This is where we will truly realize savings on your MRO storeroom. Only authorized storeroom people should have access inside the storeroom. If everyone can access the storeroom, then the next problem will be on managing your inventory accuracy. Honesty policy and trust system simply do not work in the storeroom. What goes in and out of the storeroom should be recorded and accounted for by the storeroom people. One of the integrity of a good MRO storeroom is having a high inventory accuracy level for both physically and in the system (CMMS). When too much stock-out is experienced in the storeroom, maintenance will lose their confidence in them and start squirreling parts independently.

Step 3: Surrender Parts in Your Squirrel Stores: All spares should be controlled under one roof, the MRO storeroom. Squirrel stores exist for two reasons. First, too much stock-out part is experienced. Maintenance purchase the part on an emergency basis. You ordered excess parts in the storeroom and tried to remain the keeper of the parts. Suppose your storeroom is a distance away from where you operate; it is best to decentralize your storeroom. Still, the inventory will reflect the same as in the central hub of your storeroom. The physical location will be reflected in the system. Remember that there are also other users of the parts that you intend to keep, and if they don't know that you keep this part, then the storeroom will replenish or reorder this part in their system.

Step 4: Lists All Obsolete parts and Decide What to Do with Them: Storeroom people may actually have no knowledge of assets and equipment that had been decided to be retired. The best thing to do is communicate with maintenance on which equipment had been retired for the past couple of years in the plant. Once the lists are completed, identify the parts that go with them in your storeroom and provide the lists of these parts to the maintenance. Place these parts in one location of the store and discuss what to do with these parts. There are a lot of options for obsolete parts besides scrapping them. Remember that storing obsolete parts in your storeroom cost money; it is best to free them up or simply let them go.

Step 5: Perform 5's and Housekeeping on the Storeroom: Once the shelves are in place, storekeepers can set up a day to start their housekeeping and 5s activities. Parts are labeled properly, signs are placed for easy retrieval of items. Tools are neatly organized for easy access and identification if something is missing; cabinets, shelves, drawers are properly labeled. Parking for lifters, forklifts are in place.

Step 6: Perform Actual Inventory Count: Depending on how many items there are in the storeroom multiplied by their individual quantities, the storeroom should solicit help from all departments and set a date for the inventory count for the MRO Storeroom. Parts that have completed counting should be tagged and checked on the system if it matches or not. If there is a discrepancy in the system, it should be corrected based on the actual count. Once the inventory count had been completed, storeroom people can invite people from other departments to make a random audit to choose a part at random, count them, and check this on the system. This should be a continuous activity to be done, say once a month or more.

Step 7: Secure The Storeroom: Once activities to improve inventory count are in progress, storeroom people should also improve their storeroom security and provide guidelines and procedures on the store. It should be made clear to everyone in the organization that only authorized people will be allowed inside the storeroom, and access should only be granted to the storeroom staff except for some exemptions such as audits and inventory count. Our main objective in securing the storeroom is that we want to maintain a high inventory accuracy of around 95% or more, and we want to make certain that whatever goes in and out of the storeroom will be documented and put on record.

Step 8: Locate Alternative/Substitute Parts in the Storeroom: Storeroom staff should locate identical parts in the store with different part numbers. This should be well communicated with the purchasing people since there might be the same parts delivered by different vendors, and to provide control over them, they differentiate the codification. There are also possibilities that a part can have an alternative or substitute part present in the storeroom. Once the lists are known, this should be communicated with the maintenance people to use these alternate/substitute parts in the storeroom. The part number or codification should be reflected in the CMMS system. Hence, if a part is not available, they can utilize these substitute parts in the storeroom.

Step 9: Determine Fast Moving and Consumable Items: Suppose CMMS is in place in your plant or has some form of automation for your storeroom; you need to identify all the fast-moving items and consumables. These parts can easily be known to the storeroom as they are issued or requested from your store from time to time. You can use EOQ calculation to determine how many parts to be ordered and how many times to place an order per year. A buffer or safety stock can be added to them depending on the lead time to acquire these parts. The Storekeeper must know when to reorder these parts as they are often used on the assets.

Step 10: Determine Slow and Non-Moving Parts: It is good to prepare both an ABC and FSN analysis for your MRO Spare Parts. FSN Analysis stands for Fast-moving, Slow-moving, and Non-moving items. This form of classification identifies the items frequently issued, less frequently issued for use, and the items which are not issued for a longer period, say, 2 years and longer. It is best to identify and place them on the CMMS if this part is Fast Moving, Slow Moving, or a Non- Moving item. Once the parts had been classified as fast, slow and non-moving, we can perform an EOQ calculation on those fast and slow-moving items to identify a buffer or safety stock on critical items and the correct number of quantities to stock.

Step 11: Determine a Stocking Policy for your Storeroom: While it is easy to decide to stock these fast and non-moving items, the challenge lies in whether to stock or not those parts that fail randomly or for parts that fail unexpectedly. Both maintenance and storeroom staff can review their FMEA and RCM to determine what parts frequently fail on the equipment. An MRO Decision Diagram can aid them in making a sound decision on whether to stock these parts or not. Several factors need to be considered when making decisions on what parts to stock in the storeroom. With a sound understanding of MRO Spare Parts management, we can make better decisions instead of following OEM recommendations.

Step 12: Automate Your MRO Spare Parts and Storeroom: Imagine if you are in a supermarket, and bar codes do not exist, and everything is done manually; perhaps it will take you more than 2 hours to fall in line and pay for your goods and get out of the grocery. By this time, your storeroom is ready to automate itself by using Bar codes, Radio Frequency Identification (RFID), or Electronic Data Interchange (EDI). The purpose of automating is to lessen the chances of human error, make transactions much quicker, and have more quality time with family

It is not only important to place people in the storeroom just for the sake of withdrawing parts. It is important to place people who can control and manage the spare parts list because Spare Parts Management is about having the right part at the right time when maintenance needs it most. I believe that the best people to manage spare parts are maintenance. Spare Parts Management plays an important role in any maintenance improvement strategy. Much can be saved if we can just manage our spare parts more intelligently. Each industry has its own storeroom for keeping the parts needed for maintenance work, but not every industry has control over its spare parts.

8.7: July 2014: Top Reasons Why We Need to Perform RCFA

Every industry is familiar with the word root cause, but not all do understand how it is being done. The main reason why we perform root cause failure analysis is to understand why things go wrong so that we can earn from the root cause of the failure. We need to understand the root cause so that the failure does not repeat itself. Although I do not recommend performing root cause failure analysis on every failure that occurs, I recommend performing root cause if the failure keeps recurring.

1) Failures simply won't go away by fixing them: Failures struck us right in our face every time we deal with them to keep the equipment up and running for operations to continue with their production. We encounter the same failures repeatedly, and we always perform the same thing of fixing them all the time. RCFA helps us understand why these failures occur so that maintenance can focus more on improving their assets. We can only learn from the failures themselves if we follow the evidence and perform RCFA. Without evidence, one can only guess, speculate, based it on experience, or simply address the most likely cause of the problem.

2) We perform RCFA to arrive at the correct solution to our equipment problems: RCFA is not about addressing all the probable causes of the problem, but rather failures are being looked back in reverse to determine what really causes the problem to occur. In performing RCFA, each hypothesis is verified until we have gathered enough evidence to support the actual facts that eventually lead to failure. In performing Root Cause Failure Analysis, it is important to address the physical cause, the human and latent cause of the problem. In performing Root Cause Failure Analysis, we believe that all physical failures are triggered by humans, but humans are negatively influenced by latent forces; therefore, the goal in RCFA is to identify and remove these latent causes.

3) Being proactive will give me a sense of security: Many maintenance people believe that a good backlog of maintenance work and being reactive will ensure their job security by doing a lot of overtime work. This is not the right mindset since traditional maintenance people are confined to repairs and fixing failures, but the scope of maintenance is beyond boundaries. The true job of maintenance is to improve equipment reliability, and the scope of maintenance is beyond boundaries. The best position an industry can have will always belong to the maintenance function; however, maintenance people are often treated as repair people or mechanics in most industries and are often seen as a cost center.

4) People learn from failure itself: For every failure that occurred on the equipment and that had been thoroughly analyzed through RCFA, there is a learning that we can all gain from these experiences to prevent the recurrence of the cause of failure itself. Sometimes, failures speak to us in a different language. It is like when we feel pain and when we don't remedy the pain. It will start to get worse until the equipment had finally given up. Remember that there is always a lesson that we can learn from the failure itself.

5) There are failures where the consequences far exceed the cost of failure itself: Not all failures are equal. There are failures, when it happens, will have no effect, but there are also failures that can have very severe consequences once it happens in operations. When the impact or consequences of the failure can have safety or environmental consequences, we really need to put all our collective effort into eliminating these types of failures. Industries cannot afford to encounter a failure that can paralyze their operations, even for a single minute.

6) RCFA will save us money: Unexpected failures cost us money not only on the amount of damage or cost of spare as there are other costs beyond the spare parts itself, such as downtime cost, cost of shipment delay, labor cost, secondary damages, etc. Most of the time, the majority of these causes are based on deficient organizational systems, which includes maintenance and operating procedures, policies, guidelines, and training systems since most people will make decisions based on these systems, and most of the time, we end up wasting money because we tend to spend the money on equipment and not on training our people. Investments must be spent on making our people make better decisions. Remember that it will be the people that will improve the equipment and not the other way around. This will provide us far greater benefits and returns than many equipment investments.

7) To change our people's mindset: The new paradigm is that failures must be analyzed. People can learn from failure if they take the time to take things slowly and analyze them. The true job of maintenance is not to eliminate failures and fix them all the time. While I do not recommend performing Root Cause Failure Analysis on every single failure that can occur on the equipment, but if the failure keeps recurring, it is time to slow down and perform a Root Cause. Remember that troubleshooting is no longer an effective strategy. In today's competitive world, the "Analysts" find real solutions to our problems; that is why the need to perform Root Cause Failure Analysis is a must. This simply means that if your maintenance people become so good and fast at repairing failures, something is definitely wrong with your maintenance organization because the failure simply keeps on repeating itself, and maintenance has not analyzed why it keeps on failing. Hence, let us just leave the fixing to the mechanics.

8) There are cases where the consequences of failure are more important than the failure itself: Perhaps a classic case of this will be the Challenger Disaster; what simply fail is the O-ring on the Right Solid Rocket Booster, attached to the main tank causing a leak which ignites the main tank causing it to explode, destroying the Challenger Shuttle and killing all astronauts including Christa Mc Aullife, the winner of the Teacher in Space Program which was watched by the whole wide world when it took off last January 28, 1986. The problem with industries is that most of the needs required by maintenance are being neglected or subject to cost-cutting schemes, and the industry only reacts after a painful incident occurred in which their industry was hurt and devastated. The main reason for conducting a Root Cause Failure Analysis is for industries to learn from the things that go wrong, but this can only happen if people working in this industry are brave enough to face themselves in the mirror and admit to the fact that we too are also part of the problem.

8.8: August 2014: Analytical Problem Solving Techniques and RCFA

As the Planned Maintenance teams move on to Phase 2 of Planned Maintenance which is about improvements and modifications, one of the team's requirements is to learn at least three or more analytical problem-solving tools. The most recommended tools they will use will be P-M Analysis, FMEA/FMECA, Fault Tree Analysis, Ishikawa or Fishbone Diagram, and Root

Cause Failure Analysis. They will be using these tools to understand why some parts of the equipment have a short inherent lifespan, and the challenge for the team is to lengthen the lifespan of these parts.

Five- Whys: The Five Why's have been used as a tool to identify the most likely or probable cause of the problem for many years. The approach uses a technique to search for the most likely cause of the problem. The team uses a tool by asking why five times and works on the different levels of the analysis. Once the team finds it difficult to respond to why the problem's probable cause may have been identified. As you trace the why's back to their point of origin, the team conducting the probe will find confronting issues that affect the original symptom and the entire organization. To effectively respond to the next level, why the team answering the 5 whys must always base everything on the evidence and steer away from blaming individuals. Blaming people leaves us with no option but to punish them, leaving no room for substantial change. The focus should always be on the process of the problem and not the person who is involved since frequently answering why will lead to the person who was responsible for the problem. As the team's analysis progress, the team will eventually reach the latent cause of the problem. A more comprehensive approach to why-why analysis is to perform a Root Cause Failure Analysis investigation. They need to gather the evidence before conducting a why-why analysis on their investigation.

Failure Mode and Effect Analysis/Failure Mode Criticality Analysis (FMEA/FMECA): This is a proactive tool for evaluating failure modes and probable causes. It helps prioritize critical failure modes and recommends corrective actions and measures for avoiding catastrophic failures to improve the safety, quality, and reliability of the equipment. Failure Mode and Effects Analysis discipline was first developed in the United States Military. Procedure MIL-P-1629, titled Procedures for Performing a Failure Mode, Effects and Criticality Analysis, can be dated back to November 9, 1949. It was first used as a reliability evaluation technique to determine the effect of system and equipment failures. Failures were classified according to their impact on mission success, personnel, and equipment safety. It is also commonly defined as a systematic process for identifying a potential design, process, service, or equipment failure before they occur, with the intent to eliminate or reduce the risk associated with the failure. FMEA/FMECA can be performed at a system, subsystem, assembly, sub-assembly, equipment, component, or even on a parts level. The purpose of FMEA is to identify all potential failure modes and effects on the equipment or machinery and assess the risk and criticality of each failure mode. FMECA extends FMEA by including a criticality analysis used to assess failure modes' probability against their consequences. The result highlights failure modes with relatively high probability and severity of consequences, allowing remedial effort to be directed where it will produce the greatest impact. FMECA is simply an extension of FMEA.

To perform FMECA, the analyst must perform FMEA, followed by a critical analysis. FMEA will identify failure modes of a product, design, process, or equipment and their effects, while criticality analysis will rank those failure modes in order of importance, according to the failure rate and severity of the failure. The criticality part of FMECA consists of assigning a number from 1-10 each for occurrence, severity, and a third factor relating to detecting the failure. Multiplying these three factors gives a criticality number of 1-1000. There is a problem of nomenclature and definition since in FMECA, the RPN is termed as criticality. At the same time, elsewhere and in other industries, FMEA is being used. Instead of criticality, they prefer to use the word Risk Priority Number or RPN.

P-M Analysis: P-M Analysis stands for Phenomenon and M for Mechanism. This problem-solving technique analyzes chronic losses according to the inherent principles and natural laws governing them. This technique asks the team precisely what happens when a machine breaks down or produces bad parts and how it happens. Only then can we identify and address all causal factors and thus eliminate the chronic loss. P-M Analysis physically analyzes chronic problems such as defects and failures according to the machine operating principles. It clarifies the mechanics of their occurrence and the conditions that must be controlled to prevent them in the first place. Rather than prioritize the cause of the problems, the team doing the P-M Analysis will consider logically and with an equal emphasis on all of the factors that might have probably impact on the chronic defect or breakdown, then within each factor, the team will seek out every single abnormality that they can find no matter how small the abnormality is and arrive at the most appropriate countermeasures for each problem. The team should physically analyze chronic problems such as defects and failures according to the machine operating principles. Physical Analysis is a systematic, logical investigation of phenomena such as defects or breakdowns that explains how the phenomena occur in terms of their physical principles. To analyze the problem would mean to break it down into several parts and learn their nature and relationships. Physical Analysis uses machine operating principles to clarify how various machine parts interact to generate abnormal phenomena.

Most chronic losses persist because phenomena are insufficiently stratified and analyzed. Another reason is that defects and failures are not carefully observed and stratified. People do not notice the defect pattern. Other reasons why chronic failure happens is due to the following:

• Some factors related to the phenomena are overlooked.
• Potential causes are overlooked and uncontrolled.
• Uncontrolled factors can easily lead to chronic losses
• Hidden abnormalities in individual factors are not addressed
• Failing to identify and respond to abnormal conditions
• People are more alert to address big problems than small problems
• Slight abnormalities are ignored, such as dirt, rust, vibration, and looseness

Ishikawa Diagram or Fishbone Diagram: The Fishbone or Ishikawa Diagram was developed by Kauro Ishikawa of Tokyo, Japan, in 1943. A fishbone diagram is constructed, and each bone will be assigned the 4M's and 1E. The next step is to brainstorm the most probable causes of the problem, and the team will list them accordingly. The team will brainstorm and focus most likely on the probable cause of the failure. Cause-and-effect diagrams are used to list all the different causes that can contribute to a problem or effect. A cause-and-effect diagram can aid in identifying the reasons why a process goes out of control. An Ishikawa diagram is typically the result of a brainstorming session in which members of a group offer ideas on improving a product, process, or service. The main goal is represented by the diagram's trunk, and primary factors are represented as branches. Secondary factors are then added as stems. Creating the diagram stimulates discussion and often leads to an increased understanding of a complex problem. A fishbone diagram is an analytical tool that provides a systematic way of looking at the effects and the causes that create or contribute to those effects. It is also termed as a cause-and-effect diagram. The limitation of the Fishbone Diagram is that it provides no clear sequence of events that leads to the failure. Instead, it displays all the possible causes that may have contributed to the event. While this is useful, it does not isolate the specific factors that caused the event to occur. Constructing the Fishbone-Diagram includes the following steps:

• Draw the fishbone diagram.
• List the issue or problem to be analyzed in the head of the fish.
• Label each bone of the fish as per man, machine, method, materials, and environment.
• Use an idea-generating technique such as brainstorming to identify the factors within each category that may be affecting the problem being analyzed.
• Repeat this procedure with each factor under the category to produce sub-factors and ask why it is happening; analysis may include 2nd or higher levels under each cause.
• Analyze the fishbone results after team members agree that adequate detail has been provided under each major category. Do this by looking for those items that appear in more than one category. This becomes the more likely cause of the problem.
• For items identified as the most likely causes, the team should reach a consensus on listing those items prioritized for the corrective actions.

Fault Tree Analysis: It was developed in the 1950s by BOEING Aerospace Engineer for use in the design process's development stages. It is a mathematical tool and will yield probabilities. The purpose of a Fault Tree Analysis is to predict the probability of a specific failure. Fault Tree Analysis is not actually a Root Cause Analysis tool but rather a design tool.

Pareto's 80/20 Rule: Dr. Joseph Juran, one of the pioneers of quality, worked in the US from 1930 to 1940. Juran recognized a universal principle which he called the vital few and trivial many. In an early work, a lack of precision on Juran's part made it appear that he was applying Pareto's observations about economics to a broader work scope. The name Pareto's Principle stuck, probably because it sounded better than Juran's Principle. As a result, Dr. Juran's observation of the vital few and trivial many, where 20 percent of something is always responsible for 80 percent of the results, which now became known as Pareto's Principle, the 80/20 Rule.

Root Cause Failure Analysis: This is a tool that tries to understand why something went wrong. Root Cause Failure Analysis identifies you will try to address the basic source or origin of the problem so that industries can simply learn from the failure itself. RCFA provides a methodology for investigating, categorizing, and eliminating the root cause of safety, quality, reliability, and manufacturing process consequences. Identifying the Root Cause Failure Analysis event allows us to explain what happens, how it happens, and why the failure happens. A true Root Cause Failure Analysis is not interested in who causes the problem since there are deeper causes on why people commit mistakes and errors, and the goal of any root cause will be to identify and expose these causes, which is known as the latent cause of the problem. Proper Root Cause Analysis identifies the basic source or origin of the problem. It believes that every system, spare, or component failure happens for a specific reason. There are specific successions of events that will lead to the failure. RCFA follows the cause-and-effect path from the final failure back to its origin. Root Cause Failure Analysis methodology provides a specific and solid foundation for the preventing recurrence or failure. Root Cause Failure Analysis is a tool to better explain what happens, determine how it happens, and better understand why the failure happens. Root Cause Analysis separates the facts from hearsay. RCFA is not about trial and error and seeing what works and not. While there are many techniques in analyzing a problem that provides a quick answer, it does not mean that the answer is correct every time. A true and meaningful Root Cause Failure Analysis takes the time to prove that what we say is fact and supports our hypothesis with evidence before spending our money to improve the equipment's design. When the facts are finally backed up by evidence and science, and they are separated from fiction, we now have a better understanding of the real Root cause of the problem. Both RCFA and RCA are done by determining the physical cause, followed by the human cause, and finally ending up on the problem's latent cause.

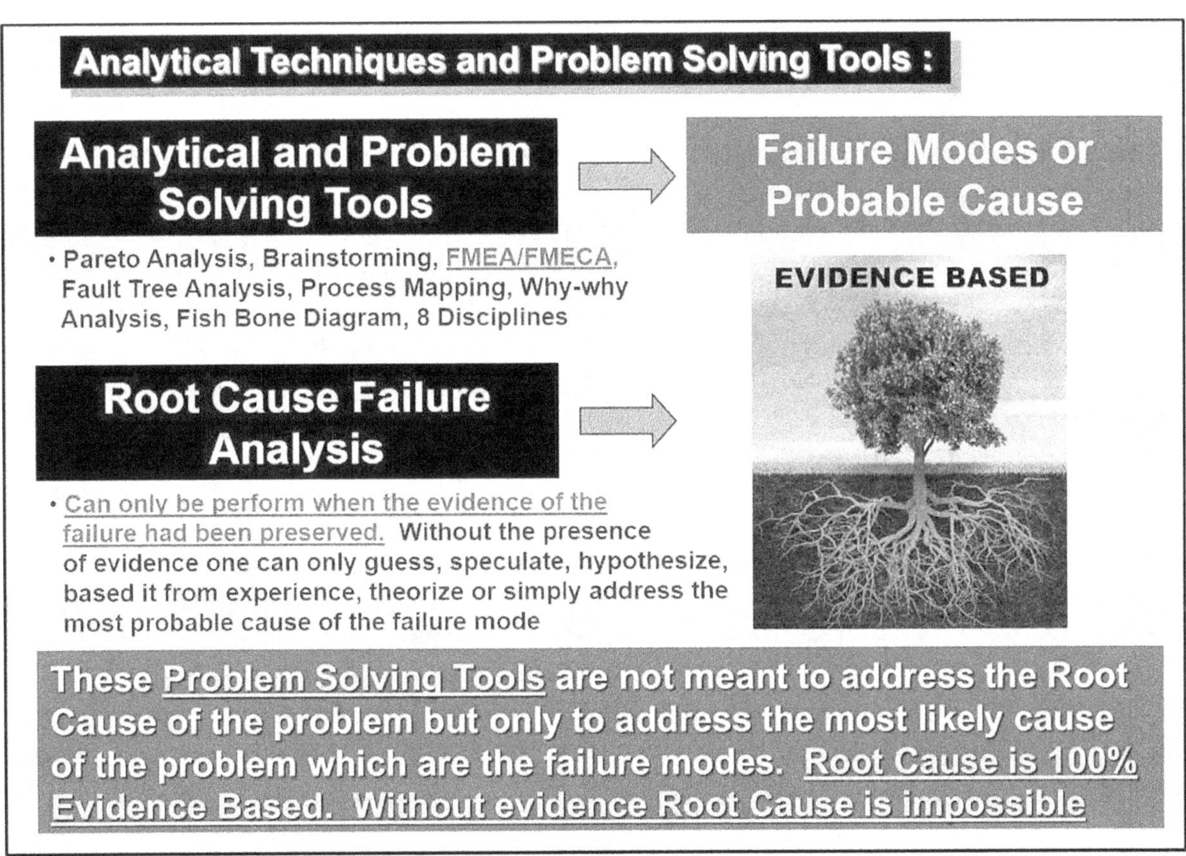

Figure 8.5: Difference Between Problem Solving Tools and RCFA

8.9: September: 2014 Step by Step Approach on How to Perform RCFA

Mini-Event RCFA is performed on mini or minor events. The failure's impact is small, but we still need to understand its cause and the need for the cause to be known. Mini-event can be performed by a single person or whoever comes in direct contact with the problem. The principal investigator will also gather the three types of shreds of evidence physical evidence, people evidence, and paper evidence. This will not require a stakeholder meeting. The analysis/investigation should include physical, human, and latent causes of the problem.

Midi – Event RCFA is performed on midi-events. They are led by an insider or someone from the affected site but as detached as possible from the specific event. This will involve a principal investigator and three evidence-gathering people. This will require a stakeholder meeting. The analysis/investigation should include the physical, human, and latent causes of the problem.

Maxi- Event RCFA is performed on maxi events or failures with a large impact on the industry. It is recommended that this event should be done by an outsider or an independent person to act as the principal investigator to avoid any form of biased. Three evidence-gathering teams will be required to perform a maxi event and will require a stakeholder meeting. The analysis/investigation should include the physical, human, and latent cause of the problem. Hence, whether the investigation or probe will be on a mini, midi, or maxi level, the following should be done to perform a Root Cause Failure Analysis on the equipment failure.

Types of Evidence

Nothing in our existence happens without it being known in one form or another. Evidence is the word we use to describe how we become aware of something. More specifically, the evidence is the way an event or condition manifests itself. The evidence contains everything we need to know about an undesired event. In fact, the evidence includes everything we need to know about life itself. Without evidence, Root Cause Failure Analysis is impossible because RCFA, by definition, is an evidence-driven endeavor. Without evidence, one can only suspect, hypothesize, theorize, speculate, guess, agree, disagree, brainstorm on the causes of an event that caused the failure to happen.

Physical Evidence or Parts Evidence: The actual collection of evidence is critical in the investigative process. Each piece of evidence collected must be handled in a way that preserves its integrity. Once an object is identified as evidence, it must be tagged. Evidence tagging helps identify the collected item. The tag can consist of as little as a sticker with the date, time, control number. Failed parts, debris, should be collected for evidence, and positions should be noted. When we try to fix the equipment, we easily replace the part that failed and throws it away. The first rule on any evidence freezing process is collecting and preserving the part that failed, taking photos of the part and equipment. Remember that we can only learn from failure if we take the time to takes things slowly.

People Evidence: An interview is a conversation intended to elicit information. Interviews are generally non-accusatory and should refrain from blaming, finger-pointing, and being biased. Interview the people who have witnessed the failure directly and indirectly. Directly means those people that have actually witness the failure, while indirect people are people who have experienced the failure in the past. Testimonies of people can strengthen and distinguish facts from fiction. In any RCFA analysis, there is always a sequence of events that eventually lead to failure. It is our goal to reconstruct what actually happened and try to connect the sequence. Once the conversation's tone has moved to an accusatory or interrogative mode, people will be defiant or hesitant to share what they know. It is, therefore, important that the investigator has a clear goal in mind when conducting the interview process. Our concern is to know what actually happened so we can put the pieces together. You need to make the interviewees feel at ease to be more willing to share what they know about the incident. No matter where the location, some amount of privacy is needed. Make sure the room is comfortable. Conduct an interview with one person at a time. The information collected will tell us a great deal about the failure itself. The purpose of the interview is to find out why things happened. The interview preparation is meant to give us a foundation for understanding what happened and gain the most relevant information from those involved. Remember that amnesty must be in place when conducting a Root Cause Failure Analysis investigation; our goal is to understand the cause of the failure so we can all learn from it.

Paper Evidence: When we speak about paper evidence, these refer to all recorded documents that we can gather at the time of failures such as history records, PM and PdM records, strip charts, manuals, operators' logbook, training records, specifications, current procedures used, equipment drawings and schematics. Paper evidence should be summarized by the person doing the probe or investigation. The investigator should collect all paper evidence as much as possible for a thorough review of the process. We should look for paper evidence to include those of operations, maintenance, and personnel-related data. What is important for the investigator is to determine which data are relevant and important at the time of failure. This will include any documents that you might find useful.

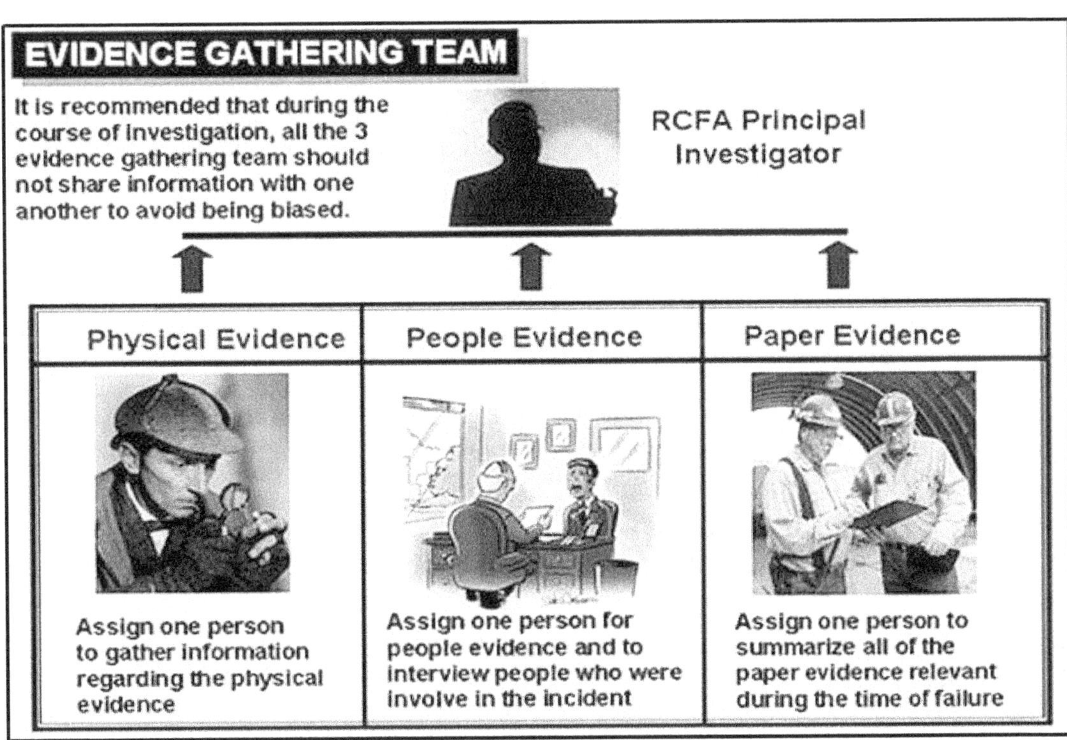

Figure 8.6: RCFA Principal Investigator and Evidence Gathering Team

Here are the lists of steps in conducting a thorough Root Cause Failure Analysis Investigation.

Step 1: Determine if the failure is a mini, midi, or maxi event: It is strongly recommended that the industry should categorize which problems and failures should be considered as a mini, midi, or maxi event. The basis for consideration can include the cost of failure, impact, or consequences of failure on the plant.

Step 2: Set up the RCFA Team: This will require a principal investigator and three evidence gathering teams for midi and maxi events. Note that the team to compose the RCFA should have been trained in the RCFA process. It is strongly recommended that an outsider or third-party group lead or act as the principal investigator for the maxi event.

Step 3: Freeze the Evidence: RCFA can only be performed on fresh failures and when the parts that failed and debris had been preserved. This will be part of the physical evidence. The three evidence-gathering teams should be independent of one another. One will be assigned to interview all people involved in the event, directly and indirectly, involve in the failure, one evidence gathering person will be assigned to collect all physical evidence, and the other evidence-gathering person will be assigned to collect all paper documents relevant during the time of the failure itself.

Step 4: Proceed with the Evidence Gathering Event: Sufficient time will be allotted to the three persons responsible for gathering the different shreds of evidence. A timeframe will be provided by the principal investigator. All evidence should be summarized independently by the 3 evidence gathering teams by determining the physical, human, and latent cause of the problem.

Step 5: Principal Investigator and Three Evidence Gathering Team Meeting: Once the evidence gathering team had completed summarizing the shreds of evidence, the principal investigator and the three evidence-gathering team will meet and perform an RCFA logic tree diagram or why-tree diagram based on the evidence collected.

Step 6: Principal Investigator and Evidence Gathering Team to Conduct RCFA Logic Tree: The evidence gathering team and the principal investigator will perform an RCFA logic tree or why-tree diagram to connect the sequence that eventually leads to the failure. If there are cases where they cannot be answered by the group, the evidence-gathering team will proceed to Step 4 again. After completing the RCFA logic tree diagram, the principal investigator and the evidence gathering team will finalize the physical cause, human cause, and latent cause.

Step 7: Principal Investigator and Evidence Gathering Team to Plan for the Stakeholder Meeting: After completing the physical, human and latent cause of the problem, the principal investigator and the evidence gathering team will name the people involved in the stakeholder meeting, and together they will set a meeting with them. Note that the stakeholder meeting will only be applicable for both midi and maxi events. Mini events do not require a stakeholder meeting.

Step 8: Stakeholder Meeting: During the start of the stakeholder meeting, the Principal Investigator will explain why the meeting and the evidence gathering team will present the summary of their investigation and why the stakeholder was involved in the first place. Stakeholder meetings would be required to determine the physical, human, and latent cause of the problem.

Step 9: Stakeholder to Determine a Smart Countermeasure: A SMART countermeasure should be generated by the stakeholder. Note that countermeasures will be required for both physical and human causes. The latent cause will not require a countermeasure since this is where a change in oneself will be required.

Step 10: Translate the Findings - The Stakeholder creates a plan to translate the findings to other people, departments, and areas in the plant that they think should also learn from the problem. The translation may be in posters, comic storybooks, and memos that people can readily read in the plant. Lessons from the failure should be disseminated to other areas where the problems are mostly to happen.

Step 11: Principal Investigator and Evidence Gathering Team Report: The principal investigator and the evidence gathering team finally conclude and generate a report on the failure and submit them to the plant's head.

8.10: October 2014: Planned Maintenance Phase 2 Flow of Activities

Phase 2: Address design weaknesses through improvement. Once accelerated deterioration has been reduced, equipment will suffer from natural deterioration. There are spares and parts of the equipment that will deteriorate naturally. Teams expose themselves to study several parts of the equipment with inherent short natural lifespan and correct design weaknesses by improving the part's dimension, the strength of materials, construction, dimensions, etc.

An MP Design (Maintenance Prevention) is usually used for this activity and later on feedback to the Early Equipment Management Pillar of TPM so that when the company decides to purchase future equipment, these improvements are discussed with the vendor's designers to be included in the new equipment purchase. A cycle must be established where MP Design improvements must be feedback to EEM. Correcting design weaknesses can prevent major breakdowns from recurring unexpectedly. Teams are trained on special tools such as P-M Analysis for a more detailed approach to chronic breakdowns. Likewise, most breakdowns are caused by human errors. Hence, both operators and maintenance must upgrade their skills to eliminate human errors. Application of Poka-Yoke solution may solve human errors but not necessarily improve the level of understanding of the mistake caused by the person involved. A much more comprehensive approach will be how to perform a Root Cause Failure Analysis and probing to the depths of the Latencies. Again for Phase 2, we are improving our equipment by addressing parts with inherent design weaknesses. Here are the details on conducting Phase 2 of Planned Maintenance.

1. Update MTBF/BDO in Phase 1 and Register the Team: Once the team passed the audit process on Phase 1 of Planned Maintenance, they are now ready to proceed with Phase 2. (BDO stands for breakdown occurrences). The team will update their indices such as MTBF and breakdown occurrences, which should dramatically be reduced as they restore their equipment in Phase 1. The team will register again for Phase 2 and will be provided a detailed roadmap on how to start and end this activity. The same team that goes through Phase 1 must be the same team that will undergo Phase 2. Usually, a sticker will be placed on the equipment, which indicates that it had already been certified on Phase 1.

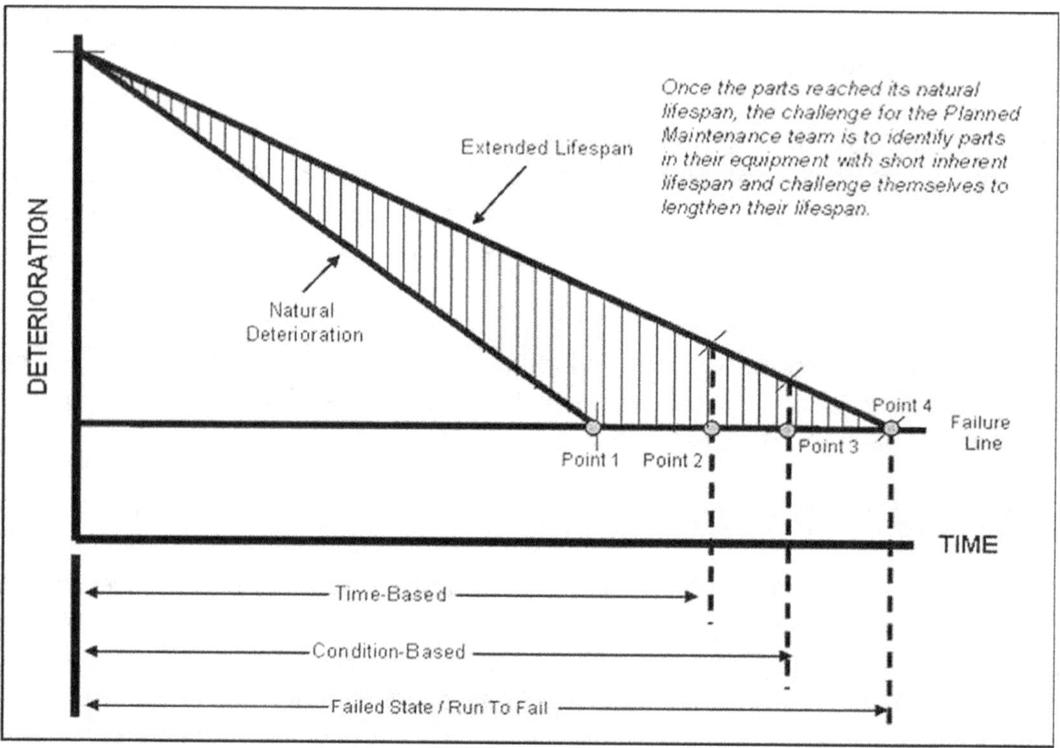

Figure 8.7: Concept of Planned Maintenance Phase 2

2. Undergo Training on Phase 2: The team on Planned Maintenance should schedule themselves to attend Phase 2 on correcting design weaknesses by lengthening its lifespan. Another important training for the team to undergo in this phase is understanding at least three or more analytical and problem-solving tools such as FMEA/FMECA, P-M Analysis, Poka-Yoke, Fault Tree Analysis, and Root Cause Failure Analysis. The team should likewise understand the concept of Life Cycle Management. Since Phase 2 on Planned Maintenance is about modification, the team needs to understand how to analyze failures before trying to redesign or modify the part itself.

3. Check recurring failures and identify parts that frequently fail: The team should analyze the equipment by identifying parts that fail frequently or those with a short inherent lifespan. This phase's goal on Planned Maintenance will be to identify these parts and challenge them to improve these parts' lifespan since no equipment is designed perfectly. There are always parts that will fail or break randomly or wear out easily on the equipment.

4. Draw the part and identify its weakness in the design: Once these parts had been identified, the team should draw or take photos of these parts and identify the weak points in the design and analyze why these parts fail easily on the equipment. These documents will also be used as a teaching aid to Autonomous Maintenance, and operators should also be aware of these parts since they will always be the first to witness the failure before the maintenance. What is important at this stage is determining why and how these parts fail and if they provide symptoms that they are on the verge of failing.

5. Prioritize and rank the list on which part to modify first according to its severity of the failure: Once these parts had been identified and analyzed, the team should prioritize which parts they will modify or redesign first. An FMEA or FMECA can be used to determine the criticality of the part so they can prioritize which of these parts will be modified first on the equipment. Likewise, the team must communicate with both purchasing and the storeroom people to control the ordering of these parts to be modified, especially if they are being stocked in the storeroom. Suppose these parts are stored in large quantities; in that case, a decision must be made to do with these parts because once the modification is fabricated and placed on the equipment. These parts simply become obsolete parts.

6. Analyze the Root Cause and perform failure analysis if necessary: When the lists had been provided to prioritize modification based on its criticality, the team needs to perform a failure analysis on these parts to determine the physical cause of the failure or simply answer how and why these parts seem to fail prematurely on the equipment. It is important to analyze the failure before proceeding with any design improvement changes to understand its failure.

7. Recommend improvement in design by changing its shape, size, or strength of the part's material: Once the team had identified the weak points in the design, they should develop a modified part prototype either by changing its shape, dimensions, the strength of materials, or construction. Remember that the goal of the team will be to exceed the lifespan of the previous part.

8. Proceed with the improved design and validate its effectiveness: Once the team had completed their analysis on the modified part, they needed to fabricate a prototype and test it on the equipment. The team should note the difference between the modified part's lifespan and the original part to be successful. Once the lifespan of the original part had been exceeded, a cost-benefit study should be done at this stage since there will be cases that the cost of the modified part might be expensive than the original part, but in the long run, it would be cheaper since the lifespan of the part had far been exceeded. Suppose the team had been successful in their design changes; they should inform both the storeroom and purchasing people to stop ordering the old part as this will now be an obsolete part of the storeroom. Also, an important decision to make at this stage is what to do with the original parts in the storeroom, if any, if they will still be used or the modified part will be used after depleting the old parts or simply get rid of them. The decision here should be supported by a cost-benefit or a feasibility study.

9. Proceed with the other identified weakness lists and follow activities 7 to 9: The team should now perform the same activities on the other parts that need to be modified on their equipment by simply proceeding with activities 7 to 9 until they have completed all the parts on the equipment that needs to be modified. An MP or Maintenance Prevention improvement design form will indicate the difference between the previous design and the modified part and looped back to the Early Equipment Management or Initial Flow Control Activities of TPM. The purpose of this is that if the plant decides to purchase equipment in the future, these modifications done on the equipment will be discussed with the designers and OEM of the equipment to adopt and add these changes before purchasing the equipment. This activity on Phase 2 will take some time, depending on the number of parts modified at this stage.

Place Company Logo	**Phase 2 ACTIVITY ROADMAP** For Pilot Equipment		QUALITY NO EXCUSE
AREA ___________ LEADER : ___________	EQUIPMENT : ___________		Teamname : ___________
OBJECTIVE : Lengthen equipment life by correcting design weaknesses and address recurring breakdowns permanently			

NO.	ACTIVITY DESCRIPTION	Code	LEV2	FORMS USED		WORKWEEK
1	Update MTBF / BDO data on Phase 1 of Planned Maintenance and register the team for Phase 2	S1-01 S1-01a	DO	MTBF / BDO Analysis Chart PM Registration Form	P A	
2	Cover the following training's : - Planned Maintenance Phase 2 - Root Cause Failure Analysis and FMEA	 S2-02a	DO	 Training (8 hrs) Training Monitoring & Skills Assess	P A	
3	Check recurrence of Breakdown and identify the sub-assembly and corresponding part	S1-01	---	MTBF / BDO Analysis Chart	P A	
4	Draw and identify spare parts that frequently fail or have a short lifespan on the equipment	S4-05	DO	Equipment Weaknesses lists	P A	
5	Prioritize improvement and rank lists according to its frequency of breakdown	S4-07	DO	Component / Spares Priority Lists	P A	
6	Analyze rootcause of design weakness, detail the analysis per part	S4-08	DO	P-M Analysis Form	P A	
7	Recommend for Improvement in Design (Change in material, shape, dimensions etc.,)	S4-09	DO	Design Improvement Recommendation MP Design Form	P A	
8	Proceed with the improved design and validate (Prepare prototype whenever necessary)	S4-09	DO	Design Improvement Recommendation MP Design Form	P A	
9	Proceed with the other Identified Equipment Weaknesses Lists and perform activities 6 - 8	-------	DO		P A	
10	Make necessary changes in procedure, PM interval, and prepare OPL or GTF for operators and maintenance	S2-10	DO	Good to Find or OPL.	P A	
11	Monitor cost impact and usage of part / component with improved design and check problems if any	S4- 13	DO	Modified Spares Parts Cost Comparison	P A	
12	Summarize Improvement Activity and update MTBF/BDO and other indices	S4-14	DO	Improvement Activity Summary MTBF/BDO Analysis Chart	P A	
13	Update PM activity board and perform self-audit on on Phase 2 of Planned Maintenance	----	DO	Step 4 Audit Guidelines	P A	
14	Proceed with Phase 2 audit	S4-18 S3-15a	DO	Step 4 Audit Guidelines Request for PM Certification	P A	
15	Identify Fan-Out to similar machine type (Note : Identify only the machine, do not replicate yet)	S4-16f	DO	Horizontal Replication Form	P A	

LEGEND : P = PLAN   A = ACTUAL   Note : Team must have Minutes of Meeting for the activities and Attendance Check for the members   Rev. 04, 03/20/01

Figure 8.8: Planned Maintenance Phase 2 Roadmap

10. Make necessary changes in procedures, PM specs and conduct training to concerned maintenance and operations:

Once all the parts with inherent design weaknesses had been modified, the team needs to adapt changes in their Preventive Maintenance since the interval or frequency of replacement had changed. A One-Point-Lesson or Good-to-find form can be generated at this stage to serve as a teaching aid for both operators and other maintenance teams and why the part had been changed or modified.

11. Monitor the impact of improvement concerning cost and summarize the improvement activity:

Allow the modified part to run for several months on the equipment and monitor the length of time it had run compared with the previous part. Monitor and observe the modified parts and update indices, especially on cost, MTBF, and breakdowns at this stage. The breakdowns resulting from Phase 1 activities had dramatically been reduced, and we still expect the breakdowns to be reduced at this phase on Planned Maintenance.

12. Horizontal Replication on other machines with the same problem:

Replication of improvement changes and modifications will only be done on the same type of asset or equipment that poses similar problems. This means that the modified part will also be adapted to other similar equipment in the plant that provides the same symptoms. Note that if there is similar equipment in the plant that does not pose this problem, it is not recommended to place the modified part on that equipment.

13. Proceed with Phase 2 Audit and Certification: The team should complete their documents for Phase 2, perform a self-audit at this stage, and schedule themselves for Phase 2 audit, where they will try to defend their validity improvements and modifications they had done on the equipment.

8.11: November 2014: Why does Many CBM Program fail in Other Industries?

Some industries invest in CBM strategies only to be abandoned because they cannot derive its full benefits. In my other first World Class Maintenance Management – The 12 Disciplines, CBM is considered in the last 2 disciplines on World Class Maintenance stage since this will require heavy investment in the plant both on the instrument and the certification of the users. Equipment basic condition must be well in place before using any CBM instrument in your plant. Here are lists of reasons why most CBM program fails in industries.

1) Failure to add a CBM group in the maintenance organization: The best option when starting a CBM program in your industry is to create a centralized CBM organization. Allot full-time and permanent manpower to handle this department. Part-time personnel is not recommended to handle CBM. It is always advantageous to have your own CBM organization rather than hire 3rd party contractors.

2) Insufficient training for CBM: As discussed earlier in this chapter, investment for CBM is not only purchasing the instruments but allowing the users of these instruments to undergo a level of certification from time to time. Some industries purchase the instruments but fail to certify their people thinking that they can save costs by providing the users with some vendor training where a little bit of principle will be discussed and explaining to the user the use of this and that button instruments. The users can only fulfill their respective functions if they are fully trained on these instruments' use. Another cost incurred in the CBM strategy is that these instruments need to be calibrated yearly or depending on the manufacturer's recommendation. For example, for certified thermographer, this is what the users will learn on each level. Remember that in CBM, there are two investments. First, the instruments, and second, the users of these instruments.

Level I Infrared Certification
• Understand the basics principles of infrared and heat transfer concepts
• Be able to detect qualitative inspections for electrical and mechanical failures
• Roofs and building envelopes
• Take accurate infrared temperatures of any targets or subject
• Can evaluate the severity of the problem

Level II Infrared Certification
• At this level, the user has mastered the science of quantitative infrared thermography and can avoid or correct measurement errors caused by emissivity, reflection, spot size, transmittance, and calibration.

Level III Infrared Certification

• Provide accurate financial documentation
• Understand and meet the requirements of OSHA, DOE, and other regulatory agencies
• Identify and specify safe practices for your company
• Prevent reoccurring problems by identifying failure mechanisms
• Integrate predictive maintenance technologies
• Quantify heat losses and calculate paybacks for corrective actions

3) Not documenting the success of CBM: Since a CBM program requires investment, it is highly recommended that success stories done by the users should be well documented to justify the group's existence. Otherwise, the Top Management team might disband or dissolve its members, thinking they are just a non-value entity to their organization.

4) Failure to identify critical equipment: Not all plant equipment will undergo a CBM strategy. It is better to run to fail some equipment if the consequence of failure is small. A standby redundancy is in place. The machine should be ranked, and parts that fail randomly should be identified, making it a candidate for the CBM strategy.

5) Top Management is not convinced about the rate of Return On Investment (ROI) on CBM: A survey by Plant Services Magazine in the US states that organizations are reluctant to invest in new manufacturing technologies because their top management is not convinced of the Return On Investment (ROI) on CBM. Many CBM programs were aborted within the first three years because a clear set of goals and objectives was not established by the CBM group before the program was implemented. These instruments are expensive, not to include the cost of training and calibrations to be incurred. Failure to compute the return of investment and the cost-benefit generated by the program will just be short-lived. It is highly recommended that industries starting to have these strategies create a feasibility study on the benefits that can be derived by using these strategies.

6) The software program of any CBM is the heart of any program: Software for these instruments is one of the most difficult things to evaluate on selecting the right brand to use before purchase. Due to stiff competition in the field, most software programs will not be compatible with each other, which prohibit transferring database from another vendor system into a different vendor of CBM. Some of this instrument's software may not even be compatible with the plant's CMMS system.

7) Top management does not clearly understand how the program works: Management was committed to the initial investment for these instruments' capital expenses but did not invest in the resources required for training, consulting support, and in-house manpower that is essential for success in any CBM strategy. Several CBM strategies had been aborted for the first couple of years, not because they failed to achieve the results but due to a lack of understanding of the concept and benefits of a CBM strategy in the plant.

8) Failure to take care of the basic equipment condition: Equipment basic condition should be well established before embarking on any Predictive Maintenance strategy. This means that the equipment should be clean, properly lubricated, no leaks whatsoever, and complete bolts. It is simply useless to purchase the top-of-the-line vibration monitoring equipment that can detect several frequencies if your equipment is dirty and lacks bolts and screws. All machinery contains nuts, bolts, and screws as essential elements of their construction. The equipment functions properly only if fasteners are securely tightened. It only takes one loose bolt to start a chain reaction of wear and vibration. As the other bolts become loose, vibration increases. All machines vibrate, but excessive vibration is destructive, as this will surely induce secondary damages on parts and components affected by the vibration. Vibration should be controlled; cracks, fractures propagate as a result of excessive vibration.

Many questions arise when trying to justify instruments for CBM, such as;
• What equipment should be permanently monitored?
• What CBM instruments should we invest in?
• What operational factors should be considered?
• Is Online or CBM the right technology to utilize?
• How much profit correlates to detecting downtime?
• How should return on investment be evaluated?

The decision to invest in Condition-Based Maintenance has always been an expensive proposition due to the large capital investment to have a strategy in place. Getting approval to purchase these instruments required heavy cost justifications. Simply writing a purchase requisition for this instrument will result in it being disapproved or denied by top management. Evaluating the return on investment is related to equipment criticality, failure mode, frequency of occurrence, and penalty cost of the downtime. This strategy is important in having maintenance, operations, top management trained and educated on Condition-Based Maintenance by qualified trainers. Source and invite resource speakers for in-house training on CBM. Invite vendors to demonstrate the capability of their Predictive Maintenance instruments in your equipment. Identify critical equipment in your plant and compute the cost of downtime and maintenance.

Prepare a feasibility study on the need for a CBM and indicate the rate of return on investment on a CBM program. Define the instruments that your plant will benefit most and start on a small scale and showcase the savings generated by these instruments for a start. Don't purchase all the instruments and software at once. Let your people adjust and appreciate the use of these instruments. It is important to document the results and success stories obtained by these instruments' users and submit a regular report to your top management to appreciate that they made the right decision.

Factors to Consider on Which Machines to Undergo CBM Strategy Critical Machines:

Machines should be ranked accordingly. Worst or critical equipment and assets will be a good candidate for the CBM strategy. Likewise, equipment subject to breakdowns that developed rapidly and have severe financial impact and consequences will be subjected to a CBM strategy. Check machines for a history of problems.

• **Personnel Safety** - If the machine or asset handles dangerous material, its condition must be carefully monitored. For example, a tank containing methyl isocyanate in one pesticide factory should be inspected and regularly checked for plate thickness and corrosion.

- **Probability of Failure is high** - Equipment operating at design limits or handling aggressive material should be monitored more closely than equipment in a routine service or those that have redundant functions.
- **Manning Level** - Equipment on remote or isolated locations or unattended normally requires condition monitoring. They will be a good candidate for the CBM strategy.
- **Cost and Effect** - If an unexpected failure is definitely high and unacceptable to the plant, CBM is required. Power plants use this strategy since there are failures that can lead to environmental consequences.

8.12: December 2014: Difference Between RCFA, RCA, and Failure Analysis

Most industries are confused about what Root Cause Failure Analysis is all about and how to conduct a thorough investigation of equipment failures. RCFA is simply based on evidence, and without evidence, it is simply impossible to conduct an investigation to determine equipment failure. One could only guess, speculate, witch hunt, or simply based it on experience. Root Cause Failure Analysis can only be done if the evidence had been preserved. While most problem-solving techniques and analytical problem tools will base themselves on data and history, Root Cause is simply based on one thing and one thing alone, and that is evidence. Here is a distinction between Root Cause Failure Analysis, Root Cause Analysis, and Failure Analysis.

Failure Analysis: If we speak about failure analysis, we stop our analysis on the parts or component level or the failure's physical analysis. We refer to this as the metallurgical aspect as to why the failure occurs. When we send a failed bearing to the metallurgical lab, they will check the raceway, perhaps conclude and write a report that the bearing failed due to fatigue. The physical cause is the reason why the parts failed technically. This is the technical explanation of why things broke or failed. This mostly explains the metallurgical factor of why the failure occurred.

Root Cause Analysis: This is much broader because this can be applied to medicine, crime investigations, etc. For industry-related issues, this can be applied to safety, environmental issues, and quality-related issues and defects. An example is that in one hospital, a failure analysis simply refers to conducting a probe or investigation on failures or breakdowns encountered on the asset or equipment. Meaning the word Root Cause Failure Analysis refers to equipment-related failures. Both RCFA and RCA will determine the problem's physical, human, and latent causes.

A patient died because the attending nurse injected the wrong blood type on the patient unknowingly. After all, the patient switched beds with another patient, but they did not switch their records. The nurse thought that the person sleeping on this bed was the one to receive the blood transfusion. Eventually, the person died after a few hours; now, this will warrant an RCA investigation approach.

Root Cause Failure Analysis: Both RCFA and RCA are performed by determining the problem's physical, human, and latent causes. The distinction is on the word failure; when we include the word failure in Root Cause Analysis, we refer to equipment-related failures. Hence, if we speak about Root Cause, RCFA is trying to understand why something went wrong. It will identify and address the basic source or origin of the problem so that industries can simply learn from the failure itself. The reason for conducting a Root Cause Failure Analysis is for industries to learn from the things that go wrong. RCFA will provide a structured methodology for investigating, categorizing, and eliminating the root cause of safety, quality, reliability, and manufacturing process consequences. Identifying the Root Cause Failure Analysis event will explain what happens, how it happens, and why the failure happens.

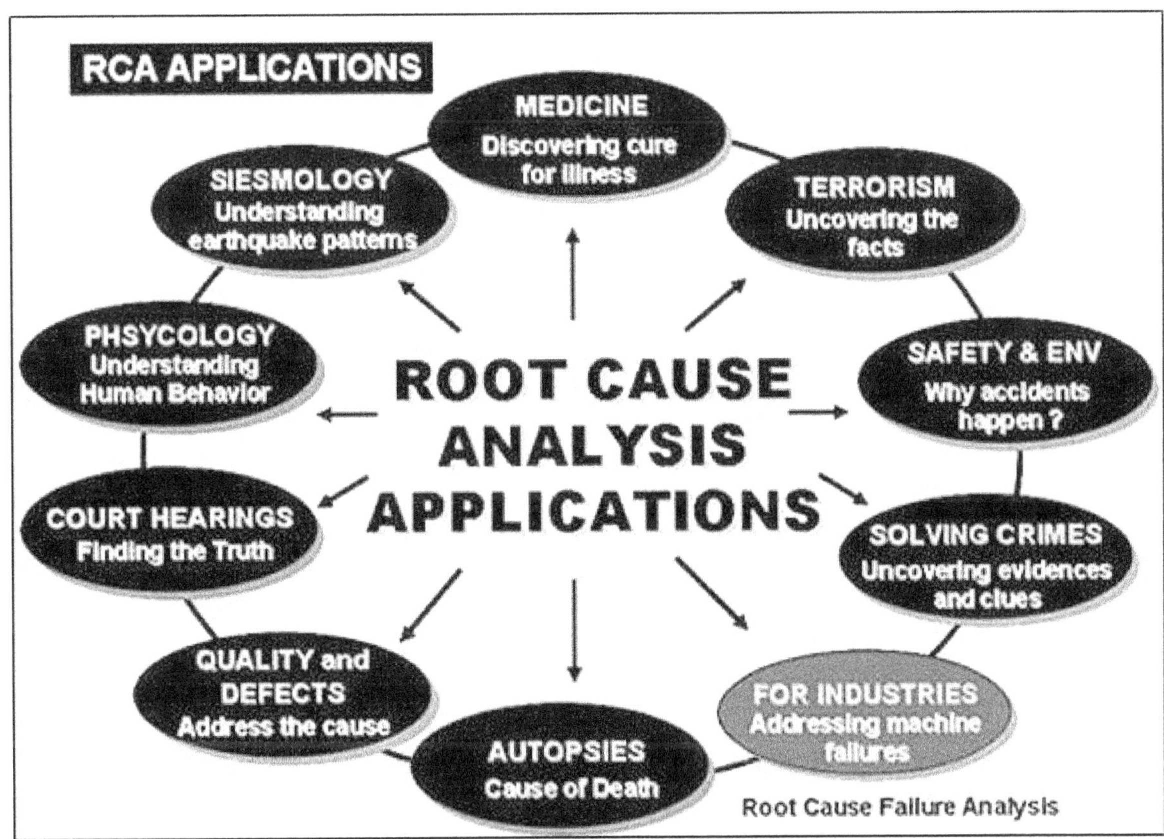

Figure 8.9: Difference Between RCA, RCFA, and Failure Analysis

A true Root Cause Failure Analysis is not interested in who causes the problem since there are much deeper causes on why people commit mistakes and errors, and the goal of any root cause will be to identify and expose these causes, which are known as the latent cause of the problem. Proper Root Cause Failure Analysis investigation identifies the basic source or origin of the problem. It believes that every system, spare, or component failure happens for a specific reason. There will always be a specific succession of events or sequence of events that will eventually lead to the failure. Failure will always have to be the last event to happen. RCFA follows the cause-and-effect path from the final failure back to its origin. Root Cause Failure Analysis methodology provides a specific and solid foundation for preventing the recurrence of the same causes of the problem or failure. It is also a tool to better explain what happens, determine how the failure happens, and better understand why the failure happens.

Root Cause Failure Analysis will separate the facts from fiction. Everything will be based on evidence, and without evidence, root cause is impossible to perform. It is not about trial and error and seeing what works and not. While there are many techniques in analyzing a problem that provides a quick answer, it does not mean that the answer is correct every time because a true and meaningful Root Cause Failure Analysis will take the time to prove that what we say is fact and supports our hypothesis with evidence before we spend time and money to improve the design of the equipment. Therefore, when the facts are backed up by evidence and science, and the causes are now separated from the fiction, we now have a better understanding of the real root cause of the problem. Both RCFA and RCA are done by determining the physical cause, followed by the human cause, and finally ending up on the latent cause of the problem.

2015 RSA Reliability Newsletter Vault Archive

> *Industries will continue to remain reactive unless both operators and maintenance remain a separate function. Remember that operators are the first line of defense on any equipment failures because they will first experience the failure first before the maintenance. If they know the symptoms, then the failure can be addressed at its earliest stage.*

9.1: January 2015: Maintenance - A Scapegoat for Operations Problems

For many industries and organizations, the maintenance department simply does not exist. Instead, they rephrase or change their organization's name to "Equipment Engineering," which is a very cool name; indeed, even though not all people in that department are engineers. When I was just starting my teaching business, I was part of a training group in the Philippines that conducts regular training in semiconductor industries. They change my maintenance and reliability training courses to "Equipment Engineering Trainings" because the trend and semiconductor industries prefer to use that term. They simply wanted to avoid the word maintenance because the word maintenance connotes a negative word to them. But the funny part is that they refer to their people as technicians, mechanics, electricians, repairmen, sustaining, which is reactive. Suppose industries understand maintenance first; they will not be calling them these names since these people's function happens right after a failure occurs, which is happening to most industries, especially in manufacturing. My apologies for being rude, but if this is a marathon race, manufacturing will always be the last in the race to reliability because their priority is on productivity. Being reactive creates more overtime for these people, which I think maintenance like because of the extra bucks, but the downfall is that pressure is intense, mostly from production people, which I think they hate. They are not proud to be called maintenance because industries perceive maintenance as evil and a neglected word and are always considered a cost center. As a result of too much fire-fighting and cost-cutting is going on. Basic equipment condition is frequently neglected on their assets. Today industries suffer from a high cost of doing maintenance because of a lack of a structured and effective Planned Maintenance system. This means that if it takes 100 dollars to manufacture a product, mining around 20 to 50 percent will be the cost of doing maintenance. Recent surveys on maintenance management in different industries indicate that one-third or 33 cents out of every dollar of all maintenance costs is wasted due to unnecessary or improperly carried out maintenance, resulting in ineffective maintenance management representing a billion-dollar loss each year or even more. Put it this way, most Preventive Maintenance is done regardless of the condition of the equipment because Predictive Maintenance simply does not exist in their plant. PM includes risks because it will replace parts and components which are still in working condition. Why, the reason is simple, "<u>Just in Case</u>."

Most industries insist that all breakdowns can either be prevented or eliminated; hence, too much focus is given to doing Preventive Maintenance on their equipment and assets. Most maintenance managers admit the fact that even with a sound Preventive Maintenance program placed in their plant, they seem to realize that failures are inevitable and do happen. They often strike without warning or when they are least expected to occur. When this happens, most maintenance managers, together with operations people, discuss what to do to prevent the failure, and the default will be to place activity in their never-ending growing lists of Preventive Maintenance. Hence, the list of Preventive Maintenance activities never stops increasing, and the pressure on doing it grows minute by minute, day by day. Most industries think that this is the right thing to do, but this is where they got it all wrong.

The traditional thinking is that all parts will eventually wear out, so the need to schedule equipment will be done for every asset in their plant. But when the schedule comes for a Preventive Maintenance, PM will be waived because operations will simply not allow maintenance to do their stuff since their line of thinking is that all that matters is just one thing alone: **output,** nothing more, nothing less. Remember that equipment is not like a television set. When you plug it in, it plays without any further need for intervention. In equipment, this will be different since mechanical parts move, and every single part that moves will be subject to stress. Once the stress exceeds the strength, then wear occurs, and since their scheduled Preventive Maintenance is missed out, more breakdowns increase and strike us right in the face at a time we least expect them to happen, so maintenance will be left with no option but to put out fire every time. They are busy recuperating and troubleshooting their equipment. If a spare part is affected, most often, the part is unavailable in the storeroom. Maintenance is left with two options: to ash advance and buy the part outside in excess. At the same time, they keep the rest of the parts to themselves or simply cannibalize other parts on idle equipment that are not used since operations are now pressuring them to get the line going at all costs. Now once again, the equipment is running with some Band-Aid fixed therapy. These people will attempt once again to get the equipment for Preventive Maintenance. Still, operations will again deny them of that chance since they need to cope with their backlog, and this humble maintenance will be working for very long hours with more pressure on them just to keep the asset running. As a result, many breakdowns and failures suffered from their assets while the cost of doing maintenance increases. The funny thing is everyone knows the reason for the high cost of doing maintenance, but instead, the boss will question no one but you on why maintenance cost was high, and they want you to provide them a very detailed answer for that. Operations people are now looking for a scapegoat, an escape for not hitting their target for the day, and there is only one thing that they can think of, which is our humble and down-to-earth maintenance people themselves. The morale for maintenance declines, and life goes on for them.

In a reactive environment, most maintenance activities happen right after a failure or simply when a breakdown happens. Repairs and fixes are done at the quickest possible time because if repairs are prolonged, there will be shadows at the back, whispering to their ear how much more time. Some planning and scheduling are done, which are well set in place, but the machine still fails even with all these done. Mostly this will always be the accepted norm of a plant's culture, and maintenance gets a pat on the back for being reactive or coming to the plant very late at night fixing the equipment where everyone has given up on how to make it run again. This is always what the incumbent maintenance teaches to their new ones. They have been doing this since the beginning of time. Training is a complete waste of time and money for these industries since they had existed with this line of culture. This is how they do things in their plant on a day-to-day basis. They do not believe in using any Predictive Maintenance instruments or any other form of non-destructive diagnostic tools since management just thinks of them as expensive things and some fancy nice to have features for the maintenance function, or simply they have existed for a very long time without these instruments. What I cannot comprehend in my thick skull is why do all international airports have infrared thermography in place to scan people, and not all industries can afford to have one? The truth of the matter is that not all failures can be prevented. For the record, not all breakdowns can be captured by doing Preventive Maintenance. In fact, increasing the number of Preventive Maintenance activities will simply increase the chances of infant mortality failures. It's as simple as that. Many failure modes will provide signs and symptoms that it is on the verge of failing. They can be predicted with precision accuracy with the use of these non-destructive Condition-Based or Predictive Maintenance instruments.

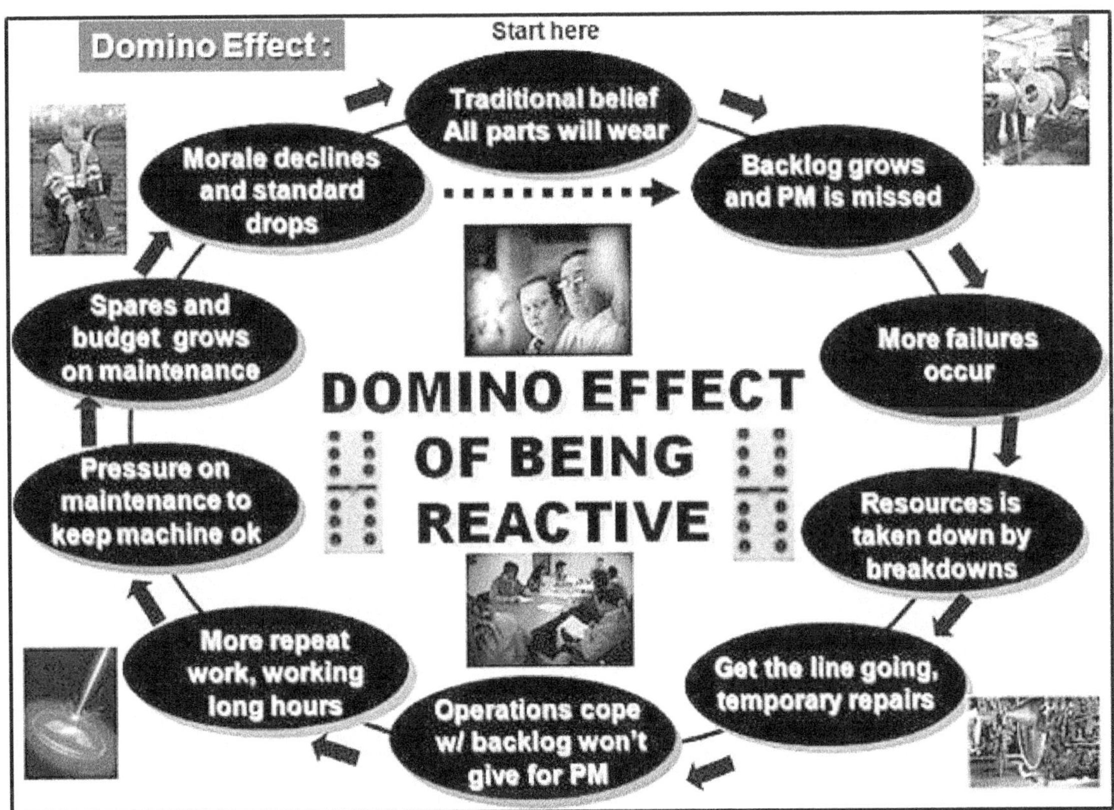

Figure 9.1: Domino Effect of Being Reactive

In reactive industries, the longer they can tolerate the pressure from operations people, the better chance for maintenance to stay longer in this type of industry. They just adopt Newton's law. Whatever they hear with their ears must come out. There are many instances where the boss will call you and ask you for the problem's root cause. You just cannot tell him that the boss is the root cause of the problem unless you still want to be employed (even if your boss is really the root cause of the problem, you need to think of other causes) and the lamest excuse for maintenance is "wear and tear." The boss will tell you to make sure it does not happen again. The default will be to add an activity in the never-ending lists on their Preventive Maintenance. The saying goes true if you cannot change them, just join them.

Changing the name of your maintenance organization to other fancy cool names to get rid of the word maintenance makes it sound really nice, but technically, it does not really make your maintenance organization more effective or proactive. I will stick to the word maintenance because this is who we really are. Maintenance should not be considered a cost center, but it can be transformed into a profit center if both operations and maintenance people truly understand what maintenance is all about and the role they play in industries. Maintenance is one area where cost can definitely be saved, and I hope that industries finally realized this message one day.

Industries that do not understand correct maintenance practices will always have one option in cutting costs: retire or terminate people. Maintenance only becomes famous when production people highlight them for a breakdown for not hitting the production of the day, even if they did not cause the breakdown. They are always considered scapegoats and will always be blamed for every single downtime on the equipment. Industries should understand that instead of blaming and putting fingers on maintenance every time equipment fails is not the right thing to do. Instead of blaming them all the time, why not just sit down with them and work in harmony on solving the problem together. Industries must understand that maintenance is not only for the maintenance people but also for operators as well.

9.2: February 2015: Industry's Procedure for FMEA/FMECA (Part 1)

Purpose of this FMEA/FMECA Guidelines: The purpose of conducting FMEA/FMECA analysis on equipment and assets is to eliminate, reduce, or mitigate failure modes' consequences, starting with the highest severity ones. This will also identify a single asset, component, system, or sub-system's potential failure modes and effects on the equipment. Although FMEA/FMECA can also be used for the design system or process, these guidelines will be limited to conducting FMEA/FMECA for the industry's equipment and assets. FMEA will not only be limited in identifying all potential failure modes and effects on the equipment but will likewise assess the risk and criticality of each failure mode by determining the severity and determining the potential causes of each failure mode. FMEA/FMECA will also be used to identify controls and design actions that can prevent the failure mode from recurring and identify possible corrective actions and countermeasures required to prevent, mitigate, or improve the likelihood of detecting failures early by providing a means for controlling or mitigating the failure modes from occurring.

Scope of this FMEA/FMECA Guidelines: This FMEA/FMECA guideline will cover only critical equipment and assets for the industry. New equipment and assets will not be covered initially in the FMEA/FMECA process as the warranty of these equipment provided by OEM may be affected due to the corrective actions and activities derived through the FMEA/FMECA process. It will be assumed that vendors and OEM recommendations will be different as OEM vendors tend to be more conservative. It is also highly recommended that equipment that is the subject of RCM Analysis will be exempted in conducting FMEA/FMECA to avoid any duplication or redundancies since the first part of the RCM process is about determining the FMEA using the RCM information worksheet in which it will derive the functions, functional failure, failure mode and effects of the failure. The equipment subject to RCM Analysis should be exempted in the FMEA/FMECA analysis to avoid redundancy or conflict in their countermeasures and corrective actions.

FMEA/FMECA Explained: Failure modes and effects analysis (FMEA) is a step-by-step approach for identifying all possible failure modes in design, equipment, process, system, service, or product. Failure modes mean the ways, or modes, in which something might fail. However, this guideline's coverage will only be for equipment FMEA as the tables on the severity, occurrence, and detection might differ, respectively. Failures are any errors, breakdowns, or defects that will affect the plant's equipment and assets. Failure Mode and Effect Analysis (FMEA) is a proactive tool for evaluating failure modes and probable causes. Note that FMEA/FMECA will be limited only in determining the most likely cause or probable cause and not the Root Cause of the failure itself. It helps prioritize critical failure modes and recommends countermeasures for avoiding catastrophic failures to improve maintainability, safety, quality, and reliability. It is also commonly defined as a systematic process for identifying potential equipment failures before they are likely to occur, with the intent to eliminate or reduce the risk associated with the failure. Equipment or Machinery FMEA/FMECA is a risk analysis tool used to identify all possible failure modes in the past and the possibility of happening in the future. The result highlights failure modes with relatively high probability and severity of consequences. Although FMEA will be using RPN (Risk Priority Number), FMECA will be using a criticality analysis. Also, it should be noted that the most critical failure mode usually has a low occurrence and may not have a high RPN number. Priority in determining corrective actions should be based upon the severity of the failure mode itself. Doing FMEA/FMECA can be conducted on a component level, equipment level, sub-assembly level, system or sub-system level, or even a parts level depending on the FMEA/FMECA Analysis team.

Difference between FMEA and FMECA: Failure Mode and Effects Analysis (FMEA) provides only qualitative information, while Failure Modes and Effect Criticality Analysis (FMECA) will provide qualitative and limited quantitative information, which means that FMECA can be measured. It attaches a level of criticality to the failure modes. FMECA is just an extension of FMEA. To perform FMECA, the team must perform FMEA, followed by a critical analysis. FMEA identifies failure modes of a product, design, process, or equipment and the effects of the failure, while criticality analysis ranks those failure modes in order of importance, according to the failure rate and severity of the failure. The criticality part of FMECA consists of assigning a number from 1-10, where 1 has the lowest criticality, and 10 will be the highest criticality of failure. The criticality of the failure mode will be done by multiplying the severity (consequences) by the occurrence (probability) and detection (confidence), which will obtain values from 1 to 1000. The detection or confidence is largely based on the ability to catch the failure mode in time. If we are doing FMECA, we estimate the probability that gives us the risk of failure. Although there is a problem of nomenclature and definition, in FMECA, the RPN will be termed as criticality analysis (CA), while elsewhere and in other industries, we understand and use the word Risk Priority Number (RPN).

Guidelines for Conducting Equipment FMEA/FMECA

The following are the steps required to proceed with the FMEA/FMECA Analysis. It is strongly recommended that people with the most extensive knowledge of the equipment and asset undergoing the FMEA/FMECA analysis should be part of the team. New employees are welcome in the team as observers and can use the FMEA/FMECA meeting as a learning ground.

Step 1: FMEA/FMECA Team Selection and Register the Team: The FMEA/FMECA team should comprise a cross selection of experienced maintenance with at least one incumbent operator. A minimum of 4 and a maximum of 8 members will compose the FMEA/FMECA team.

Step 2: Determine the Pilot FMEA/FMECA: Performing FMEA/FMECA can either be done on the most constrained or bottleneck equipment, process, system, sub-system, component, or even to a spare parts level. Note that boundaries should be defined for system and sub-system levels on what will be included and excluded in the system. It is not recommended to perform FMEA on newly purchased equipment for two reasons. First, the new equipment is still under warranty, and the FMEA failure modes derived by the team might differ from those recommended by conservative OEM vendors. Second, the impact will not be felt immediately since the equipment is new. It is highly recommended that FMEA/FMECA be done on the most critical or problematic equipment where maintenance costs are relatively high. Candidates for FMEA/FMECA analysis will include critical assets, bottleneck, and constraint equipment. Backtrack KPI and maintenance indices on the pilot equipment for the past six months to one year to provide a clear benchmark in terms of downtime, breakdown, costs of repair, Mean Time Between Failure, and maintenance costs. If FMEA/FMECA is done correctly, the cost of maintaining the asset should dramatically be reduced together with the rate of its breakdown as well. For the team to realize its benefits, FMEA/FMECA should be done only on the most critical asset or system in the plant and not on new machines.

Step 3: Team to Undergo 1 Day FMEA/FMECA Training: As a minimum requirement for the FMEA/FMECA analysis to commence, it is important to note that all team members should undergo the FMEA/FMECA one day course to provide them the basic understanding of how to conduct an actual FMEA/FMECA. This one-day FMEA/FMECA training will be the minimum requirement to proceed with the FMEA analysis. Only qualified trainers or third-party consultants will be eligible to conduct the FMEA/FMECA training.

Step 4: Have a Fixed Regular Meeting for the FMEA/FMECA Analysis: Have a fixed and regular meeting once or twice a week. FMEA meetings will last from 2 to 4 hours. It is recommended to appoint a secretary and has all relevant documents of the equipment, component, system, or sub-system to be analyzed readily available in the meeting room for convenience such as schematic drawings, OEM manuals, PM records, history records, Key Performance Indicators, Predictive and Preventive Maintenance records and so on. Provide house rules during the meeting. The FMEA analysis aims to capture the members' experiences on their assets, especially on the failure modes and effects of the failure. Members should be committed to the completion of the FMEA/FMECA Analysis.

Step 5: Write the FMEA/FMECA Header: Use the FMEA/FMECA form provided. The leader of the team should initially fill up the FMEA/FMECA header form. A description of each item in the header is explained below.

• FMEA Number: This is the FMEA reference number that can be used for tracking.
• Machinery Name: Refers to the name of the equipment and specific sub-assembly, system, sub-system, or module or component being analyzed.
• Design Responsibility: This will refer to the vendor, supplier, department, or group of people responsible for the design changes or modifications.
• Model Number: Refers to the model number and make of the equipment.
• Review Date: Enter the initial FMEA review date. The date should fall within the design and development phases of the FMEA analysis.
• FMEA Date: Enter the date the original FMEA was completed and the latest revision date. Ensure that corrective actions are already in place.
• Pages: This will refer to the number of pages of the FMEA report.
• Prepared by List of all individual team members responsible for the FMEA analysis. This may also be people who have the authority to perform the FMEA/FMECA analysis. It is recommended that all team member names, departments, contact numbers, and addresses be included in the distribution list.

Step 6: Define the asset's function being analyzed through FMEA/FMECA: Write the asset or system's primary function to undergo the FMEA/FMECA analysis. The function should consist of a verb, an object, and a desired standard of performance, which is the same procedure for writing the function in RCM. What will be required will determine both the asset's primary and secondary functions and not the equipment features. Functions will explain why the industry's equipment or asset was purchased. To answer the function, ask yourself, what does your industry want this equipment to do?

Step 7: List all possible Potential Failure Modes of the Equipment: A Potential Failure Mode is defined as how the equipment could potentially fail to meet its intended function. A potential failure mode is also the most likely cause of a system's failure, subsystem, component, or equipment. List down each potential failure mode for the particular subsystem function. The assumption is made that the failure mode can possibly occur but may not actually happen. Exclude all failure modes in which maintenance has no control whatsoever.

Step 8: Determine the Failure Effect: When this failure mode occurs, what will affect the system, asset, equipment, or machinery? Failure Effect results from a given failure on people, system, equipment, production, or the environment. Failure Effect will describe what happens when a potential failure mode occurs. Failure Effects should be as detailed as possible. It should state what must be done to repair the failure. Usually, this will range from 20 to 60 words. What is important is to write the failure effect as detailed as possible. When writing the failure effect, take note of the following:

a) How long is the asset's failure or downtime and the time to repair the failure?

b) Estimated MTBF and frequency of the failure

c) Is there any protective device or controls in place such as alarms, led that will indicate that this failure mode had occurred or is in the process of occurring?

d) Is there a redundancy or standby unit in place in the event the failure occurred?

e) Will the parts be readily available, or will it take some time to acquire the replacement?

f) Will other equipment be affected by this failure or only this equipment?

g) Is there any possibility of secondary or tertiary damages on other parts of the equipment?

h) Failure effects should be written in paragraph form ranging from 20 to 60 words.

i) Will the failure be preceded by any means such as smoke, noise, smell, increased temperature, drop in pressure, etc.? (To be continued)

9.3: March 2015: Industry's Procedure for FMEA/FMECA (Part 2)

Effects on	Criteria – Severity of Effects	Ranking
Hazardous without warning	Very high severity ranking – Affects operator, plant, or maintenance personnel, safety and/or affects non-compliance with government regulations	10
Hazardous with warning	High severity ranking – Affects operator, plant, or maintenance personnel, safety and/or affects non-compliance with government regulations	9
Very High Downtime or Defective Parts	Downtime of more than 8 hours or defective parts loss more than 4 hours of production	8
High Downtime or Defective Parts	Downtime of 4 to 7 hours or defective parts loss of 2 to 4 hours of production	7
Moderate Downtime or Defective Parts	Downtime of 1 to 3 hours or defective parts loss of 1 to 2 hours of production	6
Low Downtime or Defective Parts	Downtime of 30 minutes to 1 hour or defective parts loss of up to 1 hour of production	5
Very Low Downtime No Defective Parts	Downtime up to 30 minutes – no defective parts	4
Minor Effect	Process parameter variability exceeds Upper/Lower Control limits. Adjustment or other process controls need to be taken – no defective parts	3
Very Minor Effect	Process parameter variability between upper/lover control limits. Adjustments or other process controls needs to be taken	2
No Effect	Process parameter variability within Upper/Lower Control limits, adjustment or other process controls not needed or can be taken between shifts or at normal maintenance – no defective parts	1

Figure 9.2: FMEA Table for Severity

Step 9: Team to rate the severity of the potential failure mode and effects: Severity refers to the impact of the failure mode and its effects. These are also called consequences. Severity considers the worst scenario that can happen after a failure or breakdown takes place. It is a rating corresponding to the seriousness of the effects of a potential failure mode. Estimate the severity of the failure mode of the equipment, system, or subsystem function using the table on severity for equipment FMEA/FMECA and enter the rating in the FMEA worksheet. Ranking for severity will be 1, for the lowest severity, and 10 will be the highest severity rating. Estimate the severity of the subsystem function's failure using the table on the severity in figure 9.2 for machinery FMEA and enter the FMEA worksheet rating. A reduction in the Severity Rating index can only be effected through a redesign change or modification. The team prioritizes all failure modes with the highest severity rating obtained and not basing its priorities on the highest RPN. Remember that most critical and high severity failure modes have a low occurrence.

Step 10: Place CC for Classification if the severity is 9 to 10: Classification is used to highlight those failure modes with a Severity rating of 9 or 10 or failures with the highest impact on the equipment or system being analyzed FMEA. These failure modes will likely affect the workers' safety or lead to a breach in any environmental conditions that require design actions and modification. Classification only apples to that severity with scores from 9 to 10 by placing CC (Critical Characteristic). For the severity of 8 and below, leave the Classification column blank.

Step 11: Determine the potential causes of failure: Potential cause of the failure refers to the probable or most likely cause of the failure mode. Certain problem-solving tools such as fishbone diagram, cause, and effect, why-why analysis, Pareto, brainstorming, fishbone diagram, and other problem-solving tools can be used to determine the potential causes of the failure. This refers to the most likely cause of the failure. A failure mode's potential cause may either be a design weakness or machinery process variation that can be described in terms of something that can be corrected or controlled. The failure mode's potential causes may be the most likely cause or probable causes of the failure mode but not the actual root cause of the failure itself.

Step 12: Determine the occurrence of the failure: Occurrence is a rating according to the likelihood that a particular failure mode will occur within a specific time. It is an assessment of the likelihood that a particular failure mode will happen that resulted in failure. While severity ranking is driven by the effect, occurrence ranking is a function of the cause and is based on the likelihood or frequency that the cause or mechanism of failure occurs. A high occurrence may not necessarily mean that the failure is critical and may have a low severity and vice-versa. A rating of 10 for occurrence means that the failure happens frequently, and a rating of 1 means the failure is unlikely to happen or the occurrence is more than 10,000 hours. Almost everyone who uses FMEA is taught to use the RPN to determine which line of the FMEA works. Customers tell their suppliers they want recommended actions written for all lines with RPN over a certain value. Auditors always ask for the RPN level that triggers the need for a recommended action. Unfortunately, these auditors and those who use FMEA often look only at the RPN and ignore the severity of the failure, which leads to wasted effort on problems of little significance. High severity or consequence failure should be prioritized and addressed in the FMEA process, regardless of their occurrence. Looking only at the score with a high RPN can compromise the integrity of the FMEA process. The occurrence is also termed as a probability.

Step 13: Determine current design machine control: Design and Machinery Controls are used to control, detect, prevent, and predict the cause and mechanism or failure mode from occurring, thereby reducing their rate of criticality. This means that if a failure mode is on the verge of occurring in the equipment, these controls, which serve as a warning sign, can alert both operators and maintenance that a possible failure mode is about to occur or is in the process of occurring. Design and Machinery Controls are methods, techniques, devices, tests, or maintenance tasks that prevent the failure mode from occurring, reducing the failure occurrence rate. Design and Machinery controls can prevent the failure modes and their effects from occurring and reduce or lower the occurrence rating. Identifying these design and machinery controls should begin with those failure modes combined with the highest severity and occurrence ratings. Only those with current machinery controls will be considered to have a low detection ranking. Those failure modes which currently have no current controls but are subject to inclusion of controls in the future will still have a high detection ranking when doing the FMEA/FMECA. Failure modes with controls will have a low detection rating, and those without controls will have a high detection rating. Figure 9.5 is a sample of these controls placed on the equipment to alert operators that failure is on the verge of occurring.

Probability of Failure Occurrence	Possible Failure Rates Criteria	Ranking
Very High: Failure is almost inevitable	Intermittent operation resulting in 1 failure in 10 production pieces or MTBF of less than 1 hour.	10
	Intermittent operation resulting in 1 failure in 100 production pieces or MTBF of less than 2 to 10 hours.	9
High Repeated Failures	Intermittent operation resulting in 1 failure in 1000 production pieces or MTBF of 11 to 100 hours	8
	Intermittent operation resulting in 1 failure in 10,000 production pieces or MTBF of 101 to 400 hours	7
Moderate Occasional Failures	MTBF of 401 to 1000 hours.	6
	MTBF of 1001 to 2000 hours.	5
	MTBF of 2001 to 3000 hours.	4
Low Relatively Few Failures	MTBF of 3001 to 6000 hours.	3
	MTBF of 6001 to 10,000 hours	2
Remotely Failure Unlikely	MTBF greater than 10,000 hrs	1

Figure 9.3: FMEA Table for Occurrence

Step 14: Assign a detection ranking: Detection is an assessment of the Design and Machinery Controls' ability to detect or capture a potential failure mode from occurring on its own. To assign detection rankings, identify the current equipment controls in place for each failure mode and then assign a detection ranking for each control. A detection ranking of 1 means the chance of detecting a failure is certain, while a detection ranking of 10 means there is the absolute certainty of non-detection of the failure mode and will be quite dangerous most especially if the failure mode has a high level of consequences. This basically means that there are no controls in place to prevent or detect that a failure mode is about to occur or is in the process of occurring. This is also termed as confidence.

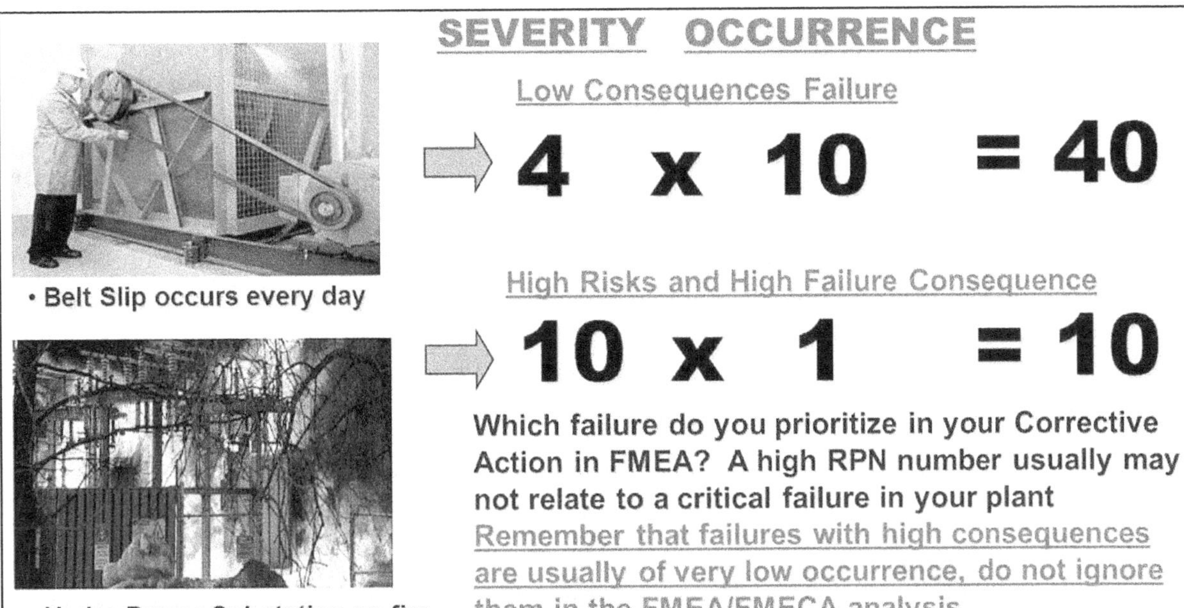

Figure 9.4: Priority on FMEA Will Be on High Severity and Not RPN

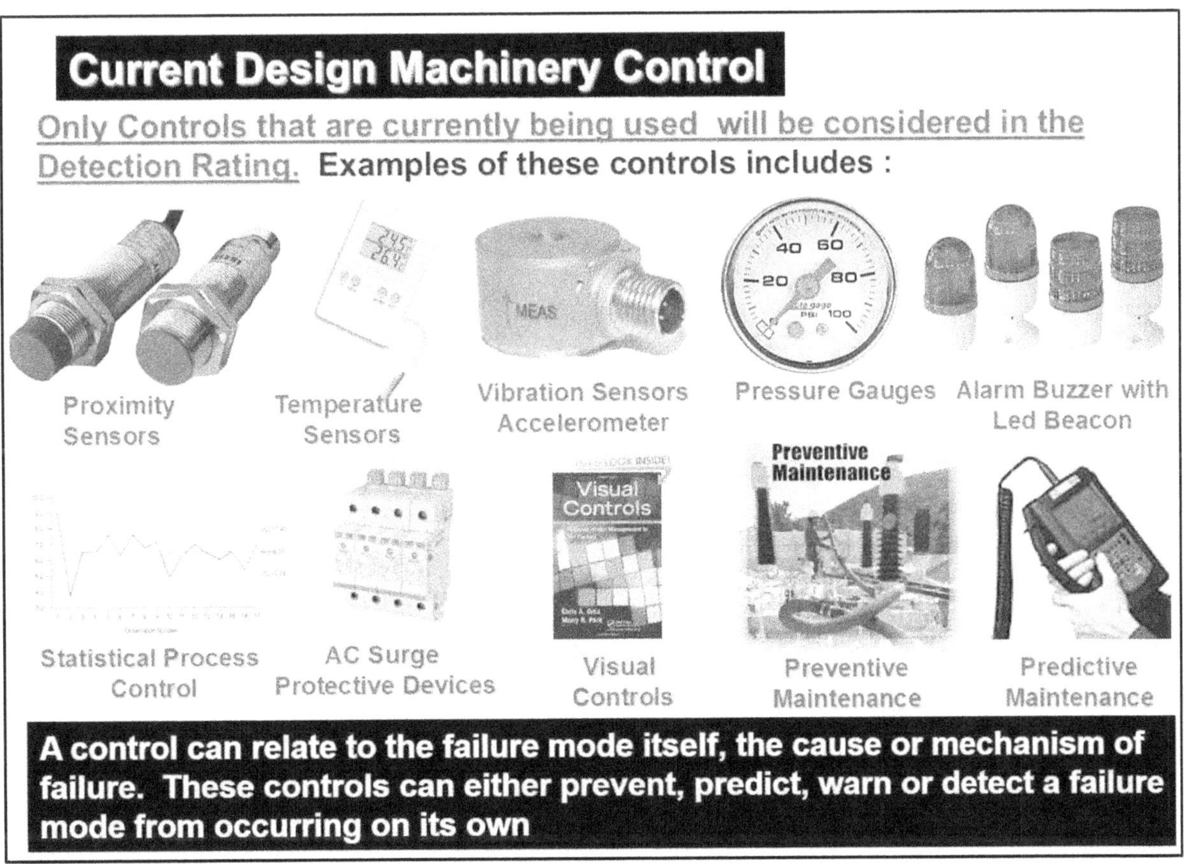

Figure 9.5: Equipment Defenses for Failures and Breakdowns

Step 15: Determine the Risk Priority Number (RPN): Risk Priority Numbers (RPN) are used to assess the failure mode's risk and effect. The team doing the FMEA/FMECA should rank the severity of each failure mode, the likelihood of occurrence for each cause of failure, and the likelihood of detecting the failure before it finally fails. RPN can be calculated by multiplying the severity by the occurrence, finally by the detection. RPN numbers should only prioritize the most critical potential failure modes to consider possible design changes or modifications to reduce its criticality or make the design more robust to reduce the failure's risks and consequences. The RPN gives us a relative risk ranking. Theoretically, the higher the RPN, the higher the potential risk, but we should likewise be very careful when prioritizing corrective actions as there are failures with very high impact and severity yet will be unlikely to occur, and there are failures that occur more frequently but are not that critical and has a low severity rating.

Detection	Likelihood of Detection by Design Controls	Ranking
Absolute uncertainty	Machine controls will not and/or cannot detect potential cause or mechanism and subsequent failure mode; or there is no design or machinery control.	10
Very Remote	Very remote chance a machinery/design control will detect a potential cause/mechanism and subsequent failure mode.	9
Remote	Remote chance a machinery/design control will detect a potential cause or mechanism and subsequent failure mode. Machinery control will prevent an imminent failure.	8
Very Low	Very low chance a machinery/design control will detect a potential cause or mechanism and subsequent failure mode. Machinery control will prevent an imminent failure.	7
Low	Low chance a machinery/design control will detect a potential cause or mechanism and subsequent failure mode. Machinery control will prevent an imminent failure.	6
Moderate	Moderate chance a machinery/design control will detect a potential cause or mechanism and subsequent failure mode. Machinery control will prevent an imminent failure and will isolate the cause. Machinery control may be required.	5
Moderate High	Moderately high chance a machinery/design control will detect a potential cause/mechanism and subsequent failure mode. Machinery control will prevent an imminent failure and isolate the cause. Controls may be required.	4
High	High chance a machinery/design control will detect a potential cause or mechanism and subsequent failure mode. Machinery control will prevent an imminent failure and will isolate the cause. Controls may be required.	3
Very High	Very high chance a machinery/design control will detect a potential cause or mechanism and subsequent failure mode. Machinery controls not necessary.	2
Almost Certain	Design control will almost certainly detect a potential cause/mechanism and subsequent failure mode. Machinery controls not necessary.	1

Figure 9.6: FMEA Table for Failure Detection

Step 16: Lists recommended corrective actions: Design actions are taken to reduce the severity, occurrence, or detection rankings. Remedial actions should be considered and prioritize failure modes with a high severity rating of 9 or 10. The priority of corrective actions should be based on the severity of the failure. Prioritizing corrective actions should also consider failure modes and effects with inherent safety as well as environmental consequences. Whenever a failure mode has a severity rating of 9 or 10, a redesign action and modification must be considered to eliminate, reduce, or mitigate the failure's consequences. Taking corrective action means reducing the RPN. This can be reduced by lowering any of the three rankings' severity, occurrence, or detection individually or in combination. Use the table in figure 9.7 to prioritize corrective actions to be done on the equipment. To be continued.

Criteria : Feasibility of Corrective Actions Implementation	Ranking
Safety/environmental problem consequences or non-compliance to Government Regulation. Unavailable necessary resources or unacceptable cost and time consumption. Having a zero chance of success and 100% probability of undesirable impact.	10
Very remote availability of necessary resources or almost unacceptable cost and time consumption. Almost zero chance of success or almost 100% probability of undesirable impact.	9
Remote availability of necessary resources or near unacceptable cost and time consumption. Have a remote chance of success and near 100% probability of undesirable impact.	8
Very low availability of necessary resources or very high cost and time consumption. Have a very low chance of success or very high probability of undesirable impact.	7
Low availability of necessary resources or high cost and time consumption. Have a low chance of success and high probability of undesirable impact.	6
Rather low availability of necessary resources and rather high cost and/or time consumption. Rather low chance of success and rather high probability of undesirable impact.	5
Moderate availability of necessary resources, cost, time consumption, chance of success and probability of undesirable impact.	4
Rather highly available resources, rather low cost and time consumption, rather high chance of success and rather low probability of undesirable impact.	3
Highly available resources, low cost and time consumption, high chance of success and low probability of undesirable impact.	2
Fully available resources, very low cost and time consumption, near 100% chance of success and near zero probability of undesirable impact.	1

Figure 9.7: FMEA Table for Prioritizing Corrective Actions

9.4: April 2015: Industry's Procedure for FMEA/FMECA (Part 3)

Step 17: Assign responsibility and target completion: Once corrective actions are finalized, assign the best person responsible for the redesign, and target the dates of completing the corrective action, redesign, or modification. The person responsible may not be the one who will do the design changes but should be held responsible for providing updates regarding its completion.

Step 18: Perform the corrective actions taken: After an action has been implemented, enter a brief description of the actual corrective action and its effective date. A thorough Equipment FMEA/FMECA will be of limited value without positive and effective actions to eliminate machine downtime or prevent breakdowns and defects. Equipment FMEA/FMECA is a living document and should reflect the latest design level and latest design actions. Besides, any machinery design changes need to be communicated well to all concerned people in the organization so that these standards, guidelines, procedures, and specifications can be updated.

Step 19: Recalculate the final RPN: After the design actions are taken into account, the severity, occurrence, and detection ratings are finally revised by the FMEA team. Recalculate and rate the revised RPN. The FMEA/FMECA team leader will review the revised RPN and determine if further design actions will be necessary. After design actions and modifications are completed, they will estimate and enter the severity, occurrence, and detection ratings. The final RPN value should go down. If no actions were listed, leave these actions blank. The purpose of recalculating the RPN is to determine if the impact of the failure mode had been. The revised RPN value should be reduced as the result of the corrective action.

Recommended Actions	Responsibility and Target Completion Dates	Actions Taken	Action Results				RPN
			SEVERITY	OCCURRENCE	DETECTION	RPN BEFORE	
- Contamination - Introduce contamination control awareness campaign to all maintenance involved - Regular cleaning and inspection	Mick Jagger 7/14/2014	- Have conducted training on contamination awareness control for all maintenance involved	3	1	10	80	30
- Lubricant Failure - improve the choice of lubricant and study feasibility of using synthetic oil.	Mick Jagger 7/14/2014	- Conduct a feasibility study on the best type of synthetic oil to be used for all motors - Standardize all lubricants for motors	2	1	8	64	16
- Misalignment - Have contractor perform lazer alignment during scheduled outage on the components	Keith Richards 12/15/2014	- Discuss with vendor to perform alignment practices whenever the components will be disassembled - Still Ongoing	8	1	10	80	80
- Contamination - redesign the system by placing pressure guages before and after the strainer and conduct a regular monitoring on the drop in pressure. To be done during the scheduled outage for the motor and pump components	Larry Watts 7/15/2014	- Pressure gauges in place	2	1	2	80	4
- Strainer Mesh too large - Modification of strainer mesh from 400 microns to 150 microns and study flow rate if it will be affected	Ronnie Woods 7/25/2014	- Strainer already modified and flow of water is normal	2	1	2	80	4

Figure 9.8: FMEA Table Comparing Before and After RPN

Tips on Conducting FMEA/FMECA

1) In doing FMEA/FMECA on the equipment, it is not recommended to perform RCM on the same equipment and vice versa as the corrective actions generated from FMEA and RCM might be different and may contradict or come in conflict with one another. Make a decision on whether to perform RCM or FMEA/FMECA on the asset being analyzed but do not perform both on the same asset.

2) Do not always prioritize failure modes with a high RPN or Risks Priority Number. Prioritize failure modes with a high severity rating of 9 to 10, especially if the failure's impact may have safety or environmental consequences. Corrective actions done should eliminate, reduce, or mitigate the consequences of the failure.

3) Remember that most critical failures have a very low occurrence but a devastating consequence, which means that critical failures seldom happen. These should be prioritized in the FMEA/FMECA analysis. Failure modes with high occurrence may have low or minor severity; hence the FMEA/FMECA team should be aware of this.

4) For equipment with current controls, it is important to perform a Failure Finding Task or Functionality Inspection on these devices to ensure that they are still functioning, especially if these devices' failure is hidden. Use the table on page 275 on Reliability Centered Maintenance to determine the correct interval or frequency in inspecting these controls and devices.

5) The output of the FMEA/FMECA team will always be subjective. Hence, it is highly recommended that the team to form the FMEA/FMECA will be the people with the most extensive knowledge of the equipment they are about to analyze, and most of the members of the team have experience the failure modes themselves.

6) For failure modes with a high occurrence rating, it is best to perform a Root Cause Failure Analysis to get to the bottom of the problem to eliminate or reduce its occurrence.

7) It is best to include an experienced operator as part of the FMEA/FMECA team since, according to RCM, operators are the best source of failure modes since they are the people that will come face to face with the breakdowns and they know the symptoms of these failures.

8) For Failure Modes that are Human Related, which caused human errors to occur, the default action will be "Redesign." Human errors should be addressed in either the RCM or FMEA/FMECA Analysis, especially when the failure itself has safety or environmental consequences.

9) Big problems and catastrophic breakdowns are just an accumulation of small problems that are always left unattended. It is likewise important to address the basic equipment condition before conducting any RCM or FMEA/FMECA analysis. These small problems, such as a leak, loose or missing bolts, dirty equipment, and improper lubrication, should be addressed at the beginning of the FMEA/FMECA initiative.

10) It is best for the FMEA/FMECA team to write down all the failure modes that they have experience in actuality based on how the equipment had operated based on its current or present operating condition in your plant rather than to use vendor or OEM failure mode which may be quite generic.

9.5: May 2015: Why Safety Cannot Be First

If you work in the manufacturing, food, drugs, pharmaceutical, automotive, electronic, semiconductor industries, the priority will always be on quality and customer service. Quality and customer satisfaction will be the main priority for these industries. Their vision will always have something to do about delighting customers or exceeding their customer satisfaction. Many people from quality will be deployed on every single part of the plant to ensure that the product they manufacture is in complete compliance and conformance to their product specification. Any small deviation from the standard, expect a non-conformance ticket from these people.

On the other hand, if you happen to work in the mining industry, shipping, construction, oil, and gas refineries, the priority of these industries will be safety. Look everywhere on the plant, and you can find a lot of signs and posters about safety. There is only one plant that I can think of that focused on reliability, and these are power plants. Suppose you're an employee of any of these plants; I am pretty sure that there will be safety, quality orientations for their employees together with their contractors and even their visitors.

Safety people teach safety orientation to their employees. Quality people teach about the importance of quality and customer satisfaction to their people, but here is the problem. Do your quality and safety people understand what maintenance and reliability are all about? While I am not against what Safety and Quality people are doing, that is a good thing for your industry. Neither am I telling you to shift your industry's focus from safety to reliability or from quality to reliability. That is not my intent in writing this book. From the bottom of my heart, what I am saying that Safety and Quality people should also be educated and understand reliability. What I am saying is that your industry should balance your priority. It should not be 100% safety first or adhere to very strict safety and quality procedures and protocols.

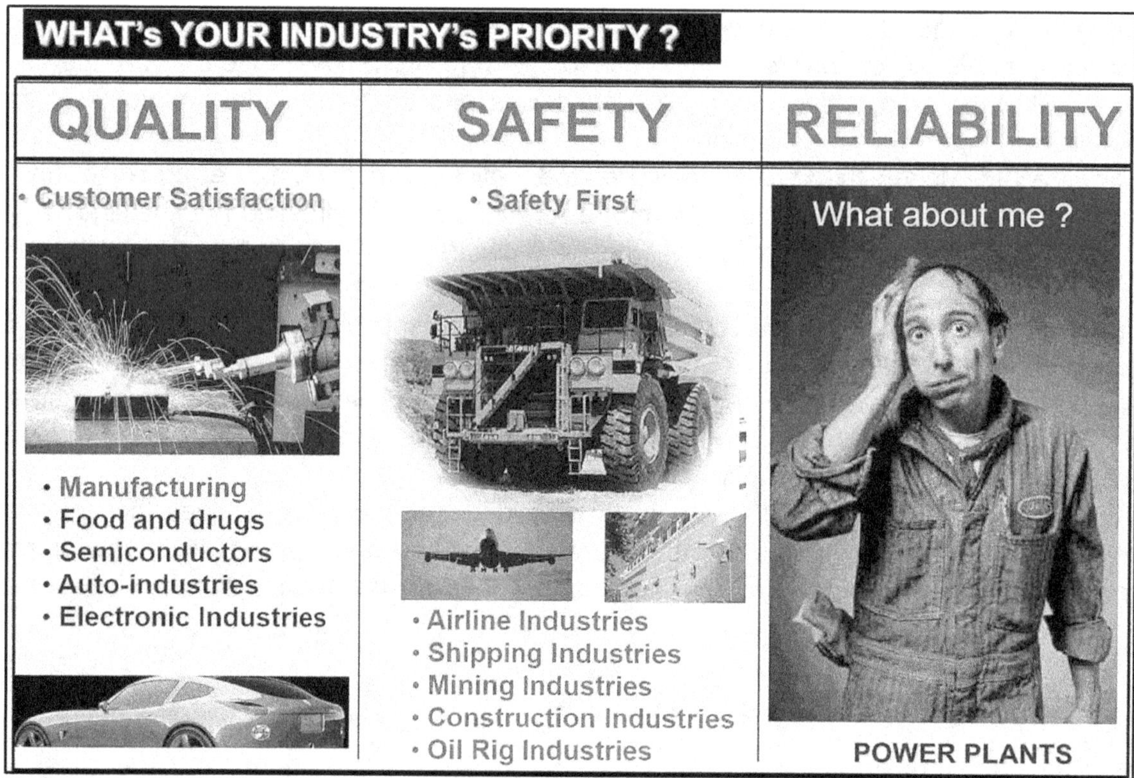

Figure 9.9: Should there be really a Priority for Industries

Assuming your safety and quality, people become so strict in implementing their policies; all they do is police and issue a non-compliance ticket for every slight deviation from their rules and protocols. There will be cases that overloading us with too many safety and quality rules and policies may affect the reliability of the equipment. Let me site this example for the record, a safety officer audited the maintenance department in their recent semi-annual Preventive Maintenance activities on their equipment. PM Compliance is low at 60% since only 12 out of 20 activities listed on their semi-annual PM checklists had been complied with by the maintenance crew. The Safety Officer issued a non-conformance ticket because of this deviation. So why did maintenance not comply with the 20 activities? First, 3 of those activities were to replace a spare part. Upon checking this with the Predictive Maintenance team, these parts need not be replaced since they are still working. Still, the PM lists indicate that it is time for these parts to be replaced. Maintenance also skipped the activity on alignment because their request for a Laser Alignment was still for approval from their Top Management, while the rest did not comply because operators seem to experience a lot of infant mortality failures if maintenance complies with every single activity on their PM checklist. Now tell me does Safety and Quality people understand what Infant Mortality is all about and why it happens?

No disrespect to safety people, but there are many cases where I find their safety rules and procedures either absurd, exaggerated, ridiculous, or just plain stupid. When using the stairs, hold the rail; if caught, a fine of 50 pesos. To me, that would be fine, but if the person in front of me keeps on coughing really hard, then don't expect me to follow that rule. Just recently, I lost a client because of one phrase I said. What was it? I told the class that Safety is not first, and it can never be first. While there was a mixed reaction from the delegates, I told my reasons. After the class, the maintenance manager approached me and told me not to mention that phrase again since their industry mindset is safety first. I told the maintenance manager that is my stand, and industries hire me so that I can tell them what is wrong with them. It is as simple as that. I decided to discontinue my training in that industry rather than take back what I said. The funny thing was that I learned from this maintenance manager previously since he was the one that picked me up in my residence for the training that he told his family not to buy their products since he knows how the product was being manufactured and most probably was thinking about the safety of his family. What an irony. In my line of profession, I only got one investment, and that is my integrity. If I lose that, I would probably stop teaching and writing. Thinking about that incident, I lost an income, and I do not mind about that as I have other industries to serve, and given that chance once more, I will do it again in a heartbeat.

Any industry's goal on the planet is to earn revenue and money, which should first be prioritized. If Safety is first, then let us make revenue for your industry a second priority. I agree 100 percent that safety is important, but like Quality and Reliability, each of them is as important as Safety, and none should be the first. What is important is that besides safety teaching their employees about safety and quality teaching about quality, we as maintenance people should start to educate these people on what reliability is all about. Quality, safety, and reliability people should work as one and not in isolation so that these people can have a better chance of dealing with human errors, which are the primary cause of the world's worst industrial accidents and disasters. Providing stricter safety rules and procedures for everyone to comply will not address human errors. I believe that safety is not synonymous with making that equipment reliable; Reliable equipment is safe for the operators. What is lacking between these folks is constant communication and utmost cooperation to optimize our equipment and assets' reliability. No disrespect to Safety and Quality people, but I believe it is time for maintenance to teach these good people what Reliability is all about.

From a failure consequences point of view again, "**SAFETY**" cannot be first. I remember several years ago when I was teaching in India since I used to teach there twice or thrice a year. Thanks to my dear friend Ronit Kapur from International Business Conference. The course I was teaching was Reliability-Centered Maintenance, and when I was discussing the consequences of failure on these slides extracted from my training materials on figures 9.10, there was someone at the back crying. Since he was disturbing my class, I told the class to break for a few minutes, and then I approach the person to find out what was the matter. He told me that he worked in this pesticide plant in Bhopal as maintenance, and he lived with his family near the plant. On December 2, 1984, he was on a business trip to meet some people in Delhi when the tragedy took place.

Once he knew from the news what happened in his hometown, his wife and two children died that night. He said to me that if only he knew how he wished that he had not left his family so that he can die with them in his arms. I held my tears and composed myself as he told me that, and I only refresh this horrifying memory and his wounds on what happened during that tragic night at Bhopal, India. It was indeed a very emotional moment. Was it a safety issue? It was not since it was a maintenance issue that led to a breach in the environment. A leak is a maintenance issue that causes this environmental disaster. An environmental consequence would have far more devastating consequences than safety consequences. What happened in Bhopal, India, is that in December 1984, 40 tons of highly poisonous methyl isocyanate gas leaked out of the pesticide factory in Union Carbide plant in Bhopal. A cloud of methyl isocyanate gas escaped from the Union Carbide plant in Bhopal. The aftermath initially killed more than 6,400 people, while many from the effects died several years after the disaster. While 30,000-40,000 were seriously injured in the world's worst chemical disaster. Over 500,000 men, women, and children have been exposed to the poison clouds, and at least six thousand people died within the first week of the disaster. The current death toll was estimated to reach over 16,000 and is still rising. Those women who were pregnant at that time were miscarried, while those who survived were stillborn. The poison affected some people's lungs and nervous systems.

Figure 9.10: Environmental Consequences in Bhopal India

Many if not all of these industry's worst industrial failures are not also safety-related, but it has more to do with the way we maintain our equipment and assets. Still, in most cases, the default will be to end up in tighter safety rules and procedures, but the way we do maintenance remains the same. Many industries do not involve operators in maintenance. These operators are the people who can sense the earliest signs of failures before the maintenance themselves. We need to address not only safety but also how we maintain our equipment and assets, and much more important is how we can manage human errors.

The first time I saw these children, my eyes were wet as tears flowed instantly. These children were deformed because of poor and neglected maintenance. Maintenance has done this to these children. Maintenance is responsible, and perhaps we are even worse than a terrorist. I cannot say if these kids were still alive today, but that is the consequence of neglecting maintenance. We have a responsibility not only to maintain our equipment and assets but also to preserve the environment. I hope and pray that this will be the last of its kind, and I rest my case.

9.6: June 2015: Is It Possible to Eliminate Human Error in Maintenance?

It is much easier to blame people for their mistakes and errors because it is simply a case of human nature. As Michael Jackson sang in his song, ***why, why, tell them it's human nature.*** This is how people react. On the other hand, people that committed the error would always have tons of excuses for not doing the right job rather than taking their time to analyze the root cause of their errors. If possible, they will always look for ways to pass the error to someone else rather than admit the error themselves. As a result, most industries have policies, procedures, and rules for people who commit mistakes. If an operator commits an error during operation or when a technician poured the wrong oil into the equipment, we reprimand or, worst, punish them instead of understanding the reason for the failure. Why? Because it is much easier, faster, and most importantly, how we do things on our plant since the beginning of time. In short, it had been part of the company's culture for many years.

Human errors can be defined as an action planned but not executed according to the plan. Research by Dr. James Reasons of the University of Manchester in England found that humans commit an average of 6 errors per week. The latest official current count for the world's total population estimate for mid-year of 2009 was estimated at 6,790,062,216 people on the planet, so that is roughly around 47.53 billion errors are committed on our planet every week, and some of these errors can be fatal or can even cause us harm or even death. A pilot error can cause the lives of the passenger on board the plane. Errors committed by physicians, doctors, and nurses can compromise the lives of their patients. We watch the news on the most recent road accidents, which are being reported almost daily by the news anchors. Errors from management decisions, flaws in systems, obsolete procedures, and policies can have a detrimental effect on its product, operations, and services that can affect our clients, customers, and services. But the most important factor in human error is learning from our own failures to become a better person. If people created problems, then people must also be capable of correcting the problem itself.

If we speak about equipment-related problems, all breakdowns are man-made. From the way, the equipment was designed, fabricated, commissioned, operated, or maintained. There is no such thing as a piece of perfect equipment. Every piece of equipment has a build-in design weakness that can be addressed by maintenance, reliability, or even the operators themselves.

Humans play an important role during the design, fabrication, installation, production, and maintenance phases. In maintenance, human error may be defined as the failure to perform a specified task that could disrupt scheduled operations or result in damage to property or equipment. While human errors had existed since the beginning of mankind, only in the last 50 years has it been the subject of scientific inquiry and studies. There were various reasons for the occurrence of human errors such as stress, fatigue, working too long without adequate break time, incorrect tools, forgetfulness, complacency, distractions, insufficient knowledge, lack of training, lack of communication, misinterpretation, ignorance, lack of standards, working under pressure, lack of awareness and the lists goes on. However, with every error that we make, there is typically an associated change or something out of the ordinary occurring in our environment. The difference between humans and machines is that people can sense change, hear, smell, see, or feel something different and take the necessary actions to correct the anomaly. In industries, human errors exist either because we tolerate them or we are just ignorant. Let me state some cases.

Try to loosen some bolts in your equipment and place a toolbox near it. Let us say that the toolbox will contain a hammer, pipe wrench, adjustable wrench, pliers, socket wrench, double-ended open type wrench, and a torque wrench. Ask one of your technicians in the plant to tighten the loose bolts and observe how he tighten them. After he has tightened the bolts, ask him to loosen them again. Try to call 4 to 5 more technicians, one at a time, and give them the same instructions: tighten the bolt and write down how they performed the tightening process. Perhaps you will observe that some have similar ways of tightening the bolt and have used the same tools, while others simply used other methods to tighten the bolts. Why? Because you just ask them to tighten the bolts, but if we are explicit in our instruction and procedure, we can instruct them to use a torque wrench with this amount of torque, or perhaps a warning is indicating not to use an adjustable wrench, hammer, or pipe wrench so that each one of them will do the same procedure on tightening the bolt.

Is it Possible to Eliminate Human Error?

There are two ways people can go wrong, they can do something they should not have done or failed to do something they should have done. We get irritated with the person who committed the error and blame the person who committed the mistake. Why? Simple because it is a matter of human nature. We love to blame and see other people's mistakes, but we do not want to see and talk about our own mistakes. We judge people quickly and avoid looking at the man in the mirror. As Leo Tolstoy once said, people love to change so many things except for one thing, and that is themselves.

Human errors will happen; **they can never be eliminated**. The good news is that although it is not possible to eliminate human errors, we can manage them more intelligently. Many maintenance organizations try to change the human themselves when they should be changing is the conditions under which the people work. Industries should treat errors as an expected and foreseeable part of both operators and maintenance work. Maintenance and operator errors not only endanger lives and assets but are also definitely bad for business. This had been among the principal causes of major industrial disasters in a wide range of industries.

Maintenance errors are not random, indeed. Most maintenance errors happen during installation, reassembly, overhauling, and repair activities or when maintenance is trying to experiment on how to deal with the repair on the failure, especially if this is the first time he experienced this kind of breakdown. We called this experience. These are the things being taught at the University of Hard Knocks. There is a risk involved in human errors. Humans are totally inclined to blame people for mistakes and errors because it is deeply rooted in their human nature. People love to blame people. According to my good friend Bob Nelms from Failsafe Network quote; the problem with industries is that blame, discipline, fear, punishment, and accountability are near the root cause of why industries never learn from things that go wrong. Although I agree that there is certainly a need for rules, regulations, policies, and procedures in our organization. Likewise, there is most certainly a need to enforce them by applying discipline to those who will not abide. Unfortunately, most organizations do everything backward. They do not consistently apply their disciplinary policies ahead of time; instead, they will wait for something awful to happen and then go after the person. This is why industries never learn from the things that go wrong.

Figure 9.11: Defenses on Human Error on Our Equipment and Assets

To cope up with these human errors, industries created a defense. There are two kinds of defenses, which include hard defenses and soft defenses. The problem is that each of these defenses has a gap, and even though how many defenses we place, once that gap is breached, then a human error occurs.

Therefore, whether we like it or not, humans will continue to commit errors and mistakes, and that is why the need to perform a thorough Root Cause Failure Analysis is a must. Failures are not really bad, because it allows us to begin more intelligently. We can only benefit from failure if we can learn from it. Perhaps we cannot change the human condition, but we can change how humans perform their work. When people commit errors, we must steer away from blaming or punishing the person, but we need to understand the reason behind the error or failure. Changing policies, procedures, and the system is not enough. It also requires a change in oneself. The person who commits the error and the people who accused them should also change because the more we blame, the less we understand the failure. When human errors occur, it is easy to blame or point fingers at someone. It is more important to understand the reason behind the failure.

All people commit mistakes, and even the most intelligent person can make the worst mistake. We must realize that most of these errors happen beyond the control of the person.

In general, there are two types of human errors. Unintended, which are actions committed or omitted with no prior thought. We typically think of these as "accidents": Examples such as bumping the wrong switch, misreading a gauge, forgetting to open a valve, spilling coffee on the floor, and so forth. If the worker intended the correct action but simply did it wrong. The error is sometimes called a slip. If the worker forgot to perform the action, the error is sometimes called a lapse.

Suppose we study and take a moment to understand human errors; most errors in maintenance are caused by either a slip, lapse, and in most cases, both. Humans make errors. The principle of error management is that even the best people can make the worst mistakes. Often very good people in established organizations keep on making the same blunders. An error can be defined as an action planned but not executed according to the plan. With all the errors occurring every day and all the ways we could destroy ourselves, life is still preserved. Why? It is because humans can sense change and break the error chain. A human error is also an inappropriate action or intention to act, given a goal and the context in which one is trying to reach that goal. It is the by-product of poor planned tasking or execution of a task resulting in a goal's failure. Maintenance errors have been the major cause of major industrial accidents and disasters in a wide range of industries worldwide. Human error is part of being human. Errors, mistakes, failures are part of being human and part of our life as well. What is important is learning from these errors; if these mistakes and failures we encounter just repeat themselves, industries never learn from their own problem.

9.7: July 2015: Maintenance Induced and Non-Maintenance Induced Errors

Maintenance Induced Failures are failures that are directly caused by the maintenance people themselves. Maintenance Induced Failures include infant mortality failures or intrusive maintenance, overhauling, and inducing early failures right after endorsing the equipment back to operators. Perhaps the maintenance reassembled the equipment incorrectly. Non-Maintenance Induced Failures are failures on the equipment which are not caused by maintenance, but other factors such as how the equipment was operated, design errors, commissioning errors, or cutting costs, and many more.

From the book of Keith Mobley, An Introduction to Predictive Maintenance, he quotes: From studies of equipment reliability problems conducted over the past 30 years, maintenance is responsible for about 17 percent of production interruptions and quality problems. The remaining 83 percent is totally outside of the traditional maintenance function's responsibility. Inappropriate operating practices, poor design, non-specification parts, and many other non-maintenance reasons are the primary contributors to production and product quality problems and not by maintenance.

Although the percentage of human errors induced by maintenance is not that big, the irony still lies behind that the worst industrial disasters that occurred globally resulted from maintenance and, of course, human errors. Human errors will have a wide range of consequences, and the problem is that we only act if the consequences of the failure already occurred. Even if we already knew the problem for as long as nothing happens, we always result in the same old ways of doing things. Here are some possible situations where the failure are caused by maintenance itself or Maintenance Induced Errors:

Infant Mortality Failures: Early failures can occur on the equipment right after a major Preventive or scheduled maintenance is done on the equipment. Operators often say sarcastically that if maintenance did not perform regular overhauling and replacement on this piece of equipment, I bet it will be running without any problems. But isn't it that the main purpose of a scheduled Preventive Maintenance is to ensure that the equipment performs as intended, but the opposite happens whenever maintenance does something. Something goes wrong. Is it profound? No! Honestly, we are just a victim of what we called Infant Mortality Failures. In almost all cases, Infant Mortality Failures are caused by human errors and intrusive or forced maintenance. In fact, many things can go wrong when we try to dismantle equipment for overhauling purposes. Human errors such as slips and lapses can occur on the part of the maintenance performing the overhauls. Infant Mortality Failures are failures that occur at the beginning of life; others refer to them as commissioning failures, start-up failures, or debugging failures. Many factors affect Infant Mortality Failures, including poor equipment design, poor quality manufactured, incorrect installation, incorrect commissioning, incorrect operation, unnecessary or intrusive maintenance, human errors, or simply bad workmanship.

The case of infant mortality failure, which is pattern F of the six failure patterns, starts off with a high incidence of early failures, which eventually drops to a constant or very slow, increasing conditional probability of failure and ending in no wear-out zone. This means that the failure can occur at any given period. If Infant Mortality Failure exists frequently, the question raised, is can we eliminate them? The answer is no, we cannot eliminate infant mortality failure completely, but it can be reduced. Reducing unnecessary PM activities we perform on the equipment can reduce infant mortality failures. In layman's terms, the more we touch, the more it fails; the less we touch, the less it fails. Suppose we review the lists of every single activity performed on the equipment; we can find out that there are maintenance tasks that duplicate itself in which it is already performed by regular maintenance yet still being performed by other contractors. There may also be maintenance tasks that are deemed unnecessary and do not address any failure modes. These should be removed from the PM lists of activities. There might also be activities that should be done, but the actors do not exist in the Preventive Maintenance lists of tasks. We can also reduce infant mortality failures through the application of Precision Maintenance.

Installation and Reassembly: Human intervention is involved during a major overhauling activity on the equipment. Although there is little or no human error when dismantling equipment, human error happens mostly when reassembling or putting the equipment back in one piece. What is important for maintenance is that the people involved in these tasks should be fully focused on completing the tasks correctly. A guided procedure on the correct way to overhaul can minimize the chances of human error. There should be no shortcuts on dismantling and reassembling the equipment and using the right tools to dismantle the equipment. Maintenance should also know that there are _do not disturb_ portions of the equipment left untouched. Maintenance must have the correct skills and the right tools to install and reassemble the equipment.

Faking Maintenance Documents to Comply with PM Specs: If we can refer to the Top Ten

Problems on Preventive Maintenance, this problem ranks third among the lists. Maintenance thinking is that it is better to fake the documents than to be incomplete, and besides, this is just a minor thing in which no one will know, or nothing happens after all if I place a check on this activity. Besides, I will be audited if I did not fake this document, or my shift is ending, and I will be late as the company shuttle bus will be leaving in 5 minutes. This is one minor problem that can lead to a dangerous situation. This is one of the major issues if maintenance will be done by maintenance 100 percent of the time. Operators are never involved because the operations manager just wants these operators to operate and produce output. You see, if Autonomous Maintenance is established in an industry, there will be checks that will be done not only for maintenance but also for operators themselves.

Not Performing a Functionality Check on Protective Devices: A protective device is placed in the equipment to serve a particular function, which is to protect something. That something is called the protected function. Protective devices should be inspected at the correct frequency because the failure of these is hidden. The only time that we know that these protective devices had failed is that if the protected function also fails. When both protective and protected functions both fail then, we experienced multiple failures. This is what we are totally avoiding. Try to think of your protective device as a parachute. If you jump off the plane with your parachute and it malfunctions, or it did not open when you press the button, do not panic as there is always a reserve parachute in place. When it is time to open the reserve parachute, and still it did not open, then all I can say is that it will be the end of your life. Protective devices are like a parachute. These protective devices, such as led alarms, sensors are strategically placed in the equipment as a defense to alert operators and maintenance of upcoming problems, deviations, and abnormal conditions.

Here is a classic case problem. April 20, 2010, Deep-water Horizon Oil Spill in the Gulf of Mexico. According to a federal investigation officer, vital warning systems on the Deep-water Horizon oil rig were switched off at the time of the explosion to spare workers from awakening up by false alarms. The revelation that the alarm systems on the rig at the center of the disaster were disabled and that key safety mechanisms had also intentionally been switched off came in testimony by a chief technician working for Transocean, the drilling company that owned the rig. Mike Williams, who maintained the rig's electronic systems, gave evidence to the federal panel in New Orleans investigating the disaster's cause, killing 11 people. Mike Williams also told the hearing that no alarms went off on the day of the explosion because they had been inhibited or switched off. Sensors monitoring conditions on the rig and in the Macondo oil well beneath it were still working, but the computer had been instructed not to trigger any alarms in case of any adverse readings since it always gives a false alarm which wakes up the people mostly early in the morning. Both visual and sound alarms should have gone off in sensors detecting fire or dangerous levels of combustible or toxic gases.

Making Maintenance a Jack of All Trades: For dynamic industries that always keep on reorganizing their people, you are the mechanical technician today; next week, you will be transferred and become an electrician. New top management was hired, and after a week of observing the plant, he decided to reorganize his staff, and you were a part of it. I have been employed in a dynamic industry where there are changes in the organization almost every month. Is this a good thing? Honestly speaking, I just can't tell. But I see a lot of wasted projects and improvements that were abandoned due to the reorganization. All I can say is that if you plan to make maintenance a jack of all trades, try to consider the anthropometric factors. While Non-Maintenance Induce Errors again are errors that were not manifested by the maintenance function and cause by others beyond maintenance. This is a much bigger problem because maintenance may have little or no control over the situation. Let me give some examples of non-maintenance Induced Errors

MRO Spare Parts is Not Managed by Maintenance: MRO Spare Parts are not controlled or managed by maintenance in many organizations today. This is a big problem as the best people to manage the storeroom are the maintenance people because they are the parts' users, and they know these parts better than any other people in the organization. While in some industries, they can be managed by Supply Chain, Warehouse, Purchasing, Finance, Admin, or even others. You see, the storekeeper's role is not only to issue and receive parts from the vendor. The role of the storekeeper is also to maintain the parts inside the storeroom. For example, belts should not be placed in a cold area, as resiliency property will be lost. Bearings that are in boxes should not be placed vertically but horizontally. Large motors and rotating components inside the storeroom should be rotated once or twice a week to avoid any chances of false brinelling, especially if there is some form of vibration felt in the storeroom. It should be regreased every 6 months. There will be a wide array of problems that will happen if maintenance does not own the storeroom. I would strongly recommend that maintenance be the ones to own and manage the storeroom as they are the users so they can be in control of the spare parts.

Purchasing Going for the Lowest Bidder: Although in most industries, purchasing will source for at least three bidders for a particular part but the end result is that they will award the part to the lowest bidder. So the Purchasing Manager together with their people, are celebrating because they have saved a few cents on the part but what they do not realize is that they make the life of maintenance much more difficult and miserable because the part breaks easily and cost thousands or even hundreds of thousands in US dollar a year not to mention the cost of downtime and repair. There were 3 vendors for this type of oil, and the purchasing awarded the oil to the lowest bidder. What purchasing do not know is that the Flash Point of the oil of this vendor is at 300 ° F or 149 ° C, which caused a lot of evaporation and tapping up of oil on the maintenance side, and now they question maintenance why this equipment consumes too much oil. There is always a temptation to purchase equipment or spares based on the lowest or cheapest source for industries. This might not be such a good idea since purchasing based on the initial costs only tells us one side of the story. The true costs can be seen based on its performance and the total life cycle costs and not on the initial costs. In my experience, this is the problem with most procurement and purchasing departments since they decide to purchase parts based on the lowest bidder or lowest possible costs. The savings that these departments claim are insignificant since both operations and maintenance can encounter many failures due to this. They always look at the part's initial cost and not the cost of problems the part may give the user in the long run. This is the problem with cheaper parts since the equipment will fail easily and will likely yield a much larger amount of cost in a period compared to the slightly higher cost of equipment or spare.

OEM Design Errors: Since God did not create the equipment, then no equipment is perfect by nature. It will always contain flaws and inherent design weaknesses. Although this can be addressed by the maintenance by deploying the Planned Maintenance pillar of TPM, which will be covered. Planned Maintenance Phase 2 will address this issue. What is important for maintenance is to observe the equipment and list all the parts that fail and break easily, and the challenge is how to extend the lifespan of these parts through modification or redesign.

9.8: August 2015: Tips on Conducting Autonomous Maintenance for Industries

1) Top Management should not only support Autonomous Maintenance but should be committed and, much importantly, own this strategy. Management can support but never commit to the strategy. Top management needs to review the team's progress from time to time for the Autonomous Maintenance team to be motivated to do their equipment activities.

2) It is highly recommended that Autonomous and Planned Maintenance should have the same pilot equipment and that initial cleaning to be done by both pillars should be done on the same day. This way, the Planned Maintenance can guide the AM team in their equipment. Autonomous Maintenance will be focusing on abnormalities mostly found on the exterior part of the equipment, while Planned Maintenance will be focusing more on deteriorations, mostly on the interior part of the equipment. There will be two sets of tags used by Autonomous Maintenance (Operators) and another set of tags used by Planned Maintenance.

3) Membership in Autonomous Maintenance should be 100% pure operators and should not include supervisors, managers, maintenance, or engineers. These people will only act as coaches, guidance, and advisor and will not be part of their team. Suppose supervisors, maintenance, engineers will be part of the Autonomous Maintenance teams; operators will lean on them and will not become empowered.

4) Management must allow operators to meet once or twice a week to perform their step activities. Management must understand that this will be a long journey, and results will not be felt overnight. If the team comprises a vertical set-up, which means that the Autonomous Maintenance members will involve all shifts, the top management team should settle overtime issues in the early stages of implementation since Autonomous Maintenance will be meeting once or twice a week to pursue their activities and roadmap.

5) Planned Maintenance should already be in place before starting up the Autonomous Maintenance activities. If the Planned Maintenance structure is weak, Autonomous Maintenance will collapse since maintenance will be the AM team's coach and mentor. This will be most important when Autonomous Maintenance reaches Step 4; training materials needed by the Autonomous Maintenance should be completed. Otherwise, Autonomous Maintenance activities will halt. Most of these training materials will be developed by their mentors, which are the Planned Maintenance teams.

6) Education, training, and transfer of technical knowledge from maintenance to the operator are key so that operators can develop their own checklist on their equipment. Once operators

understand their equipment intimately, then only the checklists will be done religiously. This is one of the main reasons operators do not regularly perform their checks. Operators should understand their equipment first and the consequences of the failures their checklists are meant to prevent.

7) Management should provide simple yet memorable recognition in completing every step of Autonomous Maintenance. This will motivate the Autonomous Maintenance team to pursue the higher Steps of Autonomous Maintenance. These tokens are meant to provide motivation and enthusiasm in carrying out the 7 Steps of Autonomous Maintenance. Remember that as the Autonomous Maintenance team moves from one step to another, the activities will be more challenging.

8) Do not imitate the Japanese; as Deming quote, companies expect miracles and look forward to a silver bullet to make things happen. It simply does not exist; the American Management thinks they can just copy from the Japanese, but the problem is that they do not know what to copy. Adopt TPM in a way that best fits the culture of your company.

9) Autonomous Maintenance activities should not conflict with other improvement initiatives in the plant for operators and should be aligned to the industry's goals. Different improvement initiatives in the industry should not be independent of each other, but rather, they should be aligned and consolidated with one another. It is important to have a main umbrella in which every single strategy will be aligned.

10) Do not perform Autonomous Maintenance when your operators are contractual. Do not expect them to get empowered when their contract is nearing an end. Autonomous Maintenance should be performed by regular employees in the plant. Remember that on an average speed that is not too fast or too slow, it will take Autonomous Maintenance around 5 years to reach Step 7.

11) Autonomous Maintenance will be meeting regularly, either once or twice a week. Management must not pull out operators during their meetings, as this will demotivate the team. What we need from management people initiating TPM is not only their support but their commitment as well.

12) If the Autonomous Maintenance membership will be on a vertical setup, the membership will be composed of different shifts. The team must agree on the time they will meet. Management must settle the issue of overtime at the very beginning.

13) Start your pilot team with the operators who complain a lot about their work and who have a voice and influence among other operators. It will be difficult initially, but once these operators are convinced, then the rest will follow.

14) If your industry is unionized, inform the union about the initiative and the benefits operators will gain when implemented. Tell them that this is not merely a transfer of work on maintenance down to the operators.

15) Let top management participate during the initial cleaning of the equipment. This will boost the morale and motivate Autonomous Maintenance start-up activities. Remember that the most difficult part will be the start-up activities, as resistance will still be felt during this normal stage.

16) Provide an activity board for both AM and PM pilot teams to showcase their activities to other operators who are not yet members of the Autonomous Maintenance team and convince other operators to join this initiative.

17) If your organization is dynamic in which workers are being transferred from one place to another, management should not separate the team from the start until the end. This will again demotivate the team completely if some of the members will be transferred.

18) Auditors for Autonomous Maintenance should never mention the word fail if the team is not compliant with each step; instead, use the word unconditional pass. This means that the team will be eligible to move to the next step to complete the initial audit process requirements.

19) If there are already many teams involved in Autonomous Maintenance, it is highly recommended that the TPM Office organize a yearly symposium where different teams share their improvements with others operators who are not yet involved and the benefits of participating in Autonomous Maintenance activities.

20) If any managers are killing Autonomous Maintenance's idea in the plant, it is best to highlight their names to your President. This is one main reason why TPM is always a top-down approach that must be initiated by the highest person in the plant, the CEO or President.

9.9: September 2015: Reducing Human Errors in Maintenance Part 1

But even if quality, safety, and reliability people team up, human errors are still inevitable. For safety, people stop calling yourself a **SAFETY FIRST**. The majorities of the World's Worst Industrial Disasters are not safety-related but maintenance-related and can be attributed to Human Errors, and the reduction of human errors can only happen if we stop prioritizing each other and calling ourselves always first. Both Quality, Safety, Reliability, and Maintenance people should not only confine themselves to their department by building stricter rules and policies and trying to police every single employee that creates a very slight deviation from the norm. We are human beings, we love to blame, and that is part of our human nature. We need to change that behavior. Management people believe that operators and maintenance who committed human errors result from carelessness or stupidity. The only way to eliminate these errors is to punish the person who committed the error outright. Reliability and maintenance people should also educate Safety and Quality people on what reliability is and what it wants to achieve. While these three departments have different goals since Safety will aim for Zero Accidents in the plant, Quality will aim for Zero Defects and Reliability together with Maintenance people will aim to reduce unplanned breakdowns and improve the reliability of their equipment and assets, each of them share a common vision which is to safeguard the integrity of the plant's existence and stay in business and remain competitive. But the problem is that they work in isolation and are independent of each other.

Working together hand in hand instead of in isolation can dramatically reduce human errors in the industry. According to James Reasons and Alan Hobbs, errors are not intrinsically bad. Success and failure spring from the same roots. We are error-guided creatures. Errors marked the boundaries of the path to successful action. What is important is that we learn from the failure as well as the error itself. Human errors will happen, and they cannot be eliminated 100%, but we can manage them more intelligently by learning from the things that go wrong in our industry. Failure is our best teacher. The most successful and well-known people in the world have failed many times themselves before becoming successful. Here are ways to reduce human errors.

1. Ergonomics: Improve the Working Condition of the People: Ergonomics is the design of equipment, operations, procedures, and work environments compatible with the workers' capabilities, limitations, and needs. It is a vital complement to other engineering disciplines that seek to optimize the hardware performance and minimize capital costs with little or no consideration of how the equipment will be operated and maintained. For example, a salvaged dial thermometer attached to a nozzle just below the third level platform might be a very inexpensive way to accurately measure the temperature in a tower. But from a human point of view, the design is flawed as both operators and maintenance need to climb a ladder, which is around 3 floors, just to get the reading and log it on their records, which needs to be done during their shift. While we always blame people for committing mistakes and errors, we should improve the operators and maintenance working environment.

2. Develop a Functionality Inspection for Protective Devices: Protective devices are defenses we placed on the equipment to alert us of an upcoming problem or error. They may be in the form of sensors, alarms, lighting arresters, overload protection, ground fault protection, overcurrent protection devices, emergency stops, over-speed devices, and so on. They should be functional 99% of the time, depending on the criticality of these protective devices. Functionality inspection or Reliability-Centered Maintenance called this Failure Finding Tasks. These will be the only maintenance tasks that can be provided for these devices to ensure that they are still functioning in the equipment, asset, or system that they are placed in. Failure of most of these devices will be hidden, and that the only time that maintenance or operators know they have failed is that when the ones they are protecting also failed, which is called the protected function. Try to assess the risks of damage. Suppose the risks are high enough and unacceptable; these functionality inspections should be carried out consistently and regularly. Everyone should be informed of the consequences and the risks involved in both the protective device and protected function failure. Remember that they are placed in the equipment for a reason, and that is to protect something. That is why they are called protective devices. The questions to ask will be what could happen in the event under consideration did occur? Is anyone likely to be hurt or killed as a result? How likely is the failure to occur, and is the risk acceptable?

3. Tidiness and Housekeeping: It is less stressful to work and operate in an environment where good housekeeping is in place rather than working in a workplace where everything is a mess. Provide adequate lighting in the workplace. Implementation of 5's will be a good starting point to address housekeeping issues and reduce the chances of slip and trip hazards. People fatigue easily if the workplace itself is untidy. Slip and trip accidents in the workplace account for many injuries attributed to poor housekeeping. This is less likely to happen if good housekeeping is in place in the organization. It also creates discipline with us, and that good housekeeping starts in our home. When I was still working in a semiconductor plant and implementing Total Productive Maintenance, my boss called me and told me to join him for lunch outside the plant, to which I said OK. While inside his car, he asked me how come we are having a hard time convincing people to do TPM and what we need to do. I sarcastically told him that why will people will follow us in his car itself is a mess. He looked at me angrily and told me to get out of the car. After that, he did not talk to me for a couple of months.

4. Simplify your Procedures: Let's look at all the procedures we used in our organization. Many outdated or even obsolete procedures are still being used and implemented until this very point in time. Perhaps these procedures were effective when it was written but would no longer hold true for now. Today, there are equipment that had already been disposed of, and some procedures still exist and are still being followed up to this point in time. Procedures should be simple, precise, and should be easily understood by the user. What is important is that if there is a way of simplifying the things being done, procedures should likewise be revisited, reviewed, revised, and simplified whenever necessary. When industries invited me to their plant for in-house training, I can already feel about how things are being done. How long security took for me to be allowed inside. Sometimes due to very strict procedures, it will take me 30 minutes or even longer to enter the plant. Although we understand their point, I am sometimes beginning to think that the bigger the industry, the more problems exist. Sometimes I told my sponsors, especially if this will be in-house training if they have coordinated my entry to their security ahead of time? I remembered in 2006, I was scheduled to go to this semiconductor plant for two days of training on Maintenance Indices and KPI; at their security, you need to enter a metal detector where you will be scanned just like in an airport.

When I entered, it alarmed, and the security told me to remove my belt, so I did; then, after removing my belt, I passed the detector again, then it alarmed for the second time. So the security told me to remove everything in my pocket, which I complied with. Once again, I entered the metal detector scanner, and it alarmed again. I told the security that I have some metal braces on my leg and told the guard on duty if I should call my doctor and removed them to enter? The security called his office, and the officer still cannot make a decision. They called the human resources. That was the only time I was able to get inside and conduct my training. The time I lost from security alone was around 45 minutes. If you ask me, what was the name of this plant? Don't bother; it's already closed as they moved to China. When you plan to visit me at my residence, give me a call or message when you come, and I'll instruct my dog that you will be coming. I can guarantee you that you will be inside my house in less than a minute. Why can't industries be like that?

5. Use of Visual Controls in the Workplace: Visual controls are placed in the workplace to expose problems and deviations easily by just looking at them. It makes deviations easily seen by operators by alerting them that something is not right. Visual controls are designed to make the control and management of the plant as simple as possible. They originated from the Japanese. They are not meant to make us dumb or look stupid. It simply entails making the problems, abnormalities, or deviation from standards much more visible to everyone in the plant. When these deviations are clearly visible and apparent to all, corrective actions can immediately correct these problems and anomalies. Visual Controls allow us to see problems more easily. They are also meant to provide instructions and convey information. These are devices or mechanisms designed to manage or control our operations process to meet the following purposes. They make the problems, abnormalities, and deviation from standards easily known to the operator and everyone in the organization. It will easily display the operating or progress status, provide instructions, convey information, and provide immediate feedback to people. Remember that the simplest solution that works will probably be the best. Good visual control management should tell the operator immediately if a process is going beyond the specified parameters.

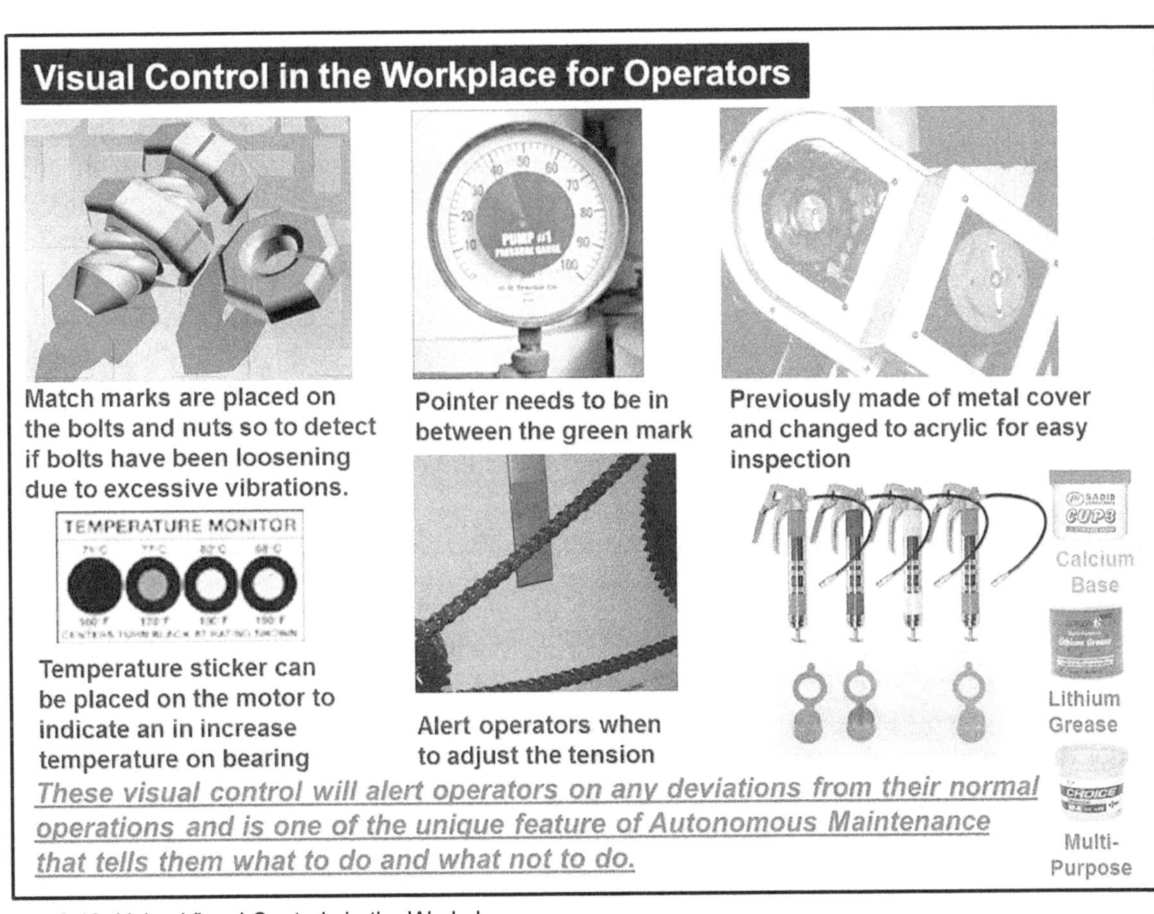

Figure 9.12: Using Visual Controls in the Workplace

Ideally, the process would be stopped automatically. Like in an airport, signs can guide the passengers to save them a great amount of time instead of asking directions most of the time on knowing where you want to go when you are inside an airport. Visual Control is also sometimes called Visual Communication. In the early 1970s, U.S. executives went on tours to different Japanese industries to see why they were being beaten so badly in quality by their Japanese competitors. They noted one specific thing: the workplaces were filled with pictures and diagrams of what to do and what not to do. With the way things were displayed in the plant, anyone passing by, whether a U.S. executive or a new employee, could sit down and do the assembly themselves without even knowing the Japanese language. The most important benefit of visual control is that it shows when something is out of place or missing.

6. Perform Root Cause Failure Analysis on Recurring Failures: Although I would not recommend that for every single breakdown or failure, a Root Cause Failure Analysis should be performed as this will be a very tedious process indeed but when the failure keeps on repeating itself, then I think it is high time to perform a Root Cause Failure Analysis investigation and take things slowly. Please note that a Root Cause Failure Analysis is different from implementing problem-solving or analytical tools such as Fishbone, Pareto Analysis, FMEA/FMECA, 8 Disciplines, and the like. The failure should be fresh when conducting a Root Cause Failure Analysis, which means it just happened. This will require a Principal Investigator and a group of people to gather evidence to shed light to see the bigger picture of why the equipment failed. Root Cause is evidence-driven. The goal of conducting a Root Cause Failure Analysis investigation is to understand why and how the failure occurred so that we can learn from the things that go wrong in their plant and finally learn from them. Conducting a Root Cause Failure Analysis will allow the investigating body to determine the Physical cause of the failure, the human cause, the system cause, and finally ending up their probe and investigation on the latent cause of the problem. While there are many techniques in analyzing a problem that provides a quick answer, it does not mean that the answer will be correct every time. A true and meaningful Root Cause Failure Analysis will take the time to prove that what we say is fact and supports our hypothesis with evidence before we spend money to improve the design of the equipment. When the facts are finally backed up by evidence and science, and these facts are separated from the fiction, we now have a better understanding of the real Root cause of the problem. Root Cause is 100% evidence-based. Without evidence, Root Cause is impossible, and we can only get the most likely or probable cause of the problem. To be continued.

9.10: October 2015: Reducing Human Errors in Maintenance Part 2

7. Apply Poka-Yoke Solution: Poka-Yoke is a Japanese term that means mistake-proofing. Poka-Yoke examples include elevator alarms, 110/220 volt or auto volt appliances, low battery indicators, fuel indicators in the car, child lock door, and so on. Poka-yoke is any mechanism mostly used by industries, especially manufacturing processes, that helps an equipment operator avoid errors and mistakes. Its purpose is to eliminate product defects by preventing, correcting, or drawing attention to human errors as they occur. It was developed by Shigeo Shingo as part of the Toyota Production System. The term Poka-yoke was applied by Shigeo Shingo in the 1960s to industrial processes in which its primary purpose is to prevent human errors. Poka-yoke aims to design the process to detect and correct mistakes immediately, thereby eliminating defects at their origin.

8. Enhance Maintenance and Operator's Training: As the saying goes that people are the companies' greatest asset is not always true. I believe that the right people are the company's greatest asset, and the wrong people are called liabilities. We can only have the right people if our people are equipped with the right knowledge to do their jobs right the first time around. But nowadays, the only training maintenance got is from the University of the Hard Knocks. As the great leader Nelson Mandela said, Education is the most powerful weapon you can use to change the world. In my first book on World Class Maintenance Management - The 12

Disciplines, I indicated that training is the foundation and backbone of any cultural change. This will be the foundation for setting up the basic maintenance strategy in the plant. Maintenance, as well as operators, should be trained and educated first. But the sad part is that training is one of the primary targets for cost-cutting and reduction schemes in most industries. While industries invest in adding hardware resources to expand their business, they forgot to invest in their human resources. Suppose industries deprived their people of attending any training; my advice is to educate yourself, read, surf the internet and join maintenance forums. Training and gaining knowledge can come both formally and informally. For the operator's training, their coach and mentors will be the maintenance people in implementing the seven Steps of Autonomous Maintenance, where the ultimate goal is to empower operators.

My last work was in the mining industry in the Philippines. It was located in Mankayan, Benguet, 16 hours from Manila by bus or 45 minutes by plane. Our main office was in Makati, but I used to teach in the plant in Benguet. When I teach, everyone was happy because of an abundance of food. They even have some take-away at the end of each training day. One day, the finance manager called me and questioned why I have a big budget for food. My answer was simple, I do not want to teach a person with a hungry stomach. After a month, I went back to the plant for another batch of scheduled training. I noticed no food during the break time, which I learned was the finance manager's directive. When I went back to the plant, I confronted him and told him that you are saving on small pebbles while the big savings will come from the people I teach. That day on, I filed my resignation and never returned to that plant the following day.

9. Develop a Risk Criticality Assessment for Failures: I believe that this is one area where both Quality, Health, Safety, Environment, Reliability, Maintenance, and operations people should sit down, relax and communicate as one team and discuss this matter thoroughly and develop a comprehensive list of all possible critical failures that can occur in their industry and assess the risks involved. Not all failures are critical; the problem sometimes or many times (I just can't tell) on maintenance is that according to them, all failures are considered critical. That is why their storeroom is loaded with excessive stocks that were not even used and ended up as a non-moving item. When the equipment failed, and the stock was not around, the boss yells at you, making the spare part critical. The criticality of the part will depend upon the consequences of the failure. Remember, for a critical failure, the risk is high, and the aftermath is unfathomable. What I am saying is that critical failures seldom happen but is possible to happen, and when they do, the impact and consequences of the aftermath will definitely hurt your industry's reputation, image, and business. The worst failure you can ever imagine will be those failures that can affect the environment. Failures that can cause harm, injury, or even death are likewise considered critical. For failures in which the boss gets mad. Would you consider this critical? I leave that good judgment to the reader.

10. Remove Departmental Barrier for Better Communication: If you are employed in an industry, that industry will be divided into different departments, and each department will have its own set of cultures, which can be different from other departments. I am not recommending getting a sledgehammer and remove the partition walls that divide every single department. For all I care, what I am saying is that each department should sing just one song. While Finance people love to sing the song of Cyndi Lauper, *Money Changes Everything.* At the same time,

Operations, Safety, and Quality want to sing the song from Police, Every Breath you take, Every move you make, Every bond you break, Every step you take, I'll be watching you. Maintenance wants to sing the song from Beatles; when I found myself in times of trouble (maintenance is troubleshooting and having a hard time fixing the repair), *Mother Mary Comes to me speaking words of wisdom, let it be.* (It seems it cannot be fixed, so let it be) The problem is they sing their songs simultaneously, and what we got is simply noise. I am saying that all departments from industries should sing one song so they can be in harmony. The only time these people communicate is when you are called for in a meeting. Sometimes your time in the plant is consumed by attending so many meetings from the morning till evening to end up with the same old problems that occurred again and again. Break down barriers between departments. People in research, design, sales, and production must work as a team to foresee their production problems.

According to Dr. Edward Deming, 14 Points of Management. Just telling people to work together doesn't do much good if the management system drives them to different behaviors. Such support for teamwork is merely a slogan without the necessary management commitment. Management can support but never commit to the initiative. We need to change the management system and the behavior of those in leadership positions in the organization. When we create incentives to optimize parts of the system, the overall system is sub-optimized. To achieve the best overall results, individual parts of the system may suffer to achieve the best overall result. When we evaluate people and provide bonuses and promotions based on optimizing a portion of the system, this creates pressure against cooperation across departments. When departments have to compete for budget and staff, this build-up barriers between departments. When departments have their budgets to protect and spend, it often creates barriers. Often, this even progresses to the point where employees are considered more a part of one department than employees of the company. If they transfer to other departments, that is seen as a sign of disloyalty. The supposition is prevalent the world over that there would be no problems in production or services if only the production workers would do their jobs in the way they were taught. In reality, the workers are handicapped by the system, and the system belongs to the management. As I said, there are no proponents for more barriers between departments as a management strategy. And there is often talk about the importance of cooperation and that we are all one team, but when the management system is structured to undermine that notion, it is not much use to tell people to work as if optimizing the overall system being desired. The management system needs to encourage the behavior the organization wants to see. Too many organizations still have difficulty breaking down barriers between departments due to the management systems they have already in place that work against their goals. Likewise, Safety, Quality, and Reliability people cannot eliminate human errors 100%. Stricter rules will only make matters worse. We need time to set aside their priorities, pride, ego and understand each other's strengths and weaknesses. There is no such thing as first. Learning from one another would make a difference instead of working in isolation.

11. Forgive but Never Forget: When someone did something wrong and apologizes to us, humans will always say that it's fine; let us forget about that and move forward. This will not be the case with human errors. Almost all industrial accidents are a result of human errors. These errors will be remembered and written down in history not to ridicule or embarrass the people involved and the industry itself but to remind us what went wrong so that industries can finally learn from the things that go wrong. Humans have flaws, and we are not perfect by nature. We will continue to commit these errors every day. Reminding us of these errors is what makes industries better. On January 28, 1986, Challenger was launched despite the Thiokol engineers' recommendations to postpone the launch because of the O-ring problems and the cold temperature that prevailed that day. Seventy-four seconds after its lift-off, everything was gone. Two days after the disaster, President Ronald Reagan addressed a grieving nation and said that the crew of the space shuttle Challenger honored us by how they lived their lives. We will never forget them, nor the last time we saw them, this morning, as they prepared for their journey and waved good-bye and "slipped the surely bonds of earth" to touch the face of God. We bid you goodbye, and we will never forget you. *Ronald Reagan*

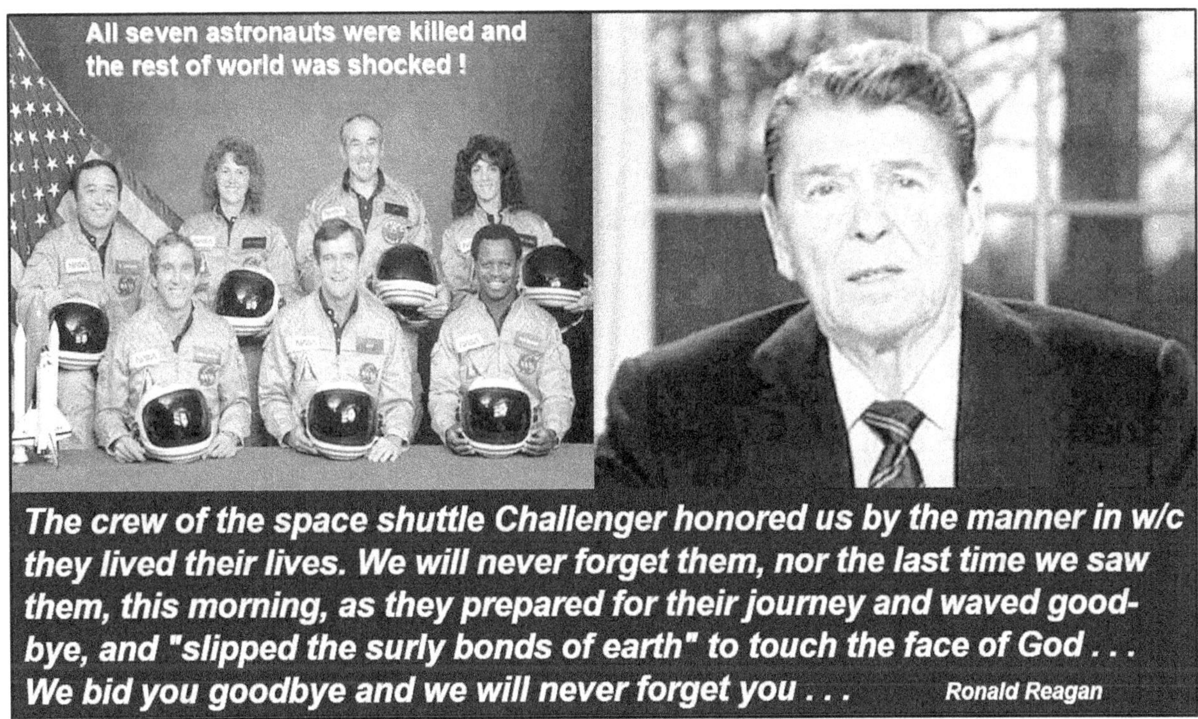

Figure 9.13: The Challenger Disaster

9.11: November 2015: Guidelines in Conducting Focused Improvement

- Focused Improvement are various improvement activities used to accomplish maximum efficiency of individual facilities, equipment, manufacturing processes, and the entire plant by thoroughly eliminating losses and improving its performance. *From JIPM Materials*

- Focused Improvement includes all activities that maximize the Overall Equipment Effectiveness of equipment processes and plants through uncompromising elimination of improvement losses and performance improvement. *From Tokutaro Suzuki*

The focused Improvement team comprises individuals who are both knowledgeable on the losses to be improved, consisting of a cross-selection of people in the organization. The respective manager spearheads the Focused Improvement team. The Focused Improvement team aims to maximize the Overall Equipment Effectiveness by eliminating or reducing the 8 Major losses encountered in their equipment. Focused Improvement teams are like an elite force of special ops in the military with a specific mission: eliminating equipment losses in their equipment and assets. These teams must be heavily equipped with several problem-solving and analytical tools to analyze the problem and address these equipment losses. The Focused Improvement team's composition will include a cross-selection of people from different departments who are familiar with the losses they are about to address.

Purpose of this Focused Improvement Guidelines: The purpose of conducting Focused Improvement on industries is to identify the different losses industry is suffering and organize a cross-functional selection of people to address these specific losses. Focused Improvement (Japanese termed this as Kobetsu-Kaizen) will include all activities that maximize the Overall Equipment Effectiveness (OEE) of the equipment by reducing or eliminating these losses and improving the asset's performance. Although not all losses will be experienced by every industry. The purpose of organizing a Focused Improvement team is to address any of the 16 big losses experienced by the industry.

Scope of this Focused Improvement Guidelines: The scope of conducting Focused Improvement activities will apply to the plant's critical equipment and assets. New equipment and assets will not be covered initially in the Focused Improvement activities as the warranty of these equipment provided by OEM may be affected due to some redesign and modification changes done by the Focused Improvement teams.

Objective of this Focused Improvement Guidelines: The objective of conducting a Focused Improvement or Kobetsu-kaizen is to assemble a cross-selection of people from maintenance, quality, reliability, safety, process, operations, and production and other departments knowledgeable on the problem or losses identified which will be addressed by the team. The Focused Improvement team aims to eliminate these losses on the equipment, manpower, and others. Note that breakdown loss will be excluded from the Focused Improvement activities since this will be cared for by the Planned Maintenance teams. Another objective is to create a loop to EEM/IFCA (Early Equipment Management or Initial Flow Control Activities) to prioritize improvements according to their importance. The reason for this is to create a loop so that when the industry decides to purchase additional new equipment in the future, these improvements and modifications will be discussed with the vendors and OEM to be included in the new equipment purchase to avoid similar problems that were previously experienced in their current equipment. Conducting Focused improvement will not only be limited to operators and maintenance but will be performed by all departments in the plant, including people from offices.

Guidelines for Conducting Focused Improvement: Below is a step-by-step guideline for conducting Focused Improvement activities to address equipment, manpower, and other losses.

1. Form the Focused Improvement Team: Membership of the Focused Improvement team will compose of a cross-selection of people from different areas knowledgeable on the problem. Minimum of 4 but not to exceed 8 members will compose the Focused Improvement team. A team should be knowledgeable on at least a minimum of three analytical and problem-solving techniques. Recommended analytical and problem-solving tools for Focused Improvement will include Root Cause Failure Analysis, FMEA/FMECA, Ishikawa or Fishbone Diagram, P-M Analysis for chronic problems, Pareto Analysis, Poka-Yoke, process mapping, and the seven QC tools. The problem-solving or analytical tool used by the Focused Improvement team will depend on the problem being addressed.

2. Decide on the Improvement Topic: The improvement topic to be addressed by the Focused Improvement team should be focused on reducing or eliminating any of the 16 big losses currently existing in the industry. Register the improvement topic with the TPM Office.

3. Understand the Situation: The Focused Improvement team should analyze and understand the current situation and identify the major losses and bottlenecks in the overall process. Losses may either be for equipment, manpower, or other losses. Provide initial measures and KPI as a baseline to set up targets. The Focused Improvement team should backtrack equipment's initial KPI and measurements to determine their success or failure. One of the measurements they will be tracking will be the OEE or Overall Equipment Effectiveness for addressing equipment losses.

4. Expose and Eliminate Abnormalities: For equipment losses, construct a picture of the optimal conditions of the equipment or process. Before applying any basic analytical techniques, the team should address all minor flaws, deviations, abnormalities, irregularities, and defects to ensure that the equipment's basic equipment condition is well in place.

5. Analyze the Causes of the Problem: The focused Improvement team uses basic analytical problem-solving tools to analyze the possible or likely causes of the problem depending on the type of losses to be addressed. There will also be cases where each loss will require two or more analytical and problem-solving tools.

6. Plan for the Improvement: During the development of proposals for improvement, formulate several alternatives. Never dismiss any ideas at this stage. For best results, do not limit participation to one or two members of the Focused Improvement team, but rather, all members should participate in this initiative. If a budget is required, the Focused Improvement team needs to prepare feasibility or economic study and calculate the return on investment (ROI) to have the budget approved by their top management. The Focused Improvement team needs to guard those improvements that may result in fresh problems on the equipment. For equipment-related losses, the team will mostly result from redesigning or modification. Refer to the questions below before proceeding with modification or redesign. If the team answers yes to all these questions, then proceed with the redesign or modification. Also, note that if a spare or part will be involved in the redesign or modification, the Focused Improvement team needs to check the actual inventory of the original parts in the storeroom and make an economic decision on whether or not to consume these parts first or proceed with the modification or redesign. Suppose the team decides to proceed with the redesign or modification; these original parts will become obsolete, and the storeroom people should be informed so that these parts will be removed from the storeroom.

• Does the failure involved cost and have major operational consequences?

• Is the cost or breakdown maintenance high?

• Are there specific costs that can be eliminated by the design change?
• Does the design have no harmful effects which can be generated afterward?
• Is there an economic trade-off study on expected cost savings?
• Is the asset to stay or be used for a long time and not decommissioned?

7. Implement the Improvement and Evaluate the Results: Once the plan had been completed, implement the improvement, and evaluate the results. If the improvement is successful, it will be replicated to other equipment with the same operating context or problem. It is not recommended to bring improvement to similar equipment which does not possess the same problem. Check whether targets have been achieved; if not, the team will again proceed to Step 5 and analyze the causes once again.

8. Monitor the Results and KPI: KPI should be based on PQCDSM. (Production, Quality, Costs, Delivery, Safety, and Morale) If equipment losses are being addressed, the Focused Improvement team should monitor the OEE of the equipment (Overall Equipment Effectiveness). If the indices and KPI improves, the Focused Improvement team needs to develop the procedures necessary and disseminate them to concerned areas and people in the plant.

9. Sustain the Improvement: Implement standardization and measures needed for preventing the recurrence of the problem. Draw up control standards to sustain results. Formulate work standards or routine maintenance to sustain these improvements.

10. Horizontal Replication: Fan-out or horizontally replicate successful improvements generated by the Focused Improvement team to other processes, equipment, system with the same operating context or problems. Similar equipment which does not possess any problems whatsoever should be excluded in the horizontal replication plan.

11. Create a Loopback to Early Equipment Management (EEM): Once the improvement had been concluded, create a loop back to IFCA (Initial Flow Control Activities) or EEM (Early Equipment Management) for the possibility of purchasing additional equipment in the future so that the improvement will already be included before purchasing the equipment with the vendor. The issue of the patent should be discussed with the OEM or Vendor.

12. Team Celebration: Suppose the team successfully reduces, mitigates, or eliminates the losses identified in their Focused Improvement activities; a simple celebration will begin. The team disbands and goes back to their original duties in the plant.

9.12: December 2015: Why are Japanese Industries Successful Compared to Other Countries?

After World War II, Japan's economy collapsed, and their nation was devastated and left in ruins. Out of humanity, the United States sent an emissary to Japan to rebuild their ruined nation. This person was invited to speak to an elite group of Japanese scientists, scholars, writers, industrial leaders, and engineers since they were in desperate need of a complete restructuring. A humble and modest American named Dr. W. Edward Deming was asked to assist the Japanese government and industry in the challenge of rebuilding their nation once again. Dr. Deming taught them about Total Quality Management, statistical control, and management. The Japanese people listened and absorbed Deming's teachings. One of his students asked him, "Dr. Deming, how long would it take to shift from the perception of the world from the existing paradigm that Japan produced cheap products and imitations to one that produced innovative quality products?" Dr. Deming told the group that they could achieve the desired outcome in five years if they would follow his directions. As time unfolded, Dr. Deming was surprised because they did it in four years. He was invited back to Japan time after time, where he became a counselor and a mentor. He was awarded the Second Order of the Sacred Treasure by the former Emperor Hirohito for his efforts. Japanese scientists and engineers named the famed Deming Prize after him. It is bestowed on organizations that applied and achieved stringent quality performance in their industry. Much of the American industry choose to ignore Deming's ideas until the last 10 years or so. Today, his ideas have been rediscovered by American industries desperate for salvation. According to Takashi Koyama, a senior economist for Global Link International, a consulting company said that for the past 40 years, Dr. W. Edward Deming has been America's most wasted domestic resource. Most of what Japanese industries have done in terms of quality improvement and greater productivity due to Dr. Deming's teachings, which was an American himself yet neglected by his own country.

Dr. Deming, Mr. Junji Noguchi, Managing Director, JUSE (at Dr. Deming's left) and other Japanese colleagues in Japan, November 1978

Lola Deming, Mr. Kenichi Koyanagi, Managing Director JUSE, Dr. Deming in the Deming's living room

Dr. Deming with Mr. Kenzo Sasaoka, President Yokogawa Hewlett Packard, in Japan, 1982

Dr. Deming and Lola S. Deming, Japan, 1978

Figure 9.14: Why Japan is Grateful to Deming

Although TPM started as one of Toyota's suppliers, Deming had greatly influenced Japanese people from a country that produced cheap products to be competitive among the very best industries worldwide. Because of Dr. Edward Deming's teachings, great Japanese quality leaders emerged, such as Shigeo Shingo, Taiichi Ohno, Kauro Ishikawa, Genichi Taguchi, of course, Seiichi Nakajima, who was later on known as the father of Total Productive Maintenance. Many learnings can be adapted from the Japanese. While US industries continue to focus more on producing mass production and big volume, they believe that the way to victory was to have tough people manage organizations that produce goods in large batches for higher profits. On the other hand, Japanese competitors do it differently; they developed a new and powerful paradigm that winning organizations listen to the voices of their customers and allow their operators to improve their products' quality and process. They design products and services that meet or exceed customer expectations and improve continuously in all areas of the organizational process that will eventually lead to customer satisfaction.

The secret of Japan's rapid economic growth is the well-tied relationship between the management and its workers. The Japanese do not believe in strikes. Japanese workers are dedicated to the work assigned to them. Interestingly, Japanese companies do not look upon the employer-employee relationship as a labor contract but as a joint membership in the same family. The firm wins the employees' loyalty by its paternalism attitude influenced more by humane than economic considerations. They listen to their people's voices rather than wait for management instructions and disseminating their supervisors for the plan to be followed by all workers in the organization. In Japan, the worker's voices matter most.

Every industry provides a specific product or service to render, but more importantly, it is the customer that spells all the difference why an industry still remains in business. To let your customer remain, the industry must satisfy three basic things, first, industries that can produce the cheapest, second, those that can produce the highest quality product or services, and third, industries that can produce the fastest delivery of them all. In today's global economy, industries' survival depends on their ability to improve and rapidly adjust to their customers' changing needs and demands. As a result, an increasing search is on improvement methodologies, strategies, and processes that drive quality, safety, reliability, and productivity. In today's fast-changing marketplace, slow and steady improvements in operations will not guarantee profitability or survival. Companies must improve faster than their competition if they want to remain leaders in their respective fields and control the lion's share of the market. That is why to survive the competition, each company is seriously adopting the best continuous improvement practices. In Japan, Total Productive Maintenance or TPM tops the list of these continuous improvement methodologies. In fact, their government is pushing this on every industry in Japan. Companies adopt TPM to improve the bottom-line results and bring out the best in their people because TPM believes that it is the people that will make or break that difference. They believe that TPM is not just a program. It is not just a strategy but rather a philosophy and a way of life for everyone. Management and workers see TPM improve the bottom line results, especially the human side. Perhaps it is quite ironic that Japan lost the war and surrendered to the United States in World War II, but it is actually the United States that lost the war in terms of its economy to Japan and the person that made what Japan is today is an unknown American by the name of Dr. Deming.

Kaizen the Concept: Kaizen is a Japanese word that means gradual, unending improvement, doing little small things better, setting and achieving even-higher standards. It is the belief that any product or process devised by man is capable of continuously being improved. It contradicts Fredrick Taylor's belief that there is only one best way to do anything. The Kaizen strategy message is simply that not a day should go by without improvement made somewhere in the company, which may improve personal life, social life, and working life. When applied to the workplace, Kaizen will involve everyone in the organization, whether they are managers or even shop floor workers. It resorts to positive thinking. It believes that we can turn each problem into an opportunity for improvement and that the starting point in any improvement initiative is to identify the problem. It is the simple truth behind Japan's economic miracle. KAIZEN resorts to positive thinking; we can turn each problem into an opportunity for improvement, and the starting point in improvement which is to identify the problem. Japanese people have a sense of loyalty to their industries. They adapt to changes incrementally or continuously, unlike innovation, which is a one-time big leap improvement effort. That is why Japanese people are known for their continuous improvement efforts. Their thinking is that they know that the best can still be made better. Here are some important things to reflect about Kaizen:

1) It reflects on gradual, unending, continuous improvement
2) Not a day should go by without improvement made somewhere in the company.
3) Quality in its broadest sense is anything that can be improved, and the means to justify improvement is Kaizen.
4) Make a continuous effort to establish a system to support the P criteria.
5) There must be a close connection and continuous interaction and communication between the design stage, development stage, production stage, and marketing stage.
6) Quality first, not profit.
7) Remember that the next station or process is your customer.
8) Kaizen must involve everyone in the organization, from top management down to the shop floor. The most important is that although culture-based, it is not culture-bound. It will work in any type of organization once you learn to accept its principles.

2016 RSA Reliability Newsletter Vault Archive

> *Many industries claim that safety is first, while others prioritize quality. Industries must learn to understand that reliability is as important as safety and quality. There is no first. Quality, Safety, and Reliability should not be in conflict with each other. The key is for these people to work in harmony as a team and not remain isolated. This way, industries can have a better chance of managing human errors and avoiding industrial disasters.*

10.1: January 2016: Strengthening Operator and Maintenance Partnership

Strengthening operators and maintenance partnerships can be done both formally and informally in any industry. To do this formally is to adopt an Autonomous Maintenance seven-step deployment, which is covered in my previous newsletter, but this will require management consent and approval. This will be part of the plant's main strategy. Doing this by yourself will be quite difficult or next to impossible as there are many walls of resistance that you will be facing. I have laid down some basic suggestions, tips, and recommendations on operator and maintenance partnerships for your industry to do this informally.

1) For any kaizen activities initiated by our very own operators, give them a slot and showcase their improvements to top management. This will definitely boost the operators' morale and convince other operators to do the same as well. It is important to recognize the efforts of these operators.

2) When the equipment fails or breaks, maintenance should not allow the operator to leave the equipment. Let operators assist them in the repair and, in return, teach these operators what went wrong with their assets and equipment. This will foster better communication between both operators and the maintenance themselves.

3) The training department needs to conduct a training needs analysis regarding what training requirements are needed by operators and maintenance to perform their job correctly.

4) For maintenance, try to visit the storeroom and ask for big non-moving parts that have never moved and use them as a teaching aid for operators. For example, a big motor that has not been a move for 5 years can be used as a teaching aid to operators. Have this cut on your machine shop so that the motor's cross-section can be seen visibly, and this can serve as a teaching aid and training kit to operators. The good point about this is that once you have

taught the operators what the inside of the motor looked like, they can actually visualize what a motor looked like from the inside.

5) Listen to the suggestions of operators since they know more about the symptoms of the failure than any other person in the plant since they are the ones who will experience and witness the failure first. Maintenance must communicate with the operators regularly. Teach them the important parts of the equipment and what happens when it fails.

6) Spend 15 minutes a day to teach the operator about a certain part or spare, its function, description, why it is important to inspect these parts, and the consequences of not performing these inspections. Do this daily in the form of One-Point-Lessons or Good-to-Find. Let operators sign the One Point Lesson themselves if they understand to grasp their learning. Remember, just one point per day and spend 15 minutes or even less.

7) If you have training for maintenance to be done in-house by subject matter experts in the plant or by third-party consultants, convince your boss to include operators in training. This will be a very good venue for operators to understand maintenance and their role in the organization.

8) If the operators have their own checklist on the equipment, maintenance should spend some time explaining each of the activities they performed and the consequences if these checks and inspections will be neglected or ignored. Ask their suggestions if they can think of a better way of performing these activities and if it is possible to adopt a visual control or visual management to spot the deviation more easily on the equipment.

9) Provide simple yet memorable recognition and rewards system for every operator's suggestion that had been implemented with success on their equipment and assets. This will definitely motivate other operators to join in the initiative.

10) For easy to restore failures, let operators participate in the repair process and assist them in doing it initially. Maintenance should be present if operators are attempting to repair their equipment the first time around. Try to assist and coach them initially, especially if this will be the first time the operator will attempt it. Guide them and thank them personally for the initiative. This is already out of their job description, but the operator got the initiative to do it. Treat them for a meal, snack coffee, and talk about what they have accomplished and experienced.

11) If Autonomous Maintenance is not in place in your industry because of restrictions, job description, resistance from operations, ask your operator personally if they are willing to learn more about their equipment. If they say yes, always remember that the white pieces move first in a game of chess, and maintenance takes on the white pieces. This means that maintenance should be the ones to make the first move on this initiative.

12) Although this is connected to item number one. Showcase minor improvements to your boss before going to your manager and convince them what you have accomplished together with your recommendation and requirements.

13) My advice is that if Autonomous Maintenance is not in your organization and some restrictions or the like, it is best to keep this between you and the operator. Ask the operator

if they are willing to learn more about their equipment and spend time doing it. The reason to keep this between you and the operator is that if the operation manager finds out, they can permanently shut you down. As I have said and based on my experience, maintenance managers will support this initiative, but I cannot say the same for operations managers as they are the main source of resistance between having an operator and maintenance partnership.

14) Start this initiative with the operators who complain a lot or those that influence other operators. It will be difficult initially, but once they are convinced of this partnership, the rest will be easy. Remember that they have an influence on other operators as well.

10.2: February 2016: Tips In Conducting Root Cause Failure Analysis

1) It is highly recommended that the Principal Investigator be assigned to investigate the incident should not come from the area where the failure incident occurred but should come from other areas to avoid being one-sided and biased.

2) It is recommended that Root Cause Failure Analysis be conducted on recurring failures with devastating consequences and not on every single failure or breakdown experienced in the plant.

3) It is highly recommended that each area in the plant have its list of Principal investigators, which can be called a plant failure incident. The minimum requirement for a Principal Investigator and the evidence gathering team is to be trained initially on Root Cause Failure Analysis.

4) If a breakdown or failure will warrant an investigation, the repair team to preserve the part that failed, take pictures, collect metal debris, and the like before repairing the failure. What is important is to preserve the shreds of evidence. This might slightly prolong the repair time, but this will be for the industry's good to have a chance to learn from the things that go wrong.

5) It is important for the Principal Investigator to choose their evidence gathering team independently. This means that the people involved in the evidence gathering team do not know who is assigned to collect the paper evidence, people evidence, and physical evidence. This means that if Bob was assigned to collect the physical evidence, he did not know that Charlie is assigned to conduct the people's evidence, and Mike was assigned to collect the paper evidence. Likewise, Charlie does not know that Bob was assigned to collect the physical evidence and so on so that they can be independent of each other. We are avoiding at this stage comparing their evidence, which can compromise the integrity of the evidence they are carrying.

6) Top Management should clarify to all people in the organization that no one should be punished due to the incident or failure if the human error committed unintentionally contributed to what they know regarding the failure. If people know that someone will be blamed and take the fall, then the people involved will be defiant, remain silent, and will not share what they know regarding the incident. They will be hesitant to speak, and there will

be little information that can be gained from the people's evidence.

7) For the area that the failure incident occurred, an operating context statement should be provided and should be given to the person that will be assigned to be the leading Principal Investigator of the incident. The operating context statement should include the following information.

  • State the actual problem as detail as possible
  • Actual date the failure incident occurred
  • Where did this event actually occurred?
  • Who did this happened to?
  • When did this event actually happen?
  • Were there any oddities or irregularities detected before the failure?
  • Were there any signs or symptoms that occurred before the failure occurred?
  • Did this failure already happened in the past?
  • Were there any parts or spare that failed?
  • Were the parts affected been preserved?

8) For the person in charge of conducting the people's evidence, have at least two sets of questionnaires used during the interview process. One set will be used for those people directly involved in the problem or those who witnessed the failure, such as operators, maintenance, their supervisors, while the other set of the questionnaire will be used for people indirectly involved in the problem. These are the people who have experienced the failure before. Ensure that the same questionnaire will be used for all directly, and another set of the questionnaire will be used for indirect people involved in the incident. It is best to conduct an interview with one person at a time.

9) When the Root Cause Failure Analysis investigation has been concluded, share the lessons learned not only on the affected area but also on all areas in the plant and other business sites. Translate the findings so that a 10-year-old boy will understand the lesson learned from this incident. Our goal is to learn from the failure and generate awareness for all people in the organization.

10) Each area in the plant should have its own Root Cause Failure Analysis kit, which will include the following: gloves, safety tapes, pen and notes, zip-lock plastic bag for collecting evidence and debris, flashlight, tweezers, camera, tape recorder, and magnet. This will serve as the Principal Investigator's initial kit in conducting the investigation.

11) It is important for the Principal Investigator not to discuss their initial findings with anyone in the organization for as long as the investigation is still ongoing and should remain confidential for the time being. What we are avoiding is for the evidence-gathering team in charge to compare their findings with others and create biased opinions, compromising the integrity investigation.

12) When the Principal Investigator and evidence gathering team is conducting the RCFA logic tree diagram, there is a possibility that there will be more than one cause that leads to the problem based on the evidence gathered. If this is the case, separate these causes by providing their own logic tree diagram. Do not combine all the causes in a single logic tree diagram, which will only complicate the investigation. Treat each of the causes separately.

10.3: March 2016: The Need for a Structures Asset Management System ISO 55001

Many industries worldwide have suffered from major Industrial disasters, environmental regulatory breaches, and recessions. Creating a structured approach is necessary to reduce, minimize, mitigate, or even eliminate the risks involved and reduce the consequences of failure if industries want to remain in business and stay competitive. Industry's assets and equipment should not only be cared for and well-maintained, but they need to develop a comprehensive and effective asset management structure that will be of utmost priority. Although both PAS 55 and ISO 55001 will provide the industry a universal framework and guidelines for managing the use of the equipment and physical assets, complying with the ISO 55001 is still not a guarantee of having a "World Class Maintenance Management Structure" in your plant. ISO 55001 standard was published in 2004 and revised in 2008, and adopted by industries worldwide to integrate and improve business practices, raise performance, and assure greater consistency and transparency. After consultation with several industry experts and professional consultants worldwide, the procedure was reviewed in 2009 and given to the International Standards Organization as the basis for a new set of ISO standards for asset management developed and approved over the last 3 years with 31 participating countries. Finally, ISO approved this as of January 2014. ISO 55001 is an international standard founded on the belief that an asset's value is defined by its contribution to achieving organizational goals. It was designed to optimize asset value and decreased organizational risks through a comprehensive asset management strategy built around its business objectives and goals. According to this procedure, asset management is defined as executing that set of processes to realize value as the organization defines it from those assets.

BS ISO 55000 Standards for Asset Management

• BS ISO 55000, Asset Management: Overview, Principles, and Terminology
• BS ISO 55001, Asset Management: Management Systems, Requirements
• BS ISO 55002, Asset Management: Guidelines for the Application of ISO 55001

ISO 55000 – Provides an overview, key concepts, and terminology for asset management. It describes both the what and why of asset management. This will require changes in policy, procedures, processes, and people. It involves a change that can only be achieved through an active commitment by senior members from the top management.

ISO 55001 is intended for assets in organizations and is principally for physical assets. ISO 55000 contains definitions of terms and key concepts.

ISO 55002 supports providing the basic guidelines and application and how to comply with ISO 55001. Both ISO 55000 and 55002 are support documents for ISO 5001.

All of these examples represent risks that must be managed. Good Asset Management entails identifying, assessing, managing, and mitigating the risks involved in the failure and the consequences of the risks. Risk management is one of the important corporate functions directly and intimately impacted by asset management activities. It is also an excellent example where the integration of asset management with other corporate functions provides substantial potential benefits to industries that adopt these standards. Good asset management can lower the risks of failure involved, reduce costs, and improve any industry's productivity and revenue. This means that before we can manage our equipment and assets, we likewise need to change the way we maintain our equipment.

A good Asset Management System is not just about having software or CMMS (Computerized Maintenance Management Software) nor having a maintenance strategy in place. Although these things will be a prerequisite to having a good asset management system in the industry. Remember that a good Asset Management system will not solve all problems, but rather it should provide a structured framework that will enable industries to make better decisions on every challenge they face. It allows industries to think more and explore things that will solve the problem they encounter instead of thinking that we have always done it this way before mindset. Good Asset Management is all about doing the right things in the right way.

Although ISO 55001 is not all about maintenance, I believe that before an asset or equipment can be managed correctly, it should be well maintained and reliable. If ISO 9000 was championed by Quality and ISO 14000 and ISO 45001 were championed by safety people in the industry, I believe that the best people to develop the ISO 55001 for industries will be the maintenance people. The use of ISO 55001 in all sectors of industries improves the way we do maintenance, but most likely, the maintenance team will be responsible to fully grasp and implement these standards to improve the way they both maintain and manage their equipment and assets in their plants.

ISO 55001 embraces the philosophy that any asset from an industry with income-generating value will deteriorate much easily if not maintained well, while actions taken to maintain different types of assets may differ from one type of industry to another. A specific asset management plan like ISO 55001 can add a much-needed dimension by allowing industries to better know their equipment and assets. Its intended result is a defined method for ensuring asset longevity, which leads to improved decision-making and better business performance. The goal of asset management is to take care of the equipment's Total Life Cycle at the most reasonable cost by having a structured plan to maintain the equipment, improve the equipment design flaws, and reduce the cost of operating and maintaining the equipment itself.

Benefits of ISO 55000

Suppose you visit an industry; you will be staying on their lobby portion of the plant for visitors. The vision, mission, policy, and objectives are all hung on the wall for everyone to see and read. It's nice to have these things around in the plant, but a few people who may no longer be connected with the plant itself drafted most of them. Try to ask any employee in the plant about their corporate values or even their mission and vision. Only a handful of these people can give what you have read on the wall. ISO 55000 must not be done to satisfy the plant's visitors or to have these credos hanging up on the wall for aesthetic purposes. The strategy for ISO 55000 must be disseminated to all employees within the organization. Yes, providing all the documents needed for ISO 55000 will give you the certification. Assuming that the purpose of certification is merely to satisfy all the required documentation needed and not for its intended purpose, which is to manage risks successfully, then everything will just end up as a big joke, and industries are fooling no one but themselves.

ISO 55000 is a top-down approach strategy where the decision-making process begins in the boardroom and aligns its asset management strategy with the organization's overall corporate strategy. An organization certified on ISO 55001 must provide a framework for cascading the entire organization's asset management plan. It provides top management with regular feedback on risk assessment, opportunities, and asset performance to inform the ongoing decision-making process. Many benefits can be derived if ISO 55000 is implemented religiously, wholeheartedly, and not merely for documentation purposes.

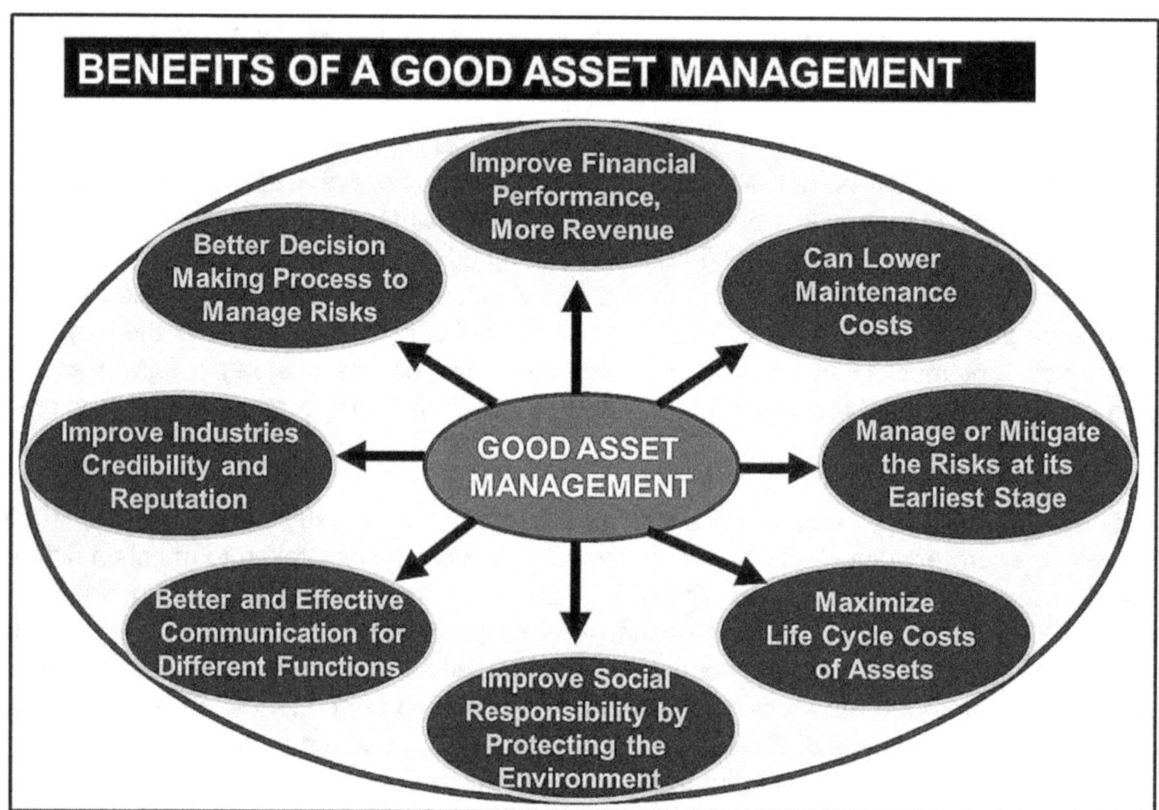

Figure 10.1: Benefits of ISO 55000

Although having an ISO 55000 in place in the organization will not totally remove barriers and silos among the organization's different functions, it would provide a good start on building effective communication with the other functions within the organization. These functions need to understand that they cannot co-exist without each other.

Good asset management in place will definitely not solve all problems. Still, it can help industries by enabling organizations to make better decisions based on the challenges they face as they come. A good asset management system entails identifying, assessing, managing, and mitigating the risks and their consequences. Risk management is one of the important corporate functions that directly can impact an asset management activity. It is also an excellent example of asset management's integration with other corporate functions that substantially benefit industries that adopt these practices. Good asset management can lower a broad spectrum of risks, reduce costs, improve production, provides decision-making, strengthen social responsibility, especially to communities, and generate revenue-generating capacity.

One more precaution I would like to reiterate in the ISO 55002 guidelines is on the sub-clause on 8.2 Management of Change and 10.3 Continual Improvement. Typically in an industry that implements a Management of Change process, any changes, improvements, or modifications made in the process must be fully evaluated, tested, validated, and approved by a group of people so it will not generate another set of new risks for the organization. The management of the change process (MOC) usually slows down the continuous improvement initiative, as everything needs to be evaluated, tested, and approved by a group of people with other priorities. Usually, those assigned to handle the MOC process are part-time people, mostly managers with other plant priorities. I know there are no industries that have full-time people for just doing the Management of Change process. They also have their main duties and obligations to attend to in the organization. If we do everything by the book, it might take several months to approve an a simple operator's kaizen improvement and take effect. This must be discussed thoroughly in the ISO 5002 implementation process to avoid these kinds of problems. I am not against the process of MOC, but perhaps it would be highly recommended that a simple decision diagram could be made if a change or improvement will undergo a MOC or be allowed to proceed. I use to consult with a plant that has a MOC system. Even incremental or small improvements recommended by operators take months or even longer to be rejected in the end, which demotivates people to further generate improvements. MOC should have a timeline for big and incremental improvements. For example, incremental improvements suggested by operators should be processed in less than 48 hours if it will truly benefit the process while innovations and modifications will take more time in the process, as it needs to be evaluated, and tested for its effectiveness to ensure that it will not provide any new problem in the process.

10.4: April 2016: Guidelines in Setting-up a CBM Organization your Industry

For those industries just starting their Condition-Based Maintenance strategy, there are detailed guidelines and step-by-step details on how to begin your CBM strategy. This discipline on maintenance requires two investments, which will require investing in the instruments, and the 2nd investment is investing in the people who will be these instruments' users. One of the key important factors is selecting the right people who will perform these measurements and diagnostics. CBM is not only about monitoring the health of their equipment but also about decision-making in allowing the equipment to stop or to continue running. Placing a CBM or Predictive Maintenance strategy in a reactive environment will not work, and the benefits will not be reaped completely. What is important is before applying any CBM, Predictive Maintenance, or whatever you call it in the plant, basic equipment conditions must be addressed first. The basic condition includes cleaning, lubrication, tightening bolts, and equipment free of leaks. Different functions in maintenance together with their c-level people must understand the role CBM will play in their maintenance activities. Predictive Maintenance requires an investment, so this strategy needs to be applied right the first time around. Do not buy these instruments if your industry will not invest in the users. This will just be a complete failure.

CBM, PM, Planners, and other maintenance functions must work hand in hand and not against each other. I recommend that the people involved in this function be under one department or under one function. In other industries, the vibration monitoring user reports to the mechanical group, while the thermographer reports to the electrical group and the ultrasonic monitoring user reports to the facilities group. I strongly recommend that all Predictive Maintenance users report under a single person and should be under one roof. For those industries starting up with their CBM or Predictive Maintenance group, here is a detailed systematic guideline on starting your PdM or CBM journey. This maintenance discipline will require three investments. The first is investing in the instruments; the second is investing in the users as they need to be certified in using these instruments, and the third will be the calibration of their instruments. One of the key important factors here is selecting the right people to monitor their equipment's performance and interpret the data. Remember that CBM is about monitoring the health of the equipment and decision-making on whether to allow the equipment to run or not and for failure prediction and analysis. Data from Predictive Maintenance can be used for Root Cause Failure Analysis investigation.

1) **Start by having basic education on Condition-Based Maintenance:** This will be done for both maintenance and operations people. Top Management must be educated regarding their support and commitment to the initial investment on the users' instruments and training to establish an effective CBM program. Understand the role of CBM in the Maintenance Strategy, as well as the benefits of PdM/CBM. It is important to understand both Planned and Autonomous Maintenance's role in the overall CBM strategy and the benefits of a CBM strategy.

2) **Define the CBM/PdM program's objective:** Make this your focal strategy in the implementation. Determine the plant's measurement of success, yardstick, KPI's and an indicator such as maintenance cost, Mean Time to Repair (MTTR), Mean Time Between Failures (MTBF), equipment availability, and downtime as well as breakdown occurrences.

3) **Conduct a Machine Ranking:** Identify critical and catastrophic equipment to be prioritized using condition monitoring. Equipment with a major impact on the following should be given utmost and top priority in CBM. Priority equipment to undergo CBM should include un-spared or stand-alone equipment, failures with safety or environmental consequences, equipment operated on unmanned locations, equipment with the high cost of failure, unplanned shutdown, frequent failure of parts, or components, hidden failures, and the probability of multiple failures.

4) **Organize the CBM group and allocate permanent manpower resources:** Assign full-time people to handle the different non-destructive and diagnostic instruments. CBM must be handled by dedicated and accountable people. Part-time personnel cannot maintain the monitoring and analysis frequency that is critical to a successful PdM program. Provide a detailed job description for each person, including their duties and responsibilities for the CBM users. Define the core vision and mission of the team. Study and determine the number of persons needed and determine if this will be an external or internal hiring. The CBM group should be centralized and not decentralized.

5.) **Define CBM/PdM instruments required for your maintenance:** If the company is tight on budget, prioritize which instrument is needed most. The most common Predictive and Condition Monitoring instruments include infrared thermography, vibration monitoring, oil analysis, and ultrasonic monitoring.

6) **Source different vendors for the PdM instruments:** Allow them to conduct some actual demonstration on your equipment. Develop the operating cost of the equipment needed. Discuss available training and possible certification processes on the instruments such as vibration, thermography, ultrasonic monitoring.

7) **Generate a feasibility or project study on acquiring these instruments:** Present them to top management, and seek budget approval. Include the following in the PdM feasibility report, computation for projected savings on PdM, benefits of using PdM and CBM instruments, case studies on PdM from other users, cost needed for the project, and the rate of return (ROI) on investment.

8) **Procure the equipment needed:** Select the best and most qualified vendor and purchase the CBM/PdM instrument required. Make sure you have evaluated the vendor regarding the following such as user-friendly software for Predictive Maintenance, the vendors' stability if they have been in the business for a long time, availability of technical support even on non-working days.

9) **Finalize the selection of people for the PdM team:** Users of these instruments should undergo Level 1 certification as a minimum requirement to successfully do the monitoring. Basic theory course on the techniques and principles of machine dynamics, heat transfer, fundamentals of vibration, and take the initial examination required for Level 1 certification of instruments handled by the user.

10) **Create a comprehensive database system management for Predictive Maintenance strategy**: This will require regular trending and monitoring to establish the P-F curve, multiple readings to develop trend statistics, defining alarm and alert limits on vibration, oil analysis, and infrared, determining the frequency of monitoring for critical equipment and regular monitoring to establish baseline trends.

11) **Define activities to be done once readings exceed the limits:** The users of these instruments must provide recommendations to concerned people from production and maintenance if a potential failure is evident.

12) **CBM/PdM group now operational:** CBM users should provide a regular report regarding cost savings generated by their instrument and update production and maintenance people.

10.5: May 2016: Reasons Why We Need to Measure KPI

To Evaluate: We need to measure KPI to determine if our equipment and assets are performing well or not. These measurements can tell us exactly that our equipment is functioning based on our expectations or not. The objective is to determine if our activities and efforts on maintenance are working or not. This means that if we are on target, we can challenge ourselves to increase the goal, while if we are way below on target, then we need to evaluate and review what is affecting our performance and how we can overcome them.

To Manage and Control: Measurements can tell us if we are in control of the situation or not. These measurements, unless manipulated, can tell us a lot about the way we do things here in our industry. The late Peter Drucker said that we can only control the situation if we are measuring it right. Every department in the organization has its own respective KPI's. What is important is that all KPI's should be aligned to the overall corporate vision and mission of the industry. This means that if the supply chain or warehouse controls the MRO spare parts of your industry, then maintenance cannot manage it because we can only manage what we can control.

To Budget: By measuring performance, we can determine which strategies and activities deserve additional budget for maintenance. Measuring performance can tell us which ones are within the budget and which are over the budget. By measuring these Key Performance Indicators, we can allocate budget wisely on the things that maintenance needed them most. The maintenance budget should not only be limited to repairs and maintaining the equipment, but a budget should be well in place for maintenance training and instruments which are needed to perform their jobs correctly, such as the acquisition of Predictive Maintenance instruments and certification of the users as well.

To Make Decisions: Developing performance measures allows an organization to determine its mission, goals, targets, and desired results. These KPI and measurements will also identify methods of measuring how well these results were achieved or not. These measurements can aid decision-makers in making sound decisions based on the outcome of their performance. What is important for every manager is to look at the results or lagging indicators and understand the process that led to the result, creating better communication with their subordinates.

To Learn: Performance measurements contain information that can be used to evaluate and learn. It is acceptable to have a low figure in our KPI for as long as we learn and do something about it. The objective of measuring these indicators and KPIs is to provide management with real numbers to manage and not blame people. If your competitor's cycle time is 24 hours, and your industry is 48 hours, we need to learn why they can deliver faster than us. What are they doing differently? If the PM Compliance is high, yet emergency repair is also high, then something is definitely wrong with the way we maintain our equipment and assets.

To Improve: But the most important reason for measuring performance is to improve the organization's performance. By knowing our performance measurement, we can ask ourselves, what exactly should we do differently to improve our performance? Industries today challenge themselves by setting goals and improving them. Industries cannot have the same goals repeatedly if they want to remain competitive and stay in business. Moreover, they should be improving at a faster rate than their competition. They do this by challenging the goals they set and continuously improving them. When I was working in Sharp Philippines, we at the Television and VHS department bought a 14 inch Samsung unit and brainstorm how come this unit sells half the price compared to our own 14-inch unit? It was a real challenge to improve and lower the cost of our own unit without sacrificing the quality of the product.

10.6: June 2016: Where MTTR Should Be Used

Some industries used MTTR to compare the repair time of their maintenance people in the plant. Others even go to the extent of including MTTR as one of their key performance indices and even included it in their Maintenance Performance Appraisal. Hence, if Joe has the lowest MTTR among all the maintenance people, then Joe will get the highest increase in his next payroll. I think that this is a terrible decision to make. Why? Many systems are repairable when the system is said to have failed in operation. We also need to consider that MTTR varies depending upon the difficulty of repairing an equipment failure. MTTR will be difficult to estimate because one must consider a variety of repairs. An engine repair will include tightening a drain plug bolt to overhauling an entire engine assembly. We cannot just compare the two. Suppose all the repairs done on the equipment are identical in every way; in this case, we can compare the repair time for different people, but that is unlikely because our equipment can fail in the real world in more than 100 ways. The repair time will vary depending upon the repair's difficulty and the person's experience performing the repair. I have a friend who repairs so fast in which if he endorsed the equipment back to the operator, it would just fail sooner than you expected it to fail.

On the other hand, this guy repairs so slow and takes his time, but when he endorses the equipment back to the operator, it is unlikely that the same breakdown will occur again. So how can we compare the two in terms of Mean Time to Repair or MTTR? What I can recommend to your operations manager is to provide them with a repair menu. If operations ask, what in the world is that? Tell operations to choose the repair they want; if they want a slow repair, we will first analyze the equipment's problem, but if they want a fast repair, then maintenance will just repair the equipment with no guarantee that the problem had been resolved. I think that is fair enough.

In most cases, when an experienced person retires or leaves the plant for good, his experience usually goes with him. Most industries have not captured nor documented these people's experience of performing this type of repair for a particular breakdown. What was his first step, then what, after that, what's next, and so on? When these kinds of things are in place in the plant, then these can be used to teach other maintenance people who are new to the plant so that maintenance will guide how to repair this type of breakdown and not to experiment all the time. I believe that this is where the true value of MTTR should be reflected. Most of us in maintenance admit that we still cannot get rid of the hero stuff where no one on that plant can repair the failure and finally when the boss called us late at night in our home telling us to go back to the plant to fix something that nobody can fix because you alone with all your sweat and tears have experienced this in dealing with this magnitude of failure, then you get a hero's welcome.

For a start, if your plant encounters a major breakdown, call on the most experienced person to repair it, then try to video the whole process of how he started and end the repair. Once the video is in place, set up a meeting with the person who performed the repair, showed him the video frame by frame, and let him discussed what he did on how he accomplished the whole repair process. Write down what he said step by step. Once the document is completed, have him review it for any more addition or deletion. Once the step-by-step repair procedure is completed, encode this in your CMMS (Computerized Maintenance Management Software). Now, suppose a similar failure happens in the future; in that case, every maintenance can print this out and have a basic guideline on repairing this type of failure if it occurs again.

Perhaps if MTTR is being monitored in the plant and in a certain group of 10 people, Charlie is said to have the lowest MTTR in performing repair and troubleshooting the hydraulic systems, while Joe has the highest MTTR in the same category. What we can do is we can define the proper procedures on repairs based on Charlie's practices, which can easily be applied and followed by other maintenance people, thereby avoiding trial and error during repair and troubleshooting, or we can either partner Joe and Charlie so that Charlie can teach Joe from time to time while Joe can assists Charlie in performing the repair. After some time, if we can see Joe repairing the hydraulics himself without Charlie, it simply means that the knowledge Charlie got had already been transferred and pass on to Joe. This is what we actually want on our maintenance people.

Finally, I do not recommend performing a Root Cause Failure Analysis for every single failure or breakdown in your industry. That will be a very tedious situation for maintenance to do, but rather what I am saying is that for failures that keep on repeating themselves, it is best to slow down and finally address the root cause of the problem to prevent the problem from recurring on its own. The saying goes that when our people become really good at fixing failures, then something is definitely wrong with our maintenance organization. Why? What is wrong? It seems that they cannot seem to let go of the problem. Try to think why they became efficient and good at repairing. It is because the failure keeps on repeating again and again, and the problem itself keeps on resurfacing. When failure keeps repeating itself, maintenance must take the time to slow down, analyze, and address the problem's root cause. If you really want to understand why the failure keeps recurring, take things slowly, and start analyzing them. I believe that to be fast, we need to take things slowly. Finally, to conclude, Root Cause Failure Analysis and Mean Time to Repair do not actually blend. Root Cause Failure Analysis is not recommended on every single failure, but all I can say is that if the failure keeps on repeating or recurring, I believe that we can set aside MTTR for a while and take the time to take things slowly and perform a thorough Root Cause Failure Analysis investigation on the problem. Having a low MTTR due to years of experience is not good because the failure simply keeps repeating.

People become good at repairing because the failure keeps on coming back. Having a low MTTR for most maintenance people due to having some repair or troubleshooting guides and procedures based on the experiences of old-timers and maintenance people from the University of Hard Knocks will be of great value. When these good people finally retire, they know that they have left a legacy in their plant. MTTR or Mean Time to Repair can never be a measure of Reliability; when people become too good at repairing, it only means one thing: the failure keeps on repeating. I would rather take my time and learn from the failure rather than repeatedly repair the same failure. Before fixing the failure, my recommendation is simple: have the restoration team take photographs of the affected part or component of the equipment and secure the part that failed. Place them on a plastic bag and any other foreign parts that they can locate near the part that failed and endorse them to the slow team, which is the team that will investigate and analyze the problem. This way, the evidence is finally preserved.

10.7: July 2016: Why is the Study of Tribology Important to Industries

Why is tribology important? It is said that 30 to 40% of the primal energy is lost due to friction, while 60% of the machine's failure can be attributed to premature wear. More than 50% of mechanical accidents and failures originated from lubrication attributed failures. This means that if industries can save at least 30 % on these frictional losses, they can provide their products and services with better reliability and quality and contribute to their country's Gross Domestic Product (GDP). The study of tribology is seriously concerned with the conservation of energy and materials in engineering design. Thus, they can sell their products and services that they produced at a much economical and lower cost with better reliability and quality. Today, the application of tribology is endless as it can apply to anything that produces contact, and its scope is still growing. According to Dr. Peter Jost, the United Kingdom could save approximately £ 490 million per year by employing better tribological practices.

Sources of Expenses	Amount	
	Pounds	USD
Reduced Maintenance Replacement	230,000,000.00	296,700,000.00
Reduced Number of Breakdowns	110,000,000.00	141,900,000.00
Extension of Machine Lifespan	100,000,000.00	129,000,000.00
Reduced Frictional Dissipation	25,000,000.00	32,250,000.00
Savings in Investment	15,000,000.00	19,350,000.00
Savings in Lubricants	5,000,000.00	6,450,000.00
Savings in Manpower	5,000,000.00	6,450,000.00
Total	**490,000,000.00**	**632,100,000.00**

Figure 10.2: Saving Generated from the Application of Tribology

Why are these interacting surfaces that are relative motion (which means rolling, or sliding, surfaces) important not only to our equipment and machinery but also to the country's economy, and why they affect our standard of living? The answer is that surface interaction usually dictates the function of practically every device developed by man. Almost everything that man-made will wear out as a result of relative motion between surfaces. Although to maintenance or those in charge of lubrication, the cost of friction and wear may appear to be small, if we can sum up all the machinery affected by this over a period and sum it up for an entire country, a very large loss of resources becomes evident.

The widely distributed incidence of tribological problems means that tribology cannot be applied solely by specialists; but instead, many engineers or lubrication people, reliability, and maintenance people should also have a knowledge of this subject. The basic concept of tribology is that friction and wear can be best controlled with a thin film layer separating sliding, rolling, and impacting bodies. There is no restriction on the type of material that can form such a film, and some solids, liquids, and gases are equally effective. Education and knowledge must precede everything. Like in the United Kingdom (UK), this should not just be a c-level initiative but also a government-driven initiative.

10.8: August 2016: Small Problems Matters Most

There is a popular short story that I usually tell my class on Root Cause Failure Analysis, which goes on like this. Before an important battle, a story is told that a king sent his horse with a groomsman to the blacksmith for shoeing. But the blacksmith had used all the nails shoeing the knight's horses for battle and was one short of a nail. The groomsman tells the blacksmith to do as good a job possible, but the blacksmith warns him that the missing nail may allow the shoe to come off. The king rides into battle, not knowing of the missing horseshoe nail. During the battle, the king rides towards the enemy. As he approaches them, the horseshoe comes off the horse hoof, causing it to stumble, and the king falls to the ground. The enemy is quick unto him and kills him. The king's troops see the death, give up the fight, and retreat. The enemy surges into the city and captures the kingdom. The kingdom is lost because of a missing horseshoe nail. This proverb was actually a true story about the Battle of Bosworth in 1485. King Richard III was stranded, horseless, defeated, and killed because of a faulty horseshoe nail. At the end of the story, I asked the class what they think was the moral lesson of this story that they find similarities in their maintenance organization?

• For want of a nail, the shoe was lost.
• For want of a shoe, the horse was lost.
• For want of a horse, the rider was lost.
• For want of a rider, the battle was lost.
• For want of a battle, the kingdom was lost.
• And all for want of a horseshoe nail.

Big failures start with small things that are being accumulated over time and left unattended. When we take care of these small things, then we have a better chance of controlling the big failures. Addressing big failures and neglecting these small problems will only create a continuance of bigger problems. We experience the same old problems that keep coming back in industries because we simply did not take care of the smaller problems. I cannot emphasize more the importance of this basic equipment condition. It will only take just one loose or missing bolt to create a chain of destruction and havoc in our equipment. All bolts must be secure and complete before advancing to any non-destructive instruments like vibration analysis. This means that it is useless to buy the most sophisticated vibration analysis instrument if your equipment lacks bolts, nuts, screws, and fasteners. These are strategically placed on different parts of the equipment for a reason, which is to minimize vibration. When we touch our equipment, we feel a certain vibration; the vibration we feel on the equipment is the summation of all frequencies and movements happening inside the equipment. If we have a bearing rotating, it will induce a certain level of frequency. It is useless to have the top-of-line vibration monitoring from SKF if most of our equipment lacks one or two bolts. Equipment will just continue to fail. Lacking and lose bolts often lead to excessive vibration, which produces secondary damages on parts affected by the vibration itself. I know several industries that have purchased these instruments and abandon them because the failure just keeps repeating themselves and blaming maintenance for these investments. Come to think of it, maintenance often remains reactive and trapped into the repair cycle because the basic equipment condition has not been well established on their equipment, and if asked why it has not been established, they really have no time to address the basics since maintenance is always undermanned and busy firefighting due to the massive amount of breakdowns experienced in the plant. When did we ask why maintenance is undermanned? In reality, they are not; the problem is that operators simply don't take part in this shared responsibility because their organization does not allow them.

Figure 10.3: Big Problems Start from Small Things

As I have stated repeatedly in this book and in my other books, operators are very important in any reliability and maintenance strategy because they will experience the failure first and not the maintenance. They know the symptoms of these failures, but the problem is that they do not know their equipment intimately. After all, they were not given a chance to learn from their equipment because their job description states that they only need to operate the equipment. One of the operators' major activities is to work hand in hand with maintenance in establishing this basic equipment condition on the equipment, but this can only be achieved if maintenance teaches operators about their equipment. I believe that majority of these industry's accidents can be avoided if operators truly understand their equipment intimately because these are the people who will sense the slightest problem, deviation, or irregularities on their equipment before it becomes a major problem. These industrial accidents happening worldwide in industries will continue to happen for as long as operators and maintenance remain a separate function. Implementing stricter safety policies and protocols will not prevent these industrial accidents from happening because almost all of these incidents are maintenance-related and have something to do with human error. Safety people need to remove these silos and departmental barriers and work for hand in hand with Quality, Reliability, Maintenance, and other departments to manage human error more intelligently and improve the way industries maintain their equipment and assets. Working together instead of being separated will have a better chance of addressing these problems from occurring in the future. I believe that it is time for reliability and maintenance people in industries to finally educate and teach people from other departments in their organization on what reliability is all about.

10.9: September 2016: Guidelines for the Autonomous Maintenance Audit

Audits are done as the Autonomous Maintenance completes each Step to check that the team complies with the activities performed on Autonomous Maintenance and if the teams are ready to move forward to the next Step on Autonomous Maintenance. Certification and audit will be done on Steps 1 to 6. There will be no audit on Step 0 and Step 7 of Autonomous Maintenance. If the team passes the audit process, the team will be eligible to proceed to the next step of Autonomous Maintenance. The audit can be done initially by the TPM Office, but as more teams join in the Autonomous Maintenance activities, TPM Office needs to develop a pool of auditors trained on auditing the Autonomous Maintenance teams. As more and more operators join Autonomous Maintenance, the TPM Office will build a pool of auditors and provide basic guidelines and criteria on what needs to be audited. Operations managers should own the audit since they will also benefit from these activities. They are also responsible for their own equipment since they are the owner. During the initial audit, the Autonomous Maintenance facilitator briefs the operations manager on what to look for. An audit guideline should be provided detailing what to check. The Autonomous Maintenance audit will compose of two parts. They will audit both the team and the equipment. Both pilot and fan-out equipment will undergo the certification and audit process.

Team Audit - The first part of the audit will be a team presentation, which can be done on the activity board together with the auditors. The team briefs the auditors on their respective activities on Autonomous Maintenance. The presentation usually takes 30 minutes to 1 hour or more, where the team will highlight their activities. To ensure the team is ready to take on the audit, the Autonomous Maintenance team performs a self-audit to gauge their confidence before the actual audit. Their self-rating should not fall below 90%, as the passing on the actual audit will be at 85%. Once they are confident enough, the team will request the TPM Office for an actual audit schedule. The TPM Office will select an auditor who will be from other departments to avoid any bias. The audit's objective is to see to it that the team takes pride in their success in reducing the number of abnormalities found on their equipment before advancing on to the Next Step of Autonomous Maintenance.

Equipment Audit - After the team's presentation is completed, both the AM auditors, together with the team, proceed to their equipment, and the auditor checks the validity of the documents concerning the actual condition of the equipment. They will also look for other unwanted items on the machine. If other abnormalities or deficiencies are found, they will be included in their pending requirements. It is best for the auditor not to mention the word fail as this can demotivate the team. Suppose the team lacks some basic requirements on a particular step; in that case, they will be given an unconditional pass provided that the team will comply with the requirements noted down by the auditors themselves on the date agreed by the auditor and the team. The team cannot move on to the next step on Autonomous Maintenance until the audit requirements have been carried out. Together with other auditors performing the audit, the TPM Office will ask the teams regarding the benefits of Autonomous Maintenance and how they will sustain these activities they had performed, especially on addressing abnormalities so that the equipment will not revert back to its deteriorating state once again. Standards being set must be performed since this will serve to sustain the equipment; otherwise, if these standards will not be carried out, then the equipment will just go back to the way it was before. Sustaining the equipment should be done at all measures.

After some question and answer portion between the team and the Autonomous Maintenance auditors, the auditor will summarize their findings and inform them. The audit is now concluded. The team will be graded for their performance. Suppose the team attains the standards and requirements as expected from them; in that case, they will deserve a passing mark, which is usually a rating of 85% or more, and they will move on to the next step of Autonomous Maintenance. If other pending items should be addressed by the team found during the audit, the AM team will receive a remark of unconditionally pass unless otherwise all pending requirements are settled on the date both agreed by the auditor and the team themselves. It is highly recommended not to provide a failing mark for the team unless they have not actually done the step requirements themselves, as this can de-motivate the team in moving to the higher steps of Autonomous Maintenance. A simple token of appreciation will be given to the Autonomous Maintenance team that passed each step on Autonomous Maintenance. Remember that audit should be done both on the team and on the equipment itself. Here is a detailed step-by-step procedure on how the audit and certification on Autonomous Maintenance will proceed.

1) The Autonomous Maintenance team should submit a self-audit in which their score should not be lower than 90% to qualify for the Autonomous Maintenance certification and audit.

2) The team should submit a hard copy of their documents together with the Autonomous Maintenance Certification Request one week before the actual certification takes place.

3) The TPM Site Office will assign auditors (minimum of 1) who will audit the team together with the Autonomous Maintenance Committee. A pool of qualified auditors for both Autonomous and Planned Maintenance will be provided to the team requesting the audit. The selection of auditors will be done randomly. It will be the AM facilitator's responsibility to find a suitable replacement if the auditor is not available on the date of the audit provided.

4) The TPM Site Office will release the time, venue, auditors, and machine to be certified on Autonomous Maintenance every week.

5) During the team audit, the auditors and his assistant introduce themselves and brief remarks before the audit starts.

6) Team presentation should be done on the activity board and last from 30 minutes to 1 hour. The whole team should be present during the certification process. The audit will start with the team and then proceed with the actual audit on their machines.

7) There will be no questions asked during the team's presentation. The question and answer portion will be done after the team's presentation and last for 15 to 20 minutes.

8) Both the auditor and team will go to their actual equipment, where the auditors will check their activities based on their actual documentation. The audit on the equipment should last for about 45 minutes to 1 hour.

9) After completing both the team and equipment audit, the auditors will reconvene and finalize their scores. Good points and points to be improved will be placed on the audit sheet. The auditors will finalize the audit sheet and discuss their findings with the Autonomous Maintenance team.

10) After reconvening, the auditors will call the AM team and announce their findings and observations. If the team passes the audit, they will be eligible to move on to the next step of Autonomous Maintenance. If they received a conditional pass, the team needs to submit the pending requirements on the dates agreed by the team and the auditor. Once these pending documents have been submitted, they can only proceed to the next step of Autonomous Maintenance. If the AM team pass, they will receive a token of appreciation from the TPM Office.

11) After the certification process, the AM team should submit to the TPM Office the audit sheet and a photocopy of the certification results. The team is now eligible to proceed to the next step of Autonomous Maintenance.

Horizontal Replication or Fan-Out of Activities

Depending on the number of equipment to be replicated or fan-out, the original Autonomous Maintenance pilot team members will break themselves into several groups and guide the fan-out teams in doing Step 1 activities to similar equipment with new members from the operators joining in. Since the original Autonomous Maintenance team has already gained some experience in performing Step 1 activities, they will be responsible for coaching new members on Step 1 for the fan-out teams. There will be cases where abnormalities found on the pilot equipment will not be evident on the fan-out equipment, and abnormalities found on the fan-out may not be evident on the pilot machine itself.

10.10: October 2016: Common Causes of Bearing Failure

Many say that failures are inevitable and can never be eliminated since there will always be the subject of parts wearing out, and when parts do wear out, then it actually had achieved its life, and therefore it had failed in its technical sense. The role of maintenance is to control the timing of failure or, eventually, to prolong the duration of the failure itself. The role of maintenance is to sustain and preserve the equipment so it will continue to deliver the functions on what the operator wants from them.

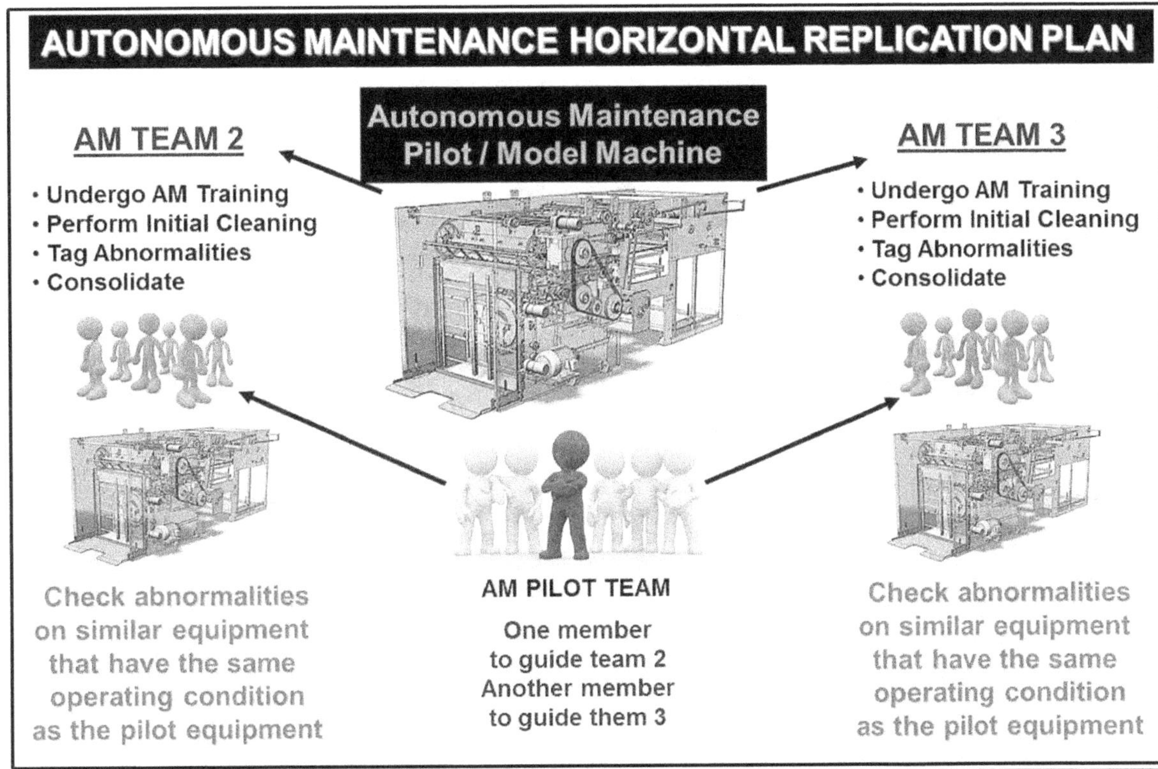

Figure   10.4: Autonomous Maintenance Horizontal Replication Plan

Not all failures are created equal, as every single failure will have its own degree of consequences. What is important is determining which parts will undergo a thorough Root Cause Failure Analysis investigation. The RCFA investigation must depend upon the consequences of the failure itself. Therefore, maintenance aims to control the timing of failure to select or perform a task before it happens. Making the equipment more reliable is about extending the life and the Mean Time Between Failure (MTBF). Our machines are not supposed to fail by chance, but if they do fail, they can tell us exactly why the parts failed. Experts say that the causes for more than 90% of all plant-related failures can be detected with a careful physical examination using a low power magnification of some basic physical testing. Every failure leave some sort of clue or evidence as to why it happens. A group of people can use some failure analysis to diagnose the mechanical cause of the failure itself. Inspecting the failure can tell us the forces involved, whether the load applied is cyclically or single load, the direction of the critical load, and its influence on the outside forces such as residual stress and corrosion. By knowing these physical roots, we can now proceed with the human and latent cause of the failure

Typically, when a bearing fails, what does maintenance do? They simply replace it like any other parts that fail. We mostly change the bearings every time it fails without understanding the real cause of the failure. That is why most failures we experience on the plant just repeats themselves. A bearing manufacturer has a dictated life, but most often, a bearing does not attain its useful or calculated life for a variety of reasons such as heavier loading that had been anticipated, inadequate lubrication, careless handling, ineffective sealing, fits that are too tight or too loose, brinelling and whatnot. A failed bearing will always leave some sort of clue on the balls, cages, and both the inner and outer raceway of the bearing for maintenance to figure out what really caused it to fail.

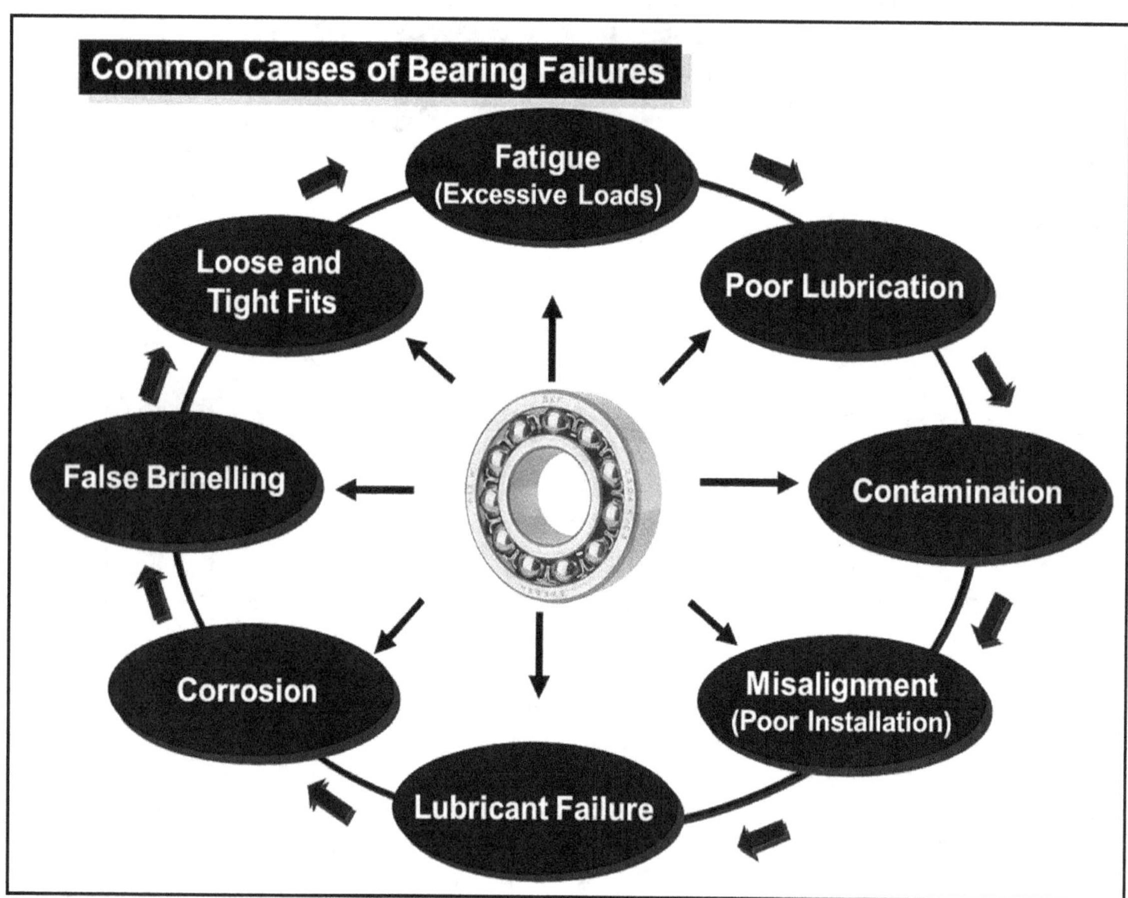

Figure 10.5: Common Causes of Bearing Failures

In most cases, when the bearing fails, the maintenance or technician will replace the failed bearing and throw away the failed bearing, which is always the case. In case the boss asks us for the root cause of the bearing, we go for the easiest excuse of them all and say that the bearing failed because of wear and tear without really understanding the wear-out process. The bearing life is determined by the number of hours it will take for the metal to fatigue, which is a function of the bearing's load, the number of rotations, and the amount of lubricant a bearing receives. According to NASA's research, it was established that ultrasonic monitoring could provide early warning for bearing failure. Different stages have been established to detect a bearing failure.

• An 8-dB gain over baseline indicates pre-failure or lack of lubrication.
• A 12-dB increase establishes the very beginning of the failure mode.
• A 16-dB gain indicates advanced failure conditions.
• A 35-50 dB gain can warn of catastrophic failure.

The mechanical movements can produce a wide spectrum of sound, and by focusing on a narrow band of high frequencies, the Ultrasonic Monitoring instrument can detect changes in amplitude and sound quality. It then heterodynes these undetectable sounds into the audible range, making it possible to be heard from the headphones. There are many reasons why a bearing fails and does not reach its useful life, but despite the cause of every failure, it can be summarized into four major reasons, according to the experts.

UNDERSTANDING BEARING FAILURE

Survey shows that the most common type of bearing failures are:

Poor Lubrication (36%)	Fatigue (34%)
• Using wrong lubricant type • Mixing lubricants • Improper lubrication amounts • Lack of lubrication • Loss or improper additives	• Bearings are overloaded • Due to misalignment unbalance • Premature fatigue cause by lubrication problems
Poor Installation (16%)	**Contamination (14%)**
• Improper installation can lead to failures such as load imbalance, misalignment, tight or loose fits	• Typical failures due to contamination may include excessive wear, abnormal surface stress, corrosion such as water and liquid contamination

Figure 10.6: Understanding Bearing Failures

Lubricant Failure: In many cases, a bearing subjected to abnormal operating conditions exhibits lubrication failure signs. The abnormal operating condition often produces excessive heat, lowering the fluid film thickness, allowing metal-to-metal contact between the raceways and the balls. Poor lubrication can lead to other problems such as overheating, contamination, fatigue, and corrosion. The evidence for attributed lubrication failures will be discoloration (blue/brown) on the raceway, cages, and balls, which are signs of lubricant failure. Excessive wear of balls, rings, and cages will follow, which will result in overheating and premature catastrophic failures. The bearings depend on the continuous presence of a very thin film between the ball and the races to avoid metal-to-metal contact. Once lubrication is inadequate or restricted, it will give rise to excessive temperature, degrade or destroy the lubricant properties, and deplete the additives. An increase in lubricant temperature can be a cause of premature bearing failure. The heat will cause the lubricant to increase viscosity, making it thick, causing more heat as it loses its ability to support the load and form a varnish residue. This will destroy the ability of the lubricant to lubricate the bearing and introduce particles in the lubricant. Many leading bearing manufacturers state that the life of the bearing oil is directly related to heat. A non-contaminated oil has a useful life of about 30 years at a temperature of about 30 °C (86 ° F) and states that the life of the bearing oil is cut in half for every 10 °C increase in the temperature of the oil.

Fatigue Failure: Fatigue in the bearing is a phenomenon leading to fracture under repeated or fluctuating stress having a maximum value less than the material's tensile strength. Fatigue fracture is progressive, which begins as a micro or mini cracks that grow under repeated fluctuating stress. When the stress finally exceeds the material's strength, then fatigue will definitely occur. Fatigue fractures are normally considered the most serious fracture in machinery parts since they can occur in normal service and may occur without excessive loads and under normal operating conditions. As the ball rotates continuously in the raceway, the outer raceway strains usually cause fatigue fractures. The ball will then forces the lubricant into the crack, which causes the metal to stand proud of the surface, which is later sheared off, forming tiny bits of metal fragments that can be detected through oil analysis. The fracture left behind changes the bearing's overall vibration characteristics and can be detected through vibration analysis. As the ball passes continuously into the crater, they make it bigger, and soon the balls themselves get damage since they are no longer rolling on a smooth surface. At some point, the bearing becomes audibly noisy and starts to get hotter until the ball disintegrates, and the bearing finally seizes. Other causes of fatigue failure include excessive loads, usually caused by premature fatigue. This is just like a small fracture on the road where the tires from trucks, cars, vans continuously run through the small crater, which makes the fracture larger, finally ending up in traffic since the vehicle needs to slow down when passing through the crater. Other cause of fatigue failures includes tight fits, brinelling, and improper pre-loading can also bring about an early fatigue failure.

4-Points in Understanding Fatigue Failure
• Without stress fluctuations, fatigue cannot happen.
• Fatigue will start at stress levels well below the tensile strength of the material.
• Where corrosion is present, the fatigue strength of metals continuously decreases.
• The cracks take a measurable time to progress across the fractured face.

Oil Contamination: Contamination is one of the leading causes of premature bearing failure. Symptoms of contamination are dents or scratches embedded in the bearing raceways where the balls or rollers result in excessive bearing vibration and wear. Contaminants may include airborne dust, dirt, moisture, or metal fragments from the bearing that usually gets trap in the clearances between the balls and the raceway. Principal sources of contamination are dirty tools, contaminated work areas, dirty hands, depleted additives, oxidized oil, and other foreign materials that should not be present in the lubricant. Contamination includes the ingression of solids or fluids into the bearing. Even with the most effective bearing seals, solid contamination can enter the bearing's cavity under an extremely dirty and dusty environment. Poor lubrication and contamination can account for many bearing failures and can be prevented by applying for a plant-wide Program on Oil Contamination Control (POCCA).

Improper Installation and Misalignment: A failure caused by misalignment can be detected on a non-rotating ring's raceway by a rolling element wear path that is not parallel to the raceway edges. Excessive misalignment can cause abnormal temperature rise or heavy wear in the cage pockets. It occurs when the centerlines of rotation between two machinery shafts are not in line. There are two kinds of misalignment: parallel and angular, which can occur simultaneously. In parallel misalignment, the centerlines of both shafts are parallel, but they are offset. With angular misalignment, the shafts are at an angle to each other. In the majority of cases, machine misalignment is actually caused by a combination of these two types. Misaligned shafts can cause an increase in the bearing load, reduction in bearing life, increased seal wear, increased vibrations, increased noise, and an increase in energy consumption. Laser alignment should be conducted regularly on the shaft for driver and driven components.

False Brinelling: Fretting damage happens if a bearing is subject to vibration when it is stationary, giving rise to what is known as false brinelling. Standby and rotating machines inside the storeroom should also be rotated slightly to change the ball's points and the bearing's outer raceway, especially if vibration forces are felt inside the storeroom. If there are standby rotating machines, it is good to run the machines alternately to reduce fretting risk. This is done for motors and other rotating assemblies with anti-friction bearings. The shaft of the motors should be rotated at least 1, and ¼ turns regularly. I would prefer a weekly or bi-weekly rotation for standby components. False brinelling can occur on standby units and rotating components inside the storeroom where vibration is present and felt. When the decision is made to stock the spare parts in the storeroom, it can be held in the inventory for many years and becomes a non-moving item. Therefore, it is important to have a program to ensure that these components remain in serviceable condition. Rotating spare parts like pumps, motors, gearboxes, blowers, and other rotary components can be damaged by the environment. Motors in the storeroom inventory over five years should be tested to ensure they are still in working condition.

When the bearing does not reach its useful life, maintenance people must take the time to perform a Root Cause Failure Analysis to determine the cause of the failure. The evidence of the failure will always be visible in the raceway of the bearing. Usually, if lubrication causes the problem, both the inner and outer raceway will be dried and have varnishes or discoloration, indicating that the lubricant had dried up.

When trying to select a bearing, determine the maximum load for the bearing. This is important for both axial and radial loads. We also need to determine the minimum and maximum running speed or RPM that the bearing will be subject to. This will help us to determine the correct viscosity of the lubricant for the bearing. Next, determine all possible environmental conditions to which the bearing will be exposed. Extremely hot or cold environments often require valid bearing specifications, affecting the type of lubricant requirements. It is also important to ask the vendor when the bearing was manufactured and how it was stored before the purchase. Ask the bearing vendor about storage and handling procedures.

10.11: November 2016: Understanding the Difference Between Poka-Yoke and RCFA

The term Poka-yoke is a Japanese term that means mistake-proofing. It is used by industries mostly for operators and maintenance to prevent errors and mistakes. The purpose of Poka-yoke is to completely eliminate human errors on product defects and breakdowns since humans are prone to errors once they stress out or reach a point of fatigue in their work. The basic concept was developed by Shigeo Shingo as part of the Toyota Production System. It was originally known as Baka-yoke, which means fool-proofing or idiot-proofing, and later on, the name was changed to a milder Poka-yoke or mistake-proofing. The word Baka-yoke was too harsh in which Shigeo later renamed it Poka-yoke. The term Poka-yoke was applied by Shigeo Shingo in the 1960s to industrial processes designed to prevent human errors. Shigeo Shingo redesigned a process in which factory workers assembled a small switch. Operators would often forget to insert the required spring under one of the switch buttons. In the redesigned process, the worker would perform the task in two steps, first preparing the two required springs, placing them in a placeholder, and then inserting them into the switch. When a spring remained in the placeholder, the workers knew that they had forgotten to insert it and could correct the mistake effortlessly. Shigeo Shingo distinguished between the concepts of inevitable human mistakes and defects in production. Defects occur when mistakes are allowed to reach the customer. The aim of Poka-Yoke is to design the process so that mistakes can be detected and corrected immediately, thereby eliminating defects at their source, but it does not actually tell us exactly the root cause of the problem. It is more ideal about addressing human errors. Here are some samples of these Poka-Yoke Devices

• Elevator Alarms
• 110 V / 220 plugs and outlets
• Auto Volt Appliances
• Low Battery Indicator
• Low Gasoline Indicator
• Child Lock Door for Cars
• Alarm Clock
• Go-No-Go Checker

Where Poka-Yoke Works Well

• Manual operations where worker vigilance is needed
• Where mispositioning can occur.
• Where an adjustment is required
• Where teams need common-sense tools and not another buzzword.
• Where SPC (Statistical Process Control) is difficult to apply or apparently ineffective.
• Where attributes, not measurements, are important.
• Where training costs and employee turnover are high.
• Where mixed model or device production occurs.
• Where customers make mistakes and blame the service provider
• Where special causes can reoccur.
• Where external failure costs dramatically exceed internal failure costs.

Although Poka-yoke is a solution, there are instances that it can be conflicting since it will not address the root cause of the problem. It provides a solution for human errors. I used to work in an assembly plant called Sharp Philippines Corporation way back in 1988 till 1994, where we assembled appliances for consumers here in the Philippines. I was assigned to the TV and VHS departments; CD, VCD, and LCD were unavailable during those times. The television sets we assembled include a CRT or a picture tube as LCDs were not yet available nor even developed, I supposed. We have one model, a 27-inch Television set that we import from

Taiwan and sell at a much higher price in the Philippines. See, I told you that the goal of any company is to make money. It was cheap, and Sharp sold it for 3x its original cost, but it was a 110-volt television set. We at engineering knew about that so what we did was we modified the carton box since the language was in Mandarin, Chinese. We convert it to English and indicate that this unit was 110 volts. Likewise, we modified the instruction manual since it was also written in Chinese and converted it into English, and we clearly indicated that this was a 110-volt appliance. After these changes were made. It passed our quality control and was now ready for marketing to the different outlets of Sharp Philippines. During the first week of marketing that unit, our Parts and Services Center was full of customers. Around 80% of them or more were carrying the same model, which was this new model of 27 inch TV that we recently released in the market. The engineering manager told us from the TV and VHS staff to do something about the problem, and so the TV staff, including myself, brainstorms for a possible solution. After several hours of brainstorming, we come up with two ideas.

• First was to place a sticker on the plug, which indicates that this was a 110-volt appliance.
• Second was to include a free 110 to a 220-volt transformer on every single unit.

And that was what we did. We went to every appliance chain outlet with this unit, opened the carton box, and placed the sticker on the electrical plug; and second, was to insert a free transformer before re-taping the carton box once more. After completing that on every chain outlet, we noticed that in the following weeks that came, the rate of customer's complaints on this unit dropped dramatically but still, there were a few people who still plugged them to 220 volts even if the sticker and transformer were already provided since that was the standard outlet for Philippine homes. Once again, the engineering manager told us to enroll ourselves in Poka-Yoke training here in the Philippines. After the training, the TV staff and I discussed a Poka-Yoke solution. We instructed all chain outlets to send us back the unit in the plant for a reworked. We modified the power supply from 110 volts and converted it into an auto volt where you can plug it on either 220 or 110 volts. After implementing this solution, there were no more customers in the Parts and Service Department holding this unit. The problem had been solved 100 percent, but on the contrary, we never knew the root cause of why people commit the mistake of plugging a 110 volt into a 220-volt outlet. We have solved the problem, but we never knew the root cause of the problem.

Poka-Yoke is a great problem-solving tool; the only drawback in using this Japanese technique is that you will never find the root cause of the problem because it was not designed to do that. It provides a foolproof solution so that the problem or mistake will go away permanently for good.

10.12: December 2016: The Common Thing RCM and TPM Both Believes

Although RCM and TPM come from different origins. TPM originated from the east, Japan, and RCM from the west, United Airlines, to be exact. These two powerful strategies have common beliefs as well as differences. One thing both these two methodologies agree on is the importance of operator involvement in maintenance.

One of the big differences between the two is how it is being approached. Since TPM is a top-down approach, RCM will be much easier to implement as this can be a down-up approach, which means that we can set up an RCM team and select a piece of pilot equipment. For a start, the team will perform the RCM analysis and implement it. Once it is successful, they showcase the results to their management team. While TPM will be done by everyone, RCM is not for everyone but for an elite few. The team that will compose the RCM team will only be the Rolling Stones. It means that the people who qualify to be part of the team will only be the people in your plant with white hair. For people with no hair, maybe 50/50 since there are young people already bald. Justine Bieber and Lady Gaga cannot be part of the RCM team. We need people with experience since RCM believes that these good old people will go out of the plant for good and never return back one day.

TPM and RCM Strengths :

Point of Comparison	TPM	RCM
1. Origin	Japan	United Airlines
2. Founder	Seiichi Nakajima	Nowlan and Heap
3. Consultant Firm	JIPM	Aladon by Moubray
4. Measure of Performance	OEE	MTBF by Component
5. Maintenance Goal	Zero BDO Zero Failures	Reduce Consequence of failure that's acceptable
6. Approach	Top-Down Approach	Down-Up Approach
7. Initial Approach	Establish BEC	Determine all FM
8. Concept	Continuous Improvement	Maintenance first before redesign
9. What it believes	Maintenance educate operators	Operators educate Maintenance
10. Implementation time	3 months/Step per machine	3 months/case/ machine or sub-assy
11. Can It be combined	YES	YES
12. Maintenance Focus	Address 6 Big Losses	Address Primary & Secondary Functions
13. Aim on Maintenance	Maintenance Prevention	Pro-Active Maintenance

Figure 10.7: Comparing TPM and RCM Strategies

What RCM is trying to do is capture almost everything they know about their equipment before they retire. RCM also recommends that an experienced operator should also be part of the team. Why does RCM require an operator in the team? Simple, for two reasons, first because RCM believes operators are the best source of failure modes since they are in direct contact with the failure themselves. Second, once the RCM analysis is completed, there will be tasks performed not by the maintenance but by operators themselves. Suppose operators are not part of the RCM team; these maintenance tasks will not be performed religiously by the operators themselves on their equipment. The good thing about the RCM meeting is that the members discussed all the failure modes they have experienced on the equipment during the meeting, which will be a good learning curve for operators. The operators will learn a lot from experienced maintenance people. But more importantly, maintenance also learns from the operators because these people sense something before the failure occurs, which will be covered in the RCM analysis. Maintenance may ask operators if they sense something before this type of failure mode occurred on the equipment. The good thing about the meeting is that the operators learn from maintenance, and maintenance likewise learns from the operators themselves.

Another main difference between RCM and TPM is that TPM is a continuous improvement program, while RCM will recommend improving the way we maintain our assets before we can even start to improve or modify the equipment. By the word itself, Reliability-Centered Maintenance means that if we center and focus on maintaining the equipment, then there is a possibility that the equipment's reliability can be improved. I have seen improvements and modifications go to waste because the improvement team forgot to recognize one thing. What was it? They forgot to ask their managers if the asset will still stay for a long time or decommissioned soon. Suppose the asset will be decommissioned in a few months; then, it is not feasible to improve or modify that asset unless we expect some miracles to happen. I think RCM had a valid point in this case. If you can see how the algorithm or RCM decision diagram was structured, modification and redesign would be at the bottom part of the decision diagram, the last resort if maintenance can no longer address the failure.

When I was still working with Sharp Philippines Corporation in 1988, I got two daughters since my son was not born yet. My two daughters shared the same room. The two slept on the floor with a mat. We called that "banig" in my native language in Tagalog. They have no bed. Every night I peek in their room and see them sleeping, and it just bleeds my heart that they do not have any bed on their own since my salary is barely enough to support my family. So what I did was I saved a little by skipping lunch and sometimes saved money I got from working overtime. Once I got enough money, I went to the mall and bought a double deck.

A double-deck is a two-bed where one can sleep on top and the other on the bottom, just if you are unfamiliar with it. I assembled it in their room while they were at school. Finally, my two children arrived, and when they entered the room, there was a joy as they were screaming with excitement. I went inside their room and told my elder daughter to sleep here at the bottom since she was heavier, while I told my younger daughter to sleep on top because she was lighter. That was my instructions to them. Around 10:00 pm, I peeked in their room that night, and both of them were sleeping on the bottom together. I did not wake them up and waited until the following morning. The morning came, and we were having breakfast; I asked my youngest daughter why she slept together in the same bed and not on top? My younger daughter answered me, father, the fan is not reaching me since they only got one electric fan. In my thoughts, I told myself, I performed good maintenance on my fan. I cleaned it every week. I oil the shaft every month.

I make sure it was working so that the user, my two daughters, will be comfortable when they sleep. But thinking about it, whatever maintenance I did will never satisfy the user because the fan's rotation was from left to right. To satisfy my younger daughter, the rotation should be from top to bottom. If you asked me if I modified the fan? Of course not. I got no time for that, and I do not know-how. What I did was I still got some few money with me, and immediately I went back once more to the mall and purchase a wall fan and mounted it on the wall. That day forward, I saw my younger daughter sleeping on top of the double deck. You see, this very simple story tells us one lesson. There are things maintenance can do and things that maintenance cannot do. If maintenance can no longer perform its function, then maintenance will walk through that thin line and go beyond maintenance, and that is to modify. Remember, maintenance people are not Superman, not even a Spiderman.

If you plan to implement TPM in your plant, I would strongly recommend performing Phase 1 first together with Autonomous Maintenance initial cleaning simultaneously on the equipment. Take care of the basics first before implementing RCM. What good will RCM do if your equipment is dirty, lacks bolts, and contains many leaks? Address the basics first to make RCM more effective when implemented. While the goal of TPM will be to establish the basic equipment condition (BEC), RCM, on the other hand, will be to lists every single failure mode (FM) that can occur on the equipment. There would be fewer failure modes if the basic equipment condition were taken care of.

Another difference between TPM and RCM is their goal of maintenance. While RCM will aim to reduce or eliminate the consequences of breakdowns, TPM will aim to zero out breakdowns. Now let me just make this statement clear to avoid any confusion on my previous books. TPM can never zero out the breakdown. What TPM is doing is to be one or two steps ahead of the breakdown, and TPM is aiming to zero not all breakdowns but what you called Unplanned Breakdowns. TPM believes that there are Planned Breakdowns. Although, as maintenance, I would prefer to agree with RCM that the consequences of failure are more important than the failure itself, especially if the damage, impact, and aftermath of the failure will be devastating or will induce safety or environmental consequences.

But despite the differences between RCM and TPM, they agree on something, and that is the importance of having a maintenance and operator partnership. These two should be inseparable, just like the Rolling Stones. TPM will be the bigger strategy, and RCM will fit perfectly in the higher phases of Planned Maintenance. These two methodologies are not contradictory, but rather they are complementary. It is only the independent RCM and TPM consultants that do the contradiction and not the methodologies. TPM will focus on eliminating the 6 major equipment losses while RCM will improve both the primary and secondary functions of the equipment.

2017 RSA Reliability Newsletter Vault Archive

Achieving world-class maintenance takes time. It is difficult but not impossible. The road is rough and bumpy. What industries need is a plan, a team, and the most important is the team should have a big heart in finishing the journey as it will take one step at a time. Remember, many industries will attempt, but only the pure and thoroughbred will finish.

11.1: January 2017: Wisdom on Maintenance - Part 2

Let me share this series of wisdom on maintenance with you, which are the guiding principles of all my training and courses on reliability and maintenance. These reflections are what I believe to be what maintenance and reliability people in industries should be focusing on and thinking. It is never too late to improve the reliability of your plant. Improving reliability is a consolidated effort not only for maintenance but for operations people as well. It is important that people from both operations and maintenance be educated and must have the correct maintenance paradigm.

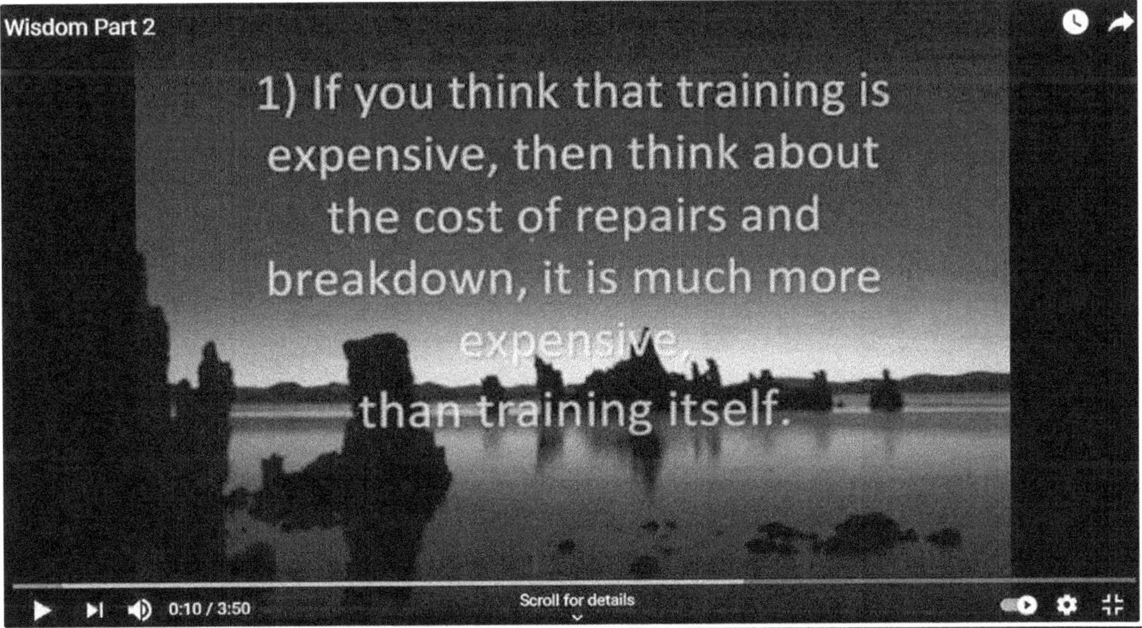

Figure 11.1: Wisdom on Maintenance Part 2

. You are free to watch and download this video on my YouTube channel and share it with your maintenance and reliability people and your management team. Here is the link: **https://www.youtube.com/watch?v=HCb4my2UyPI&t=3s**

1st Wisdom on Maintenance

If you think that training is expensive, then think about the costs of your failures and breakdowns. It is much more expensive than the training itself.

2nd Wisdom on Maintenance

Some say that changing culture should start from the top; others say it is a bottom-up approach. There is no easy and conclusive answer to this, but all I can say is that change is something that should start from within, and I think nobody says it better than the late "King of Pop," that change should start from the "Man in the Mirror." by Michael Jackson

3rd Wisdom on Maintenance

Always remember that in any reliability improvement initiative, the focus should always be given to the people. Provide them with the knowledge to do their jobs correctly. Remember, it will be the people who will improve their equipment and not the other way around.

4th Wisdom on Maintenance

In a reactive environment, we always complain about the lack of manpower resources to address equipment failures. Remember, maintenance is not measured by how fast they repair but how they could analyze the failure themselves.

5th Wisdom on Maintenance

TPM is 80% people and 20% equipment. Focus on the people, and the people will focus on their equipment.

6th Wisdom on Maintenance

The level of achievement that we have at anything reflects how well we could focus on it because the only thing that is holding you back is the way you're thinking. By Steve Vai

7th Wisdom on Maintenance

The message of Root Cause Failure Analysis is simple and straightforward: we can only learn from the things that go wrong if we are brave enough and look ourselves in the mirror and admit that each and every one of us is also part of the problem.

8th Wisdom on Maintenance

All failures are not created equal. Every failure has its own degree of consequences. Being Proactive has something to do about reducing or eliminating the consequences of failure rather than eliminating the failure themselves.

9th Wisdom on Maintenance

The most important reason we maintain our equipment is that there are times when the consequences of failure are far greater and more important than the failure themselves.

10th Wisdom on Maintenance

People are not the company's greatest asset. The right people are the company's greatest asset, and the wrong people are simply called liabilities, and we can only have the right people if they are equipped with the right knowledge to do their jobs right the first time around.

11th Wisdom on Maintenance

In any reliability and improvement strategy, we do not need management support. What we need is management commitment and ownership. Remember, management can support but never commit to the initiative.

12th Wisdom on Maintenance

Knowledge and education are what maintenance needs to fulfill its duties and function. As a maintenance myself, I humbly ask, let us not deprive maintenance of education because that is all they ask for.

11.2: February 2017: Most Common Operator Indices

As mentioned previously, most operators' indices will be intangible, but they are still important in boosting the morale of operators, especially if they are performing Autonomous Maintenance in seven steps. These indicators will be tracked at the very beginning of their activities on their pilot equipment.

Number of Suggestions: Autonomous Maintenance will be done in small group activities, perhaps with a membership between 4 to 8 members. As they complete their initial requirements for Step 0, the team proceeds to Step 1 of Autonomous Maintenance, the Initial Cleaning process. Operators will be exposing problems and abnormalities on their equipment. Their suggestions will mainly be focused on addressing these abnormalities, addressing contamination sources, improving their workplace, minimizing their cleaning, lubrication, and inspection time, making it hard to clean and hard to reach areas in their equipment easily accessible. But the real challenge here is not only about making a suggestion but rather if the suggestion will be implemented successfully or not. This will be the start and will be one of the main activities that will boost the team in pursuing the higher steps of their Autonomous Maintenance activities. I have seen operators that reached step 6 and even step 7 of Autonomous Maintenance with my own eyes. Operators that reached these steps have changed dramatically in how they deal with their work. It is no longer a routine for them, and believe me when I tell you that they can decide independently. Operators know their equipment better than anyone else in the organization, even the maintenance, because they are the people day in and day out that come face to face with their equipment all the time. They know the problems, and they can provide improvements together with the maintenance function. The problem with most industries is that they have not been allowed to voice out their suggestions on how to improve their equipment because they are only limited to operating the equipment from the start till the end of their shift.

Number of Question Lists Generated and Answered: As the teams start their Autonomous Maintenance activities, they raised questions about their equipment. Usually, the team will answer these questions raised by their fellow operators. If no one can answer a question raised by a team member, they will refer to their coach, the Planned Maintenance people. This is part of the learning curve for operators on doing Autonomous Maintenance. The goal is for the operators to know their equipment intimately. If they think that this question is important, either the operators or maintenance will generate a One-Point-Lesson to provide a clear answer to their questions. As they complete each of the 7 steps of Autonomous Maintenance, every single question generated by operators and those answered will be recorded and placed on their activity board. No matter how hard or easy the question is, it will be listed and answered by the team leaders or solicited from the maintenance function. The purpose of generating questions by the operator is to generate awareness. For example, asking a name of the part will not be part of the question lists, but if I have a pressure gauge and the pressure goes beyond its limit, an operator may ask maintenance what will be the effect of going beyond this limit will now be included in the question lists generated.

Number of Abnormalities Detected and Corrected: As Autonomous Maintenance Step 1 activities are initiated, operators will tag their equipment for any abnormalities, slight or major deviation that they can see during the initial cleaning process. Maintenance will also have its own tags when they performed the Planned Maintenance Phase 1 restoration stage. Phase 1 and Step 1 of both Planned and Autonomous Maintenance aims to bring the equipment back to its original basic equipment conditions. The distinction between Autonomous Maintenance and Planned Maintenance tags is that maintenance will be focused more on the interior part of the equipment, including deteriorated parts that need to be restored, while the Autonomous Maintenance team will focus more on the exterior part of the equipment. As much as possible, it is best for the operators who locate the abnormality to correct it themselves if they can. Although big abnormalities in which the operators cannot correct themselves will be passed on to the Planned Maintenance team. It is highly recommended that when Step 1 Initial Cleaning is performed on the equipment, both operators and maintenance should perform their initial cleaning and restoration activities on the same day so that operators tagging abnormalities can be assisted by the Planned Maintenance teams. Those abnormalities found by operators that cannot be corrected can be passed on to the Planned Maintenance team. One thing important to note is never ever passed all abnormalities to maintenance. There will be abnormalities that can easily be corrected by operators. The Autonomous Maintenance team should provide a before and after picture of the abnormalities they have corrected. Hence, when doing step 1, let us say that 180 abnormalities were detected, and 90 of them have been corrected so far, which means that their accomplishment rate will be at 50%. This percentage is what the members of Autonomous Maintenance will be tracking. It is also highly recommended for Autonomous Maintenance and Planned Maintenance to have the same pilot equipment at the start. Planned Maintenance can easily focus on supporting the operators' needs instead of having different pilot equipment for operators and maintenance. One more important point is that Planned Maintenance activities should be ahead before Autonomous Maintenance is implemented. This means that Phase 0 or the Preparatory Stage of Planned Maintenance should already be completed since Planned Maintenance will be performing a machine ranking on all equipment in the plant. The operator's pilot equipment should belong to the Rank A (worst) category with the possibility of large replication or fan-out.

Number of One-Point-Lessons Generated: A One-Point lesson, or a single-point lesson, provides us specific information on a single topic. Usually, this will contain just one topic. This will include a picture, chart, graph, or diagram and a one-point explanation. Usually, it is best to place these One-Point-Lessons on Flip Charts. Maintenance or the operators themselves will discuss these One-Point-Lessons with the Autonomous Maintenance team, where they spend around 10 to 15 minutes for each One-Point-Lesson generated. Once they understand the message, then all team members will sign on to the One-Point-Lesson. Suggestions, questions generated, improvement, and Kaizen activities can be included in the One-Point-Lesson. The important thing to consider is to discuss just one point at a time. A One-Point-Lesson is a learning tool for communicating standards, problems, and improvements in the work processes and equipment. Workers and supervisors use One-Point-Lessons to provide key information about their everyday work and improvement opportunities. Thus, a one-point lesson may contain information on a wide range of topics. In other words, whenever a worker needs key information to perform their jobs, One-Point-Lessons can be an effective tool for delivering this information. OPL is a good tool for sharing knowledge among the team. It translates knowledge into practical information that the Autonomous Maintenance team can use to correctly perform their jobs. This will be part of the operator's continuous learning process.

WHAT AUTONOMOUS MAINTENANCE TEAMS SHOULD TRACK

WHAT TO TRACK	STEP 1 :	STEP 2 :	TOTAL	PERCENTAGE
1. Abnormalities Tagged	145	+ 25	170	Percent = 72.8 %
2. Team Suggestions	25	+ 20	45	Percent = 67.0 %
3. One Point Lesson	20	+ 25	45	Percent = 100.0 %
4. Questions Generated	20	+ 25	45	Percent = 100.0 %
5. Difficult to Clean Areas	5	+ 6	11	Percent = 100.0 %
6. Contamination Sources	4	+ 0	4	Percent = 75.0 %
7. Kaizen Activities	0	+ 25	25	Percent = 75.0 %

Figure 11.2: What Autonomous Maintenance Should Track?

Number of Kaizen Activities: The team will also be tracking the number of improvements or Kaizen activities they have generated to motivate the team and inspire other operators to participate in the Autonomous Maintenance implementation in the plant. These kaizen activities will focus on how they addressed contamination sources, hard-to-reach areas, hard-to-clean areas, how to reduce their cleaning, lubrication, inspection time, and how to improve the workflow process. The team will generate kaizen activities to address quality problems and defects on their equipment on the higher steps. These indices for operators will be placed on their activity boards and updated by the operators themselves.

11.3: March 2017: Tips in Setting-up Maintenance and Operator's Indicators

Decide what indicators should be measured by both operators and maintenance. There is no single KPI or indicator that will tell you everything. Have a minimum of at least 5 KPIs, especially for maintenance. Since industries differ from their products and services they produce and rendered, their KPIs will not be the same all the time. Maintenance needs to sit down and discuss the key indices that are important to them that they need to measure. This should be done at the very beginning of any reliability and maintenance strategy.

Maintenance Indicators and Measurements

No.	Measurement Indicator	Code	Formula	Trend
1	Machine Downtime	MDT	n.a.	The lower the better ⇩
2	Repair and Maintenance Costs	RNM	n.a.	The lower the better ⇩
3	Breakdown Occurences	BDO	Frequency of Breakdowns	The lower the better ⇩
4	Mean Time Between Failure	MTBF	$= \dfrac{\text{Available Time - Machine DT}}{\text{Breakdown Occurences}}$ or $= \dfrac{\text{Loading Time - Machine DT}}{\text{Breakdown Occurences}}$	The higher the better ⇧
5	Mean Time To Repair	MTTR	$= \dfrac{\text{Machine Downtime}}{\text{BDO}}$	The lower the better ⇩
6	Overall Equipment Effectiveness	OEE	$= \text{Availability x Performance Rate x Quality Rate}$	The higher the better ⇧
7	Availability	AVAIL	$= \dfrac{\text{Available Time - Machine DT}}{\text{Available Time}} \times 100\%$	The higher the better ⇧
8	Mean Time Between Assists	MTBA	$= \dfrac{\text{Operating Time}}{\text{Frequency of Assists}} \times 100\%$	The higher the better ⇧
9	Ratio of PM Compliance vs Corrective Maintenance	PMCM	n.a.	PM Compliance ⇧ Corrective Maintenance ⇩
10	Ratio of Maintenance to Operating Cost	MC:OC	$= \dfrac{\text{Total Maintenance Cost}}{\text{Total Production Cost}} \times 100\%$	The lower the better ⇩
11	Failure Frequency	FF	$= \dfrac{\text{Total Breakdown Occ.}}{\text{Loading Time}} \times 100\%$	The lower the better ⇩
12	PM Achievement Rate	PMAR	$= \dfrac{\text{PM Tasks Completed}}{\text{PM Tasks Planned}} \times 100\%$	The higher the better ⇧

Figure 11.3: Maintenance Most Common Indicator and Indices

Whether these indices will be generated by software or tracked down manually, it is important to determine the frequency or period of generating these indices. This means if these indices will be tracked on a daily, weekly, or monthly basis. What is important is the trend of each of these indices. For maintenance, determine which will be your leading and lagging indicators and measure them consistently. Management should not only look at the results but at how the results were achieved. What is important is that maintenance indices and measurements should be aligned to their corporate goals and targets.

Never manipulate these key performance indices to look good to management, as we will never know the problem itself. What is important is that if the goals were not met, maintenance should evaluate the situation, determine why these goals were not met, and do something about it. These indices should be deployed to all maintenance levels, from the highest down to the hands-on people on maintenance. Everyone in the organization should be aware of the company's goals and targets. These KPI should be deployed down to the supervisory level and all involved in maintenance down to the shop floor level. Goals and targets should be visible to everyone.

It is recommended that teams that contributed to the goals should be recognized. It is highly recommended to recognize the team and not the individual. Recognition may not necessarily be in monetary value, but recognizing the people's efforts will motivate them to improve the plant. It is a good feeling to work in an industry where Top Management and Decision-makers value their people.

Although the operator's indices are mostly intangible and difficult to assess, they contributed to the lagging indicator. Allow these operators to showcase their improvements to their top management. This is to let them show that they are part of the improvement process of the plant. What is important in any indicator maintenance is measuring the trend to indicate if they perform well based on their efforts and expectations.

Identify the losses on your equipment to determine the correct Mean Time Indicators for them. Suppose your equipment is suffering from many breakdowns and failures; in that case, a good indicator will measure the MTBF of the equipment. For that equipment with a high frequency of assists and errors, a good indicator to measure will be MTBA or Mean Time Between Assists.

If you plan to track the number of breakdowns in your equipment. Maintenance and other departments need to be consistent on when to declare equipment failed or not. A safety officer may declare equipment failed if it leaks oil severely for safety purposes to avoid accidents even if the equipment is still running. If this will not be done, then each and everyone will have their own definition of when to declare an equipment breakdown, affecting other indices such as MTBF. What is important is to be consistent on when to say that a breakdown actually occurred or not. Example, Case 1: A motor attached to a pump burnt, the standby motor and pump activated automatically without any manual intervention. Question do we have a breakdown? If we speak about the component level, then we have a breakdown even if there was no downtime; however, if we speak on a system level, the system was not affected, then we have no breakdown. Case 2: A compressor overheated and failed in which that it did not provide compressed air to five pneumatic equipment. If we speak about the five pneumatic equipment that stopped, do we have a breakdown or not? If we speak about the compressor in the facilities, then we have a breakdown, but if we speak about the five pneumatic equipment that stopped, there was no breakdown, but still, we have a downtime caused by the compressor in the facilities or utilities.

11.4: April 2017: Wisdom on Maintenance - Part 3

Let me share this series of wisdom on maintenance with you, which are the guiding principles of all my training and courses on reliability and maintenance. These reflections are what I believe to be what maintenance and reliability people in industries should be focusing on and thinking. It is never too late to improve the reliability of your plant. Improving reliability is a consolidated effort not only for maintenance but for operations people as well. It is important that people from both operations and maintenance be educated and must have the correct maintenance paradigm.

1st Wisdom on Maintenance

Maintenance is not synonymous with repair and troubleshooting equipment failures, but rather it has something to do about preserving our equipment and assets and to do that, we need to adapt the different disciplines on maintenance.

2nd Wisdom on Maintenance

For Decision Makers who do not understand maintenance and reliability, expect them to cut costs on people either by terminating them or cutting their benefits, which are just pebbles or a small fraction of forced savings, but what they need to understand is that the biggest savings will always come from improving the way we do maintenance on our equipment and assets. These are the big rocks for savings.

3rd Wisdom on Maintenance

Industries must learn to understand that they will continue to remain reactive for as long as both operator and maintenance functions remain a separate function because reliability is a shared responsibility for both operators and maintenance.

4th Wisdom on Maintenance

Operators will always be the first line of defense on any equipment-related failures and breakdown since these people will encounter the failure first before maintenance; hence operators in industries need to understand the symptoms of these failures before they occur.

5th Wisdom on Maintenance

The true job of maintenance is to be ahead of failure. Maintenance can prevent, predict, prolong, control, or manage the failure before it occurs. Remember that it is less expensive to correct the failure rather than react to the failure when it occurs.

6th Wisdom on Maintenance

In Japanese industries, they do not understand the word best. They only understand the word better because they know that the best can still be made better; that is why continuous improvement was born since they believe that the older the equipment gets, the better it runs.

7th Wisdom on Maintenance

Let us not blame maintenance for all the failures and breakdowns happening in our industry. Remember that 83% of equipment failures are non-maintenance induced. It is important for management to understand and provide maintenance what they need to perform their true function.

8th Wisdom on Maintenance

Many industries claim that Safety is first, while others prioritize Quality. Industries should understand that Reliability is as important as Safety and Quality. It is important for these people not to work in isolation but rather in harmony as a team. This way, we have a better chance of managing human errors and avoid industrial accidents and disasters.

9th Wisdom on Maintenance

The only way to improve reliability is to accept failures. There is nothing to be ashamed of for as long as people accept their mistakes and learn from them. Remember that errors and mistake is a sign of being human, and that is who we are.

10th Wisdom on Maintenance

A maintenance leader's role is to combine all the talents of your maintenance people and move in one direction as a team and not in isolation. If this can be done, then expect maintenance to move mountains.

11th Wisdom on Maintenance

Achieving world-class maintenance takes time. It is difficult but not an impossible task as the road is tough and rough. We need a plan, a team, and the team should have a big heart in completing the journey. Making this journey on World Class maintenance will take one step at a time and will start with the basics. Remember that many industries can start, but only the thoroughbred will finish.

12th Wisdom on Maintenance

In maintenance, we aim high, but nevertheless, we keep our heads down and humble ourselves. Remember, humility is one of the traits of true-blooded maintenance, and that is who we really are.

11.5: May 2017: Wisdom on Maintenance - Part 4

Let me share this series of wisdom on maintenance with you, which are the guiding principles of all my training and courses on reliability and maintenance. These reflections are what I believe to be what maintenance and reliability people in industries should be focusing on and thinking. It is never too late to improve the reliability of our equipment and assets. Improving reliability is a consolidated effort not only for maintenance but for operations people as well. It is important that people from both operations and maintenance be educated and must have the correct maintenance paradigm.

1st Wisdom on Maintenance

Improving reliability cannot be achieved by cutting costs on maintenance but rather reducing maintenance costs can be achieved by improving reliability

2nd Wisdom on Maintenance

Industries that value their people have a greater chance of changing their culture than industries that believe their people are expendable. Changing the culture, in this case, will either be the most difficult or impossible to achieve.

3rd Wisdom on Maintenance

All physical failures are triggered by humans, but humans are negatively influenced by latent forces. Therefore, in any Root Cause Failure Analysis investigation, the goal must be to expose and identify these latent causes.

4th Wisdom on Maintenance

We can only learn from failures if we take the time to analyze them. It is an opportunity for us to begin more intelligently.   Therefore, whether from the lessons of life or equipment-related problems that industry experiences, failures tell us something. It tells us to take things slowly so that we can understand their causes, but in most cases, we just fix the symptoms; that is why failure keeps on repeating itself again and again.

5th Wisdom on Maintenance

There is no question about having highly skilled people in our maintenance organization. Technically speaking, the goal of MTTR is to reduce repair time, but the real goal of maintenance is to analyze failures by addressing the root cause of the problem, therefore when the failure keeps on repeating, forget MTTR and start performing a Root Cause Failure Analysis on the problem.

6th Wisdom on Maintenance

Most failures start from small things left unattended and neglected on our equipment and assets; Autonomous and Planned Maintenance emphasize establishing these basics on their equipment, including cleaning, lubrication, and having complete bolts needed for the equipment. Once these simple basic conditions had been addressed, their impact will be felt immediately.

7th Wisdom on Maintenance

When TPM says that failures and breakdowns can be zeroed out or eliminated, it simply means that the Mean Time Between Failure was lengthened and that breakdowns will still happen, but this time on a longer period or interval. Technically speaking, the breakdown was not eliminated; the MTBF was only prolonged. Zeroing out breakdowns is simply not possible in its real sense.

8th Wisdom on Maintenance

Maintenance must understand that Infant Mortality failures are introduced during overhauls when the equipment is not properly put back together. Overhauls and replacements should be done by skilled craftspeople with the right tools and skills. If you have the slightest doubt about putting the equipment back again in one piece, think twice or even thrice before dismantling it.

9th Wisdom on Maintenance

Maintenance is not about shifting from Preventive to Predictive Maintenance nor transitioning from a Reactive to a Proactive Maintenance stage but simply understanding when to use these different strategies simultaneously depending on the consequences of failure. This is done with the aid of a maintenance algorithm or decision diagram.

10th Wisdom on Maintenance

Improving our equipment's reliability does not mean retiring or firing maintenance people in their organization but understanding that new doors will open for maintenance when equipment reliability improves.

11th Wisdom on Maintenance

Reliability-Centered Maintenance is not a replacement for Preventive Maintenance. PM will always have its place in the overall maintenance structure. The message of RCM is simple, PM cannot silence all failures. It needs the help of other maintenance tasks to prevent, predict, prolong, control, or simply allow the failure to occur if the consequences of failure would have minimal consequences.

12th Wisdom on Maintenance

In implementing TPM at the very beginning, expect a lot of resistance from members. This is normal; what is not normal is if they do not show any resistance at all. TPM is a long and painful journey with its ups and downs, but once you reaped the benefits, then your efforts will be well rewarded. Remember, you are not only improving the equipment but changing the lives of people too. If people change, then the rest will follow.

11.6: June 2017: Wisdom on Maintenance - Part 5

Let me share this series of wisdom on maintenance with you, which are the guiding principles of all my training and courses on reliability and maintenance. These reflections are what I believe to be what maintenance and reliability people in industries should be focusing on and thinking. It is never too late to improve the reliability of your plant. Improving reliability is a consolidated effort not only for maintenance but for operations people as well. It is important that people from both operations and maintenance be educated and must have the correct maintenance paradigm.

1st Wisdom on Maintenance

The maintenance department's real mission is to provide reliable physical assets and excellent customer support by reducing unnecessary, forced, and intrusive maintenance. Do not confuse maintenance as synonymous with repair; these two are entirely different.

2nd Wisdom on Maintenance

In a Proactive Environment, MRO Spare Parts Management can be defined as simply acquiring the right part, at the right place, during the right time. While in a Reactive Environment, Spare Parts are those parts that are available in plenty when not needed and those not available when needed.

3rd Wisdom on Maintenance

Maintenance is not synonymous with repairing, fixing, and troubleshooting failures all the time. Remember that maintenance is a noble profession, and the best positions an industry has will always belong to the maintenance function, be proud that we belong to the maintenance function.

4th Wisdom on Maintenance

Predictive Maintenance is never a replacement for Preventive Maintenance. PM will always have its place in the overall improvement strategy on Maintenance Management. What we are saying is that there are instances that checking the condition of the equipment is much better than overhauling the equipment itself.

5th Wisdom on Maintenance

OEE or Overall Equipment Effectiveness is not a perfect KPI; in fact, there is no such thing as a perfect KPI. There should always be a series of indices that maintenance needs to monitor on their assets. Remember that OEE only refers to the asset's primary function. There are cases that sometimes a failure in secondary functions poses more threat than the failure of the equipment's primary function. In short, both the primary and secondary functions of the equipment should be functioning at all times.

6th Wisdom on Maintenance

Although Autonomous Maintenance is the pillar in TPM with the most population involved, Planned Maintenance should be the strongest pillar in any TPM implementation. Why? Simple because if the Planned Maintenance structure is weak, then Autonomous Maintenance will collapse. Planned Maintenance will be the pillar that will support and guide the Autonomous Maintenance teams.

7th Wisdom on Maintenance

As the late Peter Drucker quote, you cannot manage something you cannot control, and you cannot control something you cannot manage. MRO storeroom and spare parts management are some of the major strategies in improving the reliability of any industry, and the best people in the position to control and manage the storeroom are the users of the parts themselves, which are the maintenance people.

8th Wisdom on Maintenance

CBM is not an isolated group and can never be a replacement for Preventive Maintenance. Both will have their roles in the maintenance function. Lines of communication between the two should always remain open.

9th Wisdom on Maintenance

The message of RCM is that a holistic maintenance strategy should be to develop a maintenance strategy that will address infant mortality, random and age-related failures. Remember that Preventive Maintenance cannot address random failures since PM is only designed to address age-related failures. The more Preventive Maintenance tasks you perform, then expect more problems with the equipment.

10th Wisdom on Maintenance

Maintenance is not about transitioning from reactive to preventive and from preventive to predictive, but rather maintenance is about understanding when to use the different maintenance tasks. This will be possible with the aid of an algorithm or decision diagram.

11th Wisdom on Maintenance

In a reactive environment, we always complain about the lack of manpower resources to address equipment failures, but when reliability starts to improve, our people become more visible. We always wonder where they had been. In reality, maintenance is not really undermanned; operators do not take part in Reliability's shared responsibility.

12th Wisdom on Maintenance

Industries should strengthen their Condition-Based or Predictive Maintenance activities and lessen their Preventive Maintenance for the simple reason that Predictive Maintenance will capture more failures than Preventive Maintenance with precision accuracy.

11.7: July 2017: Recommended MRO Storeroom Measurement and KPIs

You cannot manage something you cannot control, and you cannot control something you cannot manage. *From Peter Drucker, 1909 - 2005*

Maintenance must know where they are and how they are performing. Although no single one KPI will tell us everything, maintenance must have a minimum of around 5 and a maximum of 8 KPIs to measure regularly and what is most important is that this should be known not just by the top but down to the shop floor. Maintenance KPI should be aligned to their corporate KPI. What is important is the trend of their KPI to understand if they perform well or below expectations. Success or failure in managing your spare parts depends entirely on the things you measure in your MRO Storeroom. My stand remains firm that MRO Spare parts and Storeroom should be managed by the maintenance department since they are the users of these spares. We can only measure what we can control. The success or failure of the things we control can only be known by these KPIs and measurements.

MRO Storeroom Inventory Accuracy: One of the critical success factors for Spare Parts Management is achieving a high level of inventory accuracy. Suppose the actual inventory is lower than the system recorded; in that case, the risk is high that physically an out-of-stock condition occurs, while if the actual inventory is higher than the system recorded. Parts are overstocked even when it's not needed. When a certain stock or part indicates a certain amount of quantity in the computer, this should reflect the same physical amount of quantity in the storeroom. The Storeroom Manager can invite people from other departments monthly as part of their routine activities to conduct a random audit at any part of the storeroom and conduct a physical count to check them out in the system. Small items in large quantities can be weigh-in instead of determining the quantity. Having an activity like this can assure high inventory accuracy for both physically and the system. Having high inventory accuracy can minimize the chances of a stock out. Inventory accuracy should be 98% and above.

Number of Stock-Out: The storeroom people should keep the stockout log to determine the number of times an item is requested but is not currently held in stock or has run out of stock. A good target value for stockout items would be around 0.5% each month, but the ultimate goal is to have a zero stock-out. A stock-out condition occurs when the inventory reaches zero before the replenishment stock arrives. A safety stock level and its lead time should always be known to avoid stock out. Frequent stock-out can make maintenance lose their confidence in the storeroom and start to keep the part themselves. Although it is not a recommended practice to keep every single part in the storeroom, neither reducing inventory by creating large stock-outs. Storekeepers should also be aware of the consequences of a prolonged downtime because the part is not available when needed. There are three causes of stock-outs. First, the holding parameters may have been incorrectly set for the expected or incorrect demand for the current level. Second, there is some delay in the demand or supply chain, including the internal processes associated with purchasing, including reordering and lead time. Third, the part is not stocked. Stock-outs can mean losing confidence in the storeroom people, which makes maintenance keep the parts themselves in their squirrel stores.

Measuring Obsolete Parts: When equipment is decommissioned or retired, the spare parts that go with them also become obsolete because the equipment has no other similar equipment in the plant. When a spare part is modified since it has an inherent design weakness, the original parts kept in the storeroom become obsolete. Engineers doing redesign should communicate with storeroom people to discontinue ordering these parts. When deciding to retire or decommission equipment, maintenance should communicate beforehand with storeroom people on an asset that will be decommissioned to identify the parts that go with the equipment. Storeroom people should be patient in making continuous attention to obsolete parts so that maintenance can make their final decisions on disposing of these parts, which are no longer used. Disposing parts that are no longer needed not only saves on carrying cost but also space is occupied in the storeroom.

Carrying or Holding Costs: This is the accumulated cost on parts held in stock. These costs begin to accumulate the day the item is put in the storeroom. Carrying costs will include borrowing interest rates, taxes, insurance, obsolescence, shrinkage, rent, depreciation, electricity, and so on. Most maintenance thinks that there is no cost of holding the part in the storeroom and does not want to get rid of the part by hanging on to it on the assumption that it might be needed someday; besides, it does not cost anything to keep it. This is where they got it all wrong because all parts in the storeroom have a cost on holding them, which is called the holding cost. This is the cost of holding the item in the storeroom, which is usually around 20 to 30% of the item's value.

Costs of Non-Moving Items: Most of these items are classified as insurance spares in which a spare part that you hold in your store that you would not expect to use in the normal life of a plant and equipment, but if they are not available when needed, would result in significant losses. This differentiates the item from capital spares because capital spares, by definition, are not inventory because they are capitalized as plant assets and equipment. One key aspect of insurance spares is that they have a long lead time and are not available on short notice. Think of insurance spares like the insurance of your car or house. You don't want to claim them, but you don't want to be without them either. Insurance spares are the ultimate 'just in case' inventory. Others termed this as Rarely Used Inventory. These items are difficult to obtain. Their lead time to acquire usually takes several months, even years, either because most of these parts are unavailable or imported from other countries. Most of them were recommended by the OEM since the part is classified and deemed as critical. Today, they sit on the storeroom shelves for years and remain untouched.

Recommended Measurements For Your Storeroom

Measurements	Frequency	Trend
Total Line Items and Cost of Inventory	Weekly	The lower the better
Number of Stock Outs	Monthly	The lower the better
Inventory Accuracy in Percentage	Weekly	The higher the better
List of Obsolete Parts and Cost	Weekly	The lower the better
List of Non-Moving Parts and Cost	Weekly	The lower the better
List of Fast-Moving Parts and Cost	Weekly	The lower the better
Lead Time for Critical Parts	Weekly	The lower the better
Average Waiting Time for Spares	Weekly	The lower the better
Carrying and Holding Cost of Spares	Yearly	Should be equal

Supply Chain | Warehouse | Logistics | Finance | Purchasing | Maintenance

Figure 11.4: Important KPIs and Indicators for MRO Spare Parts and Storeroom

Performance measurement is one of the methods at the heart of propelling an organization towards breakthrough performance. This generally takes the form of key performance indicators and measurement programs, all designed to focus on various performance areas. The old adage is, "if you can measure it, you can manage it." Before you think about measuring it, first work out what you want to manage. It is important to understand what we want to measure to determine where maintenance is currently headed in the future.

11.8: August 2017: Consolidate All Improvement Initiatives in the Plant

Have you ever experienced coming to work in the plant, starting with a morning meeting with operations, then you need to attend another meeting and another? At the end of the day, all you have done was to attend all these meetings? One day the manager called you and discussed problems about housekeeping on the plant, and on that day forward, you were called the 5's Champion. You will create an ad-hoc committee composed of different people from different plant departments, and at first, you and the committee members will be excited about this initiative. You and the team agreed to meet every week to determine the progress of the 5's initiative. The following morning there was a near-miss accident, and the safety officer called a meeting on each department and asked for a champion in each department to ensure the safety of the people in the workplace. The Quality Department initiated a Quality Improvement Circle composed of operators, maintenance, and process people to ensure the quality of the product, and these improvement circles will be meeting once a week. The President of the plant wants the OEE to be tracked and monitored regularly on every single piece of equipment and assigned you once again to spearhead an OEE committee, and again there will be a weekly meeting because of this.   The maintenance manager attended a course on RCM and was convinced that he needed RCM to be implemented in the plant, so he organized the team, and the team will be meeting twice a week.

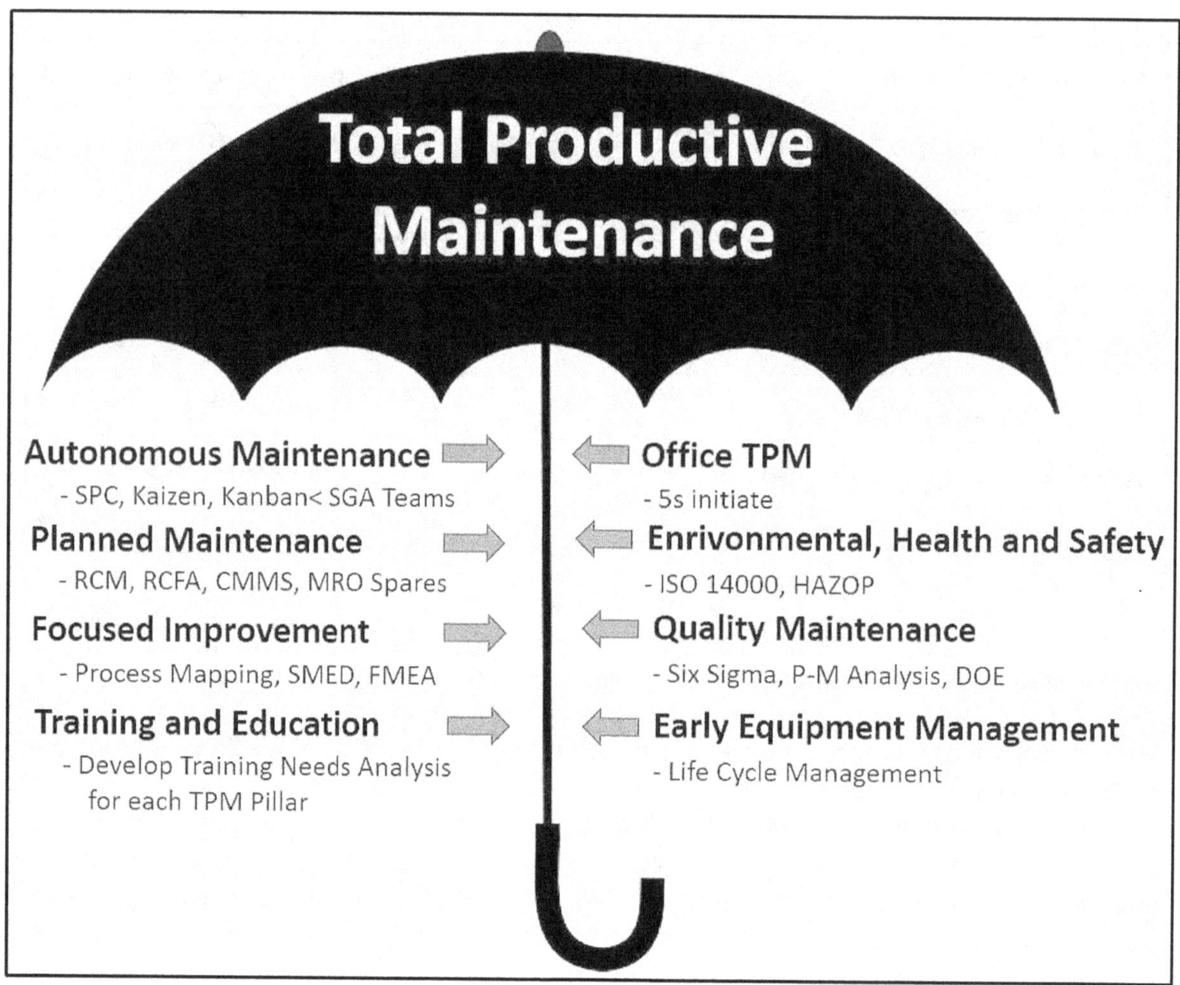

Figure 11.4: Consolidate All Improvement Initiatives in the Plant

Hey wait, you still have the human resources, and they got this cool initiative which is called the employee of the month award, and you were lucky or perhaps unlucky to be selected as one of the committee members to derive the procedures and criteria in which the bottom line is that you need to attend this once a week meeting. Not to mention the daily operations meeting, toolbox meeting, and supervisors meeting, which will only end up having many action items to complete? Are we employed just to attend these meetings? One manager I knew went to Japan and attended a training with Masaaki Imai on Gemba Kaizen. After the training, he was so convinced of its philosophy, and after returning to the plant, he assigned a champion to head the Gemba Kaizen project, but wait, isn't there was already a 5's champion? Wait, we still got the Karate people from Six Sigma Yellow, Green, and Black belt, and I can go on and on, and I hope you got my point now. The bottom line is that you have accomplished nothing for the day because your time was consumed by attending too many meetings.

While their initiative is good, the problem is that the same members will be part of the team or committee or whatever your industry called them. The main problem is that these plant initiatives were never consolidated. What will happen is that many corrective actions will be generated due to these initiatives wherein you finally forgot your purpose and why the industry hired you. The more meetings you attend, then the more problems that industry got. What is important in this case will be to list down all the initiatives taken by your industry and try to consolidate everything, just like a barbeque or satay. I love to eat that when I am in Malaysia. Have the main umbrella, and that all improvement initiatives will be under one umbrella. The next will be to assign which department or group is best to handle that initiative.

Do not confine your industry with so many improvement initiatives as there may be cases where some of them may try to conflict or even contradict each other. I remember when I was still working in the semiconductor industry, they organized a group called the BKM group or Best Known Method, in which the role of this group was that if a modification was performed on one of the equipment, it should be fan-out or replicated in all similar equipment in the plant since there are different departments with the same types of equipment. I truly contradict this because these people do not understand what operating context means. RCM contradicts this. If the problem is not present in other equipment, even though they are identical, then the modification should not be placed there to induce new equipment problems. This means that if the equipment is stable, even if they are of the same equipment type. The best thing to do is not to disturb stable equipment. This group will ensure that the modification will be carried out to all similar equipment in the plants, even if other similar equipment poses no problem. Disturbing stable equipment may induce a new set of problems in the long run. When the equipment is stable, the best thing to do is to leave it alone and run.

In manufacturing, quality people do is to fan out improvements/modifications to similar equipment, assets, or machines. They called this process the Best Known Method or BKM. If we have, let's say, 20 similar equipment in the plant and modification have been done on one equipment with a problem, only that equipment should be modified as well as those that emit the same problem, but for other similar equipment which is stable and poses no problem, then such modification should be discarded. Let us say that 2 out of the 20 pieces of equipment have an existing problem, then only these two should be modified and not the whole 20 pieces of equipment, which is my point. But for Quality people, they will modify the 20 pieces of equipment, which I think is wrong from a reliability point of view. My point is that if a machine poses no problem at all, then let it run.

Answer Posted by Joel Levitt (Maintenance Book Author): Rolly, This is a deep question that I've never thought about. I think of the airplane makers, and when they find an improvement, they spread it out to the entire fleet. Also, when an RCM analysis turns up on a particular failure mode with a dire consequence, all the units of the same type should have the modification. I'm happy being wrong on this but would side with quality.

Answer Posted by Nathan Wright (Maintenance Book Author): Hello Rolly, I agree with you that it is not always a good idea to spend money and make changes just because similar equipment needs it. Equipment is like children; they need individual attention and be held to a standard that meets their needs. Thank you for your email.

Answer Posted by Keith Mobley; book Author: Rolly, Your point is well taken, but consider the following. If the identified problem is an inherent design deficiency common to all 20 assets, then the modification should be made to all 20. However, suppose it is not an inherent design deficiency, and the failures are limited to one or two of the assets in specific areas or applications; in that case, there is no value and a high potential for creating more problems in modifying all 20. As a general rule, you determine whether the problem is unique to one asset or all common or similar assets when problem-solving. In the former, the problem is unique to that asset. If all similar exhibit problems, then the problem is systemic, caused by something common to all similar assets. I would suggest that you not apply one rule for all cases. Instead, determine to the best of your ability whether the problem and modification are unique to one or a few assets or systemic and exhibited by all. Quality's rollout practice is common, and many consider it best practices, but only in cases where the modification is typically warranted, the root cause is an inherent design weakness common to all assets in the family or a systemic (infrastructure) problem.

11.9: September 2017: The Physical, Human, and the Latent Cause of the Problem

One of the biggest confusion in performing a thorough Root Cause Failure Analysis is understanding how deep we should probe our analysis or simply stated where do we stop our investigation in performing a Root Cause Failure Analysis? Going too deep will lead us to the Christian bible, Timothy 6:10, _for the love of money is the root of all evil_, and going too shallow will allow the problem to recur once again. Although a lot of analytical tools are currently being used by industries today. Are they really meant to uncover the root cause or simply what we termed the Physical Cause of the problem only? One industry found out that John Doe was responsible for closing the valve that wrecked one of the turbines. John Doe was sentenced to be suspended from work for 1 month without pay. The lawyers and managers were happy that the culprit was finally condemned. One year later, the same problem occurred and this time by a man named Johnny Thorr, and he quote, I just use to sweep dirt around here, but the supervisor instructed me to close the valve, and I don't know which valve he was talking about so I close them all. The question being raised here is John Doe and Johnny Thorr, the Root Cause of the problem? Are we certain if John Doe was severely punished, will the problem will be gone for good?

Before we begin any further analysis, we need to ask ourselves why we need to perform a Root Cause Failure Analysis. Do we simply do it to comply with customer's requirements, such as manufacturing plants? Do we perform a Root Cause Failure Analysis because our management wants to know what happened? Do we perform a Root Cause Failure Analysis because we want to know what caused the problem to occur? Do we learn from failure when we punish someone, or we just expand the gap between our people and ourselves?

Physical Cause of the Problem: This is the physical reason why the parts failed. This is the technical explanation of why things broke or failed. This refers to the Physics of the incident, which usually explains how the failure had occurred. For example, a bearing failed due to fatigue. This mostly explains the metallurgical factor of why the failure occurs. This is simply the same as Failure Analysis, in which the probe or investigation will end up on the component level or on the part the eventually fail.

Human Cause of the Problem: Root Cause Failure Analysis believes that all failures are caused by humans, and all humans are prone to commit mistakes and errors where either someone did the wrong job or simply did the job wrong. People commit slips and lapses. When I tell you that even with the best procedures industry got, people are likely to commit errors. All people commit errors. Errors can be classified into slips and lapses. Many factors can be attributed to the cause of these errors, such as fatigue, pressure, environment, inattention to details, which are just to name a few. We must understand that most human errors are not necessarily the fault of the person who committed the error. We need to understand that either the error was caused by external circumstances far beyond our control or by flawed rules and procedures that need to be changed. We also need to understand the reason why the error was committed so industries can learn from them. But the most important lesson of all is, do we really learn from failure by punishing people?

System Cause of the Problem: These are the management system weaknesses, including lack of training, outdated policies, and procedures. People make decisions based on these, and if the system or procedure is flawed or outdated, then the decision will be in error and will be the triggering mechanism that causes mechanical failure to occur. Usually, most system causes are addressed by updating their current procedure or standard operating procedures for the industry. Still, the root cause should not end here since there is a much deeper cause that eventually leads to the failure. Unless the latent cause of the problem is not completely exposed and addressed, the problem will continue to resurface once again. Usually, this will be procedural, and the corrective action or countermeasure is to update or the procedure or process.

Latent Cause of the Problem: Underneath every problem lies a deeper cause called Latent Causes. These are the concealed and hidden causes that eventually cause a human error to be committed. The only way to address these Latent Causes is to expose them, and we can only expose them if we truly understand what it is all about. Latent Cause Analysis is not just about system flaws and procedures that eventually led a person to commit the mistake. It is not just about organizational management weaknesses. It is not just about flawed management decisions, but rather Latent Causes means understanding that we are also part of the problem. Physical, human, and system causes will require corrective action, while latent causes will not require any corrective action. The only way to address a latent cause is for people to change. Learning from the things that go wrong is not as easy as we think it is. The only way we can learn from the things that go wrong is to see ourselves in the mirror and admit that we are also part of the problem. Collectively, we all have our share that eventually caused a person to commit a mistake. The sad thing about this is when this happens; we isolate the person who commits the mistake and put all their fingers and blame him. We always wanted a fall guy. Before we can truly understand the latency behind the problem, we need to ask ourselves the following:

• What is it about the way we are that contributes to the problem?
• What is it about the way I am that contributes to the problem?

If we are part of the problem, we must be responsible for being part of the solution. Engineers, technical people, maintenance can easily arrive at the physical cause of the problem. Suppose some mechanical component such as a bearing failed; in that case, these people could dig up evidence that will eventually lead to its physical cause, but digging deeper into the latent cause is the most difficult part as we need to look ourselves in the mirror and admit to the fact that we too are also part of the problem.

Figure 11.6: Mr. Sirivat, From Riches to Rags and Bouncing Back

Mr. Sirivat from Thailand is a former millionaire, tycoon, stockbroker, and well-known investor in Thailand. When the stock market crashed in 1997, he was left with no money and large debts to the banks. All his assets disappeared, and he was left in large amounts of debt. One day sitting at a table with his wife, he asked his wife what to do, and his wife told him, let us sell sandwiches. He followed his wife's advice, removed his coat and tie, swallowed all his pride and ego, and walks down the streets as a peddler and started selling sandwiches, but only this time he was not wearing an expensive coat and tie but an ordinary T-shirt. Mr. Sirivat and his wife started making and selling the first 20 sandwiches on April 20, 1997. It took them six and a half hours to sell them. Today, Sirivat owns several coffee shops and still sells sandwiches in the streets. He had been a motivational speaker and an inspiration to Thailand in which he is better known as Thailand's Sandwich Man.

Can each and every one of us be this kind of person? Is it difficult? Latency is not about system causes but rather understanding how we contribute to the problem and learning from it since we are taught that we serve its best interest. I consider Latency to be a higher form of human cause. Again, exposing these Latent Causes is the only way to understand a true and meaningful Root Cause Failure Analysis. Hence, to answer the question, where do we end our probe on Root Cause Failure Analysis? The answer is when we reach the Latent Cause of the problem. Latencies are all about looking at "The Man in the Mirror.

11.10: October 2017: What to Do with Obsolete Parts in your Storeroom

There are large quantities of obsolete parts inside the storeroom, which will no longer be used since the equipment was already retired or decommissioned, and there is no other similar equipment of this in the plant. Frequently, there are many parts held in the storeroom that do not belong to any other equipment in the facilities since both maintenance and management had already been remove or retire several types of equipment in the plant, but the problem is that maintenance forgot to include the storeroom people in the communication loop. The facility's equipment may be retired and no longer in sight, yet, the parts of that equipment are still inside the storeroom. And worst if the storeroom personnel may still be re-ordering these obsolete parts since some of these parts are either slow or fast-moving, and their reordering is automated. This is like having a patient died in a hospital, and you still pay the bills even if the patient is already dead. All parts inside the storeroom have a cost of holding the part called the carrying or holding cost. It is like renting a condominium or apartment. Usually, the cost of holding the part may range from 15 to 30% of the value of the spare parts per annum, then that is the cost of storing the part in your storeroom per month. Let us say that I have a big spare motor inside the storeroom, which is already obsolete, and the cost of the motor says it is $ 100,000.00. Therefore the cost of holding the motor in the storeroom will be around $ 15,000.00 to $ 30,000.00 per year or divided by 12, then this will be per month, and this is just one of the thousands of parts inside the storeroom. The problem is that most storekeepers do not provide a report that on this; that is why maintenance thinks that it is ok to hold these obsolete parts in the storeroom. Most maintenance thinks that there is no cost of holding the part in the storeroom and does not want to get rid of the part by hanging on to it on the assumption that it might be needed someday; besides, it does not cost anything to keep it. They got it all wrong because all parts in the storeroom have a cost on holding them. These obsolete parts eat up space from the storeroom; hence, if you have several plants, the department and storeroom can notify other departments if they have similar equipment; otherwise, the best option for these obsolete parts will be to sell the part as scrap.

Why Spare Parts Become Obsolete: First, the equipment will no longer be needed, and the company decided that this equipment should be decommissioned, retired, or scrapped. Second, a part or item had some design weakness and had been modified by an improvement team without looking into the current inventory of the part in the stores, and the improvement team decided to use these parts in the equipment. If that will be the case, then the storeroom's existing or original parts will become obsolete. Third, some parts have a shelf life, and the part's life had already been reached. When equipment is retired, the spares associated with it must be identified so it can be freed up from the storeroom. This will free up space inside your storeroom. Storing obsolete items cost money and space in the storeroom. A plant can terminate plans to expand its storeroom by freeing up space consumed by these obsolete parts. This is one common problem in industries with their storeroom because their storekeepers do not report directly or indirectly to maintenance; hence, they are frequently forgotten in the communication loop.

Difference Between Non-Moving and Obsolete Items: Although Non-moving parts are another big problem with maintenance, we shall cover them in a separate newsletter. The difference between non-moving and obsolete parts in the storeroom is that the machine or asset is still being used in the plant, but the part is simply not moving and remains inside the storeroom for years. An analysis of 100 MRO stores tells us that 60% or more items had no usage for the past 2 years. Yet, most of these items should be on hand when needed because the failure's impact and consequences are merely unacceptable. Imagine a spare tire in your car; how many times have you used them, but yet nobody wants to drive their car without one. I am just wondering why do motorcycles and scooters do not carry a spare tire too. The general thinking here is that industries stock them "just in case"; we might use it someday.

When a part becomes obsolete, the best thing to do is to identify them and inform the head of maintenance that you need to get rid of them; if your company had multiple plants, you could share or sell these parts, the option is geographically practical. The most preferable way to do this is to sell the parts back to the supplier you bought. Most vendors agree with this idea, provided that the part is in its original wrapping or package. The Internet can be a great source for reducing your obsolete parts in your storeroom, for as long as it is in good condition. Your purchasing can auction them on eBay or sell them online as a surplus. A better approach might be to use these obsolete parts as tools in your maintenance training program. If you are stuck with obsolete bearings, you have an opportunity for all of your maintenance and millwrights to practice bearing installation. The removal of obsolete items from the storeroom should be an ongoing activity. The ultimate and long-term approach to these obsolete parts in the storeroom is having a good spare part list for the equipment. Remember that there is a cost in carrying obsolete parts in your storeroom. When equipment is removed from operations, the part that goes with the equipment should also be removed from the storeroom. CMMS should be able to tie apart the equipment on production. Remember that storekeepers have no information on what equipment is being decommissioned unless they are informed.

There is also the possibility of human error that the part was purchase by mistake, and the vendor had a no return policy. In this case, the company shoulders the burden for the mistake. The worst thing that can ever happen to obsolete parts is that if these parts are considered fast-moving items, Purchasing might still be ordering these parts; they do not know that the equipment was already retired replenishment is fully automated and has not been adjusted. This is like having a patient die in a hospital, where the patient's family still pays the bills. Unlike non-moving parts, there are many options on what to do with these obsolete parts. Listed here are just some of them.

- List the obsolete items accordingly and place them in the newspaper for bidding
- Ask Purchasing to find a vendor where these obsolete parts can be sold as scrap
- Some of these obsolete items can be used for training purposes for the operators and maintenance. For example, have the motor's cross-section cut and use it as a teaching aid for operators.
- If the plant has other business units, they can sell these obsolete parts to them.
- Try to seek Purchasing's help to locate similar industries that are still using the equipment you have retired and sell directly to them at a lower cost.
- Sell them back to the OEM at a depreciated cost.
- Join online maintenance forums or bulletin boards and post about the equipment you retired and spare parts you will no longer use. Once you found someone, get their contact details, and let your Purchasing handle this case. Your trash and garbage can be someone else treasure.
- Trade them back for something that you can use.
- Nowadays, the internet is the greatest source of information. During my time, we got the Britannica Encyclopedia. You just need to type the correct keyword on what you are looking for, such as vendors or traders who buy this stuff. Some vendors buy scrap and obsolete items.
- Sell them at e-bay, sulit.com, Amazon, and other online platforms.
- You can post the items for sale or auction on maintenance forums or LinkedIn.
- Seek vendors in the same category and sell to them at a depreciated cost.
- If none of these items work, they can be totally scrapped or just sell them in a junk shop.

Remember that obsolete parts now become garbage to your industry. Still, there is a saying that your garbage can be someone else treasure. It is also important that the software include what equipment the spare parts belong to at the very beginning of automating the MRO Storeroom tasks. Suppose a part belongs to several machines, and one of the equipment is retired; in that case, the parts that belong to this equipment are still active and not yet considered obsolete. The spare part only becomes obsolete if all the equipment utilized is finally retired from the industry.

Usually, storeroom people may actually have no knowledge of assets and equipment that have been decided to be retired. The best thing to do is to talk and communicate with maintenance on which equipment had been retired in the past to generate lists of spare parts inside the storeroom. Once the lists are completed, identify the parts that go with them in your storeroom and provide lists of these parts to the maintenance. If your industry has other business units with the same equipment, then you can contact them about selling these parts to them since these parts will be of further use to your industry. Place these obsolete parts in one store's location temporarily and discuss what to do with these parts. Remember that storing obsolete parts in your storeroom cost money; it is best to free them up or simply. Hence, if you are the storekeeper in your plant and you have these tremendous amounts of obsolete parts inside your storeroom, just remember the song of Frozen. Let it Go.

11.11: November 2017: Skills We Need for Autonomous Maintenance Operators

In industries today, what we need are operators who can detect, correct, and prevent equipment abnormality from occurring. These include understanding the importance of proper lubrication, including correct lubrication methods, and checking lubrication points in their equipment. Simplify cleaning inspection and the use of visual controls for inspection. Operators will always be the first line of defense on any equipment-related breakdown or failures since they will be the people who will experience the failure first before maintenance. What is important for operators is to know the signs and symptoms of these failures while they are still in their earliest stage. Remember that it is 3x more expensive if the failure happens first and maintenance reacts to it.

• Operators' ability to understand these basic functions on their equipment. In most industries, since operators are not technical people, maintenance finds it hard to translate what operators say when they are being asked something about what went wrong with their equipment. Operators' basic human senses must be enhanced to detect any signs of potential failure or symptoms in their equipment and their ability to detect causes of abnormalities.

• Operators can conduct simple analyses on the causes of equipment failures by using conventional analytical tools and problem-solving techniques such as brainstorming and why-why analysis.

• Operators' ability to conduct minor repairs, conversion, adjustments in their equipment. For a start, when their equipment fails, operators should be present at their equipment site whenever maintenance performs repair. Operators can assist them while maintenance can provide operators key points on what to detect in their equipment and the name of the part that failed. Operators can ask questions regarding the failure. The line of communication is open, and operators learn much more about their equipment. This creates a better understanding and communication between both operators and maintenance.

• Operators who know how to operate their equipment. These people can make a sound decision on whether to shut down the equipment in the event of breakdown or quality defects. The key is for the operators' ability to understand the relationship between quality and productivity. They can detect and predict problems using their basic human senses.

• Finally, the goal for any Autonomous Maintenance initiative is to have empowered operators in the workplace that not only take care of their equipment but also can transfer basic equipment knowledge to other operators as well. This is what we want to achieve in our Autonomous Maintenance journey.

Autonomous Maintenance is the pillar of TPM with the most population and number of members. Perhaps this is the missing link in any maintenance reliability improvement and initiative. Maintenance always struggles to cope with their day-to-day problems and failures just to end the day feeling exhausted. The following days will be no different. Nothing will ever change if the operators will not be involved in accepting shared responsibilities with their equipment. This can only happen if maintenance starts educating operators. Education is not a one-time event but a continuous process. Education can be in any form, whether formal or in an informal way; when maintenance repairs the equipment, operators can assists maintenance on repair and maintenance; on the other hand, maintenance can teach operators regarding the failure experience on their equipment. This will be the start of better communication between the two. The line of communication between operators and maintenance should be open at all times. We saved the parts that failed in our equipment during our TPM days and explained them to operators.   Worn-out critical spare parts are not thrown away. Instead, they are mounted on a whiteboard, which later can be used by the maintenance people as a teaching aid to their partners who are the operators, a brief description of the part, the part's function in the equipment, and why it failed. We can spend around 10 to 15 minutes every shift teaching operators on why the part has worn out or why it failed, and its criticality to the equipment, as well as possible signs of potential failure or symptoms that can be detected by operators that the part is already on the verge of failing.

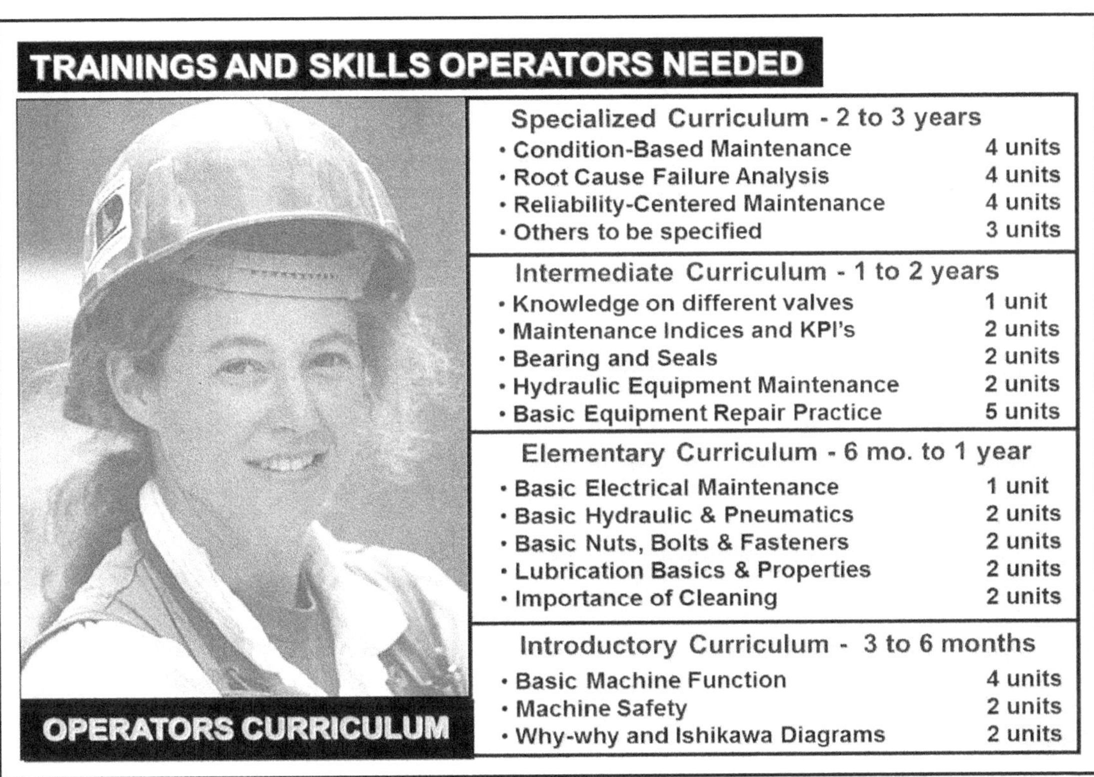

Figure   11.7: Training and Skills Operators Need

During our TPM days, I asked some of my Planned Maintenance members what support they have provided to their operators. One of the members said, "When I repair the equipment, I make sure that the operator is beside me while I'm repairing; I tell her what this part is all about and its function in the equipment. "I said that's good." He said that he made sure that they ate together during lunch breaks, and as they ate, I told her all about what she needed to know about her equipment. Again, I said, that's good. During the shift, I told her how to perform light repairs on the equipment that I believe she can handle it. But that is not all; at the end of our shift, since we are on the same shift, we go home together with the bus shuttle (since our company offered a free bus shuttle as part of its employee's benefits). It's an hour-long journey, most especially during rush hour due to traffic, so instead of being bored and just sit down on the bus, I tell her stories about my experiences in the line, and from time to time, I still discuss with her about what I think of her equipment. Today, they are now married and have 3 kids. Anyway, so much for that back to Autonomous Maintenance.

Autonomous Maintenance focuses on eliminating dirt, dust and address sources of contamination in the equipment. Operators need to understand why they need to keep their equipment clean and what contamination and dirt can do to their equipment. Operators need to expose all abnormalities they can find. An abnormality can be defined as any deficiency, disorder, slight irregularity, defect, bug, or flaw, or any unwanted condition that can lead to other equipment problems. Operators must be keen on their senses in exposing abnormalities in their equipment. They begin correcting flaws they can find, such as deterioration, deformation, cracks, fracture, and so on. When serious damage is discovered, such as metal fractures or broken parts, they seek maintenance people's assistance in addressing them. The key is finding problems when they are still in the early stages.

11.12: December 2017: The Roots of TPM

In 1945, the imperial of Japan, Japanese Emperor Hirohito, finally announced its surrender to the United States after the atomic bomb was dropped in Hiroshima and Nagasaki. The Japanese officers formally signed on September 2, 1945, their surrender ending World War II and avoid further destruction.

Japan Surrenders

Figure 11.8: Japan Surrender to the United States

After World War II, Japan's economy collapsed, and their nation was left in ruins. Out of humanity, the United States sent an emissary to Japan to rebuild their ruined nation. His name was Dr. Edward Deming. This person was invited to speak to an elite group of Japanese scientists, scholars, writers, industrial leaders, and engineers right after Japan's economy collapsed after the war since they were in desperate need of a complete restructuring. The United States sends a humble and down-to-earth person by the name of d Dr. W. Edward Deming. He was asked to assist the Japanese government and its industries in the challenge of rebuilding their nation once again. Dr. Deming taught them about quality, statistical control, and management systems. Japanese people listened and absorbed Dr. Deming's teachings. One of his students asked him, "Dr. Deming, how long would it take to shift from the perception of the world from the existing paradigm that Japan produced cheap products and imitations to one that produced innovative quality products?" Dr. Deming replied and told the group that if they would follow his directions, they could achieve the desired outcome in five years. As time unfolded, Dr. Deming was surprised because they did it in four years. Japan invited Dr. Deming back time after time, where he became a counselor and a mentor. For his efforts, he was awarded the Second Order of the Sacred Treasure by former Emperor Hirohito. Japanese scientists and engineers named the famed Deming Prize after him. It is bestowed on organizations that applied and achieved stringent quality performance in their plant. Deming taught statistics at the New York University when he was invited to visit Japan in 1950 to run a series of speaking engagements with Japan's business leaders. Other Quality gurus also went to Japan, such as Joseph Juran. Juran's visit to Japan provided the industry leaders in creating a mindset on quality in which, through the years, Japanese factory management and leaders created a paradigm on the importance of quality in their products and services. And as the years, passed many well-known Japanese Guru's emerged, which all started out from Deming's teachings.

Although TPM started in one of Toyota's suppliers, Nippondenso, Deming had been the key to Japan's restructuring and indirectly influenced Japanese people from a country that produced cheap products to a highly competitive country that can compete with the best industries in the world. Because of Deming's teaching's, these great Japanese quality leaders and gurus emerged, such as Shigeo Shingo, Taiichi Ohno, Kauro Ishikawa, Genichi Taguchi, and of course, Seiichi Nakajima, who is later on known as the Father of Total Productive Maintenance (TPM) who recently passed away last April 2015 at the age of 96 years.

Today, many learnings were adapted from the Japanese, even if Dr. Deming, an American who taught the Japanese people how to restructure their economy. While US industries continue to focus more on producing mass production since they believe that the way to victory was to have tough people, manage organizations that produce goods in large batches for higher profits. On the other hand, their Japanese competitors do it differently. They developed a new and powerful paradigm that winning organizations listen to the voices of their customers. They design products and services that meet or exceed customer's expectations and improve continuously in all areas of the organizational process that will lead to customer satisfaction.

Perhaps we have known, and heard of big industries before, mostly in the US, such as Polaroid, Kodak, IBM, and many more who no longer exist. Today, only around 52 US companies have made it to the Fortune 500 companies since 1955. This means that only 10.4% of these companies have remained operating today after 64 years since 2019. More than 89% of these companies have either gone bankrupt, merged, or acquired by other industries or have fallen out from the original Top 500 Fortune Companies in the United States. Many of these companies were big shots during their time in 1955. Today, they are now unrecognizable, forgotten, absorbed by other industries, or completely gone and unknown. In general, the most fruitful success strategy is, to begin with, leadership tools, starting with hiring the right people, envisioning the change in place, defining their roles in the organization, developing their objectives, measurement, and control systems.

Every industry provides a specific product or service to render, but more importantly, it is the customer that spells all the difference why an industry still remains in business. In today's global economy, industries' survival depends on their ability to improve and rapidly innovate according to their customers' needs and demands. As a result, an increasing search is on for methodologies, strategies, and improvement processes that drive quality, costs, productivity, and reliability. In today's fast-changing marketplace, slow, steady improvements in manufacturing operations will not guarantee profitability or survival. Companies must improve faster than their competition if they want to remain competitive in their respective fields and control the lion's market share. That is why to survive the competition, each company is seriously adopting the best continuous improvement practices. In Japan, Total Productive Maintenance or TPM tops the list of these continuous improvement methodologies. Companies adopt TPM to improve their bottom-line results and bring out the best in their people because TPM believes that people will be the ones to make or break that difference. They believe that TPM is not just a program. It is not just a strategy but rather a philosophy and a way of life for all employees. Japanese Management and workers see TPM as a means of improving the bottom line and, most especially, the human side.

By definition, according to Ginder and Robinson, TPM is a plant improvement methodology, which enables continuous and rapid improvement of the manufacturing process through employee involvement, employee empowerment, and closed-loop measurement of results. It is a production-driven improvement methodology designed to optimize equipment reliability and efficiently manage the plant's people and assets by eliminating losses. Every single employee in the plant will be involved in TPM. TPM covers all departments, including maintenance, production, finance, development, marketing, sales, quality, safety, reliability, human resources, warehouse, and administration. It involves all people in the organization, from top management to frontline workers. It aims to achieve zero losses through overlapping small groups of activities. In TPM, everyone will be involved.

2018 RSA Reliability Newsletter Vault Archive

> *When equipment is preserved well, sustained, and breakdowns are reduced, it does not mean that jobs will be lost and people will be fired from their organization. What it simply means is that if industries becomes less reactive, then this is an opportunity for them to allow new doors to open for the maintenance function.*

12.1: January 2018: How to Initially Implement CMMS in Maintenance

I have provided an easy and simple user guide for those still planning to acquire their CMMS in their maintenance operations.

Step 1: Organize a team to study the need for CMMS: I would strongly recommend organizing a cross-selection of maintenance and other departments that will be part of the CMMS process and study what the plant needs to automate in maintenance. For example, suppose spare parts and ordering systems will be part of the CMMS structure; in that case, we need to include the Spare Parts and Purchasing people to compose the CMMS project. Although the team can meet twice a week for the project, someone must work on this full time in preparing all the details, legwork, meeting agenda, minutes, roadmap, and other relevant matters in this study. The team must define why they need to adopt a CMMS system in their maintenance department and align this with its business goals and objectives. Make a survey with your people and interview some key people in the maintenance department. This will provide you with the initial requirements you need. Study likewise the capability of your IT or MIS Department in developing this software for maintenance and include them in the team formation. Make a list of details and categorize them. If you belong to a large organization with several departments, include key people in the other departments. Remember that the role CMMS will play will be confined to your own department and the whole plant, as well as other departments that may have different requirements from yours. The team should also draft a proposal to their management regarding the feasibility, plans, benefits, costs, and ROI of the said project. Management buy-in is required at this stage for a go or no go on the project.

Step 2: Decide what to include in the CMMS: Ask what needs to be included in the CMMS program. Since we have covered almost all of the maintenance disciplines in this book, ask what we want to automate in this respective discipline and its reasons. CMMS is not just limited to having a list of equipment and work orders for routine maintenance. It is important to understand what the user wants out of it to benefit most from the software. The usefulness of the CMMS system depends on the amount of information we put into the system. Ask what indices should we include and what type of reports and documents need to be automated. Customize the requirements according to your plant needs, whether the CMMS will be developed in-house or purchase from a third-party vendor or both. The vendor must adjust to his clients' needs and not just merely purchase the software as it is. If the vendor's CMMS software was design for manufacturing operations, and your sector belongs to the oil and gas, then some features of the CMMS software will not apply to the oil and gas sector.

Step 3: Decide Access Levels for CMMS: While CMMS is designed to be used by the maintenance, not all maintenance will have access to everything. For example: if part of the Maintenance 201 file will be the training and details of every maintenance person in the plant, including his appraisal results, access should be limited only to the right and authorized persons. Sensitive information should be accessed by authorized personnel only. While there are access levels for a common purpose such as repair procedures, maintenance measurements, indices, equipment 201, spare parts inventory, portions on the CMMS are for authorized eyes only.

Step 4: Select the Most Suitable Vendor for CMMS: If your IT department cannot provide you with all the requirements you need, it is time for you to select a vendor. Each company has a unique set of requirements, and the vendor should adjust to your requirements and not you adjusting to them. It is vital to ask the CMMS vendors how long have they been in business. Choose a vendor who has been in business for several years. CMMS software must be easy to use and learn. It must be user-friendly. Technical support must be available, especially for periodic and regular updates, to keep technology current in this fast-paced digital age. Determine the support available and see if it satisfies your requirements. Likewise, check the compatibility of the software with other software you have in your plant. Can your Predictive Maintenance software communicate with the CMMS software you are planning to buy? One of the most important factors to discuss with the CMMS vendor is the cost of the software. Even if all the features of a CMMS are available at hand, if the company's budget is way below the cost, it will be useless. Compare at least three vendors and evaluate which of them best fits your requirements regarding cost, services, customer support, and features.

Step 5: Install the CMMS and start encoding: Once your organization has found the most suitable vendor for your CMMS, purchase it, and install it in your plant. The process of installation should also include the initial populating of important and relevant information. CMMS can only provide us what we put or encode into the system. The benefits and usefulness of any CMMS system depend on the information we supply to the system. If what we populate in the CMMS system is irrelevant and garbage, then what goes out will just be garbage and irrelevant to the maintenance department. Encoding relevant information will take some time, depending on the amount of data collected and encoded. Do not just confine this to one person. This is like commissioning the equipment for the first time and debugging the system for flaws. Below is the list of some of the information that initially needed to be encoded into the CMMS system.

• Asset of Equipment 201 file and history records.
• Maintenance 201 file, training, scheduled leaves, asset assigned to

- Asset hierarchy
- Downtime record for the past year
- Maintenance measurements and indices for the past year
- Equipment safety functions and features
- Autonomous Maintenance, regular cleaning and inspection standards and frequencies
- Lubrication, checks, frequencies, and lubrication inventories
- Asset common repair procedures (Expert System)
- Daily, weekly, monthly, quarterly PM and shutdown procedures
- Spare parts lists, bin, location, classification (fast, slow or non-moving), and inventory
- Parts vendor 201 file and records
- Previous reports, relevant documents, drawings, schematics, and machine modifications
- Previous vibration, thermography, oil analysis records

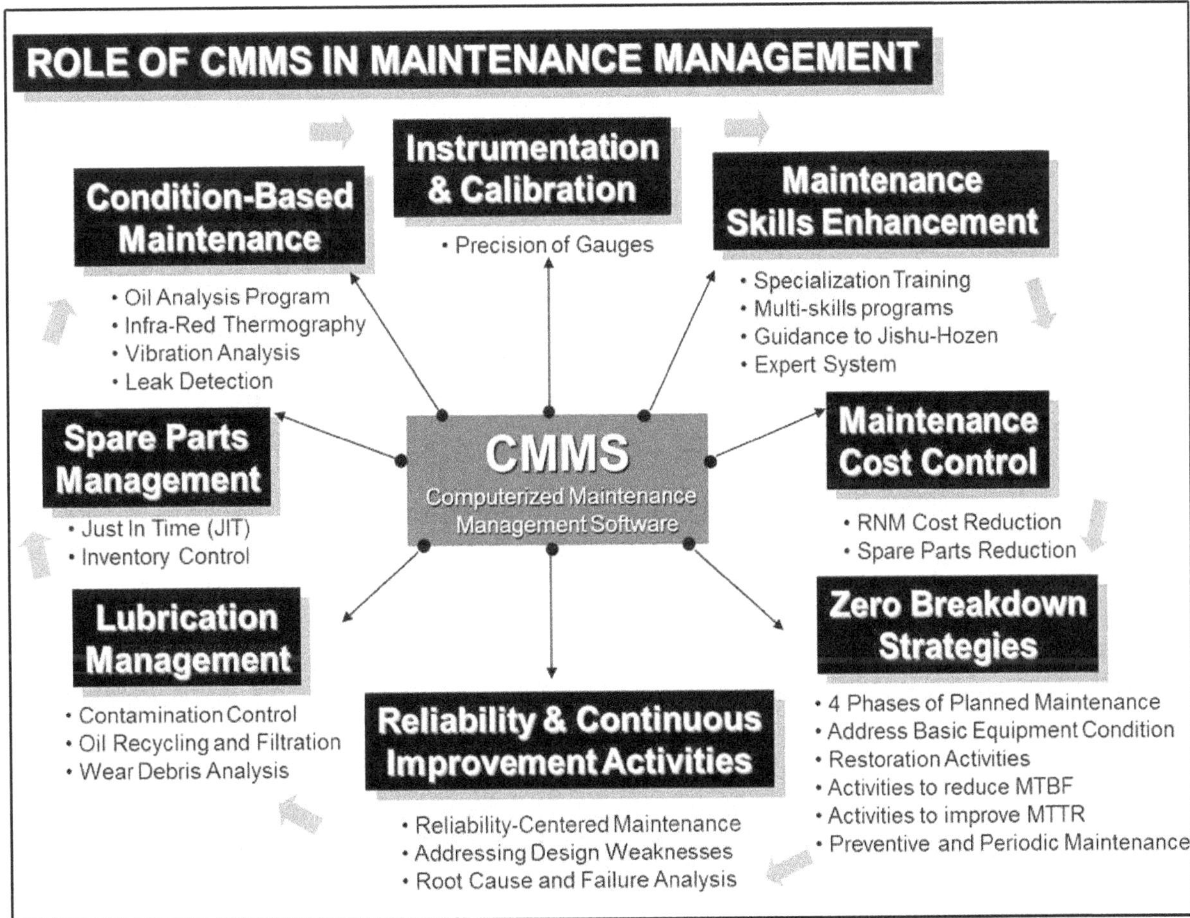

Figure 12.1: CMMS Software

Step 6: Train the Users of CMMS: The reason for encoding the data first before proceeding with the training is that it will be much appreciated by the users if the samples and data from the CMMS are the actual data instead of make-belief data. The users' training should be done hands-on, meaning that they should be using the CMMS. This is also to understand how the users will react in making every bit of information online against their current practice. The users can provide the best feedback on what parts of the CMMS they find difficult and need improvement. On the lighter side, they can inform us which part of the CMMS they enjoy doing most and find it easy to do. Remember that the goal is to streamline and automate the process and remove as much paperwork as possible. If it takes the maintenance an hour to provide an MTBF report for the month, a few clicks will do the trick with the CMMS, which can be ready to send to people via email. In 1988, when I was working in an assembly plant, the internet and email did not yet exist or still at an early stage; it took me the whole day just to photocopy the reports, arrange, sort them out, and send them to at least 30 people from different departments in the plant. With the emails around, maintenance reports can be generated in excel for preparing graphs and sending them to concerned people. With the CMMS, this will be a lot faster since the graphs are automatically generated. All we need to do is provide some highlights regarding the trends in the KPI's. Once data have been encoded, the report is completed, it can be automatically forwarded to a selected group of people in your department.

Step 7: Train all the others on the use of CMMS: After the CMMS users have been trained, it is best to coordinate this with your training department and provide training to other departments such as operations, quality, purchasing, and management regarding the use of CMMS's in the organization. Grant them special access to menus that they will be using, and it is likewise important to create some interface and communication with their existing system. For example, HRD and training should also have access to the CMMS Maintenance 201 file regarding their people's training needs and updates. If a separate program is being used by the training department, the CMMS system should be compatible and can communicate with their existing system. Likewise, the Purchasing Department must also have access to CMMS Spare Parts Management Menu for purchasing stocks and parts for maintenance. While the Training Department is busy educating its people on the new CMMS, all people involved with the software, the IT department, and the CMMS vendor should finalize debugging and improving the system software based on the feedback from the users CMMS on Step 6.

Step 8: Close the CMMS Roadmap and make the final report: Once the CMMS training and debugging have been completed, let the software run for a month and interview or survey people directly or indirectly using the CMMS. Summarize the interview and prepare a final report regarding the project. The CMMS roadmap will now be closed, and it is time for the cross-selection of team members for this project to go back fully to their original work. Provide the CMMS team with the final report. Updates, revisions on the CMMS should be provided to the responsible department initiating the revision, whether this is a modification, redesign, or Root Cause initiative. For RCM case studies completed, the current state of Preventive Maintenance should be revised, conforming to the findings generated by the RCM team. The RCM derived activities generated weekly, monthly, quarterly, and so on should be revised on the CMMS to conform to the newly derived RCM worksheet; the RCM team should be responsible for this initiative.

Step 9: Assess your CMMS performance: After a year of implementation using the CMMS software, evaluate and assess CMMS performance's current status if the users are satisfied with the outcome or if other features need to be added to the system. Does the current CMMS live up to our expectations? Is there anything we need to improve, or is there any feature that must be added, which is definitely needed by the user? Have we captured all the needed information and features in the software?

12.2: February 2018: Two Types of Improvements

There are two types of improvement, **kaizen and innovation**. The kaizen improvement originated from Japan, the east, while innovation is mostly preferred by the west. The Kaizen methodology is the simple belief that any product, process, service, or anything devised by man can continuously be improved. It contradicts Fredrick Taylor's belief that there is only one best way to do anything. It is the simple truth behind Japan's economic miracle, the reason the Japanese have become masters of flexible manufacturing. The KAIZEN strategy message is that not a day should go by without some small incremental improvements done in the company. Kaizen refers to small but consistent and continuous improvement.

On the other hand, we have innovation, which is more of a breakthrough improvement in which its impact could be felt. Kaizen requires a mindset; that is why all employees must contribute and suggest something, which is a way of life for most Japanese manufacturing firms. It does not matter whether its impact is big or small. It's a part of their day-to-day culture and philosophy. I was employed many years ago in a Japanese firm, and as far as my poor memory can recollect, they would prefer to use the word "better" than the word "best." Their reason is simple because they challenge themselves that the best can still be made better.

Comparison	Kaizen	Innovation
• Effect	Long term and lasting	Short term but dramatic
• Pace	Step by step, incremental	Great leap, a big step
• How often it is done	Many times continuously	One time
• Timeframe	Continuous and incremental	Intermittent and non-incremental
• Change	Gradual and constant	Abrupt and volatile
• Involvement	Everyone	Few selected with the capability
• Approach	Team approach	Individual's creativity
• Mode	Maintenance Improvement	Rebuild, redesign and modification
• Spark	Conventional know-how and state of the art	A technological breakthrough, new invention, and design theories
• Ideas	Information is open and shared	Classified, confidential, trade-secret
• Requirement	Requires little or no investment but great effort to maintain	Requires large investment or capital but little effort to maintain
• Effort Oriented	People	Technology

Figure 12.1: Comparison between Kaizen and Innovation

Perhaps kaizen, also known as continuous improvement, has something to do with the Rolling Stones' famous song, "I Can't Get No Satisfaction," as the band itself, the older they get, the better they become. Kaizen's true meaning is somewhat difficult to absorb and comprehend since this is about mindset. Most Japanese operators and maintenance will do small improvements in their equipment or asset since their belief is that the older the equipment gets, the more it should run smoothly. This mindset is very different from the Philippines and other countries in Asia (as I cannot speak for the rest of the countries) since the older the equipment gets, the more problematic it becomes.

Innovation, on the other part, refers to something new. It is considered a change made to an existing product, service, idea, or field. We can say that the first telephone invented by Elisha

Gray and Antonio Meucci was an invention; the mobile phone, today with so many apps and features, can be considered an innovation. Not everyone can innovate, unlike kaizen.

Let me share this short story about a Japanese manufacturing soap factory to explain the difference between simple kaizen and an innovation. The final process in this soap manufacturing will be packaging the soap and placing it in a box distributed to the retailers and supermarkets. The problem starts when one of the customers who purchase this soap only contains the box with no soap inside. A few days followed, and several customers also complained of the same problem. One customer even posts it on his Facebook Social Media about the box without the soap inside and naming the company, which went viral with more than one million views. It really damaged the reputation of the company. The factory's management was alarmed and called for a meeting with the head of engineering, quality, and process on all those big heads in the plant to address the problem. After several days of brainstorming, the factory decided to invest in an expensive x-ray machine, which will be included in the process. One operator was also assigned at the x-ray to inspect all soaps 100% on her shift. The plant operates for three shifts, so there are three operators assigned to do the task. There will be one operator per shift assigned to the x-ray machine, just like in the security area before you board your plane. The soap factory management team also purchased an automated weighing system that includes all the soap in the box that should be measured with + 5 /- 5 grams. The operations team again assign three people, one operator per shift, to do the tasks.

The plan was executed with the x-ray machine and automated weighing system, but still, there are a few customers who purchased the soap without the soap but only the box. Although the complaints were reduced, it was not 100% foolproof. The six operators assigned were doing their tasks, which were assigned to them even if it bored them to death. But, hey, they also need to break and eat their lunch. This means that both the x-ray and the weighing system were unmanned when these operators eat their lunch. Finally, one operator proposed his suggestion to his supervisor to borrow one of the blowers from their facilities and run it continuously, blowing away those empty boxes that contain no soap inside. When this was implemented, the six operators were relocated to other parts of the production, and both the x-ray machine and the automated weighing scale were never used again. It was just a simple suggestion from an operator which they implemented, and the problem of the missing soap was finally solved. In Kaizen, they make their improvements as easy and as simple as possible.

• When NASA began the launch of astronauts into space, they found out that pens wouldn't work at zero gravity since the ink would not flow down to the writing surface. It took them one decade and an estimated amount of US $12,000.00 million to solve this problem. The NASA Engineering and scientists developed a pen that worked at zero gravity, up-side-down, underwater, on practically any surface, including crystal, and in a temperature range from below freezing to over 300 ° C. What did the Russians do? They use a pencil.

12.3: March 2018: World Class Maintenance Management Book

My intent in writing this my first book is to provide a strategic framework on what it takes for an industry to achieve a World Class Maintenance Management structure and let each one understand the role of maintenance in our industry. The role of the maintenance function is not about eliminating failures but understanding that each failure has its own unique and distinctive consequences. Maintenance is not about repairing and scheduling equipment for repair. It is much more than that. Each of us who belongs to the maintenance function should stand proud because maintenance plays a major role in the success or failure of any industry. Achieving an excellent maintenance structure is not only about having the best Predictive Maintenance instruments and CMMS software in town, but it has more to do with the people, their culture, and how each maintenance thinks.

This book is divided into three parts. Part 1 covers the basic discipline that has often been neglected in our equipment. Part 2 refers to the intermediate discipline, which refers to the different strategies, which must be adapted on maintenance to improve and sustain our equipment reliability, and Part 3 covers the more advanced disciplines.

Part 1: Basic Discipline – Back to Basics
- Discipline 1: Training and Education
- Discipline 2: Set-up Maintenance Indices and KPI's
- Discipline 3: Autonomous Maintenance
- Discipline 4: Addressing Basic Equipment Condition
- Discipline 5: Preventive Maintenance

Part 2: Intermediate Discipline – Maintenance Strategies
- Discipline 6: Spare Parts Management
- Discipline 7: Life Cycle Management
- Discipline 8: Lubrication Strategy
- Discipline 9: Root Cause Failure Analysis
- Discipline 10: Reliability Initiatives (TPM and RCM)

Part 3: Advance Discipline – Specialized Strategies
- Discipline 11: Predictive Maintenance
- Discipline 12: Computerized Maintenance Management Software (CMMS)

It is not easy to change a maintenance system that had been in place in the industry for many years, but every company can transform from a firefighting to a world-class maintenance structure, and this is the objective of writing this book. I have detailed every step so you can follow them thoroughly. Let us face the facts about your system of maintenance so we can learn from them. Many companies are struggling to survive together with their competition. Maintenance is one big factor where a company can save on cost if one has the right tools and knowledge to convert its maintenance task into an excellent maintenance structure. Reliability cannot be achieved by cutting costs, but improving reliability will reduce our maintenance and operating costs.

There is no such thing as plant overnight success or one-day transformation. It will take several years to achieve a state of world-class maintenance structure. Being world-class is a long-term and not a short-term approach. It is a long journey, yet the time spent will be very rewarding. I recall a movie I watched in 1991 about a reporter and an arrogant congressman.

The reporter interviewing the congressman said that he wanted to confirm that the cases pending in their justice system are very slow based on his sources. The congressman laughed at the reporter and said that their justice system is fast and that his sources were unreliable. To prove his point, he invited the reporter to come with him to the justice hall to observe. And so they went to the hall of justice, and a court hearing is currently taking place. So both of them sat down and listened behind. It was a case of two mothers claiming custody of a baby, so the judge called the two mothers, and someone said that both the mothers were already deceased. The judge then asked for someone to represent each of the mothers, but no one came forward. Finally, the judge asked for the baby boy's presence, and an old man went up to the stand. The judge looking curious asked the old man who he represented and what his purpose was in going to the stand. The old man replied that he was "baby boy." I just hope that changing your culture would not take this long.

Achieving a world-class maintenance structure is not about being the best but giving it our best and doing it better. Having a world-class maintenance structure is not about improving our equipment but improving the people maintaining our equipment. The focus will always be on the people. You need three things to make this happen, a plan, a team, and a big heart. This book is about the first part: having a plan; I leave the other two elements with you: a team and a big heart. When your people unite into a common goal, you can only realize that your maintenance people can move mountains. Make these changes now, not because you can sleep soundly at night without your phone ringing and asking you to report to your plant due to an unexpected breakdown but do it for the right reasons. Millions of dollars can be saved on maintenance by adopting a change in your maintenance system. Investments should have to be made in the people because people will improve the equipment. It is not the other way around, and this is not where we should cut costs. The first step in every change initiative is getting yourself and your people trained and educated. Industries today adopt one law, and that is the law of the jungle: survive or be left behind; that is why they need to adopt the best maintenance strategies is a must.

This book is about the sum of my previous experiences with Amkor Technology, where I worked with TPM for seven years handling the pillars of Planned Maintenance and where we have actually implemented and reaped its benefits. Here is also where I learned the process of Reliability-Centred Maintenance and have it applied in our Facilities section. This book also includes my knowledge of the best practices from both the eastern and western parts of the world, their conflicts, and how they can complement one another. This book also covers my visits, seminars, and consultation I have conducted with various manufacturing and non-manufacturing industries such as shipping, oil, and gas, metals, automotive, mining, and power plants, where I have come to realize where the best standards in maintenance exist, my continuous communication with other reliability and maintenance experts from around the world who consider me not as a competitor but a friend, and ally. This book also covers my previous training on reliability I have conducted and implemented, such as Lubrication Strategy and Condition-Based Maintenance during my employment with Lepanto Consolidated Mining Company, and my principles and passion for understanding what it takes for an industry to achieve a state of world-class maintenance management structure.

This book depicts the life and struggles of maintenance in seeking better ways to manage and maintain their equipment and assets. The author shares his passion and experiences about the day-to-day struggles in the life of maintenance. The book is easy to absorb as it is structured into three parts: the Basics, the Strategies, and the Advance Disciplines. The Twelve Disciplines are grouped accordingly into these three parts. Maintenance often time seeks advanced ways to deal with their everyday problems and issues. This book's message is that there is no better way to start by going back to the "basics" and addressing these very small problems we have in our plant. Big problems, unplanned breakdowns, and catastrophic failures are just an accumulation of small problems that have always been ignored and most neglected. The author strongly emphasizes the importance operators' play in addressing this basic equipment condition and is considered a partner with maintenance on this shared responsibility they have towards their equipment. It is very difficult or impossible for maintenance people to transcend from a reactive to a proactive mode if operators are not involved along the way. When the Basics had been set and well established, then maintenance can move on with the different maintenance and reliability strategies, explained in detail in this book. Each chapter covers a specific maintenance discipline. Chapter 14 of WCM book covers an implementation plan on how to proceed with these 12 disciplines by understanding where the current maintenance is and bringing the maintaining herd in the direction they want to go. Finally, the author would like to share his message on the conclusion part of this book and share his actual experience on what it takes for the industry to reach World Class Maintenance Management Stage.

12.4: April 2018: Maintenance - Roadmap to Reliability Book

This book on **Maintenance Roadmap to Reliability** depicts the life and struggle of maintenance in seeking better ways and means to improve the equipment and assets' reliability. The author shares his experience on how to achieve such a feat. Transitioning from a reactive to a proactive maintenance stage is not an easy task, but it is not an impossible task. The author believes that the key to everything is educating the maintenance people on what maintenance is all about. Training is where we acquire knowledge to develop the skills required to do our job right. This book contains real-life stories, struggles, and actual experiences by the author in his maintenance and current Reliability and Maintenance Consultant career.

The message of this book is simple and straightforward. There is no better way to start the journey to reliability than to go back to the basics and address these very small problems in our plant. Big problems, unplanned breakdowns, and catastrophic failures are just an accumulation of small problems that have always been ignored. Maintenance is always shared responsibility for operators and maintenance working together in complete harmony. It will be difficult for maintenance to transition from a reactive to a proactive mode if operators are not involved in maintenance since maintenance is always shared responsibility for operators and maintenance. This book explains how to proceed with the 4 Phases of Planned Maintenance and integrate RCM into the TPM process. It also covers the importance of Autonomous Maintenance and Spare Parts Management, which is believed to be the missing link theory on any reliability and maintenance strategy. Chapter 11 of this book is a classic case study on what maintenance can achieve if there is a clear roadmap to follow. The last chapter states that maintenance is just human like you and me. It is important not to blame them for every failure in the plant but for both operations and maintenance to work together.

Improving the reliability of your equipment and assets cannot be done overnight. This is when you need to change from being a manager/engineer to being a leader. This is where you need to turn these negative things into positive thoughts. This is where you need to command respect regardless of your position. Most importantly, this is where you need to realize the virtue of extending your patience. It is your job to motivate these maintenance people to follow you. Nelson Mandela unites South African people through a sport known as rugby by motivating his players to win the world cup. Mahatma Gandhi inspired his people through the use of non-violent means. He was a stubborn man who would starve to death by fasting unless his principles of non-violence would be heard and followed. As a result, he was not only heard in his own country but his non-violence principle was heard worldwide. Remember that maintenance people are much inclined to do the same old things day in and day out. If what you say is new to them, expect a lot of resistance, which is quite normal initially. People assigned to handle this kind of initiative should have a strong conviction and passion that what they are doing is the right thing and for the benefit of the whole maintenance in your plant.

Initiating a reliability and maintenance improvement is all about bringing the whole maintenance herd in one direction, and this can never be an effort of a single person. It can never be a one-man strategy since you need to take the whole maintenance workforce to work together towards a common and united goal. Although equipment and assets in industries vary from one another, much of what is said in this book can be applied, whether on manufacturing, mining, oil industry, power plant, process plant, automotive, or whatever type of industry as long as you have assets and equipment to maintain. The strategy covered in this book can be done for individual equipment, component, process, system, or sub-systems. What is written here can be applied and adopted. So let me provide the reader with some basic guidelines on starting a plant-wide maintenance improvement in your industry.

The first thing you need to do is to develop a roadmap for your maintenance structure. Start by conducting a maintenance survey and have a random group from across all divisions answer the survey. The survey should tell what areas of maintenance need to focus on and improve. This will serve as a guideline on how to start or kick off the activities on maintenance. It is important that before trying to embark on any reliability and maintenance improvement journey, maintenance should have a specific and crystal clear direction on where they are headed forth. Provide a time frame on how long it will take for them to complete each set of roadmap. Provide specific indices and Key Performance Indicators on what is important for them to measure since this will spell the difference between their initiative's success or failure.

After you develop the roadmap, try to meet with your top management team for a buy-in. Explain to them what will be required initially in undertaking this effort and make sure that they understand the benefits that can be derived from initiating this maintenance strategy in your plant. Seek not only their support but also their commitment as well. Remember that there will be some investment initially, especially when restoring the equipment and setting up the Predictive Maintenance strategy.

Once you got their nod, you need to organize a core team for maintenance. The membership will depend on how large your plant is. Also include the facilities/utility department, as they play a specific and major role in maintenance. If you belong to a manufacturing plant, the consequences of breakdowns and failures from facilities/utilities will be more devastating than equipment failures on the production floor, and they will have a major impact on your plant. Each member of the maintenance core team should represent his area of responsibility. Make him aware of his roles and responsibilities for this initiative. Once the maintenance core team has been organized, meet regularly. Conduct some gap analysis between the current status and what the team wants to achieve in the future. Develop a master plan projected on a 3 to 5-year scale. A good starting point can be derived from the book of Stephen Covey on Principle-Centered Leadership by asking the following questions:

First: Determine where we are. This will define your current situation and baseline on where the maintenance organization structure currently is. Make a selection of people and start with a survey or you can select a cross selection of maintenance people and have them conduct an internal survey on the different disciplines on maintenance. Knowing where you are at the starting point on any continuous improvement and reliability initiative and effort.

Second: Where do we want to go? This will be the direction in which the team is going. It is important for maintenance to know the destination where it is headed in the future, similar to a vision and mission. Aim for a high or ideal vision and set goals so that people will continuously improve. Set a time frame for achieving it.

Third: How are we going to get there? This will be the path or map to be done to implement these disciplines. These are the activities and steps to be done to implement the different strategies and disciplines on maintenance. Maintenance must understand that these strategies are long-term and not on a short-term basis. Generate a master plan for each of these disciplines and have them review regularly.

Fourth: How shall we know we have arrived? The first and last step is defining and measuring indices that tell us that we have arrived but are headed in the right direction. It is important to measure these performance indices to determine if maintenance has reached its goals and targets. Measurements should be defined at the very beginning of implementation and must be reviewed regularly. The success or failure of all these efforts depends entirely on these measurements and KPIs.

Once the improvement areas have been determined, determine the initial training that must be initiated initially. For example, suppose your focal point of improvement is on lubrication; your people need to take a full-blown course about Lubrication Strategy and Oil Contamination Control. If you want to improve your current Preventive Maintenance structure, you need to have the maintenance people trained on Reliability-Centered Maintenance. Remember that training should be considered a long-term initiative, which means that this will be required by maintenance from time to time.

Once the indices improve, there will be less resistance, and other people will follow the path. It is important to let these people realize that they are initiating this roadmap on reliability to ensure that the industry remains competitive and is here to remain. Lastly, people want to be recognized for their efforts. Provide some very simple yet unique recognition to teams that successfully deployed this initiative with success. Rewards and recognition may not necessarily be in the form of money or raising their salary but something they will remember and make them proud of as part of this initiative. Remember that this is a team approach and not an individual performance, so credit should be given to the team and not to a single person.

I have been in training and consulting maintenance and reliability courses for twelve consecutive years, and I have been privileged to have taught many maintenance people in industries. My understanding further deepens the real problem maintenance people face in their day-to-day work. It allows me to meet different people from maintenance working in different industries, countries, races, and cultures. Despite the type of industry you are in, the maintenance problems are just one and the same. Almost all of them are man-made. Being reactive and the lack of training tops the list, but I still see hope for maintenance people in industries despite all of these. I believe that the way we do maintenance has something to do with the people doing it. Maintenance is all about people, nothing more, and nothing less. The saying that the company's greatest asset is its people is not always true. What is correct is that the right people are the company's greatest asset, and we can only have the right people if they are equipped with the right knowledge and skill to perform their job correctly. Training will always be the backbone of any cultural change. Industries cannot remain the same if they want to exist in the future. It is like the law of the jungle where only the fittest survive. If you ever thought that training is expensive, then think about the errors and mistakes we do in maintenance; it is much more expensive than training.

There is no rocket science or silver bullet solution that can transform a plant's reliability overnight. Everything will start by educating the maintenance workforce with the right knowledge and providing them with the right courses on reliability and maintenance. If we provide these humble people the chance to learn and gain knowledge of their equipment, they will improve the reliability of their equipment and assets. Remember that people will be the ones to improve the equipment. It cannot be the other way around. It will always be the people who will determine the culture of maintenance. And this aligns with my mission, which is to reach out to industries to discover ways to improve their maintenance human resources. All I can say is that despite the day-to-day pressures at work, maintenance will always remain humble and down-to-earth human beings. (I speak for the majority.) On a typical day, these people mostly work in shifts, and when equipment fails, they attend to it and do their best to fix it. Like a good captain, they never abandon their post and see that the equipment is up and running before they call it a day's work. Maintenance people are like soldiers. They never leave a good man behind. Whatever heroic deeds they have done for the day, they know that nobody will ever thank them for doing a good day's work because they know that maintenance is simply a thankless job, and they will always remain a complete unknown. They know that they are paid to do the job. This book will benefit all maintenance in providing them with a clear guide and roadmap on improving their equipment reliability. Lastly, I would just like to state for the record that improving the reliability of your assets and equipment cannot be done overnight. It is like playing the guitar the first time around. It takes time to become like the legends.

12.5: May 2018: Reliability - A Shared Responsibility for Operators and Maintenance Book

The message of my third book is that maintenance is not for maintenance alone to perform. It is for both operators and maintenance to perform, but the two functions remain separated in most industries. Operators are important in any reliability and maintenance strategy since they are the people closest to the asset during failure. It means that they will be the ones that will experience the failure first and not maintenance. They are our first defense on equipment failures and breakdowns and what is important is that they can sense the failure from happening at the earliest stage possible if they know their equipment well. The majority of failures and industrial accidents worldwide can be prevented if the operator knew the earliest stages of failure. This book strongly emphasizes the bond between operators and maintenance and creating a permanent partnership in maintaining the equipment. The purpose of writing this book is for industries to realize that operators will also play a major role in maintenance and that maintenance can "never" escape the vicious cycle of being reactive if operators will not be involved with maintenance itself. I believe that creating a partnership between operators and maintenance will always be difficult during the start-up. It does not matter whether it will be done through Autonomous Maintenance or other means. It is important to break down the departmental barrier and for operators and maintenance to develop a solid partnership for both of them to start preserving their equipment assets. Empowering operators requires a culture change. This cannot be done overnight; it will take years to build an empowered workforce since changing the people and the organization's culture. The role operators play will be critical since they can sense the earliest stages of failure, leading to a catastrophic breakdown. These failures can be prevented if industries will give a chance to operators on being not just switch flickers but having a thorough understanding of the equipment and assets they operate.

This book is composed of twelve chapters in which I include a quiz at the end of each chapter for the reader to answer to grasp the level of understanding they got from reading each chapter. Chapter 1 discusses why operators are important in any maintenance and reliability strategy. As our equipment continues to be upgraded and automated, we need operators besides switch-flickers that can operate the equipment, but we need operators who can sense if something is wrong with their equipment at its earliest possible stage. Dealing with a small problem is less expensive than waiting for the failure to come. The breakdowns and failures we experienced on our equipment are just merely an accumulation of small problems that had been neglected so far. When these problems were small, nothing had been done to correct them until another small problem emerges and another and another in which finally the equipment can no longer bear which ends up in a breakdown. And when the machine fails, then that is the time we react.

Chapter 2 explains what maintenance is all about. What it can and what it cannot do. Maintenance is simple, but often industries complicate matters. For example, Preventive Maintenance is one of the strategies for maintenance. This is a very good strategy indeed as its role is to extend the lifespan of the asset instead of doing maintenance on a reactive or crash basis, but the problem is that most industries misuse, abuse, or overuse this strategy ending up with more breakdowns instead of the other way around. Chapters 3 discuss human errors. This is a very important topic as most of the world's industrial accidents, and disasters worldwide were mostly caused by maintenance and human errors. Suppose operators are taught by maintenance on their equipment and know-how to sense the smallest signs of their equipment problems; in that case, most of these disasters can be avoided. Although technically speaking, there is no way to eliminate human errors since this is part of being human. Human errors have many origins, and even the best and smartest employees we have can commit the worst errors and mistakes, but the good news is that human errors can be managed more intelligently. What is important is to understand the risks and consequences involved in the error. Chapter 4 discusses the most common Key Performance Indicators recommended for maintenance and operators, although the operator's indices will be more intangible than maintenance indices, which are more tangible and measurable. The important thing is that before starting any maintenance strategy, it is important to create a starting point, just like when we want to gain or lose weight. This KPI serves one purpose: to tell us if we are successful or not in our maintenance initiative. Chapter 5 discusses key points in strengthening operators and maintenance partnerships. A good way to start is to assess both the operators and the maintenance function in your industry to determine both their needs. I have included a classic case of an operator and maintenance assessment, which I conducted in the Philippines to determine where your industry is from the worst to being world-class in terms of maintenance. While industries claim they are world-class in terms of their services, quality, safety, facilities services, and so on, achieving World Class Maintenance Management is a different story and perhaps the most challenging.

The second part of this book covers Chapters 6 to 10, which discuss how to adapt the different strategies for both operators and maintenance. Chapter 6 has written detailed guidelines on how to implement Autonomous Maintenance for the operator's function. Chapter 7 is a detailed guideline for implementing TPM Planned Maintenance in 4 Phases. Both Autonomous and Planned Maintenance will start by implementing the Preparatory Stage, Step 0, and Phase 0, respectively. Phase 1 deals with restoring the equipment to bring it back to its original basic condition, while Phase 2 will address inherent design weaknesses in the equipment and how to move forward in implementing these phases. Chapter 8 provides detailed guidelines for implementing Phase 3 of Planned Maintenance, which can be done through Reliability-Centered Maintenance for industries using the classical seven basic questions on RCM. While there may be many variations on how RCM is implemented, I think that the best way to implement RCM is to do it based on the original version of Stanley Nowlan and Howard Heap, which are considered the pioneers of the RCM methodology.

Chapter 9 discusses how to implement a Root Cause Failure Analysis investigation for equipment-related breakdowns and failures that maintenance can perform on recurring failures and those in which the failure consequences are unacceptable. Chapter 10 is a detailed guideline for implementing equipment FMEA/FMECA on equipment and assets. Although the readers of this book may come from different industries, RCM and FMEA should be done based on how the equipment is being set up in your plant. For example, if you are working in an oil and gas industry where equipment and components are interconnected with one another, it is best to do RCM and FMEA on a system or sub-system level, while if you are working in a manufacturing plant in which your equipment is independent of one another, the best approach will be to do RCM or FMEA by selecting the pilot equipment which should be the worst in terms of breakdown and cost in your plant. If you work in a process industry, the best approach is to do RCM or FMEA by selecting the bottleneck or constrained process. If you have very big equipment with so many sub-assemblies, I recommend creating a ranking to determine the worst sub-assembly in that equipment and prioritizing that on the RCM or FMEA process. I have written the nitty-gritty details on implementing these strategies. Chapter 11 is a detailed guideline on conducting the Focused Improvement pillar of TPM in your industry. This chapter discusses how a Focused Improvement team is organized, what problems they need to address in their industry, and the requirements needed from the team before starting their activities.

Finally, Chapter 12, the last chapter of this book, calls for unification between operators and maintenance. This book's message is simple, creating a permanent partnership between operators and maintenance. While creating a partnership between these two may not be accepted by many industries of today, the reason I wrote this book is for industries to realize what they are missing, which is something big if these two remain separated. I hope that industries will change for the better in the next generations to come. More and more industries adopt a permanent partnership for both operators and maintenance as they cannot co-exist without one another. We may be of different races, cultures, ethnicities, religions, or beliefs, but all I can say is that whatever industry you are working in right now, if you are still employed, the problems on maintenance, regardless of industry, are just one and the same. This book's message may be ignored by industries for now, but I believe this book needs to be written to serve as a wake-up call for all industries to realize their biggest mistake, separating operators and maintenance. For as long as these two remain separated, maintenance in industries will always remain reactive.

12.6: June 2018: Why Is Cheap Actually, Expensive?

Perhaps you and your friends went out and bought some personal things for yourself. The Department Store had some big discounts on clothes' sales, and it definitely suited your tastes. Finally, you decided to purchase some clothes you wanted to wear for certain occasions. You looked in the mirror, and the pants fitted you perfectly. However, when the pants were washed, they shrunk by 3 inches, and now you cannot use them anymore, but they fit your younger sister, that is, if you have one.

Ten years ago, I decided to paint my roof and hired a painter to do it. The painter advised me to purchase some primer paint (red color), so I went to the paint shop to purchase some paint. There were varieties of paints to choose from. Some of them were even shown on TV commercials, but those shown on TV were higher than some unknown paints. The paint difference is quite a big margin since low-cost paint costs around Php 245.00 per gallon or roughly around USD 5.21. The cost of some leading brand paint shown on TV was around Php 390.00 or USD 8.30 USD. (Conversion used here is 1 USD is equal to 47.00 Php). I've decided to purchase the cheaper paint since I will need around 5 gallons of it. When I reached home, I contacted the painter, and he went to our house. I contacted him and paid him daily to paint the roof. He said that I should have purchased a more reliable brand to the last longer, but I guess the paint was already there, so he just started painting and completed the work, which lasted for a week. A year later, I had some problems viewing my TV reception since the antenna's orientation changed due to the wind. I decided to go to the roof to fix the antenna; when I saw the roof, almost all the primer paint faded. I might have saved on the paint's initial costs, but since I needed to repaint it again, the paint's cost and the labor have now been doubled and is much costly than the leading brand of paint. If I considered both the paint's economic part, perhaps I would be saving more if I purchased the paint with an initial higher cost. That was a very poor decision on my part.

There is always a temptation to purchase equipment or spares based on the lowest or cheapest possible source for industries. This might not be such a good idea since purchasing based on the initial costs only tells us one side of the story. The true costs can be seen based on their performance. In my experience, this is the problem with most procurement and purchasing departments since they decide to purchase parts based on the lowest bidder or lowest possible costs, leaving the problem to maintenance. The savings these departments claim are insignificant since both operations and maintenance encountered many failures due to this. They always look at the part's initial cost and not the cost of problems the part may give the user in the long run. This is the problem with cheaper parts where equipment that fails every day and will likely yield a much larger amount of cost in a period compared to slightly higher equipment. The equipment's true cost can be felt when the equipment was commissioned to the plant until decommissioned or disposed of. Decisions about purchasing or acquiring equipment and spares should not only be based upon the tag price or initial cost of the part or equipment itself but rather on its Life Cycle costs. Some of the questions that need to be raised:

• How will the equipment or part perform in the long run?
• What will be the cost of operating and maintaining this equipment?
• What are the critical parts of this equipment?
• How frequently should the part or equipment be replaced?
• How much is the cost of spares and downtime?
• How long will the equipment be down in the event of a breakdown?

Purchasing cheaper parts and equipment may actually mean spending more in the long run. A true and meaningful cost reduction effort can be achieved not by purchasing the cheapest costs but by having a thorough understanding of what Life Cycle Cost is all about. Many plants are initiating a cost reduction program in their plant; the best way to reduce costs is to understand that it is more important to look into the equipment or part's running cost rather than the initial or procurement cost of the equipment or part itself. When you buy a shirt in a department store, it has a tag price. In the study of Life Cycle Costing, this is what we called the initial cost of the shirt. However, once you purchase the shirt, there will be a cost to spend, such as washing the shirt, the labor of the person who washed it unless you that wash it, detergent consumed, ironing the shirt, electricity, water bill, and so on. These things are what we called the running costs so that we can re-use the shirt once more. In the study of Life Cycle Costing, it is also important to look not only at its initial costs but as well as the running cost since there is a possibility that purchasing the cheapest equipment may not be the best alternative unless we have an idea of its running costs which may end up more expensive than purchasing a slightly higher cost from a different vendor. You see, the initial costs of the equipment are just one side of the story. When industries purchase new equipment, there will be costs in commissioning the equipment, such as the installation cost, transportation costs, warranty costs, debugging costs, contractor's costs, especially if they maintain the equipment for the first few months of operation. And when the equipment can now be operated, we will still have a series of costs, including its maintenance cost, spare parts costs, downtime costs, energy costs, facilities costs, modification costs, training costs, operator and maintenance costs, and others. My point is that in making decisions on purchasing new or additional equipment, in the future, the industry should focus on its initial costs and running cost. The equipment may be the cheapest in terms of its initial costs but not exactly its running costs.

Sometimes a higher performance product costs less than a commodity type product, even though it is higher. To gauge whether this will be the case, we should look at the product life cycle cost rather than its purchase price or initial cost. Life Cycle Costing is a way of analyzing equipment purchase choices. If the decision was based on several factors rather than its initial costs, we would make our selection based on the least amount to own over its entire life. This is all about Life Cycle Management. Equipment must not only be inexpensive in terms of its initial cost (also termed as procurement cost, fabrication cost) but also in terms of its running cost. This brings us to the key concept of Life Cycle Cost and Management. Therefore, our goal is to develop and design equipment with the lowest possible Life Cycle Costs. LCC must not only be looked upon entirely on purchasing new equipment but rather should also be used in selecting the right spare or component to use. Most equipment designers emphasize equipment initial cost over its running cost, but today more designers are now thinking about LCC. They have now learned that the surest path to profitability lies in minimizing equipment LCC. Here is one of the many actual cases that an item may be expensive in its initial costs, but the real savings can be felt in the long run.

Details	Before	After	Unit
Filter Size (L x H x W)	24 x 24 x 6	24 x 24 x 12	inches
Cost per Unit	$ 47.00	$ 60.00	USD
Total Filters per Air Handling Unit (AHU)	16	16	pieces
Filter Costs per AHU	$ 752.00	$ 960.00	USD
Total Air Handling Units	7	7	units
Average Life in hours	1800	3600	hours
Average Life in years	0.2055	0.41096	years
Cost for 7 AHU per year	$ 25,615.57	$ 16,351.96	USD
Yearly Savings in USD	$ 9,263.62 per year		USD
Year Savings in PHP	PHP 472,444.62 per year		PHP

Figure 12.2:   Modifying the Width of AHU Filters

Case Study 1: Resizing the Width of AHU Filter from (24 x 24 x 6) to (24 x 24 x 12)

- Division: Facilities
- PM Step: PM Step 4
- Team Leader: Arman Jusay

• PM Committee: Arman Jusay
• Model Machine: AHU-401
• Impact: Lengthen the life of the air filter from 1800 to 3600 hours
• Note: 1 USD = 51 PHP

12.7: July 2018: The Importance of having a Strategic Planning for Maintenance

It is important for maintenance in industries to have a yearly strategic planning. This can be done before the year ends or in the very first month of the New Year, which is January. The objective of this strategic planning for maintenance is to provide a clear direction and strategy" in achieving a close to perfect high-reliability equipment and safe workplace. This is done by identifying all obstacles and pitfalls in achieving a "World Class Maintenance Management structure. Maintenance needs to understand why they need some time to reflect on their previous activities on things they need to change for the better.

Why the need for a Maintenance Strategic Planning

Know where we are and where maintenance is going. Without Strategic Planning is like sailing to the sea on a sailboat without a rudder but rather at the mercy of the wind or driving a car in which the windshield is painted black. Maintenance needs to look back and reflect on the past. This is where they ask themselves, have the goals been met? What are the things that maintenance should improve upon? What is important for maintenance is developing both a short-term and long-term strategy for the year by asking, am I satisfied with our current maintenance system? Ask ourselves, are we ready for the change, or can we just work the way we were last year? Is my current maintenance organization equipped with the right tools and tactics to carry out these changes? Do we need to change now? Do we need to change things here? Was this a good year for our industry?

Figure 12.3: Actual TPM Planned Maintenance Strategic Planning 1999

Culture Defined

Culture is the way of doing things around here. It is about people's common values, shared beliefs, shared things, shared sayings, shared doings, and shared feelings. Culture is the pattern of basic assumptions that a given group has invented, discovered, or developed in learning to cope with its problems of external adaptation and internal integration that has worked well enough to be considered valid and therefore taught to new members as the correct way to perceive, think and feed about these problems. (Schein 1983). Culture is like a fingerprint as there are no two persons with the same fingerprint, and each person has 10 fingers with different fingerprints. Like industry, there are no two industries with the same culture, and every industry has different departments in which their culture might not be the same, just like our fingers having different fingerprints. A culture change is possible if we can alter or change the way we think and do things around here. People are willing to make changes in the organization for the better. What is important is for people to provide ownership of the initiative for change. Members should know their role in the change initiative by making amendments to previous rules, policies, and procedures for the better. It is important to understand the roadblocks and hindrances in these initiatives and recognize people for their efforts. Cultural change is possible, but don't expect it to happen overnight. A strong culture is embedded by having a common vision and mission, a sense of inclusion among its members, and pride in the workplace. People need to have a sense that they belong to their organization.

Why do Improvement Programs fail In A Company?

- Failure to create a powerful mandate for change. In our experience, the biggest barrier to change is that the organization is simply not ready for it.
- Failure to deliver early, tangible results. If we are not showing tangible benefits in six months, expect support to cut in halve, barriers to double, and an increase in resistance to change. Most managers always look at the results, but the better manager should not only look at the results but at the process of how the results were achieved. This creates better communication between their people.
- Measurements and indices are not aligned with the company's corporate goals. If our improvement goals and measures are different from the company's goals and objectives, we can be sure that the program will be treated separately.
- What's in it for me is unclear? Although this is a management prerogative, management must seek ways to recognize the team; rewards may be in the form of recognition. A guaranteed job, career growth, and other financial rewards. The old thinking simply states that everyone is waiting for a miracle man who will change and do everything; it won't happen this way.

It is important to clearly state the reason for the change and do it right the first time. Otherwise, any initiative for change will treat it "as another program of the month." Some factors are essentially and equally important; if one is missing, changing the maintenance culture will be impossible. These are having the right tools and a heart to make the change and having a system in place, and the right attitude to do the job. Many people do not want to improve since their thinking is that improvement programs are fads, and they aim to reduce the workforce in the long run. Programs fail because they are poorly planned, developed, and implemented. Most improvement strategies and programs look good on paper. People view these programs as the latest in a series of fads that come and go. When a program takes too long to implement, management will abandon it, and they're back to their old ways.

How Do We Change the Maintenance Culture?

I think the first step is to form a steering group or core group. For union plants, be sure to use a joint union/management selection.

Define the Planned Maintenance Vision and Mission: Once the member of Planned Maintenance has been organized, they need to define their maintenance vision, mission, policy, and goals, which should be aligned to the overall vision and mission of the company. This will be their credo that the maintenance organization will be aiming for shortly. This will be their direction, which they will aim for in the future. Be sure that this is observed and that all activities being performed on Planned Maintenance must be aligned to their corporate mission and vision.

Conduct Machine Inventory and Ranking: A group of people can be assigned initially to conduct an inventory on all active assets, machines, equipment, systems, sub-systems used currently in their plant. This will be their official machine inventory list. Only equipment being used should be included in the list. Exclude assets that are planned to be decommissioned shortly. After completing the list of equipment and assets, conduct a machine ranking for all assets and equipment. Rank A will be the worst, Rank B getting worst, and Rank C should be tagged as good machine. Priority for Planned Maintenance activities will be given only for equipment and assets categorized as Rank A or worst equipment, followed by Rank B. After the inventory has been completed, set up criteria for ranking equipment and rank them all accordingly. Perform this on all assets and equipment lists.

Select the Pilot Machine and Backtrack Breakdown Data: Determining the first Pilot Machine to undergo Planned Maintenance must belong to Rank A or the worst machine in breakdown and costs. For manufacturing, that pilot machine should have similar equipment types belonging to Rank A or B for fan-out or horizontal replication purposes. After deciding which equipment will be piloted, backtrack data on a breakdown by checking history logs and records, backtrack BDO (Breakdown Occurrences) for the past 6 months or more as this will be the benchmark and reference as the team's measure of performance on their activities. The team must monitor the data on the breakdown and provide a form on its occurrence.

Determine the Basic Training Requirements for the Pilot Team: The Planned Maintenance facilitator must see that all team members are trained on the basic requirements needed to undergo each step/phase activity on Planned Maintenance. This training should be conducted in-house for a wider participation of maintenance delegates. This will be the basic training for the teams as they journey their way to implementing their Planned Maintenance activities. Training should be a continuous activity to upgrade the knowledge and skills of maintenance in industries. These training do not include specialization training on their equipment as training may differ depending on the industry they are working in.

Generate a Master Plan of Completion: A Planned Maintenance Master Plan is where the team projects completing their activities on Planned Maintenance for all their equipment and assets categorized as Rank A and B. It will indicate the total number of equipment and projected completion of each phase or step activities on their Planned Maintenance activities. This will serve as the Planned Maintenance organization's overall plan, usually projected from a 3 to 5-year timeframe or more depending on the number of assets or equipment they have in the plant.

12.8: August 2018: Preventive Maintenance Explained

Preventive Maintenance is a basic maintenance performed on the equipment and facilities. The main goal of performing a task on a scheduled basis is to extend the equipment's life and assure its capacity to support the plan's goals and targets. In Preventive Maintenance, the basic law to consider is that the cost of performing Preventive Maintenance must always have to be lower than the cost of not doing it, which eventually will result in failure. PM is also a series of tasks performed at a defined frequency dictated by the passage of time, the number of machine-hours, the mileage that extends the asset's life, or detects that an asset had a critical wear about to fail or break-in operation. Preventive Maintenance is planned maintenance of a plant designed to improve equipment life and avoid any unplanned activity. PM includes cleaning, lubrication, adjusting, component replacement, and overhauling to extend the life of equipment and facilities. Its purpose is to minimize breakdowns and excessive depreciation. In its simplest form, preventive maintenance can be compared to the service schedule of a car. (By William Worsham Senior Consultant and Trainer, Reliability Center Inc.,) The amount of Preventive Maintenance needed at a facility varies greatly. It can range from walk-through inspection of facilities and noting equipment deficiencies for later correction up to shutting down the equipment after a certain number of hours or after a certain number of units had been produced. The care and servicing by personnel to maintain the equipment and facilities in satisfactory operating condition by providing for systematic inspection, detection, and correction of incipient failures before they occur or before they develop into major defects. It includes tests, measurements, adjustments, and parts replacement, overhauls performed specifically to prevent faults from occurring. (Source: from Federal Standard 1037C and from MIL-STD-188 and from the Department of Defense Dictionary of Military and Associated Terms)

While preventive maintenance is generally considered worthwhile, it is important to note that there are risks such as equipment failure and human errors when performing Preventive Maintenance. Preventive maintenance is conducted to keep equipment working to extend the life of the equipment. Corrective maintenance, sometimes called repair, is conducted to get equipment working again. The primary goal of PM is to anticipate the failure of equipment before it actually occurs. It is designed to preserve and enhance equipment reliability by replacing worn components before they actually fail in operation. Preventive maintenance activities include equipment checks, partial or complete overhauls at specified periods, oil changes, lubrication, etc. The workers can record equipment deterioration, so they know to replace or repair worn parts before it causes system failure. The structural integrity of Preventive Maintenance includes:

Engineering: The tasks have to be the right tasks, done w/ the right technique at the right frequency. Many PM systems are being performed, yet the breakdown occurs anyway because wrong things are being looked at in the wrong frequency. PM is not about eliminating the failure. This is an impossible feat to accomplish. PM is about reducing or eliminating the consequences of failure. All activities being performed on PM are geared towards anticipating and planning for the failure itself.

Economics: The tasks must be worth doing. One measure of worth is that doing the tasks furthers the business goals of the organization. The economic question is very critical, and most of the time, does not exist in the PM program. Hence, spending $1000.00 to maintain an asset worth $500.00 is usually a waste of time and resources. Understanding PM's economic point of view will allow us to understand that PM is not just a cost center but rather done correctly; it will yield profit.

People Psychological: People doing PMs have to be motivated so that they can perform their tasks properly. A good PM inspector can work alone with little or no supervision. He should be interested and trained in advanced PdM technology, knows how to review history, and, most of all, is proactive. He can decide to act on prediction rather than react to a situation where failure is already in place. Sad to note that many industries reactively regard heroic work, yet little attention is given to inspectors, such as people who grease and lubricate equipment regularly.

Management: PM has to be built into the system and procedures that control the business, and these systems must be designed so that good PM will result. Dr. Edward. Deming (quality guru) said that quality must be built in the production system and not in the individual effort Information on PM had to be integrated into the flow of business information.

<u>Activities on Preventive Maintenance Includes:</u>

<u>Routine Inspection and Checking</u>: An inspection is usually performed on operating machinery using the basic senses of sight, hearing, sound, and touch. Unless specific parameters have been identified, the inspection quality depends on the inspector's sensitivity in detecting potential equipment problems. Associated checklists used on inspections are often just a list of equipment, the symptomatic conditions to be looked for, and a location on the checklist where a "checkmark" is entered. The use of Visual Management can aid in simplifying inspection techniques.

<u>Overhauls and Replacement</u>: PM involve periodic replacement of disposable components, such as air, oil, or fuel filters. Replacement also includes replace-in-kind for inexpensive but critical components and parts. Often, such replacements are economically justified over more frequent checking. Overhauling means dismantling the equipment based on a scheduled time interval. Care should be taken in overhauling the equipment since this causes problems such as infant mortality failures or start-up problems when endorsing when not properly executed. Note that overhauling and replacing parts should only be done for age-related failures and not for random failures.

<u>Planning and Scheduling</u>: To be effective, all system users must be well trained in background theory so that there is a good common understanding of not only how but also why certain processes and tasks are necessary. If maintenance is asked what is planned for the next shut, they will frequently display a schedule. By way of clarification, a planning system is defined as the total process set up to ensure that the right resources and the right materials arrive at the right place, at the right time, to do the right job in the right way. Planning has a macro meaning when reference is made to the array of shuts for the year, the business plan, the marketing plan, budget, etc. Planned Work is defined as work (skills, resources, and materials) that have been identified, documented, estimated in terms of time and cost, and is ready for scheduling.

Similarly, scheduled work is planned work that has been placed on a formal worklist to be undertaken at a prescribed time. In summary, Planning answers the question of how and scheduling is when.

Managing Spares and Inventory: When a machine fails, it is mostly caused by a part that fails to fulfill its function. Mechanical parts wear and need to be replaced. What is important is to keep the downtime to a minimum. A storeroom is a place to store parts that we need to keep our equipment running but managing the parts simply means how fast we can respond to the right part when maintenance and operations needed it most. One of the storeroom functions is to provide parts, tools, and supplies for the technicians to perform PM Tasks. It will be a good idea for the spare parts personnel to have access to the PM schedule. PM kits are prepared in advance before the PM schedule starts. This saves time in acquiring the parts to be used for PM activities.

Figure 12.4:   Preventive Maintenance Activities

PM replacement and overhauling should be applied only for age-related failures. Aging is a summary term for a set of processes that, over time, contribute to the deterioration and wear of parts and components. Any process that contributes to an age-related decline in performance, productivity is a component of the aging process that deserves our attention and intervention. Intervention in this context can include preventative strategies. This traditional view assumes that most items operate reliably for some fixed period, after which the probability of failure starts to increase rapidly, and that analysis of failures will allow us to predict the remaining useful life of an item and take scheduled action to overhaul or replace it before it reaches the "wear-out zone" where the risk of failure now becomes unacceptable.

12.9: September 2018: Selecting the Correct Interval for PM

In most industries, replacements and overhauls are based on mere guesses, but in most cases, they guess it wrong all along. As said previously, PM uses the concept of JIC or Just in Case. This means that replacements and overhauls are done because our thinking is that it might fail again sometime if we do not do the tasks now. If a part is still working and replaced, then we are throwing something that is still useful. It means that we are throwing money away from our company. This is because most industries are highly too conservative most of the time, in which industries try to select intervals that are way too short or doing maintenance that is intrusive or not needed. This means that we might overhaul equipment after one year, in which the correct overhauling interval should be done after 5 years. These activities only cause our PM efforts to be more costly than ever. According to a recent survey, around 33 cents out of every dollar is wasted due to unnecessary maintenance being performed on the equipment.

While most industries trust to use the maintenance interval dictated in the equipment manual prepare by the Original Equipment Manufacturer or OEM, the problem is that, in most cases, the interval may be too conservative, or the conditions in which the equipment is currently operating may not be exactly the same as dictated in the operating manual. I have nothing against OEM, but in many instances, PM replacements are done too soon, especially if the spare part will be coming from them. If I have a mechanical seal that will last 5 years, it is unlikely that the OEM will recommend replacing the seal every 5 years; perhaps they will recommend that the replacement be done every six months or even less. I think the reason here is obvious. Another case is that if the equipment operating environment is hot and humid like the Philippines, then the greasing frequency and the grease to be used might not be exactly the same grease written on the OEM manual. While it will be easy to derive the maintenance tasks for a failure that has a symptom of what we called the Failure Development Period, we can perform the maintenance tasks less than the Failure Development Period. This means that if the Failure development period is 1 month, then we can perform the task every week or every two weeks for inspection purposes. We want to maximize the useful life; hence, the PM replacement will be done when the life reaches around 80 to 90% of its useful life. If the lead time to failure will be 30 days, then the actual replacement will be done either on the 24th until the 27th day period. In this case, we have more or less maximized the useful life, and the failure has not yet occurred, but the thing is, there will also be failures that will just happen without any form of signs or symptoms.

There are also calculations and formulas to determine the interval or frequency, such as Weibull distribution analysis. However, most industries have insufficient data, and only a handful of maintenance knows how to use them; in this case, we need lots of data to determine the correct maintenance interval. There are two ways that we can go wrong with selecting the interval for specific maintenance tasks. We might be doing the tasks either too soon or too long. Suppose we are doing the maintenance tasks too soon. In that case, we are just wasting precious operating time for the equipment to operate and the maintenance resources. On the other hand, if we are doing the tasks too long, then there is a probability that we might have actually missed the failure. It is difficult to precisely tell when the best interval for doing these tasks, but there are basic guidelines and recommendations on selecting the correct interval for the different maintenance tasks.

In most cases, a car service center will recommend the owner for a car service every six months or 5000 miles, whichever comes first. Depending on the car's brand, the scheduled semi-annual maintenance will include a change oil and filter, inspection of brake pads, checking fluid levels, tire rotation whenever necessary, an inspection of wiper blades, and others. This will be the standard that most service centers offer. Although car service center checklists may differ from one brand to another or from one service center to another. For example, a Toyota Camry maintenance checklist usually includes a tire rotation and resets the tires to the correct pressure. Scheduled Toyota Camry service may also include an oil change and oil filter replacement. For Toyota Camry owners who prefer to use synthetic oil, they will recommend a longer oil change interval of around 12 months or 10,000 miles, whichever comes first. They will also check and adjust the fluid levels and check the brake pads for both the front and the rear parts of the tire.

Before we answer why we need to perform these maintenance tasks on our equipment and assets, what is important for industries is that every single maintenance task included in the lists should address a failure mode. If certain maintenance tasks do not address a particular failure mode, they should be removed entirely from the lists. Second, the tasks should be precise and well understood by all maintenance whose job is to perform these tasks. According to John Moubray, a common problem with mature maintenance programs that, if not correctly designed, states that around 40 to 60% of the PM tasks serve very little purpose and, therefore, evaluating our current Preventive Maintenance system will lead us to the following conclusions.

• Many tasks duplicate other tasks.
• Some tasks are done too often, while others are not enough.
• Some tasks serve no purpose whatsoever.
• Many tasks will be intrusive (forced) and overhaul-based, whereas they should be Condition-Based.
• Some tasks cost more to do than the failure it is meant to prevent.
• Maintenance is costly by replacing perfectly good parts since we base replacement on Time-Based instead of Condition-Based.

What should be the correct frequency or interval for these maintenance tasks? Although RCM Question number 6 on what should be done to predict or prevent the failures only provides the initial interval, the question is, does this initial interval the correct interval? How did we come up with this interval? Did we just cut and paste the interval from our previous PM tasks? Doing the maintenance tasks too soon will be a complete waste of time, while doing it too late can cause the failure to occur even before the maintenance tasks can be carried out. One of the mistakes I often see in doing the RCM analysis is to cut and paste the interval from their existing Preventive Maintenance into the RCM analysis. In some cases, the team is conservative or just follows whatever the OEM recommends them. Once the maintenance tasks are identified to address every single failure mode possible to occur on the equipment or asset, the next question is to determine the maintenance interval.

Determining the correct frequency of replacements and overhauls for Preventive Maintenance activities should be based on useful life and not on the average life of the part, which inhibits an age-related pattern. The RCM team needs to identify the occurrence of the failure before dictating the interval. Suppose the equipment had been restored as we have, we can expect the failure to last longer because the basic equipment condition had been well established. Parts that fail randomly do not have a useful life and should not be replaced on a time-based dominated frequency because their pattern is different. Only parts that will wear out because of stress-induced should be the subject for PM replacement and overhaul. This simply means that when the stress finally exceeds the part's material's strength, then it is time to replace or overhaul that part or item. Aging is the process where certain parts of the equipment deteriorate because of the stress induced upon the part over a given period. Any part that conforms to this process will be the subject of Preventive Maintenance intervention. The people that derived the interval should review their frequency every 6 months to determine if the intervals still hold true or need to be adjusted.

Using our experience to guess the task interval is the only option in the absence of data and historical records. One proven method can be adapted to determine the correct task interval for overhauling and replacement, called the age-exploration method. When we first try to overhaul or inspect the equipment's condition on parts were aging and wear-out are thought to be possible and reveal that no wear-out signs exist, we can automatically increase the interval by 10% based on its original schedule. Repeat this process until we can finally see a normal wear-out pattern on the part being inspected. This means that if the equipment is overhauled after one year and reveals that there are still no traces of wearing or aging on suspected parts upon thorough inspection, we can automatically increase the interval by 10% during the next overhauling schedule. This means the next PM overhaul will happen on the 13th month. If there are still no signs of wearing out, we increase it again by 10% until a wear-out is evident on the parts.

12.10: October 2018: Steps in Adapting an In-house Oil Analysis Laboratory in Your Plant

It is important to sample oil regularly. The decision on whether to adopt an oil analysis in-house or have your sample offsite has both its pros and cons. We should likewise consider the economic and feasibility point of building an in-house laboratory inside your plant. If your plant consumes a large amount of lubrication like mining industries, you definitely need to consider having an in-house laboratory. The disadvantage will always have to be on the initial costs; that is why it is important in the following steps to compute the payback time or rate of return on investing in these instruments. An in-house laboratory's advantages lie in the results; the plant will no longer have to wait for several days to have the report in its hands. This is the problem if the oil will be sampled offsite. Remember that it is not only your industry that they are catering to, and your sample will fall in line with the rest of the sampling bottles. In oil analysis tests, there is what we call a tier or importance of sampling.

A good oil analyst should know these things that sampling oil is not on a first-come, first-served basis, but rather it must be done on the level of importance. For example: when Joe arrived at their in-house laboratory, he bought with him a sample of oil where he wants to check its viscosity. Ten minutes later, Bob arrived, and he wanted to test his oil for moisture. Ten minutes later, Pete arrived and wanted his bottle sampled for particle count. Since there is only one person in the lab, who do you think the laboratory technician would prioritize? It should be Pete because particle sampling is the most sensitive oil analysis test. When the bottle is opened, it can induce dirt. Once the dirt is induced in the sample, it will be counted. The particle counter will not give an accurate representation of the system it is representing. Likewise, if it takes too long to test the particle count in an independent Oil Analysis Laboratory, the contaminants will subside and settle on the bottom part of the oil and might stick with it, even if the oil contaminants are dispersed to put them in suspension in the oil. Others have already settled at the bottom of the bottle. Testing oil for particle count should be done while the oil is still warm. The second is moisture. Again, once the bottle is opened, there is a tendency for some of the moisture contents to evaporate. The less sensitive is the viscosity test. This means that even if Joe was the first to arrive in the laboratory, he should wait for the most important test to be finished. That is how oil analysis tests should be done. However, if the sample would be sent offsite to an independent laboratory, all of these factors will be lost, and again your bottle falls in line together with the rest of the samples from other industries.

Below is a detailed step-by-step approach to setting up an oil analysis laboratory in your plant. These instruments are expensive; hence, just placing them on some purchase requisition and wishing for everything to be approved will always result in your request being denied. These things should be justified, and a feasibility or project study should be conducted and defended to the management team since they are the approving party in this case.

Step 1: Data gathering and documentation
- Determine how much cost your industry is spending on spare parts attributed to lubrication failures.
- Determine your plant oil consumption every month.
- Determine the costs of lubricants your industry consumes.
- Specify current problems your industry has with lubrication.

Step 2: Conduct basic training on Oil Analysis and Contamination Control
- Provide in-house training on basic lubrication and oil analysis program.
- Provide benefits and usefulness of adopting an oil analysis program in-house.
- Compare current practice with having an oil analysis program.

Step 3: Decide what type of oil analysis test is highly needed and recommended
- Decide what types of oil test best suits your industry.
- What equipment will be sampled, and how frequently?
- Provide a procedure on how to get a proper sampling from the unit.

Step 4: Source vendor on Oil Analysis laboratory instruments
- Develop the operating cost of the test equipment needed.
- Decide what test must be done in-house and off-site
- Discuss training package with the oil analysis dealer.

Step 5: Prepare a management presentation on Oil Analysis
- Comparing current practice and propose practice on lubrication.
- Determine the cost needed for the project.

• Determine if the project is a go or no go.
• Seek approval of the project
• Assign a champion to spearhead the oil analysis.

Step 6: Locate the laboratory site and manpower needed

• Study feasibility and strategic location of the laboratory inside your plant.
• Create a centralized oil analysis laboratory.
• Provide a detailed job descriptions and responsibilities for the oil analyst.

Step 7: Conduct Project Study and Feasibility Report On Oil Analysis Project

• Compute for the rate of return on investment and payback period
• Compute for projected savings with an oil analysis project
• Finalize feasibility and strategic location of the laboratory

Step 8: Procurement of the equipment needed for the Oil Analysis laboratory

• Select the best and most qualified oil analysis vendor.
• Purchase the equipment for oil analysis.

Step 9: Installation and Commissioning of Oil Analysis Equipment

• Hands-on training on the equipment with the vendor.
• Attend dealers were training on the use of oil analysis.
• Set meetings with other departments regarding procedures on sampling.

Step 10: Implement an Awareness Campaign on Oil Contamination

• Conduct brief training on all maintenance on oil contamination control
• Adopt by-pass and offline filtration to improve contamination levels.
• Monitor progress and results, especially on the reduction of cost on spares and lubrication.
• Provide savings generated on your in-house Oil Analysis Laboratory

12.11: November 2018: Benefits of Oil Analysis

Many benefits can be derived from a successful Oil Analysis program for your industry. The figure below reflects the reduction in costs in adapting an Oil Analysis strategy in industries. Here are some of the benefits that we can derive from this application:

Prolong the oil change interval: Oil analysis provides users with a strategy to detect wear and contamination in their equipment. Oil becomes a working history of the machine itself. The report tells us if additives are still present and oil needs to be changed or not. Since we base our maintenance on the oil condition, it is more feasible and economical to follow the oil analysis report's recommendation instead of replacing the oil based on running hour's frequency. This means that if Preventive Maintenance activity includes changing the oil on one of the equipment and the Oil Analysis indicates that the oil is still fit for use, then the decision must be not to replace the oil. This lessens the burden of Preventive Maintenance people in replacing the oil and disposing of the used oil.

For deciding to purchase used equipment. A regular Oil Analysis performed on equipment can prove to be a great tool when selling a piece of used equipment. This will show your buyer how well you have maintained the equipment throughout the years. On the other hand, buyers know if the equipment they are going to buy is still in perfectly good condition. Oil Analysis makes it possible to peek inside an engine, transmission, or hydraulic system by monitoring the oil condition and oil contaminants present. When buying a second-hand car, check the oil before buying it.

Early detection of problems: One of the distinct advantages of the oil-analysis strategy is anticipating early problems in the equipment and advising Preventive Maintenance to schedule maintenance work to avoid further downtime during critical operation. Since failures can be predicted in advance, this can aid maintenance in making the right decisions about their equipment.

Cost reduction: An oil analysis program can save us not only in lubricant cost but also in the cost of spare parts, repair and maintenance, oil disposal, energy, labor, and another cost. The figure below says it all. Industries seeking serious strategies in reducing the cost of doing maintenance can further study the benefits of having an oil analysis program in their plant.

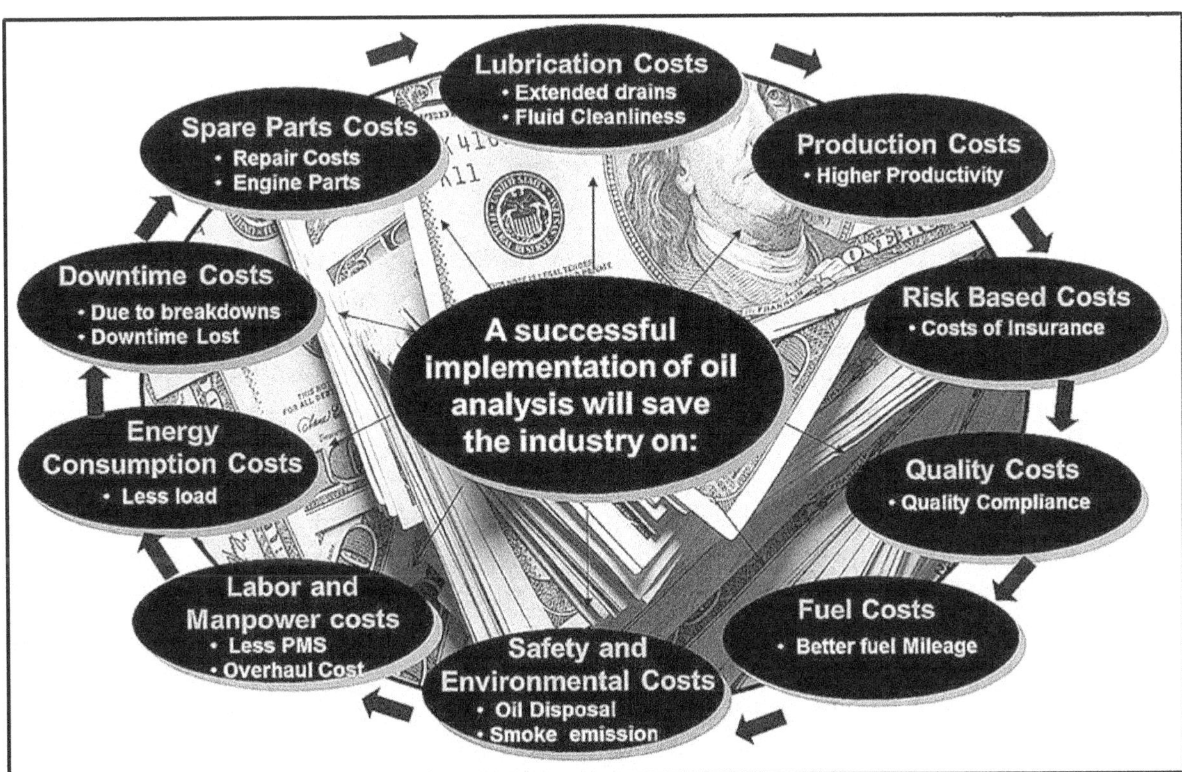

Figure 12.5: Benefits of Oil Analysis in Terms of Reducing Cost

Prolong the equipment life: Oil Analysis tells us what types of contaminants are present in the oil. Through this, maintenance can monitor the trend of contaminants present. Maintenance can also strategize ways of removing these particles in the oil to prolong the equipment's life. As stated previously, there is a direct relationship between the number of contaminants and the rate of wear inside the equipment. The fewer the contaminants, the longer the life of the equipment will be. Every industry needs lubrication for its equipment to function, yet only a few industries understand the role lubrication plays in their maintenance structure. Applying a Lubrication Management strategy should start with how we contain and store our lubricants. We can only remove contamination from the source if we understand how contamination occurs in the equipment. Many people are inclined to think that oil should be changed based on the running hours of the equipment, which is true, but what is more important is that oil should be changed based on the number of contaminants and the condition of the oil itself.

Oil, like any human, is the lifeblood of any equipment. It can provide useful information about the condition of the oil and the equipment itself. This can only be made possible if we analyze the oil in our equipment. There is a direct relationship between the rate of wear inside our equipment and the number of contaminants present in the oil. When the oil is maintained clean, we can prolong both the oil and the parts' life. A study conducted by the Canadian National Research Council found that contamination was the leading cause of wear in various industries investigated. In fact, 82% of all wear was found to be caused by contamination itself. New oil from the drum or barrel is not necessarily clean oil. New oil from the drum contains contaminants that are not fit for use in hydraulic and engine systems. The best way to remove them is to use offline filtration in transferring fluid from the source to your equipment. Remember that clean oil is not synonymous with new oil. These two have different meanings. It is like saying that a fat child is also a healthy child.

12.12: December 2018: The sinking of RMS Titanic - A Classic Case of Human Error Part 1

One of the classic human errors I can think of is the RMS Titanic sinking, which I covered in my Root Cause Failure Analysis class. I think most of us know this classic story. On April 10, 1912, RMS Titanic set sail from Southampton, England, on her maiden voyage to New York. At that time, she was the largest and most luxurious ship ever built. At 11:40 PM on April 14, 1912, she hit an iceberg about 400 miles off Newfoundland, Canada, despite several warnings from other ships. Although the crew had been warned about icebergs several times that evening by other ships navigating throughout that region, she traveled at a near top speed of about 20.5 knots when one grazed her side and met her faith. There were 2,223 people on board, and out of those, 1,517 people died, and there were 706 people who survived. The overall survival rate for men was 20%. The majority that survived belong to the upper class or rich people. For women, it was 74%, and for children, it was 52%. Most of the people that died on the Titanic were caused by hypothermia. People can only survive for a few minutes at a cold, freezing temperature. Many of those who didn't get into the lifeboats were alive when they went into the water, but the water was so cold, which was at a freezing temperature, that they could not survive very long after being immersed at this temperature. Here are a lists of human errors based on my research that actually led to the sinking of the RMS Titanic.

Human Error Number 1: Lack of Lifeboats on RMS Titanic: It was a fact that had the Officers filled the lifeboats per their specification, an additional 470 people could have been saved. Most lifeboats were not full and not even half because the board officers were unsure how many people each lifeboat could carry. This was very unlikely for an officer of a ship. Should the officers knew exactly how many each lifeboat can carry, then an additional 470 people can still be saved. Just imagine that human error. Titanic carried 20 lifeboats: 14

wooden boats, each with a capacity to carry 65 persons; 2 wooden cutters, each one being able to carry 40 persons, and 4 collapsible boats with a 47-person capacity. Altogether, the 20 lifeboats were built to carry 1,178 people, but around 705 to 708 people were saved and could make it to the lifeboats. Officers on board a ship should know the emergency procedures and precautions, especially if the passengers' lives are at stake. Although the RMS Titanic faith was already doomed that night, a total of 470 people can still be saved if each of these officers on board this ship exactly knew each lifeboat's capacity. You might also be wondering why the RMS Titanic carried a few lifeboats instead of having enough lifeboats to accommodate all the passengers on the ship; well, that will be another set of human errors discussed as you try to read along what went wrong with the RMS Titanic.

- Total passengers on Titanic   2,223
- Max. Lifeboat Capacity   1,178
- Actual Loaded   708
- Passengers that can be saved   470

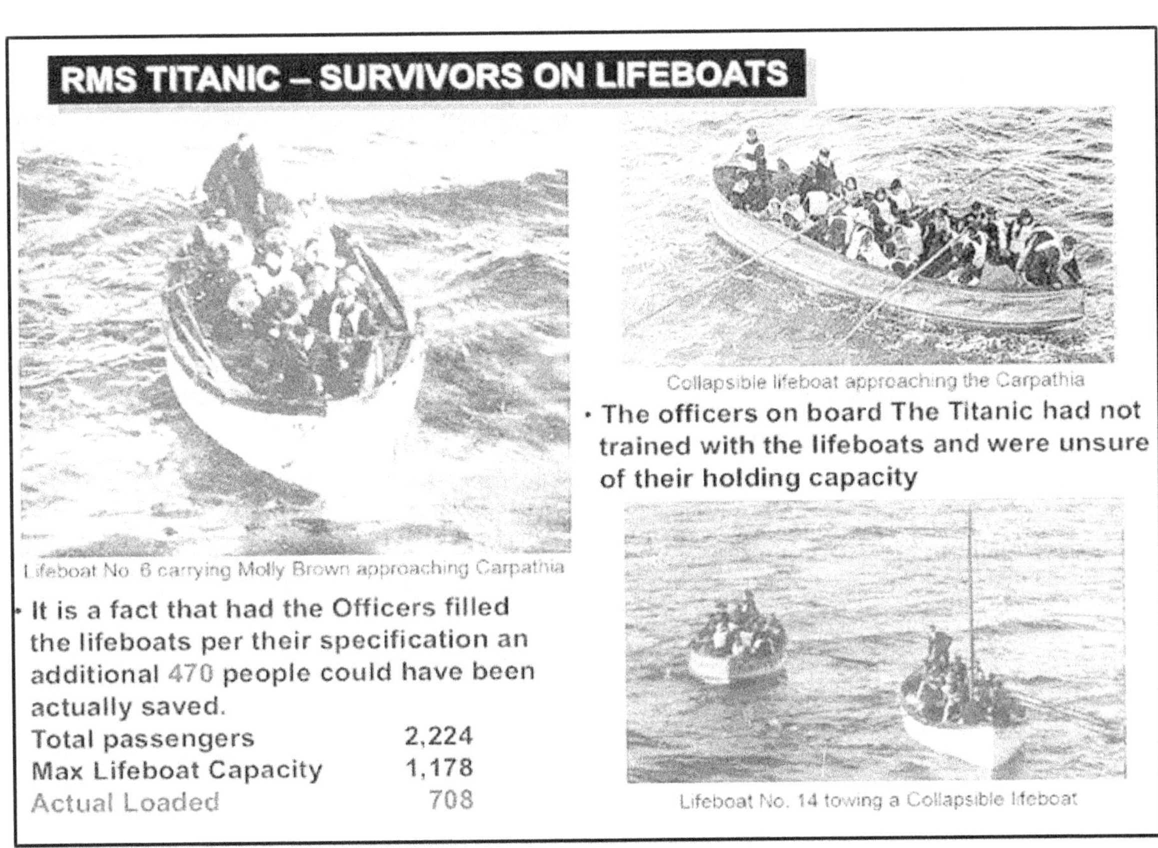

Figure 12.6: Limited Lifeboats on RMS Titanic as it is not required by Law

Human Error Number 2: The Law Itself was the Problem: When the lifeboat needs were finalized, the general feeling was that the modern ship was engineered and built so well that even if a ship was in a situation where it might sink, there would be plenty of time for other ships' to rescue. Why were there so few lifeboats on RMS Titanic? Well, believe it or not, the Titanic actually had exceeded the number of lifeboats required by the Board of Trade at that time. The regulations, ratified in 1894, applied to ships of 10,000 gross tons or larger, indicate as ships increased in size over the years, the lifeboat requirements stayed the same. The

Titanic was designed to carry 48 lifeboats, but the White Star Liner, which was the owner of the RMS Titanic, decided that passenger comfort was more important. They believed that an increase in the number of lifeboats (beyond 20) would have cluttered the decks and taken up valuable space. The law was only changed after the sinking of the RMS Titanic.

Human Error Number 3: The Lookout Did Not Have Any Binoculars: The lookouts in the crow's nest did not have binoculars. Having binoculars might have prevented the RMS Titanic from hitting the iceberg. Fredrick Fleet was the person on duty as the lookout. He had no binoculars that night. If the lookout had binoculars with them, they could spot the iceberg from a much further distance. Remember that it was a cold night on April 14, 1912, with a freezing temperature. Looking with your naked eye might be quite a painful experience. In a test done to determine the stopping distance of the RMS Titanic, the ship was accelerated to 20 knots, and then the engines were reversed at full power. Reversing the engines will be the only way to stop a ship. The distance required to stop the Titanic was about half a mile. When the iceberg was spotted by the naked eye, the distance between the ship and the iceberg was less than a mile.

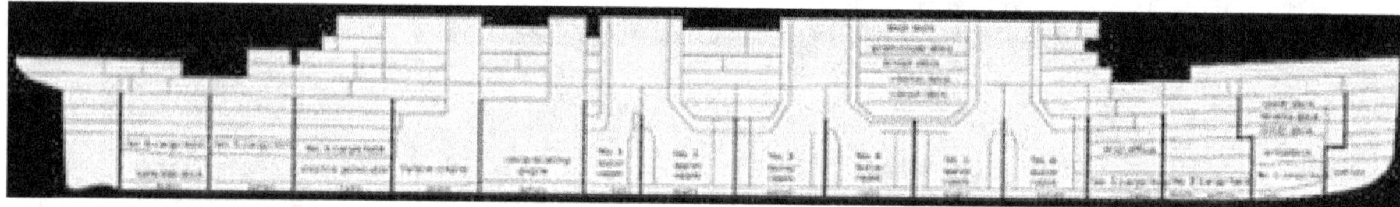

Figure 12.7: RMS Titanic 16 Compartments

Human Error Number 4: Design Error on the 16 Compartments: The Titanic design was that the door was watertight, but each compartment was not completely watertight. For the compartment to be watertight, each of the six sides must be completely sealed. The RMS Titanic design problem was that only five out of the six compartments were watertight. The Titanic design extends to the lower decks, making the bulkhead or the upper ceiling not watertight. Making it watertight was either to remove some decks for passengers or make the passengers walk on stairs to the other side, which is not a practical idea. The RMS Titanic had 16 watertight compartments, and the ship could stay afloat with up to four of these compartments flooded. After hitting the iceberg, water began flooding the Titanic's forward six compartments. To be continued.

<u>2019 RSA Reliability Newsletter Vault Archive</u>

> *Maintenance is not about shifting from Preventive to Predictive nor transitioning from Reactive to Proactive Maintenance. In maintenance, there is no transition; it is about understanding when to use the different maintenance tasks simultaneously depending on the consequences of failure. This can only be done with the aid of a decision diagram or algorithm.*

<u>13.1: January 2019: The Sinking of RMS Titanic - A Classic Case of Human Error Types Part 2</u>

Human Error Number 5: Radio Operator Ignored Warning of Icebergs: On April 14, the day it hit the iceberg, the Titanic received seven heavy ice warnings, including one from the Californian less than an hour before the fateful collision with the iceberg. The message said we were stopped and surrounded by ice. The RMS Titanic radio operator sent back a message that said, shut up, we are busy. It was at 7:15 pm where the first iceberg warning from the Baltic was posted on the bridge. First, Officer Murdoch then ordered a sailor to secure a hatch, from which a light glow might have been obscuring the view on the crow's nest. At 7:30 pm, the Californian radioed Titanic and reported ice. An additional ice warning from the Mesaba came in at 9:40 pm but was never taken to the bridge. Should the message have reached the Captain or the officers on deck, they can be more prepared and have slowed down the RMS Titanic, traveling at 20 knots per hour. It was also known that Jack Phillips prioritized delivering the messages from first-class people instead of attending to the iceberg warnings from other ships.

Figure 13.1: Titanic Radio Operator Jack Philips

Human Error Number 6: Command Error by First Mate William Murdock: The Titanic sank 2 hours and 40 minutes after hitting the iceberg. It probably took the RMS Titanic about 15 minutes to sink the ship to her final resting place beneath the ocean floor. That means that the Titanic sank at a rate of 10 miles per hour or 16 km/hr. The Titanic hit the iceberg on the starboard or right side of the bow. Immediately right after, the lookout spotted the iceberg and informed the deck. First Mate William Murdock immediately commanded an astern or reversed the engine. Reversing the engine will cause the ship to turn at a slower phase. Should the officer did not order a reversal of the engine, then the ship can turn to the port side or left side at a much faster rate and might have avoided the collision of the iceberg by a few meters. It has also been speculated that the Titanic may have suffered only from minor damage and minimal loss of life and will not sink if it had hit the iceberg head-on with the engines at full reverse. It has also been suggested that the Titanic may have completely avoided colliding with the iceberg had the bridge not requested that the engines be reversed (Full Astern) before steering the ship to the left (Hard-a-port). This action would have decreased the forward momentum of the Titanic, causing it to turn at a slower rate. William Murdock committed two human errors, which was about not maximizing the people in the lifeboat because he did not know exactly how many people each lifeboat can carry, which was very unlikely for the ship's officer where he was the second in command. He was believed to have committed suicide at the time of the sinking.

Human Error Number 7: Bruce Ismay Pressuring the Captain: Bruce Ismay was the Managing Director of the White Star Liner that owns the RMS Titanic. He was also on the RMS Titanic's maiden voyage. It was believed that Bruce Ismay was pressuring Captain Smith to light the remaining boilers so that the ship can run faster in the icebergs' vicinity. The reason for pressuring Captain Smith was due to the stiff competition in the shipping business, and Bruce Ismay does not just want the RMS Titanic to be the biggest and most luxurious ship during that time, but he also wanted it to be the fastest ship of his size. Another human error Bruce Ismay did was that one of the radio operator's messages was given to him. He forgot or intentionally did not give the message of radio warnings to the officers on duty. The message of iceberg warnings remained in his pocket.

Human Error Number 8: Mistake by Captain Smith: Captain Smith allowed the ship to run at top speed on icy waters. He also allowed himself to be pressured by Bruce Ismay to light up all the boilers so that the ship can run much faster to reach its destination much earlier than expected. No practice drills for emergency preparation were done. Captain Smith skipped an emergency drill. As a result, there was chaos after the ship hit the iceberg and nobody knew what to do or how to use the lifeboats. Captain Smith ordered the ship at full speed. This was despite the reports of icebergs and a clear dark sky that evening. No proof had been found that Mr. Ismay had previous discussions with the Captain to have the ship go full speed. Titanic Passengers and crew died of cold or hypothermia. The temperature was below freezing. Suppose Captain Smith ordered to return the ship back to the iceberg and docked the passengers on the iceberg; there could only be fewer fatalities. It can take hours for people to freeze, but it would just be minutes before they can die when submerge in an icy cold freezing temperature on the water.

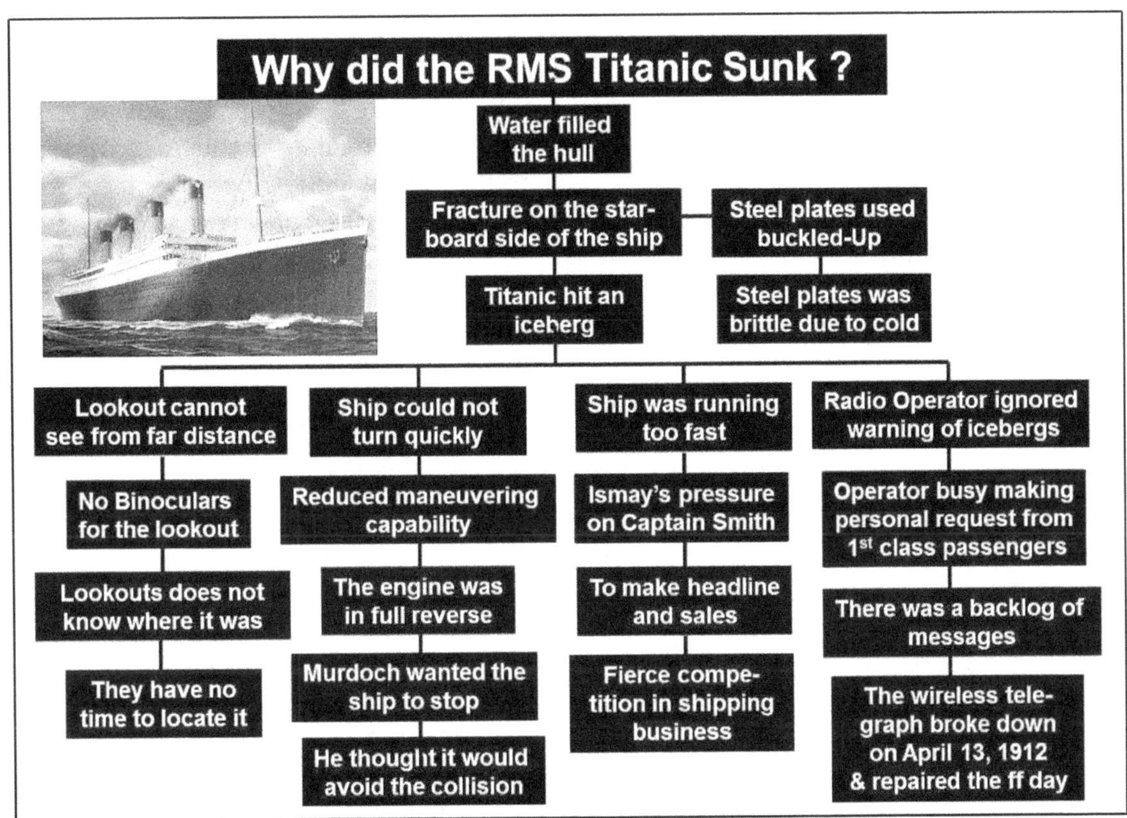

Figure 13.2: Why Did the RMS Titanic Sank

Figure 13.3: Last Remaining Survivors of RMS Titanic

As of this point of writing, all survivors from the ill-fated RMS Titanic had died. The last known survivor was Milvina Dean, who died recently last May 2009 at 97 years old. She was the youngest survivor of the Titanic. May their souls finally Rest in Peace. But perhaps the biggest human error was thinking that RMS Titanic was unsinkable because of its size and grandeur and defying Mother Nature laws. Although technology is advancing as time goes by, nature shall remain undefeated. Let us not be too arrogant with the current technology we have and preserve our nature.

13.2: February 2019: Switching Redundancy and Standby Tasks

Duplicating the system or component is done so that operations will not be affected in case failures and breakdowns occur. Failure is allowed or being tolerated through some form of redundancy and standby or when the asset has some form of duplicated function. The presence of standby or redundancy, which is an alternative means of production, is a feature of the operating context that must be considered in detail when defining the asset's functions in its present operating context since the maintenance requirements for the duty and the operating asset will be different, even if they are of the same maker and model. In other words, standby units can have a different maintenance task as the duty or running unit, and most failures for standby units are considered hidden.

The presence of redundancy or alternative means of production is a feature of the operating context that must be considered in detail when defining the asset's functions in its present operating context. In the aviation industry, many items have some duplication or redundancy level, allowing failure to happen. They ensure that the failure is fail-safe if ever it happens. When the aircraft is airborne, back-ups are used to experience failure when the plane is already airborne. Failures are then allowed to happen for the aircraft industry. However, run to fail may not be the best option for the land industry if a system, sub-system, or asset has a standby or redundancy is present. It is much better to create a change-over or a switching pattern to ensure both the duty and standby are functioning. The problem will allow the duty pump to fail can create a problem where the standby may also be in a failed state or tock up since it was not used for a definite amount of time.

The limitations of a run to fail are that it causes unplanned downtime and production delays, resulting in revenue losses and extending the maintenance to work overtime to fix the equipment. Repairing the equipment after it fails usually creates the possibility of secondary damage, which can increase maintenance costs. Allowing failures to occur can be applied to the asset if the consequences of failure and repair cost are minimal and acceptable to both the user and maintenance. There will be no scheduled replacement on the item or part that will fail since maintenance will just wait for failure. The only strategy for a run to fail is to keep a stock of spare in the storeroom to ensure that downtime will be minimal. The only issue is if the failure will provide some secondary or tertiary damage to the equipment or asset. For a run to fail, there will be no interval.

Selecting Switching Interval for Redundant Components: While in most industries, an alternative of switching the standby and duty is done weekly. The problem with this case is that the number of start-ups in operation influences the number of seal failures. This is because seals that are running are well lubricated and generally show low wear rates. The worst time for a seal to fail is during the start-up condition. During start-up, the seals are dry, and it will take a bit of time to build up the fluid film between the clearances. Frequent start-ups such as weekly is a major cause of wear in seals. By reducing the number of start-ups, we can reduce the number of failures for the seal. Another issue will be the impeller if the switching is done evenly; this means that the rate of erosion on the impeller for both the duty and standby will happen at the same time, which means that there is a possibility that both the duty and the standby pump can fail at the same period.

The most important part of a standby pump is to start on-demand or when needed if the duty pump fails. Depending on historical failure rates relating to this failure mode, a test start can be organized at a suitable frequency to ensure its desired availability. The solution is to run the test for a reasonable length of time, say 8-24 hours, with the duty equipment shut down. In this case, the duty pump can be run either between 3 to 5 months, and the standby equipment runs for 1 month. What is important is the duration of their operating time should not be the same. This policy's advantage is that it produces the lowest number of start-ups for both pumps in a year while allowing a longer duration of the test run of 1 month for the standby and a frequency of 3 to 5 months for the duty equipment. Besides, note that even if both duty and standby pumps are of the same type and make, their maintenance tasks will be entirely different. One maintenance to perform on the standby pump is to rotate the shaft manually weekly around 360 degrees + 60 degrees or 720 degrees plus 60 degrees to avoid the chances of false brinelling occurring in the raceway of the bearing. Just make sure that the bearing balls will not end up in the same exact position as the outer raceway when it is rotated. This bearing problem often occurs on stationary equipment since vibration is also present in the standby component. If a bearing is subject to vibration when stationary, fretting damage occurs at the points of contact, giving rise to what is known as **false brinelling**. Standby components inside the storeroom should likewise be rotated once a week if the vibration is heavily felt inside the storeroom.

When the duty pump runs, the vibration forces are not just confined to the duty pump but also to the standby pump. This means that the balls inside the standby pump and the bearing's outer raceway are vibrating continuously. Suppose the ball's position concerning the bearing's raceway remains unchanged; in that case, this will create some fretting and can cause early fatigue fracture once the standby pump is used.

13.3: March 2019: Aftermath Additives as Seen on TV

I used to remember the commercial from Duralube oil additive many years ago when I was still working and employed, where they extract the oil from the engine and just place the Duralube additives in the car, and the engine runs. After that, they freeze the engine on ice and place the Duralube oil additives. The engine runs once more, where the audience was mesmerized by what they saw and were clapping with disbelief. In fact, I was so intrigued by that commercial that I said to myself that I need to place that in my car. The following day, I bought the Duralube additive and placed it in my car, an old Nissan 90 box type model. Honestly, I can see no changes in the performance of the engine. There was a time where the water in my radiator leaked, and my car overheated. I told myself that this is not possible as I got this Duralube inside my engine. My advice is, if you are so intrigued or spell-bounded (like I was) by their television commercial that you intent to buy one yourself, I think you need to check the website of FTC or Federal Trade Commission of the United States before purchasing one. Most of these aftermath additives have cases with the FTC in which they have already lost and agreed on a settlement with affected consumers. Although they have already stopped manufacturing their products in the US, it does not stop them from selling in other parts of the globe. If you are in the Philippines, try to visit a mall, and they have large hardware called ACE hardware, which still sells these aftermath lubricating oil additives up to this point of writing. Perhaps Asian countries are still ignorant about the FTC cases imposed on these additive manufacturers, so they are still allowed to be sold in Asian countries.

- The Slick 50 Oil Additive in $ 20M lawsuit finally lost to FTC. Blue Corral, the manufacturer of Slick 50, has agreed to pay upwards of $ 20M in damages to affected customers. In the United States (FTC), the Federal Trade Commission found that company claims of increased performance and reduce engine wear were unsubstantiated. *Source from Max Power Magazine, March 1998*

- Prolong challenged by Consumer Reports, lose to FTC. Consumer Report attempted to reproduce the no oil test. All the oil was drained out of an engine and treated with Prolong, where the engine ran for 13 seconds and seized. *Source from Consumer Reports, October 1998*

Figure 13.4: Aftermath Additives as Seen on TV

- Duralube additive was also challenged and lost to FTC, where the same test was conducted by the Consumer's Report, where the Duralube additive was added in the engine. During the

TV commercial, the engine was running smoothly without any signs of problems, but when it was actually tested, the engine lasted 11 seconds and then seized. According to FTC, the ads claimed by Dura Lube were false and unsubstantiated. In the latest law enforcement initiative targeting ads that use deceptive performance claims to tout motor oil additives, the Federal Trade Commission has charged the marketers claims of Dura Lube Super Engine Treatment and Dura Lube Advanced Engine Treatment with making false and unsubstantiated advertising claims, in violation of federal law. Dura Lube is one of the largest engine treatment companies in the country and the fifth to be charged by the FTC with making false or unsubstantiated claims.

• In February and March 1997, an independent laboratory performed two CRC L38 tests of the ZMax oil additive for the Speedway and Oil Chem. During the tests, motor oil is treated with ZMax produced more than twice as much bearing corrosion as the motor oil alone. The complainant also states that the defendants fabricated the results from the two test reports, eliminating the results of bearing corrosion and all other negative test results, and then used it in the official laboratory results as part of their marketing claims.

The FTC has also previously halted allegedly deceptive ads for Valvoline Engine Treatment, Slick 50 Engine Treatment, STP Engine Treatment, and Motor Up Corporation, which was sued by the FTC on April 14, 1999. These additive companies have already lost to FTC. The FTC states that Dura Lube did not have a reasonable basis to substantiate these claims.

Ashland, Inc. has agreed to settle with the Federal Trade Commission charges that ads for the Valvoline Company's Teflon-containing TM8 Engine Treatment products were false and unsubstantiated. These are the latest findings in a series of FTC cases involving unsubstantiated or false claims for automotive additives and high-octane fuels. The Federal Trade Commission has also filed suit in the U. S. District Court seeking to stop false and misleading advertising for zMax auto additives and has asked the court to order refunds to consumers who bought the products. In addition, the FTC agency states that enhanced performance claims for the products are unsubstantiated. The tests cited to support performance claims actually demonstrated that motor oil treated with zMax produced more than twice as much bearing corrosion than the motor oil and that the three different products and engine additive, a fuel line additive and a transmission additive, were all actually mineral oil. zMax is manufactured by Oil-Chem, a wholly-owned subsidiary of Speedway Motorsports, Inc., which is based in Concord, North Carolina. They also operate NASCAR race tracks in the South and in California, in addition to marketing the ZMax products.

Most of these aftermath additives contain a compound called Polytetrafloeraethylene (PTFE), more commonly known by the trade name Teflon, a registered trademark of the DuPont Chemical Corporation. The Teflon compound problem is that it is a solid additive, and it expands rapidly when exposed to heat. PTFE or Teflon in oil additives is a suspended solid. It means that the oil filter's job is to remove suspended solids; therefore, it would seem logical to follow that if the oil filter is doing its job, it will collect as much of the PTFE as quickly as possible. This can result in clogging the oil filter and decreased oil pressure throughout your engine. One of the important things about lubricating oil is that they already contain additives based on the type, function, and application used by the oil. There is also the problem of compatibility issues when mixing additives of different brands, as they may or may not be compatible with each other.

13.4: April 2019: The Evolution of Maintenance

If we go back in time and trace back to history during World War II, equipment was simple in design and not that complicated. The downtime did not matter at all, and when the equipment failed, it was not a big issue. When the equipment failed, maintenance will just repair it. There was no need for a systematic maintenance strategy on the equipment. Some common and basic inspection and lubrication tasks were the best courses of action when maintaining the equipment. This was because demand during that time was low; hence, breakdown did not matter much.

After World War II, things changed dramatically. There was an increase in the demand for products, goods, and services globally. This time, equipment downtime was a real big deal for industries, and something must be done about it. This led to the concept that equipment failures were not good for business. Breakdowns must be prevented and are unacceptable, and so the concept of Preventive Maintenance was born, which sort of scheduled equipment for maintenance. Big and slow computers started to emerge on the market as a result. Traditional thinking was that there is a direct relationship between the rate of wear of an item and the time it operated. To cope with this issue, scheduled repairs, overhauls, and restoration were done to ensure that equipment would be available whenever operations needed them. Uptime needs to be high to deliver the goods and services to its customers on demand. The expectation on maintenance was to provide equipment with higher availability at a lower maintenance cost. Still, as we all know, the cost of performing maintenance also begins to increase and take its toll. During the '80s, greater expectations were now being realized on the maintenance function. As equipment becomes more complex, diversified, and automated, more and more industries now realized that not only should equipment provide higher availability with a low maintenance cost, but this time, equipment must be reliable, safe to operate, can produce goods with higher product quality and most importantly, equipment failures must not damage the environment whatsoever. New techniques, strategies evolved, such as Predictive Maintenance and software on maintenance, circulated on the market throughout the years. Equipment failures were monitored by checking the equipment's actual condition, which has been part of the maintenance strategy.

Predictive Maintenance instruments developed as a result. Today, we are now facing the digitalization age; others termed this as Maintenance 4.0. Industry 4.0 refers to the fourth industrial revolution, where equipment and machines are often amplified with wireless sensors and technology, connected to a system that can visualize the entire production line and make decisions independently. Industry 4.0 is the trend towards automation and data exchange in manufacturing and non-manufacturing industries. However, oil industries and power plants have already used this technology originally. This includes the Industrial Internet of Things (IIoT), cloud computing, wireless sensors, artificial intelligence, robotics, machine learning, and you name it. They are now the current trend in automation for equipment and assets.

The truth is that as our equipment becomes more complex and automated, yet the way industries do maintenance is still in fire-fighting mode. To make matters worse, the cost of doing maintenance continues to increase as time goes by. Before we can start to improve at anything, maintenance should be educated on the correct way of doing maintenance on their equipment and assets. There are training, tools, skills, knowledge, instruments that maintenance needs to perform their true function, but the sad part about this is that in most cases, these things maintenance need are deprived of them as part of the industry's cost-cutting measures. The problem is how on the planet can maintenance function and do their jobs correctly. I shared my sentiments and frustration in the class sometimes. I told them that it is hopeless teaching industry, and I am thinking about shifting to another job. When the delegates ask me what job I will be shifting to, I told them that I plan to go into politics. I told them that if I am elected, one of the bills I will pass will be for every industry to have their own set of Predictive Maintenance instruments, and the delegates would smile. I also told them that one of the bills I will propose is to increase the salary of all maintenance people on the planet (with a high tone) because industries should understand that no industry on this planet can run without the maintenance function and that is how important we are. Then I get applause from the class. They told me that they will back me up and vote for me if I ran for politics. Still, I told them that politics is not in my blood, and on second thoughts, I'll stay with maintenance because that is who we are, and I'm damn proud of that.

13.5: May 2019: The Roots of RCM

RCM originated from the airline industry. In the **1950s**, the commercial aircraft fleet's size became bigger, and the airline industries posed two major problems. The first is that during this time, the accidents per million take-offs were high. This KPI indicates that for every 1,000,000 planes that took off from the runway during this time, how many will fail and crash? The second reason is with the development of the 747 aircraft, which is three times as many passengers as the 707 aircraft or DC8, with its structure, avionics, engines that are much more extensive, the cost of operating and maintaining an aircraft of this magnitude would not be much profitable under the current circumstances. This is if they will carry out the same Preventive Maintenance requirements and procedure. In **1958**, after several years of study, research, and evaluation, the FAA (Federal Aviation Administration) finally concluded that it was impossible to control the failure rate of certain types of unreliable engines by any feasible changes in either the content or frequency of scheduled overhauls. This means that the amount of failure that could occur has nothing to do with the interval or the activities dictated by their PM. It simply means that even if the overhauling interval or frequency was shortened, the chances of failure to occur are not guaranteed and will not be captured. The review group was focused on why Preventive Maintenance was done, how it should be done, and when it should be accomplished.

In the **1960s,** the FAA organized a new task force composed of several FAA representatives and the airline industries to further investigate the limitations and capabilities of Preventive Maintenance. Bill Mentzer, Stanley Nowlan, Howard Heap, and Thomas Matteson, the Vice President of the Maintenance Planning for United Airlines, were the pioneers of this effort. It is highly noted that Thomas Matteson was essentially the major groundbreaker and pioneer among the pioneers behind the detailed development of this methodology used by the Maintenance Steering Group (MSG-1) to produce the Preventive Maintenance requirements for the 747 aircraft. A simple decision diagram was developed in **1965** and revised in June **1967.** A paper on this was presented at the American Institute of Aeronautics and Astronautics (AIAA) regarding commercial aircraft design and operations meetings. The MSG-1 document was published in **1968**. It was used to develop the scheduled maintenance for Boeing 747.

The MSG-1 was further refined, which led to the establishment of a second document called MSG-2. MSG-2 was used to develop the scheduled maintenance for Douglas DC10 and Lockheed L1011 aircraft. In **1972**, the Department of Defense (DoD) contracted United Airlines to provide its Navy military aircrafts S-3 and P-3. In **1974**, it was also used for the Air force Mc Donnell F-4J. Later on, it was also adapted to several aircrafts such as Airbus A300 and Concorde. The dossier's objective for MSG-1 and MSG-2 was to provide a scheduled maintenance program that will provide maximum safety and aircraft reliability at the lowest possible maintenance costs. A major saving on scheduled Preventive Maintenance derived from the MSG-1 and MSG-2 document was realized compared to their traditional PM program for the Douglas DC-8 aircraft, which required a scheduled overhaul of 337 items in which they dramatically reduce it to 7 items for a much bigger aircraft such as the DC-10. It was a remarkable breakthrough by the pioneers of the group. One of the items that were no longer subject to overhaul was the aircraft's turbine propulsion engines, which led to a major reduction in their spare parts by as much as 50%. Another major accomplishment on MSG-1 was reducing more than 4,000,000 man-hours for smaller aircrafts such as DC-8 on structural inspection to 66,000 man-hours before reaching 20,000 hours of flight travel. Such remarkable changes in their PM program were achieved with no increase in the aircraft's maintenance costs and provided much lower accidents per million take-offs.

However, the MSG-1 and MSG-2 document was further refined and improved since its decision logic tree diagram was merely focused on the maintenance tasks rather than the consequences of failure. The basic goal of MSG-3 is to identify these maintenance tasks, both effective and efficient, in enabling new aircraft to be designed and operated to achieve a satisfactory level of safety and reliability throughout its entire lifespan. In **1975**, the Department of Defense (DoD) directed that the MSG-3 concept be named Reliability Centered Maintenance and directed that it should be applied to all major military systems. On December 29, **1978**, the document was finally completed by Stanley Nowlan and Howard Heap (495 pages typewritten) and given to DoD. Thomas Matteson was not included as one of the book's main authors because he already retired from United Airlines. The Nowlan and Heap report on RCM was the basis for the MSG-3 concept. This year, the US Navy has also applied the RCM concept, Stanley Nowlan and Howard Heap quoted in their RCM document.

- *One of the underlying assumptions of maintenance theory has always been a fundamental cause-and-effect relationship between scheduled maintenance and operating reliability. This assumption was based on the intuitive belief that because mechanical parts wear out, any equipment's reliability is directly related to its operating age. It, therefore, followed that the more frequent equipment was overhauled, the better protected it was against the likelihood of failure. The only problem was in determining what age limit was necessary to assure reliable operation. In aircraft, it was also commonly assumed that all reliability problems were directly related to operating safety. Over the years, however, it was found that many types of failures could not be prevented, no matter how intensive the maintenance activities. Designers were able to cope with this problem, not by preventing failures but by preventing such failures from affecting safety. In most aircraft, essential functions were protected by redundancy features, ensuring that the necessary function will still be available from some other source in a failure.*

In **1983**, the Electric Power Research Institute (ERPI) in San Diego, California, USA, started RCM pilot case studies on their Nuclear Power Plants. The success in their implementation led further to the adaptation of the RCM to nuclear facilities in North America; Électricité de France (EDF) is the country's main Electricity Generation and Distribution Company, which manages around 58 power reactors and finally adapted to other industries worldwide and the rest was history.

Although the RCM concept's origin was first applied in the airline industry, two people are responsible for promoting RCM for land industries. First, we have Anthony (Mac) Smith, a friend of Thomas Matteson, and the late John Moubray. Stanley Nowlan continued his research on RCM and, in **1983,** started his collaboration with John Moubray, who was the founder and CEO of Aladon until he died in **1984,** to adapt RCM to land industries.

The classical RCM approach simply tells us that the right way to do RCM was to do it the way it was originally done in the 60s, developed by United Airlines Industries. However, a set of standards on RCM evolved, which are SAE JA1011 and SAE JA1012. The purpose of this document is to set a minimum requirements for what is and what is not, able to be defined and considered as RCM since today there so many streamlined versions and wannabe gurus on RCM. Their selling point is that their version is much faster.

• SAE JA1011 - Evaluation Criteria for the RCM Processes
• SAE JA1012 - A Guide to the Reliability-Centered Maintenance (RCM) Standard

What C. H Waddington Discovered about Scheduled Maintenance

Conrad Hal Waddington was born on November 8, 1905, until his death on September 16, 1975. He was a British biologist, paleontologist, geneticist, philosopher, poet, and even a painter who laid the foundations for system biology epigenetics. In 1943 during World War II, the Royal Air Force Coastal Command acquired C. H. Waddington services. During World War II, he was involved in operational research with the Royal Air Force and became a scientific advisor to the Commander in Chief of Coastal Command from 1944 to 1945. His job was to provide recommendations to the British Military on how they can defeat the German U-Boat submarines, where he made his stunning recommendations in their favor.

One of Waddington's recommendations was to repaint the RAF bomber planes from black to white. Based on studies conducted, this was because these German U-boats cannot be seen until they were 20% closer to the target, which resulted in an additional 30% of its sinking. Another recommendation by Waddington was that the depth that the bombs be released was changed from the depth from 100 to 25 feet, which resulted in more accuracy of their target and increasing the number of German U-Boats destroyed. Although at first, the military was hesitant to agree with Waddington, in the end, they realized that his recommendation was indeed correct and effective.

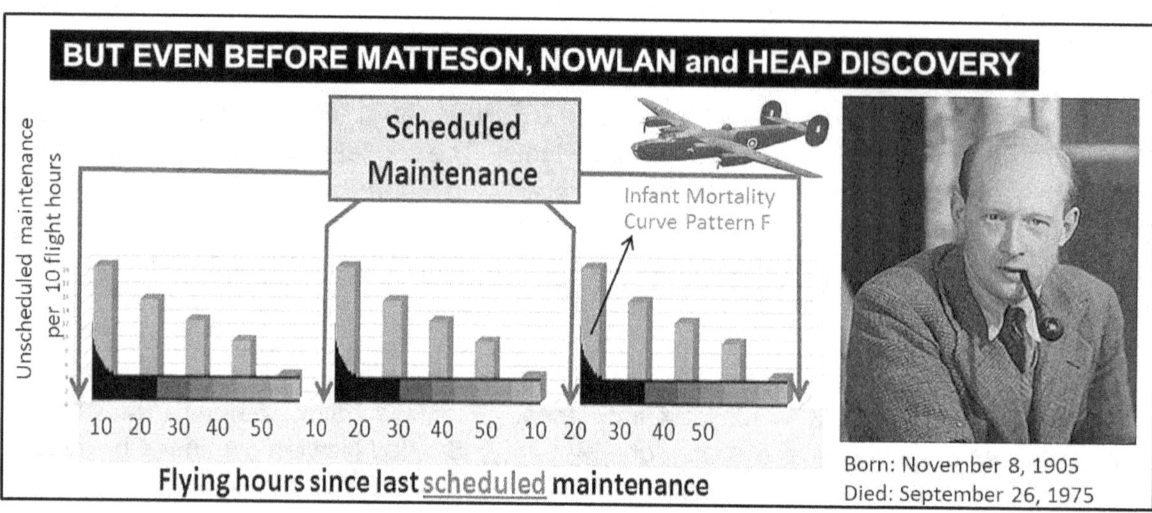

Figure   13.5: Waddington's Discovery of Early Infant Mortality Failures

The Waddington Effect: Waddington observed that British bomber squadrons whose mission was to sink German U Boats consisted of around 40 bomber planes, but during the mission, only half of them were available while the rest were either in a failed state only after a short 50 hour flying time. The reasons for the unavailability varied from one bomber to another, but a trend is clear that after a scheduled Preventive Maintenance performed on the RAF-Bomber planes, most of them were unavailable. Waddington tried to study the data and plotted all these downtime events concerning the last Preventive Maintenance, expecting no relationship. What he discovered was that unscheduled repair and downtime increased immediately after performing maintenance. He observed that the scheduled Preventive Maintenance performed on these bombers provided them with more harm than good.

Waddington's recommendation was to increase the time interval and duration of scheduled Preventive Maintenance on these RAF Coastal Command bomber planes. He removed all intrusive and unnecessary maintenance performed, and the results were an astounding increase of up to 60% on their availability of these RAF bomber planes. This was known as the Waddington Effect, and the principle behind it is the very core principle of Reliability-Centered Maintenance.   Therefore, the less invasive the Preventive Maintenance is, the better the outcome, according to the Waddington effect. As the Waddington effect shows, more is not always better from a maintenance point of view; that is to note that Waddington was neither maintenance nor a reliability engineer but a poet and a painter.

13.6: June 2019: Different Awards on TPM Given by JIPM

TPM or Total Productive Maintenance has already made its mark worldwide, where many industries globally are implementing its philosophy and principles. It is important to implement TPM so that the company's culture can accept it and not entirely implement TPM the Japanese way. There are several industries certified by JIPM (Japan Institute of Plant Maintenance) outside Japan. The TPM Awards consists of six categories and given to industries that have matured, reaped the results, and benefitted their TPM implementation in their plant. I also noted that India has its own TPM Club India, which I believe is a joint venture from JIPM that promotes TPM and provides training, references, consultation, conferences, and research among TPM industries and strengthening its ties with industries implementing TPM. TPM has a very strong foundation in India, as there are many plants involved in doing TPM. India has the second-highest number of TPM awards from industries outside Japan. I have been teaching in India since 2009, and what I admired about India is that my classes on maintenance are always full house with a large participation of not only maintenance managers but top management themselves. I have several cases where a CEO or a Senior Vice President is sitting in my class for a couple of days and listening to what I am teaching. I believe that their mindset is that before my people should know, I should be the first to know. I just cannot say for the rest of the countries I teach. What bothers me is how top management people from India have time to attend this training, while most countries don't.

The TPM Awards was established in 1971 to recognize the development of Total Productive Maintenance (TPM) in Japan and encourage its development. The first TPM Award outside Japan was given in 1991. Japan Institute of Plant Maintenance (JIPM) promotes the TPM Awards. Approximately more than 2,000 plants worldwide have already received the TPM Awards in Japan and outside Japan since its inception.

Award for TPM Excellence, Category B
- Must have a minimum of 2 years or more of achievement activity after the introduction of TPM
- Must have deployed activity based on the 5 pillars of TPM focusing on the production site
  (Focused Improvement, Autonomous Maintenance, Planned Maintenance, Training, and EHS)
- Must have completed Step 4 for Autonomous Maintenance activity
- Must have completed infrastructure development for TPM activities with both tangible and intangible achievements obtained

Figure   13.6: JIPM TPM Awards Category

Award for TPM Excellence, Category A
- Must have a minimum of 3 years or more of achievement activity after the introduction of TPM

- Must have deployed activity based on eight pillars of TPM by all staff members of the plant
- Must have completed Step 4 for Autonomous Maintenance activity
- Must have completed infrastructure development for TPM activity with both tangible and intangible achievements obtained

Award for Excellence in Consistent TPM Commitment

- Must have received the Award for TPM Excellence (Category A or B)
- Industry seeking this award must have approximately 2 years of achievement activity after receiving the Award for TPM Excellence Awards
- Must have deployed activity based on the 8 pillars of TPM by all staff members of the plant/factory
- Must have maintained and enhanced the results achieved at the time the Award for TPM Excellence was received and have established measures for their maintenance and continuation
- Must have completed Step 4 of Autonomous Maintenance
- Must have deployed activity based on the 8 pillars of TPM

Special Award for TPM Achievement

- Must have received the Award for Excellence in Consistent TPM Commitment
- Must have a minimum of 2 years or more of achievement activity after receiving the Award for TPM Excellence in Consistent TPM Commitment
- Must have deployed activity based on 8 pillars of TPM by all staff members of the plant/factory
- Must have maintained and enhanced the results achieved when the Award for Excellence in Consistent TPM Commitment was received. Award for Excellence in Consistent TPM Commitment is required for the applicants who wish to apply for the Special Award for TPM Achievement.

Advanced Special Award for TPM Achievement

- Must have received the Special Award for TPM Achievement
- Must have a minimum of 2 years or more of deployment activity based on the 8 pillars of TPM after receiving the Special Award for TPM Achievement and show significant improvement as a result
- Must be deploying TPM activities after establishing important items and showing results on management, and other contents are independently determined in line with the business category of the eligible site.

Award for World Class TPM Achievement

- Must have a minimum of 2 years or more of deployment activities based on the 8 pillars of TPM after receiving Special Award for TPM Achievement or Advanced Special Award for TPM Achievement.
- Must have deployed distinctive and creative TPM activities and showing results

13.7: July 2019: What Empowerment can do to Operators?

Autonomous Maintenance takes around 5 years to achieve, on average. Step 7 is the last step on Autonomous Maintenance. Not only have we improved the equipment in terms of reducing breakdowns and defects, but also we have finally built a highly skilled, self-directed, and empowered operator. At this last stage, operators become independent, skilled, confident, and well-motivated. They can perform simple repairs and equipment restoration by training them in repair skills. Operator's improvement skills are raised as they join improvement initiatives and projects in eliminating the six big losses in their equipment. In Step 7, these knowledgeable operators conduct autonomous supervision and follow the standards set by themselves on the orderly shop floor, where they can easily detect any deviation from the normal or optimal operating condition at a glance.

I have been privileged in my previous work to see operators who have reached Step 7 on Autonomous Maintenance. They work and think differently. There are different from the traditional operators who just operate their equipment. They spend more time with maintenance than their fellow operators do. An empowered operator has no limits, and their work is no longer routine. They can make decisions on their own.

What a Self-Directed Empowered Operator Can Do

One day walking into the plant, Tom Marshal is bothered with thoughts about one of their suppliers who are not meeting the standards because they almost lost an important client due to quality issues. As he approached the production floor, he was met with Linda, who had problems with temperature fluctuating on one of their equipment, and told Tom about the problem. Tom and Linda went to the machine to point out some possible causes for Linda to check. Tom knows how tricky these adjustments are to a new operator because he was exactly on the same spot when he was new to the plant. At 10:00 am, Tom glances at the production figures and is pleased to note that things are going well. He looked at his watch, and it was time for him to go for a production meeting. First up on the list was to decide on the best applicant, and the managers asked Tom about his opinion since he was the one who interviews them. Next on the list was the supplier who had problems with quality. Tom suggests meeting with the vendor and, likewise, train them on Statistical Process Control or SPC, and the management agreed with Tom. After the meeting, he went back to the line and started operating the controls. You might think that Tom is a manager or a supervisor, which he is not, but Tom is just part of an empowered operator and self-directed Autonomous Maintenance team leader.

Just imagine the caliber of this operator. Imagine if even 1/3 of your operators are like this. Achieving this level of Autonomous Maintenance will take a very long journey to reach. Breakdowns and quality defects will tend to revert back as frontline people, engineers, maintenance, including plant managers, are being replaced every year; what is important is to sustain the team's enthusiasm on every step that the team will undertake.

Empowerment means power, control, authority, or dominion. The prefix "em" means to put on to or to cover with. Empowering is the passing of authority and responsibility. It occurs when the power goes to the operators, who then experience a sense of ownership and control over their jobs. An empowered operator knows that their jobs belong to them. Given a say in how things are done, these operators feel more responsible. When they feel responsible, they show more initiative in their work, get more done, and most importantly, enjoy their work more. TPM believes that constant improvement has to come from the individual and that empowerment is the only way to get the people to adopt constant improvement as a way of life in the individual business. In other words, empowerment means is to create a working system based on principles that will enable people to take charge of the situation. TPM believes that the person closest to work knows best how to perform and improve their day-to-day operations.

Most employees want to feel that they own their jobs and are making meaningful contributions to the organization. Teams provide possibilities for empowerment that are not available to individual employees.

A self-directed work team is an intact group of people who normally work together on an ongoing day-to-day basis. It is not a group brought together for a special purpose, such as a quality action team or a quality circle, or a product launch group. An empowered workforce can plan, control, and improve their own work process. They set their own goals and inspect their own work. They review their own performance. They can prepare their own budget, coordinate with other departments, order materials, monitor their inventories, and even deal with suppliers. They are responsible for improving their workplace and equipment and acquire the necessary training they need. In short, these operators can decide on their own with little or no supervision.

Empowered teams provide a vehicle for employees to take on the responsibility typically reserved for managers or supervisors in the workplace. Organizations that acquire a capable workforce will offer a culture that matches a new workforce. Teams will offer greater participation and feelings of accomplishment. Industries with self-directed teams will retain the best people, while most will not. These empowered teams are designed to facilitate job-sharing and cross-training, but perhaps the biggest reason for the movement towards an empowered workforce is that it actually works. In the long term, they will reduce costs, boost morale, and improve productivity.

Reflections on Autonomous Maintenance

- Most equipment failures start from small things that have been left unattended and neglected. Autonomous Maintenance emphasizes the importance of establishing this basic equipment condition on their equipment. This basic equipment condition includes cleaning, lubrication, having complete bolts with the correct torque, and machines free from leaks. Once these basic conditions have been addressed, their impact will be felt immediately. Always remember that this is always a shared responsibility for both Autonomous and Planned Maintenance.

- Our equipment is a shared responsibility for both operators and maintenance working together in harmony. Once operators learn and accept their responsibility, they can only advance to any equipment improvement efforts and escape from a fire fighting or reactive mode of maintenance.

- When a failure occurs on the equipment, operators will encounter them first, not maintenance, since they are the people closest to the asset most of the time. Frequently operators do not perform an inspection on their equipment. In a traditional industry, maintenance develops the checklist, but operators do not perform the checks. This seems to be the main complaint of the maintenance people. Operators do not resist change; they just do not want to be dictated by anyone. The secret is to let the maintenance guide them into understanding their own equipment. Once they learn, they will take care of their own equipment. The key is to let the maintenance understand that operators are capable of developing their own checklist.

Remember, if operators develop their own checklist, they will be guaranteed to perform the checks themselves religiously.

13.8: August 2019: Precision Maintenance Explained

According to the dictionary, precision means the state or quality of being precise or being exact. It is the ability of a measurement that can be reproduced consistently. Precision Maintenance is the act of maintaining a consistently copied manner. This means that the maintenance performed is done the same way and delivers the same exact results regardless of the person who performs the maintenance tasks. It is also a method of performing maintenance tasks that must always be consistent, precise, and accurate. Not only do we need to be precise, but also we need to execute the tasks in the shortest possible time precisely.

Precision Maintenance involves performing maintenance work in a consistent, precise, and industry-accepted way. If properly implemented, maintenance should yield the exact same results, no matter who performs the tasks. Precision Maintenance is much more than merely having procedures on PM. Precision Maintenance must be the correct culture of the organization. The good thing about doing Precision Maintenance is that it can even reduce the chances of Infant Mortality Failures seen during the start-up of assets right after a Preventive Maintenance shutdown or Scheduled Overhauls are performed. It involves performing maintenance work accurately and precisely, which means that the maintenance should provide the exact same results no matter who is performing the work, whether the work is done by the most or the least experienced maintenance craft in the plant.

When different maintenance people perform the same PM on the same equipment, variations occur. These variations cause problems since they do not meet the requirements for the process. Precision maintenance ensures that equipment will be maintained to the highest possible standard so that the variations that can cause defects and failures can be reduced or eliminated, enabling equipment and assets to run at their optimized and peak level of reliability. Precision maintenance rebuilds machines and equipment to the highest standards so that fewer problems can occur during operation. It is a matter of ensuring that equipment and machinery health are done correctly and accurately.

History of Precision Maintenance: The first real evidence of Precision Maintenance skills and techniques was pioneered by NASA by Dr. Wernher Von Braun and his rocket scientists' team in the 1960s. He was a German and later American aerospace engineer and space architect. He was the leading figure in rocket technology development in Germany and a rocket and space technology pioneer in the United States. They discovered that for every 20% decrease in vibration, the life of the bearing is doubled. Further reductions produce an exponential increase in the bearing life and all other expensive components that usually are damaged during bearing failure causing a dramatic reduction in costs. In his 80's, Ralph Buscarello (also known as the Vibe-Father) taught thousands of people in more than 34 plus countries about vibration analysis, including innovations in-phase and resonance analysis. He pioneered the principles of practical prevention through precision and collective teamwork. He is said to be the father of Precision Maintenance. Ralph Buscarello preached that common assembly errors, mistakes, and omissions of essential field installation and rebuild details, along with simple equipment specification issues at purchase, are significantly the sources of destructive vibration that fatigues the bearing's metallurgy casing early, causing premature bearing failure. Ralph coined the phrase Precision Maintenance in the 1960s and spent the rest of his life trying to spread the word.   Like Dr. Braun, Buscarello is convinced that the lesser the equipment's vibration, the lesser the maintenance cost.

Figure   13.7: Pioneers of Precisions Maintenance

Ralph Buscarello preached about the relationship between an equipment vibration and the direct impact on costs and introduced the Precision Maintenance role in increasing machine life and productivity. The following is an extract from his paper on "Proven Benefits of Precision Maintenance," presented by him as his keynote address to the Vibration Association in New Zealand in May 1998. He quotes that in over 46 years of teaching on vibration reduction and conducting seminars in over 34 countries, I could produce hundreds of outstanding financial case histories on Precision Maintenance benefits, many involving millions of dollars. However, for those who remain skeptical, I suggest you accept none of my figures. Instead, start with the first idea of measuring the vibration levels taken at the worst point on about 50 ordinary but common machines, such as motor and pump assemblies. Plot the trends of their previous year's maintenance costs on the same graph as their vibration amplitudes. Then make up your mind as to whether reducing vibration to precision levels pays off or not.

Machine Type	Highest Velocity mm/s	Dollars Spent in USD	Lowest Velocity mm/s	Doing Precision Maintenancee	Percent Savings
Single Stage Pumps	5.6	$3,200	2	$650	80%
Multi Stage Pumps	4.8	$6,100	1.5	$1,100	82%
Major Fans & Blowers	9	$900	2.8	0	100%
Single Stage Turbines	3.8	$8,200	1	$2,000	76%
Other Machines	7.8	$11,850	3	$3,700	69%

Figure   13.8: Effects of Implementing Precision Maintenance

13.9: September 2019: Requirements for Precision Maintenance

The key to Precision Maintenance lies in the knowledge and skill of the maintenance craftspeople. Precision Maintenance will take time to achieve, but the benefits that can reap off are worth doing. Doing Precision Maintenance will not only reduce downtime, but it will likewise reduce the cost of performing maintenance. By training maintenance to perform, the correct tasks on the equipment will yield outstanding benefits for the plant. Organizations that can build a Precision Maintenance culture will reap the rewards of a better bottom line, a safer workplace, and more productivity throughout the facility.

First, a skills gap is necessary to conduct, assess the difference between the skills required to perform a specified task and the skills maintenance currently possesses. Precision Maintenance describes the maintenance culture in a facility. This means that the maintenance should yield the exact same results no matter who is performing the tasks. A scoring system will help determine the knowledge and skill maintenance currently has. With these scores, a sensible skills training plan can be developed to transform the maintenance workforce into a highly skilled team trained and qualified to perform these tasks precisely every time maintenance tasks will be performed on the equipment.

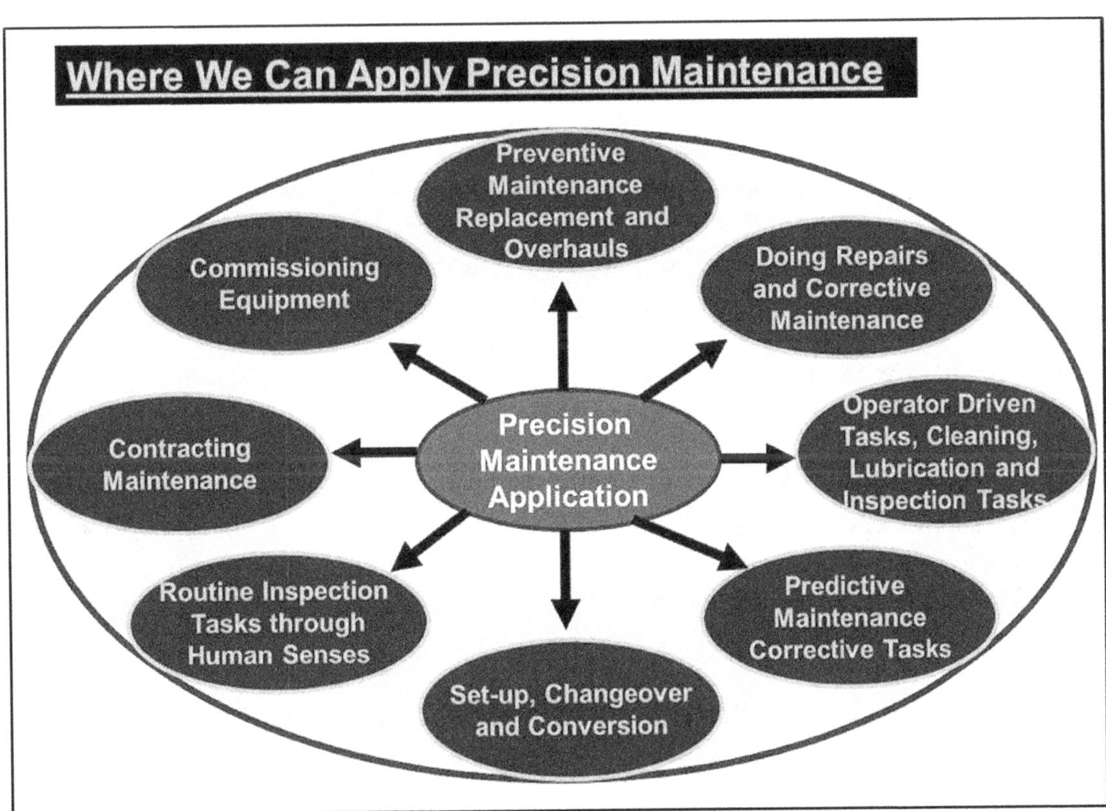

Figure   13.9: Application of Precision Maintenance

Achieving Precision Maintenance implementation, success, and sustainability requires a full commitment and advocacy from top management, senior leadership, and supervisory and management roles. There should be a clear focus on setting written expectations for maintenance, engineers, planners, MRO Stores, and all contributing roles, including operations in achieving Precision Maintenance and reducing human errors, especially during Preventive Maintenance activities.

Knowledge is the key requirement for Precision Maintenance. Knowledge is where we develop our skills. Essential to any industry are competent people who understand their assets and equipment intimately. Training is the backbone of any cultural change. It is mainly the missing link ingredient in any change or continuous improvement effort. Skill is the capability to do one's job correctly, apply knowledge and experience in all kinds of events over an extended period. Skill is the product of personal motivation and thorough training in which the end result is mastery. When mastery is achieved, then maintenance can be capable of transferring their knowledge and skills to others. This is what we want to achieve in our industry, and to enable us to achieve this, companies must develop the most effective training methods. This refers to the skills needed to perform the required work correctly. Define the gap between the current and what is expected from our people. Precision Maintenance involves performing maintenance work in a consistent, precise, and industry-accepted way. Here are some requirements needed to establish a successful Precision Maintenance program:

- Maintenance must be given the training and skills to perform their maintenance tasks correctly to do their job right the first time around. Likewise, they must also have the correct tools and instruments needed. The tools and instruments used for PM should be calibrated regularly.

- Maintenance should have knowledge of lubrication as well as oil contamination. Oil and lubricants must be properly stored and free from contaminants. Maintenance must understand the effects of storing lubricants in an open and contaminated area. For grease, maintenance must understand that grease should not be mixed with other types to avoid grease incompatibility issues. This problem will shorten both the life of the grease and the part it is greasing. According to Svante Arrhenius, a Nobel Price physicist, the life of the oil is cut in half for every increase or rise in its operating temperature. This means that if the lubricants have exceeded their base activation temperature, the oil will degrade or oxidize twice as fast for every $10°C$ rise in its temperature.

- Maintenance procedures and tasks must not only be written but should be precise and in detail. Use simple words that everyone can understand. Provide drawings, visuals, photos, or, if possible, videos whenever needed. Adopt visual controls to easily spot deviations from normal. The scope of work or maintenance tasks must be made crystal clear to the people who will execute the tasks.

- If bolts are removed during a Preventive Maintenance disassembly, the correct torque and tension should be known and checked during reassembly and installation. For rotating equipment being dismantled, precision alignment tools such as laser alignment should be performed before reassembling both the driver and driven components. It is important to check the vibration frequency level before and after conducting PM on rotating machines.

- Besides the written procedure regarding the details on doing maintenance tasks, it is more effective to provide videos on how to correctly perform the tasks to understand instructions.

Many videos on the Industrial Internet of Things (IIoT) can be shared with maintenance. Browse youtube.com regarding the correct way to grease bearings, bearing installation, alignment, dismantling, reassembly, and the like. There is free software available on the internet on how to download videos from www.youtube.com. Share these videos during your toolbox meeting, or much better if maintenance can create their own videos and include these as part of their training program for all maintenance in the plant. Just make sure that the person you are watching on youtube is an expert as there tons of wannabe gurus on youtube.

- Talk to purchasing regarding MRO spare parts being supplied to them. Only original spare parts should be used on the equipment unless a replicated part has passed an engineering evaluation that complies with the requirements that the part supplied has the same dimensions, tolerances, measurements, and material strength.

- Only the correct tools specified are used in performing the tasks. Measuring tools should be calibrated before using them. Any deviations from the standard should be corrected and calibrated immediately.

13.10: October 2019: Different Types of Lubricating Film

Oil provides a thin layer or coating called film. The oil film's function is to separate two moving parts from coming in contact with one another. In layman's terms, the film can be compared to the paint in the car. When it is peeled off, we can visibly see the metal. The paint serves as protection from rust and corrosion. For oil, what we are avoiding is metal-to-metal contact. Once this happens, heat is generated and will shorten the useful life of both mechanical parts and the lubricating oil itself. When we rub both our hands, we experience heat after a few seconds of continuously rubbing. When we place something between our hands, perhaps a thick cardboard, the heat generated is minimized. The cardboard acts as a separation to minimize the friction and heat caused by the two hands rubbing continuously. Similarly, lubricating oil separates the two metals through a thin film or coat to minimize friction by avoiding the two metals to contact each other relatively.

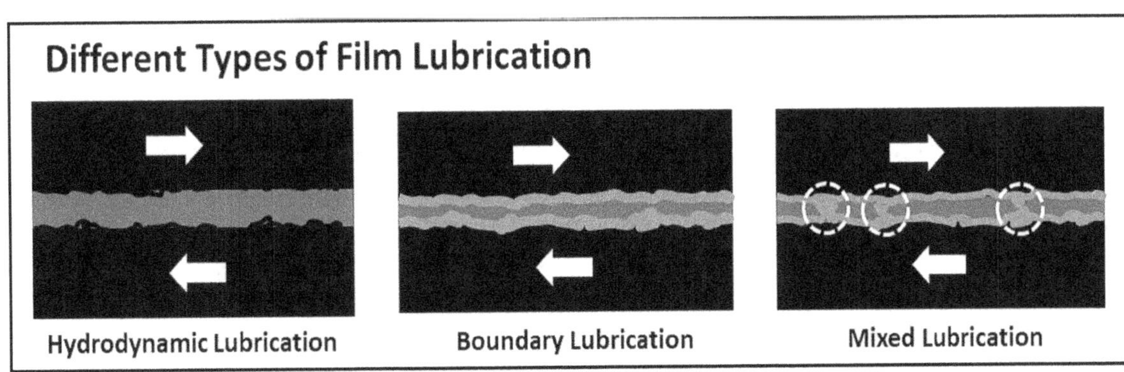

Figure 13.10: Different Types of Film Lubrication

Lubricating oil's primary purpose is to form a film or coating to avoid metal-to-metal contact between two metal surfaces' clearance. There are different types of film that the lubricant can provide, which includes the following:

Hydrodynamic Lubrication (HL) is also called full-film or fluid film lubrication. Hydrodynamic lubrication is where the film is thick enough to separate interacting surfaces to minimize friction. This type of film usually occurs at maximum viscosity with a minimum load but fails on shock loading with low speeds at high torque operation. In physics, the torque, which is also called the moment of a force, is the tendency of a force to rotate the body to which it is being applied, just like when we tighten the bolt.

Boundary Lubrication, In this case, the lubricant film becomes too thin to provide total boundary surface separation. The fluid film is squeezed out because of the high load as the two metals come closer to each other. With boundary lubrication, it is essential to use a lubricant with Extreme Pressure (EP) additives where the asperities will be coated with the oil's film itself for the opposite asperities to slide instead of adhering to each other and create a fracture. This film has a low shear strength but a high solid to liquid temperature. Boundary lubrication is affiliated with metal-to-metal contact between two sliding surfaces as the asperities come in direct contact with one another. This mostly happens during start-up and shutdown.

Mixed Lubrication: This is also referred to as partial lubrication, which is present in internal combustion engines. Both elastohydrodynamic lubrication and metal-to-metal contact occur in mixed lubrication. The load is supported partially through a fluid film and partly by its surface asperities in which it combines the properties of fluid film and boundary lubrication. Elastohydrodynamic lubrication (EHD) applies to extremely high contact pressure. Due to the bearing material elasticity during operation, the bearing metals usually deform to produce a very small area in which hydrodynamic film is formed. The film that separates the component's load is very small, less than 1 micrometer in thickness, while the pressure is extremely high, as much as 500,000 psi.

13.11: November 2019: The Use of Barcoding for MRO Spare Parts and Storeroom

The end goal of managing spare parts effectively is to ensure that spare parts are available when required for repairs or scheduled Preventive Maintenance in the plant. Without some sort of centralized system to track data, it is seemingly difficult to do things manually and may take you a lot of time just doing everything manually

If we want to know what the total cost of inventory in the storeroom at this very moment, or who is the vendor that can deliver this part within the day or other important questions, would it be possible to know this without sitting through a year's worth of emails, dossier, or purchase orders? If the person in charge of these documents is on leave, now what. Suppose a critical asset went down, and we need a critical part to remedy the problem immediately. Would you know if it is in stock in the storeroom, and if not? Which of the vendors could ship it with the shortest lead-time at the lowest possible cost? Implementing CMMS or EAM software to manage spare parts can provide information to answer these questions to have the tools needed to manage our MRO spare parts more efficiently and effectively. The direct integration with maintenance work orders, history, purchase, and even approval system can streamline the work process and easily give us visibility into the information we need.

Barcoding: **Most CMMS** and EAM software come with barcoding capability, which can be included, or as an add-on feature of the software. Storeroom people can use barcoding technology to link work orders, purchase orders, spare parts, and equipment. They can also account for parts, tools, or supplies that leave the storeroom and bill those materials back to the user. The use of barcoding in the MRO storeroom has not been fully implemented well by many industries. Although software such as SAP, Oracle, and other EAM or CMMS software available has capability or interface, which is available to accept bar codes. Bar code applications are slowly being adopted by maintenance organizations to manage the spare parts and MRO inventory. Grocery stores were the first to take advantage of this technology. Just imagine yourself if groceries do not have any barcoding technology. How long do you think you will queue to pay your grocery bill?

Benefits of Barcoding:

• Accurately records all aspects of every transaction and eliminates time-consuming data-transcription errors common with paper records
• Eliminates manual data entry and mistakes for parts return, receipt, and reorder.
• Records transaction histories for parts movement, cost, and charge-back
• Streamline purchasing orders and receiving practices.
• Uses transaction history to help determine minimum levels and reorder quantities
• Provides a rapid Return on Investment (ROI), which can generally be less than a year

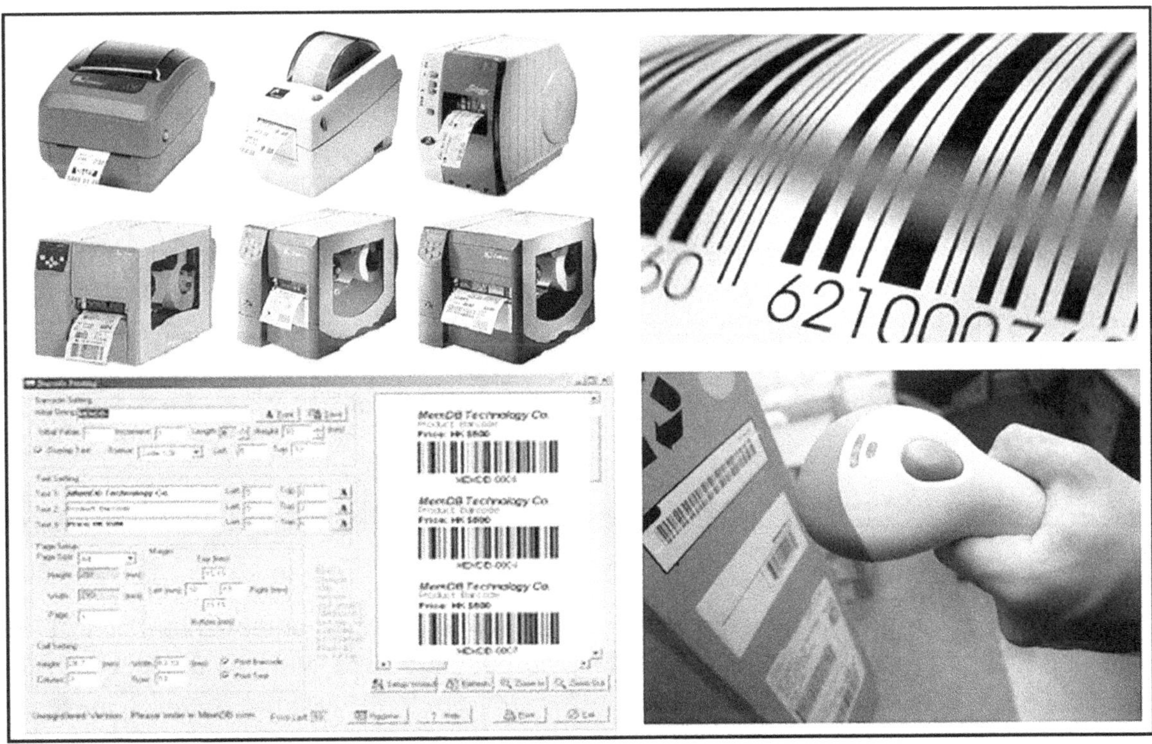

Figure   13.10: The Use of Barcoding on MRO Storeroom

One option for the storeroom is to use Bar Coding to reduce data entry errors and save manual entry. It is said that the human error rate ranges from 1 in 100 for manual data entry, while a bar code reader has an error rate of 1 in a billion. Just imagine that. There are many types of bar codes available on the market. Most barcodes comprise of a vertical dark or black bars and spaces. Information is encoded by varying the width of the bar and spaces. The height of the bar is usually of equal height. The barcode sticker label can be placed either on the bin location or attached to the part's plastic cover for the storeroom. Once they are retrieved, they will be read by a barcode reader, the same thing you see when you buy items in a supermarket, mall, or department store.

Barcodes were developed in the 1950s, and all barcodes conform to the standards developed by American National Standards Institute or ANSI and Automatic Identification Manufacturers (AIM). Barcodes are just a different way of labeling products, allowing us to a quicker checkout than doing everything manually. So, how does a barcode work? Once an item is run across the scanner, it picks up the barcode, which is done by reading the bars and spaces between them. As the scanner picks up the bars, it transfers the information on the software to a computer. The computer finds the record associated with that barcode, containing the information as the price, stock on hand, location, quantity on stock, and other relevant information of the part. By using a barcode, the margin of human error will almost be negligible to none. It also helps a company know exactly how much of an item they have in stock and replenish or reorder it. It also can save us time and money for data entry errors on prices.

Barcode accurately records all aspects of every transaction. It eliminates time-consuming data transcription errors common with paper and manual records and eliminates manual data entry and mistakes for parts checkout, returns, receipt, and reordering, which is popular among maintenance because of the ease and speed in which they can check out the parts, compared with doing it manually.

Another investment needed for barcoding will be a printer. There is a wide selection of barcode printers. Barcode Microsoft Windows application interface consists of a DLL set, which can create custom-made applications that run on the latest Windows or on any Windows NT platform. Whichever printer software will be used, a printer is necessary to print a barcode. Thermal, dot matrix, and laser printer would be best in the printing of bar code. Inkjet printers currently used in the offices are not recommended for printing barcodes as the ink can smear off easily. The laser printer will provide the best option for printing bar codes where the printing quality can be 600 to 1200 dpi. The cost of the barcode printer can range from $ 400.00 to 3,500.00 US dollars.

13.12: December 2019: New ISO 55010 Alignment of Financial and Non-Financial Functions

In many industries, there seems to be a grudge between finance and maintenance. I have seen this happen in many industries where finance people interrogate maintenance for every single cent of purchase maintenance people did and why they needed them. I have also experienced several industries with opposite and contradicting views or conflicting paths, thinking that their paths are the correct ones. I have mentioned in my other books that I resigned in one industry because finance cut back my training expenses. Why on this planet did you hire me if I will have no budget for training? I told the finance manager, are you stupid or something? The following day I resigned. One person from maintenance requested a tire for his service pick-up truck in one of the plants I served. He told me that finance called him, asking him why he needs four sets of tires instead of only one; his reply was because all four tires are in contact with the road. Lol. ☺.

I have a client in the Philippines. They have three power plants in different cities. Although the three power plants have different capacities, they also have similar equipment. The finance people did have a separate stock item (others called this part number, stock number, item code, or whatever) for each spare part for the three power plants. If I have a common oil filter for a diesel engine, the stock item or part number will be different for the three power plants. My thinking was that if plant A has an excess of these filters, and plant B and C need these filters, then they can just borrow from plant A and return it at a later date. That would not be possible because the three plants have a different stock item or part number for this filter, even if it is the same part. Although both departments are keen on controlling their company's cost and budget, they contribute to the cost itself. People from Finance must also understand that there are better and more proven ways to reduce costs by doing maintenance correctly. Just a piece of advice to industries, do not try to restrict or cut off your industry's training budget, especially maintenance.

Organizations are typically structured along hierarchical lines for different organizations or departments within a company, leading to barriers, red tapes, bureaucracy, and silos. In September 2019, a new ISO standard was published. This standard provides guidance on the alignment between Asset Management functions such as Finance, Accounting, and other technical functions within an organization. These non-technical and technical people are commonly the key people responsible for delivering asset management within their organization. However, in most cases, the two functions developed their own way to manage and control risks, creating conflict. A new ISO was established to address this problem.

ISO 55010 specification was written to help all functions within an organization to align both their financial and non-financial, or what we consistently called technical and non-technical issues, to maximize value from their assets. ISO 55010 provides guidance to bridge the gap between an organization's financial and non-financial functions to help them resolve their conflicts with one another. However, we all know that the finance voice is bigger in an organization than technical people such as maintenance since they are near to God, I mean the CEO. Over several years of discussions involving several hundreds of experts worldwide, ISO 55010 specifications were now approved for publication. Having this alignment can enable the organization's different functions to better share and utilize their information and improve their organizational objectives.

Industry's brand and reputation are what we are paying for when industries aim for these certifications and awards. Just imagine drinking your favorite soft drink only to find out that a cockroach is at the bottom of the can. Will you drink this brand again? Will you tell someone or anyone about your experience with this soft drink? I bet you will even post this on your social media or FB account, and maybe it will go viral. Once a brand had been damaged, it is very difficult or impossible to bring it back once more. It is done. Likewise, it is the organization's responsibility to value both its internal and external resources. While industries become more aware of their customers' value, these awards will help them win more business and clients. That is why top management and executive decision-makers must fully understand the reliability and how it affects their business because its reputation is at stake. Reliability should be given the same priority as safety and quality. Industries must understand that reliability is responsible for reliability and maintenance people, but it is everyone's responsibility in the plant.

2020 RSA Reliability Newsletter Vault Archive

> *Knowledge and education are what maintenance people need to fulfill their duties and function. As a maintenance myself, I humbly ask, let us not deprive these good people of education because that is all they ask for.*

14.1: January 2020: What Maintenance Need is Knowledge and Not Experience

The objective of doing maintenance in our equipment is to take care of the equipment's Total Life Cycle at the most reasonable cost. Industries purchase equipment and assets to earn revenue to ensure that their equipment delivers what will be expected from them at the lowest possible Life Cycle Cost under optimal conditions with compliance to quality, safety, and the environment. To do this, maintenance must, therefore.

• Preserve and sustain equipment reliability
• Find ways to reduce the cost of maintenance
• Minimize or reduce equipment losses, breakdowns, and failures
• Identify weak equipment points in the design and improve them
• Have the needed knowledge and skills to correctly maintain the equipment and assets

If you read the last one, it said to have the needed knowledge and skills to correctly maintain the equipment and assets. I am not speaking about the experience here, but having the correct knowledge to build their skills is significantly important. Having the correct knowledge is different from experience. Maybe have you wondered why there are no college degrees or technical courses on Maintenance and Reliability in most countries? Actually, there is, in some well-developed countries such as Australia, the United Kingdom (England), and the United States. There are courses on Maintenance and Reliability that you can enroll. Here, in the Philippines, and for the rest of the world, the only university we got is the "University of Hard Knocks," where everyone learned maintenance because of their day-to-day pressures, stress, tension, difficulties, pain, despair, frustration, anger, disputes, and struggles that we now called "experience."

There is a saying that people are the company's greatest asset. I totally disagree with that since not all people are assets. If I will re-phrase this, people are not the company's greatest asset, but rather, the right people are the company's greatest asset, and the wrong people are called liabilities. We can only have the right people if they are equipped with the right knowledge and skills to do their jobs right the first time around. Knowledge comes from training. My point is that top-management people must not deprive their key maintenance and technical people of training. Today I work as an independent trainer, preacher, and teacher on Reliability and Maintenance (I hated it when they called me consultant, perhaps a teacher will do). I have taught industries, but they also hire me to provide recommendations and advice on how to perform the correct maintenance on their assets and equipment. From my observations, 95% or more of the industries I taught in the past perform maintenance differently, if not the opposite of what I have been teaching them.

In my class, I usually asked a simple question which goes like this, if you were tasked to grease a bearing, how would you know when to stop pumping the grease gun? How would you know you have provided the correct amount of grease to be placed in the bearing? The common answer I got is when the old grease is coming out, and their answer, I guess, was based on their experience. Let's touch a little about the subject of lubrication. There is only one thing that matters, if you over-lubricate or under-lubricate, this will be detrimental and cause problems with the equipment. It means that when the old grease is coming out, then we are over-greasing the bearing. Over-greasing the bearing in motors could damage the seals and allow the grease to leak into the windings or other parts, causing further complications.

On the contrary, lack of grease can result in oil starvation. It can cause overheating and increase friction, wear, and contamination. There are two ways to grease a bearing correctly. The correct way to stop pumping the grease gun is when the grease has actually reached the inner and outer raceway of the bearing. This will be possible if we have an ultrasonic monitoring instrument. This instrument will try to measure the ultrasonic frequency emitted by the bearing. Through a process called heterodyning, it will translate these frequencies into the audible range. The process of heterodyning is the mixing of two frequency waves. This will produce both the sum and difference of the original frequencies, which allows the shifting of a high-frequency sound to the audible or sonic range that the human ear can hear. For example: let us assume that we have a bearing emitting a frequency signal of 35 kHz. The human ear can hear only up to a maximum of 20 kilohertz. This means that 35 kHz will not be audible. Mixing a 30 kHz constant-frequency wave with the signal, we will get the following by adding a constant of 30 kHz will give a sum of 60 kHz, which is not audible to the human ears. When we subtract this frequency, we will get a frequency of 5 kHz, which is audible. This will be retained in the filter since it is audible. Heterodyning consists of a mixer, local oscillator, and a low pass filter. With today's technology, there are already ultrasonic grease guns that are available on the market. The second way is to determine the diameter and width of the bearing by making some mathematical calculations. There are grease guns on the market with a digital counter, which indicates the grease being pumped by the grease gun, either in grams or in ounces.

After hearing their feedback, I thought that maintenance should know these things a long time ago since this is part of their basic activity. Yet, deep in my heart, I do not blame them since they were never given a chance to attend this important training on maintenance. This knowledge maintenance currently has been passed on by their ancestors ever since the beginning of time. It will continue to be passed on to the newer generations when the time comes for them to retire. Training is a venue where we acquire knowledge. This knowledge is used to correctly perform our maintenance tasks to do our jobs right the first time around. I have been teaching for 16 years independently to different industries from different countries, cultures, races, and religions. During this period, I have experienced more than a dozen times where maintenance will talk to me most during break time and tell me in a calm voice that how they wish they have met me ten years ago or even longer. If I asked the person, what makes you say that? The person would reply that he is retiring next month. My heart bleeds for this guy. I feel sad that his company never invested anything in this person until his very last day on the plant. What a shame. I know that there are many industries like that, and all I can say is that if management thinks that attending maintenance training is expensive, then all I can say is look then at the cost of all your failures and breakdown; I believe they are more expensive than my training.

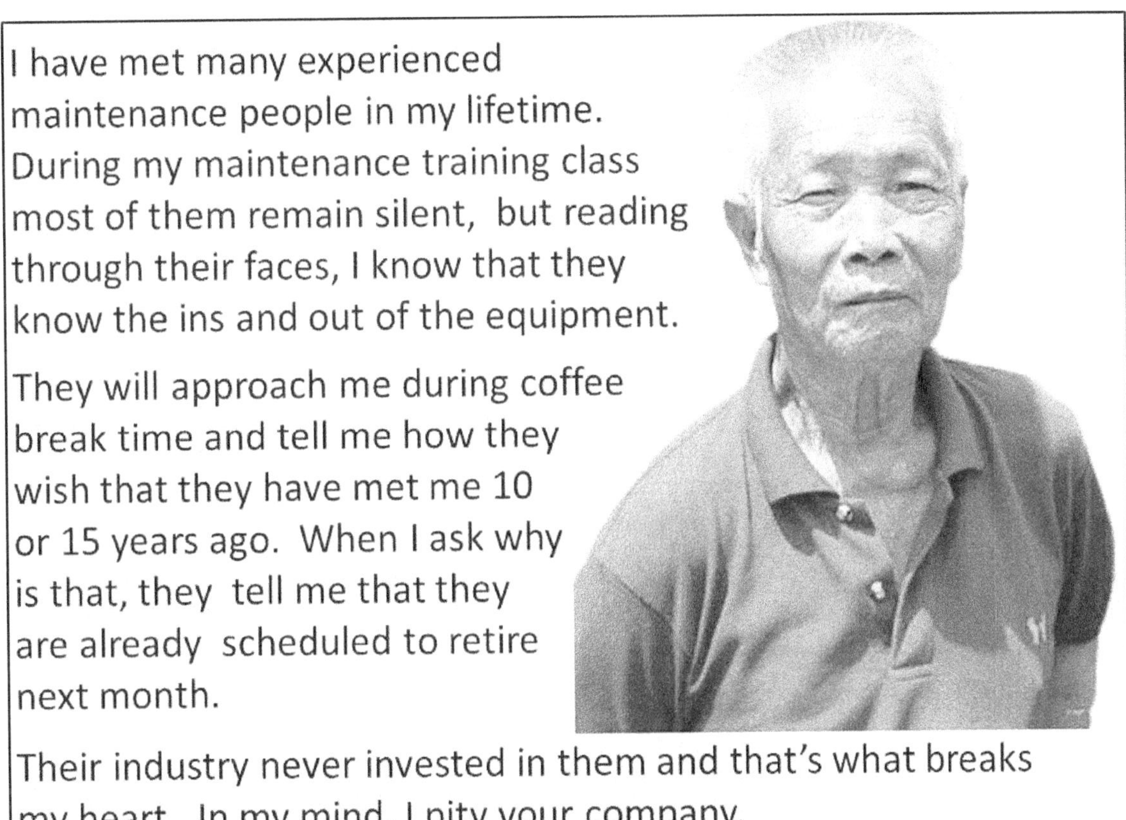

Figure 14.2: Let Us Not Deprive Maintenance of Training

14.2: February 2020: Industrial Internet of Things (IIoT) and Maintenance 4.0

The industrial internet of things (IIoT) uses wireless sensors and actuators to enhance manufacturing and industrial processes by hooking them on the equipment and providing data directly on your computer. It tries to leverage the power of smart machines and provide real-time data that can provide information on how the equipment is performing in real-time, which was not commercially available during the past decade, especially during my time. The concept behind IIoT or the Industrial Internet of Things is that these wireless sensors can capture and analyze data in real-time. They communicate these data directly to the user and alert them for any deviation from their normal operation, allowing both operations and maintenance to decide whether to continue operating or stop the equipment to correct the problem.

The Industrial Internet of Things (IIoT) is about the connection and interaction between humans and machines. Using wireless sensors, cloud computing, artificial intelligence (AI), and the internet of things, these machines can be connected with each other and can transmit data to make monitoring easier without even bothering to go to the equipment itself. Traditionally, a Predictive Maintenance user will carry his vibration-monitoring instrument, place some probes, sensors, accelerometers, or whatnot, and extract data from the equipment. This data will be downloaded on the laptop, and the user will plot the data and compare it to his previous readings and check if everything is well within limits or not. In today's technology, wireless sensors are ready to be mounted on the equipment that will transmit these data, such as vibration, temperature, pressure, and other information directly from the equipment to the laptop. This means that you no longer need to go to the equipment and extract the data manually.

Technology has changed in a very rapid mode. I remember when I was still studying that when the teacher asked us to take notes, each of us will get our notebook, pen and start writing whatever the teacher writes on the board. Those were the times. Those were my time. When I asked students to take down notes during my training class, then this is how they do it today in figure 14.2. Perhaps I can adopt these technologies during my training as I can only speak Tagalog and English. There is already a technology existing that can be used for training that will be available soon. This technology can transform you into a hologram where you can teach anywhere in the world in the comfort of your own home and space. You can be in Japan speaking their language without even bothering to know the Japanese language. I can actually be in Japan teaching maintenance without even boarding a plane because, physically, I am at home. What a cool technology. Well, I only got one problem here, what if they also pay my check-in hologram. I still prefer to teach the old-fashioned school style of training with myself physically present in the training room, having a great time cracking jokes with my students.

Figure 14.2: How Students Take Notes Today

Maintenance 1.0 to 3.0 includes sending highly trained Predictive Maintenance users to place some probes or sensors on the machine. The user will collect machinery vibration readings on the rotating equipment and download them to their laptop. With today's digital technology with Maintenance 4.0 will include a wireless vibration sensor connected to a cloud server and platform. It will try to use artificial intelligence to analyze complex patterns simultaneously, provide automated recommendations, and advise the end-user of the asset. This is like Kasparov playing chess against Deep Blue, the computer. I remember that game where Grandmaster Chess champion Garry Kasparov defeated IBM's Deep Blue in their final six-game match. However, Deep Blue was quick to learn, and the following year, a rematch took place, and this time, Deep Blue goes on to defeat Kasparov in a heavily publicized rematch.

Figure 14.3: Kasparov Losing to Deep Blue in their Rematch

In Maintenance 4.0, the vibration analyst will no longer go to the equipment to extract data as the data will be readily available automatically. They will no longer be having the problem of a Vibration Analyst level 4 resigning from work since another company seeking the same position recently pirated him. Perhaps the user was not available because he was on leave or utilizing his Vacation or Sick Leave as part of his benefits with the company. Like Deep Blue, these sensors will provide the needed data 24 hours a day. These wireless sensors do not need any sick leave, vacation leave and will not even be bothered with this **covid 19** pandemic. They will provide data 24 hours a day. With this kind of automation, it will connect machines, data, and people with ease. These wireless sensors and software today will allow humans to share this information with each other and can provide maintenance a complete platform to track, collect, exchange, access, and analyze large amounts of data chunks at a glance with ease. The insights obtained from this data can be used intelligently to decide whether maintenance tasks are needed to be performed on the equipment or not. These data can provide us with a decision-making process on whether we need to continue running the equipment or stop it from performing a maintenance task. However, I see here that there will be lots of data for failures that really do not matter much. These are failures that will not be a big issue if they occur on their own since there are no safety nor environmental consequences or may cause some very minor operational consequences, but there will be very little data for failures that truly matters most, and this is where these wireless sensors need to be prioritize and mounted on. Remember that machine learning needs lots of failure data to do its learning. There will not be very many instances that matter if these sensors are placed on the wrong equipment.

14.3: March 2020: Reliability is Everyone Responsibility in the Plant

I have been active in LinkedIn for a year now, and you can read many posts daily about reliability, but what I have never read so far is, what is the definition of reliability to them in the first place? Let me start with my own definition and interpretation of reliability. Reliability is the collaborative activities and efforts of every organization's function to ensure their customers and stakeholders that the products and/or services they produce are trustworthy, reliable, and dependable to the people they served. Now you might be thinking that this is about quality and not reliability. Let me put it this way, reliability is the input; it has something to do about the physical assets we have, such as equipment, while quality is about the output or the product or services in which the equipment is providing.

Reliability is not just about maintenance since everyone in the organization will play a role in this. We have defined reliability as being capable of being trusted or trustworthy in the previous chapter of this book. This means that reliability is not limited alone to our physical asset, which is our equipment, but it has more to do with the reputation and brand and its ability to be trusted by its customers. All the things about maintenance discussed in this book are merely a fraction. Every function in the organization has its own tasks and responsibility to contribute to reliability. Reliability is beyond MTBF and Asset Management. Let me give a very simple example of what I am trying to dig up here. Just imagine the following

• A can of soft drinks with a cockroach inside
• An airline industry that has a couple of consecutive crashers recently
• A TV set that lasted a couple of days
• A brand of noodles with some metal fragments inside the cup
• A telephone company that is always out of order
• A city with unlimited blackouts
• A well-known cruise ship that sank, killing 4000 people
• A chemical plant that exploded
• A pharmaceutical company that have cases of several people killed
• A well-known car brand with sudden unintended acceleration

How would you feel about that if you were the victim of these? Your company's reputation, brand, or logo is damage. One of your industry's names has been impaired, then it would be difficult to gain the trust of your customer once again, or worse, you can close your business forever. Your industry is no longer reliable, and you have lost reliability. Building industries that last can be possible if we can break down these departmental barriers, silos, bureaucracy, and red tapes. Each function of the organization can communicate effectively on what works and what does not work for them. The most important key to reliability is not about technology or the latest fancy Predictive Maintenance tools or software, but everything boils down to having the right people on board the bus to take these challenges and maneuver these bumps along the away. Only the c-level people can make this happen. Industries whose c-level executives value their people have a greater chance of survival in the long run.

Figure 14.4: Reliability Can Only Work if Silos are Removed

Unlike maintenance, deploying a reliability strategy in the plant will involve bigger plant-wide participation, including every single function in the entire organization. Each of these functions must be linked together by having a common voice, a common direction, and every single one of them singing just one song. The problem right now is that most industry's function is open only to themselves and remain closed to others. We need to break down these departmental barriers so we can communicate effectively with others. The reason why c-level people must be the pioneers of reliability is that they are the only people with complete authority and control on this matter. The first thing that we need to do is generate an awareness and understanding of all our people in the organization on what reliability is and how every function can directly or indirectly impact the bottom line, which is the company's reputation. All people in the organization must understand that these failures, breakdowns, and losses are not solely caused by maintenance. The only way to address this problem is for everyone in the entire organization to understand how their function affects its reliability. Here is an excerpt from R. Keith Mobley's message on the forward of my book on Cutting-Edge Maintenance Management Strategies

- *Without reliable business and work processes and employees who execute them, physical asset reliability is impossible. Instead, the volatility and instability of the unreliability of the infrastructure, not the assets, determine their reliability. Understanding the true magnitude of reliability defines a reliability leader. It is also what differentiates reliability leaders from their peers and enables them to lead their organizations into a brighter future, which would not have happened without their leadership. Reliability is an essential, foundational part of a viable enterprise's life cycle asset management process, but it must be more than just reliable assets defined by MTBF. Capital or tangible assets are the heart of any enterprise that relies on its ability to produce or manufacture products or, as in the case of facilities, to maintain conditions that occupants require. They are the engine that drives the revenue stream but can quickly become a demanding drain on the enterprise's cash flow and net operating profit when not effectively managed. Reliable processes, procedures, and practices are the framework, the infrastructure that binds the entire enterprise operations, including asset management, together. It provides stability, consistency, and dependability that is crucial to all aspects of the organization.* R. Keith Mobley

Finally, each function within the organization must also understand that we can transform maintenance from a cost-center into a profit-center that can contribute to the company's revenue big time if this can be done. Today, most industries still remain reactive, and they will continue to stay that way while the rest will blame everything for maintenance. Nothing will ever change if c-level people will not be involved with this. There are a strong connection and correlation between the company's revenue and equipment downtime. This means that if we can reduce downtime, then this can add up to the bottom line, but this will only happen if everyone in the organization realizes the fact that reliability is everyone's responsibility in the plant. All organization functions, whether from Finance, Purchasing, Warehouse, Maintenance, and all the others, must sing just one song; if not, then nothing matters, and our industry will just be at the mercy of the wind. I have never seen even one industry that educates all its people about reliability, just like what safety and quality people are doing. Perhaps industries in this generation are not yet ready for this, but I do hope that the next generations to come can find ways to make this happen on their plant for them to survive. Remember that all industries follow the law of the jungle, survive now, or be left behind, and I think that is all I have to say about that.

14.4: April 2020: Are Industries Ready for Maintenance 4.0?

Last night (as of this writing is April 16, 2020), I was listening to the Philippine President Duterte's nation's address about the **covid 19**; he was quite disappointed as the number of cases in the Philippines and deaths continue to rise, which is the same likewise with the majority on other countries globally. He was mentioning about China, that everything started in China, and after 59 days, they are now open, meaning the majority of businesses are now open in China. The President said that they are now open because the people of China have complete obedience. It struck me for a while, and in my mind was the people of China have discipline, but perhaps in the majority of the cases in other countries, the Philippines included, they have none. I absolutely agree on this one. Technology is traveling at warp speed, but the way we do maintenance is still in the fire-fighting mode, which is true. Many industries still remain reactive up to this point in writing this book. If you asked me why, just like China's **covid 19** pandemic case, the maintenance people have no complete obedience to the disciplines on maintenance.

One oil analysis vendor on LinkedIn was promoting his top-of-the-line oil analysis instrument with the capability to provide data on the oil every 2 minutes. That will be around 720 data per day or 262,800 data per year. Do we actually need this massive amount of data to understand the problem? If you think that by adopting these wireless sensors, artificial intelligence, IIoT, or whatnot, our problems are solved, and they will do wonders in maintenance magically like Shin Lim, I am not so much optimistic about that. Remember, there is no silver bullet solution to all problems on maintenance. My limited knowledge simply tells me that a clean motor that is well lubricated and totally aligned can last 10 years or even more than a dirty motor, which can last for 6 months or even less. Equipment with a couple of missing bolts will increase its vibration and end up with a premature failure, which will happen sooner than we think. Equipment with leaks also contributes to energy loss, and I can continue on and on. What am I telling you here? Even if we place these wireless and smart sensors, as they called it, on all our equipment and asset, the equipment will still fail prematurely because the basic equipment condition has not been done on the equipment. These smart sensors can tell you, hey, it's getting hot, but that's about it. These sensors are not magic wands that will do their wonders and maximize the lifespan of parts when placed on the equipment. Equipment that is poorly lubricated, which has excess or inadequate lubrication, can build up friction again, ending in an early random failure. What happens is that even if I installed these sensors on the equipment, then it will just fail. The sensors can advise us that the temperature is increasing, and we can perform maintenance to address the problem without the part reaching its lifespan. If a part is supposed to have a life of 5 years but failed after 2 months, it will be known, and that's about it. What is my point here? My point is simple: if you buy these sensors and your plant is reactive, you cannot benefit from these technologies.

As R. Keith Mobley quote, there are no random failures, just our inability to look deep enough to find the cause. Failures do not just happen, nor are they random in nature. Most are caused by our failure to design, install, operate, and maintain our assets properly. We cause failure through acts of commission or omission. We are the pattern. We are the forcing function that resulted in these so-called random failures. If you digest this statement, I find it very deep. If we can decode the last statement, he said, we are the forcing functions that result in these so-called random failures. The word "we" means you and me. It means that you and I are human, and I believe that humans cause all failures. According to James Reason, an average human being commits around six errors a week. This also coincides with what I have learned in Latent Cause Analysis from my friend Bob Nelms that all physical failures are triggered by humans, but humans are negatively influenced by latent forces. The goal is to expose and identify these latent causes. This means that for every failure that happens, there is a corresponding human error committed. There will always be a cause-and-effect relationship for every human error, whether it was the person's fault or the deeper cause of it happen. All random failures have a cause as to why they happened.

Industries will find little or no value in implementing Maintenance 4.0 if their maintenance is reactive and their equipment lacks the basics. Fully automating equipment will not eliminate the chances of random failure from occurring. I believe in technology, but implementing this in an environment with many random failures will not change anything. Breakdown and failures will just continue to happen. If I place some smart sensors on a piece of rotating equipment, sure, it will tell us that it is vibrating beyond its limit or the temperature is increasing by the day, but still, the failure is just waiting to happen. If we can take care of the basics and provide care for this rotating equipment by supplying the correct lubrication, making it clean, and applying the correct alignment practices, then this is where these technologies can benefit our industry. If these basics are in place, then these cool technologies will work smoothly in our equipment.

One of the questions I asked myself is, what if these smart sensors and software go haywire. Figure 14.5 is a true story that happened on October 7, 2008. Quantas Flight 72 (QF-72) was on a routine flight from Changi Singapore Airport to Perth, Australia. The flight captain was Captain Kevin Sullivan, with 13 592 flying hours of experience and a former Top Gun pilot. The first officer was Peter Lipsett. This is what happened.

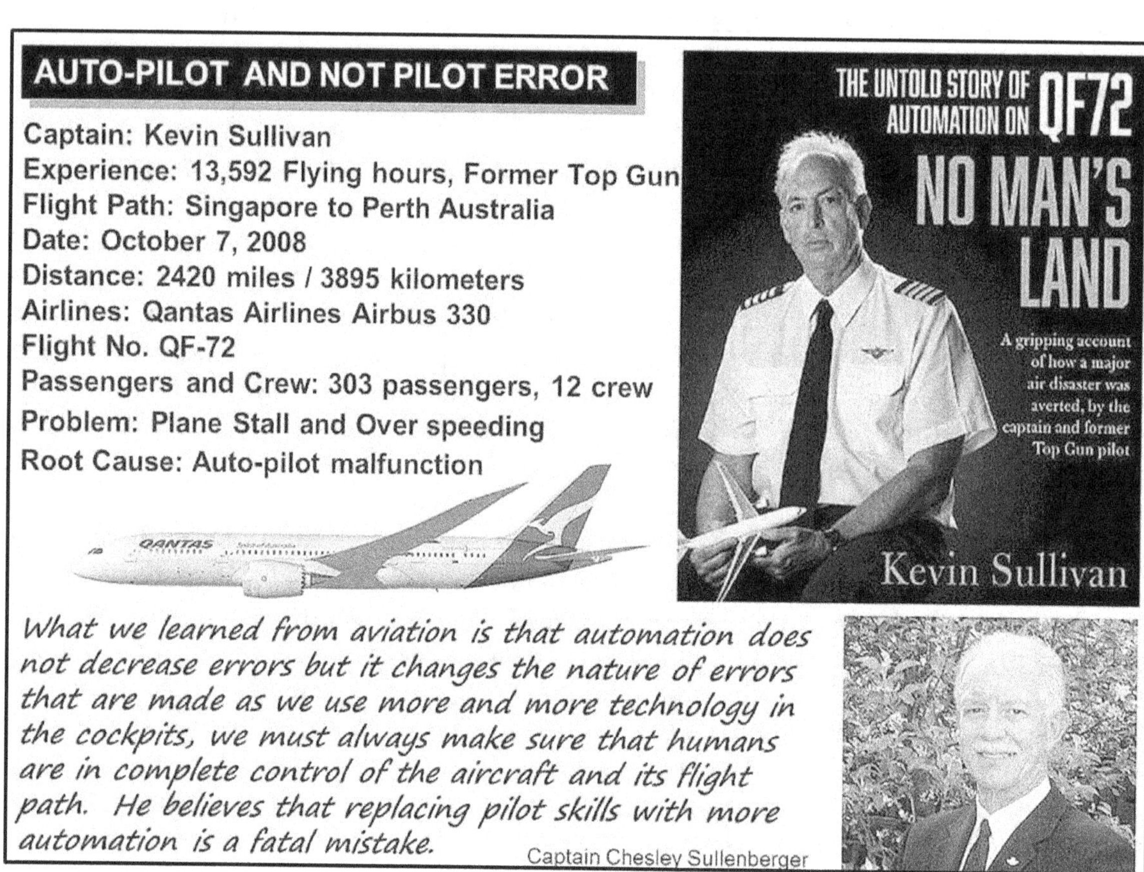

Figure 14.5: Automation is Not 100% Fool-Proof

- At around 12:40:26, one aircraft's three air data inertial reference units (ADIRU) started providing incorrect data to the flight computer. In response to the anomalous data, the autopilot disengaged automatically, and the captain took over. A few seconds later, the pilots received an electronic

message on the aircraft's ECAM, warning them of an irregularity with the autopilot and inertial reference systems. The pilots were receiving messages that the plane was stalling and overspeeding.
- At around 12:42:27, the aircraft made a sudden, uncontrolled pitch down maneuver reaching around 8.4 degrees pitch down and rapidly descending at 650 feet (200 m). Those that were not wearing their seat belt bumped their head on the ceiling and were injured, including one male crew.
- At around 12:45:08, the aircraft made a second uncontrolled maneuver of a similar nature, causing an acceleration of a 3.5-degree pitch down and a loss of altitude of 400 feet or around 120 meters.

The pilots radioed the tower and declared a state of emergency. The flight made an emergency landing at Learmonth Airport, Western Australia. Although the pilot landed the plane safely on an emergency landing, there were no recorded fatalities among the 315 people on board the plane, but 119 people were injured as they bumped their heads on the ceiling during the pitch down motion of the plane. The final report concludes that the incident occurred not because of pilot error but due to a combination of design limitations in the flight control primary computer software of the Airbus A330/A340 and a failure mode affecting one aircraft's three air data inertial reference units (ADIRU). The design limitation meant that multiple spikes in the Angle of Attack (AOA) data from one of the ADIRUs resulted in commanding the aircraft to pitch down in a very rare and specific situation. In layman's terms, the software provided the error that made the equipment stall and pitches down. Thanks to the experience of the pilot, no one died in this incident. However, other recorded instances of similar events like this, but unlike Quantas Flight 72, the other planes crashed where all people died in the mishap.

What Captain Sullenberger is stating the automation changes the nature of errors means that with these new technologies, automation, and digitalization around, let us not be too complacent about technology as we might face new problems in the future what we have not yet experienced before and industries should be ready to accept the risks to face them.

14.5: May 2020: Risks, Criticality, and Consequences (Part 1)

Risk is the effect of uncertainty on objectives, and an effect is a positive or negative deviation from what is expected. Criticality is a measure of how important an asset or equipment is to your process. The more critical the asset, the more impact it will have when a failure happens. Criticality is based on what could happen if the failure occurs. The consequences of failure are the aftermath of the failure. In question number 5 on the Classical RCM, it asks us how failure matter. This means that if the failure happens, what will be its consequences. When a part is unavailable when needed, then what would possibly be the worst thing to happen? In this case, both risks and criticality are based on the possibility of whether it may or may not occur. The consequences will be the final outcome of the failure.

We have a pump used as a cooling for a steam turbine in a coal power plant. In this case, let us assume that the pump has a redundancy. If either the pump or motor failed in an operation, the steam turbine would overheat. However, the standby pump and motor will be activated, and the turbine will be running, where power is still capable of being generated. In this case, the other pump was already in failed state mode, and we only have one pump running at this point. If we again assess the risks and criticality of the failure with only one pump left, then the consequences are more likely to happen. The risks and criticality are now higher when both the running and standby pumps are in place. This is on the assumption that the previous failures of the motor and pump have not yet been repaired. We now know at this point that if the standby pump that is now running failed and the two pumps are both in a failed state, then the consequences will be a city blackout, which is an environmental consequence, which will more likely, pissed off some crooked old politicians in town that will be affected. Let us assume that the bearing failed, and it was available in the storeroom; the maintenance withdraws the bearing and repairs the pump; in this case, since the standby pump is running, there were no consequences of the failure whatsoever since the failure was averted. If both pumps failed and the city experienced a blackout that lasted for 5 hours, that would be the consequence of the failure. If the standby pump is maintained well, even if it is not used and the lead-time to acquire the bearing will just take a few days, then, in this case, I will not keep a stock of the bearing. However, if no maintenance was done on the standby pump, then there is still a greater probability that the standby can also fail after operating it just for a few days or even for an hour.

One of the things I learned from my students is that they often provide the same running interval for running and the standby pump. This means that if the running pump runs for one week, it will be stopped, and the standby pump will be switch on and run for 1 week, then stop and switch back to the original one, and so on. I would not do that if I were you, both running and standby pumps should not have the same running interval. The rule of thumb here is that one should run longer than the other should. If the two pumps run on the same interval, let us say a one-week's interval, then there is the probability that the two pumps will fail at the same interval. The second point, the more startup then, the more chances of failure. In this case, I would recommend a switching interval of 3 to 5 months for the running pump and 1 month for the standby pump to run. The advantage of this is that it produces the lowest number of start-ups in a year while allowing a long duration for the standby pump to run. In this case, since the standby pump will be idle for 5 months, it is advised to rotate the shaft around 360 degrees plus add 60 degrees to avoid the chances of false brinelling on the bearing. This must be done on the standby pump at least once a week or every two weeks. This means that since the other pump is running, it is vibrating. These vibration forces are not only confined to the running pump but the standby pump will also be absorbed. This means that inside the bearing's raceway, the balls or the rollers are slightly vibrating concerning the inner and outer raceway of the bearing, which produces some fretting marks, mostly on the outer raceway of the bearing. If a bearing is subject to vibration when stationary, fretting damage occurs at the points of contact, giving rise to what is known as false brinelling. Standby components likewise in the storeroom are no exception and should be provided the same treatment. The pumps and motors should be rotated slightly once per week to change the points of contact of the ball to the raceway of the bearing. Just do not rotate them exactly at 360 or 720 degrees, as the ball or rollers will end up exactly in the same location with the outer raceway of the bearing. Where there are installed standby machines, it is a good practice to run the machines alternately with a different interval to reduce the risk of fretting, especially when vibration is heavily felt in that area.

Let me state another classic case regarding risks, criticality, and consequences, which is the story about the Challenger Explosion that happened on January 28, 1986. It took just 74 seconds after lift-off for the challenger to explode, and there were no survivors from the crew.

Here is the chain of events that happened before the explosion. On January 27, 1986, the weather forecast indicated a temperature below freezing of around -6.67 ° C the following day during the launch date. The Morton Thiokol engineers raised their concern to their managers about launching at a cold, freezing temperature. That evening, one day before the actual launch, a conference call (Skype and Zoom were not yet available) between NASA, Morton Thiokol, and Marshall Space Flight Center took place to make a very important decision whether to launch the Challenger into space or not. I hope you don't mind if I expand a little bit so the reader can follow the story. The Challenger Mission is the most exciting shuttle mission from NASA because it was the first time that they will be sending an ordinary citizen into space together with the astronauts. This caught the attention of millions and millions of viewers worldwide. NASA launched a contest called the Teacher in Space Program, where around 11,000 teacher applicants from elementary and high school applied. The selection was finally trimmed down to 100, and then finally to 10 candidates. Out of the 11,000 Applicants, Christa McAullife was finally chosen as the winner of NASA's Teacher in Space Program.

14.6: June 2020: Risks, Criticality, and Consequences (Part 2)

Marshall Space Flight Center was responsible for all rocket and engine systems of the Challenger shuttle, and their plant was located in Alabama. During the conference call, George Hardy, Deputy Director of Science and Engineering, Judson Lovingood, Deputy Manager of Shuttle Project, Ben Powers Engineering Structure, and Propulsion.

Morton Thiokol is the manufacturer and OEM of the Solid Rocket Boosters. Their plant was based in Utah. During the conference call were Jerry Mason, Senior VP of Wasatch Operations; Carl Wiggins, Vice President and General Manager of Space Division; Joe Kilminster VP Space Booster Program; Robert Lund VP of Engineering, Roger Boisjoly Engineer Structures Section; and Arnold Thompson, Supervisor of Rocket Motor. NASA (Kennedy Space Center) was responsible for all space shuttle launches and was located in Florida. Present during the meeting was Allan McDonald Space Booster Project (from Morton Thiokol, but was currently in NASA during that time), Lawrence Mulloy, Manager of SRB Project, Stanley Reinartz Shuttle Project, Jack Buchanan Manager, and Cecil Houston Resident Manager.

Morton Thiokol in Utah manufactured the Solid Rocket Booster. It was transported by rail to NASA in Florida with a distance of around 2000 miles. The SRB (Solid Rocket Boosters) was done in segments, as in figure 14.6. The purpose of the SRB is to provide the explosive lift of the shuttle from the ground. The Challenger shuttle was attached to the main fuel tank during lift-off. The left and right solid rocket boosters were transported by rail in segments and were assembled in Florida. Each segment of the SRB had two O-rings made of rubber, acting as a seal to prevent the flammable fuel from leaking. Due to the tremendous amount of vibration during liftoff, there will be a bit of play from each left and right solid rocket booster segment. The outer shell of the SRB is made in segments, and two rubber O-rings are sealed, each joint between the segments. After both the main tank and solid rocket boosters separate from the space shuttle, the shuttle will proceed with their mission, while the fuel tank and SRB will fall back to earth and land on the sea through a parachute. A ship will recover the solid rocket boosters (SRB) as they land with their parachutes on the sea and will be returned back to

Morton Thiokol for inspection for any traces of wear, especially erosion, so that it can be re-used again during the next shuttle flight mission. That was the protocol from NASA and Morton Thiokol. NASA and Morton Thiokol had data from earlier flights showing that hot gases had burnt and escaped the first O-ring seal on a previous cold-weather launch; luckily, the second O-ring caught the leak. Morton Thiokol engineer Roger Boisjoly inspects the solid rocket booster during the previous flight and discovers a massive amount of erosion on the O-ring on the previous Flight 51C. The primary seal did not work, and hot gas escaped.

The conference call was made to answer one important question: Do we launch the Challenger at a temperature below freezing? Note that Morton Thiokol, which was the manufacturer of the SRB at that time, recommended to NASA to abort the mission. Morton Thiokol can only recommend it, but the final decision will still have to be from NASA. Although this had always been the case, NASA had no history where the OEM was overruled by NASA. This means that NASA always followed their OEM recommendations, just like in most industries. As the evening on January 27, 1986, slowly passed by, the lives of the seven astronauts hang on one single question: how will the O-rings react to a cold, freezing temperature launch. This question was answered by Roger Boisjoly, one of the main engineers at Morton Thiokol, and he said;

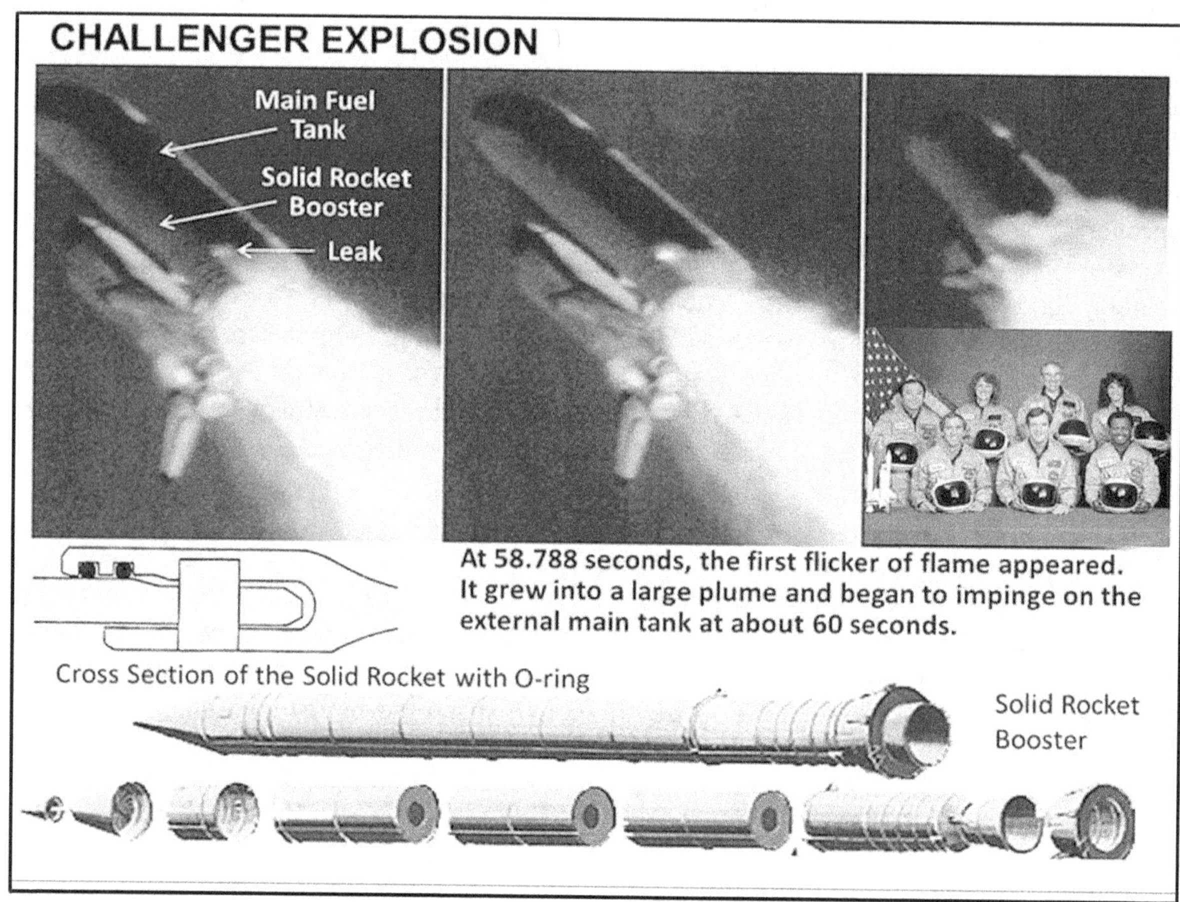

Figure 14.6: Why Did the Shuttle Challenger Exploded?

• At low temperatures, the seals are not so elastic. It won't seal properly. It will be like putting a brick versus a sponge. The grease, too, will be ineffective. It will be thicker and not a slick. We will be having a higher O-ring actuation time. If that happens, then we will approach a threshold of secondary seal pressurization. This means that we have a situation where neither the primary seal nor the secondary seal will function. *From Rogier Boisjoly*

Roger Boisjoly said that if we launch tomorrow at a freezing cold temperature, the O-ring will not function since rubber loses its resiliency during cold. This means that the rubber O-ring will not be so elastic. When the rubber seal does not function, then hot gases from the Solid Rocket Booster will leak. If we analyze the risk involved in this situation, it asks what harm it can do if the risk happens. This means that the failure had not yet happened; Morton Thiokol, NASA, and Marshall Space Flight were still in the conference call meeting and discussing the outcome.

Risk Number 3 (Low Risks): The solid rocket booster can leak since the rubber O-ring will not fully become functional. No one knows where the leak will occur, which can be on the left or right solid rocket booster. It may be on just one segment or on several segments. The leak may not be directly facing the main tank or in the opposite direction. During the meeting, Morton Thiokol Roger Biosjoly only knows what will happen to the O-rings function, which could or could not destroy the entire Space Challenger and pose a danger to the entire crew. If the leak happens opposite of the main fuel tank, their commander chief Dick Scobee will know that since the shuttle will lose fuel slightly, which will be indicated in the fuel gauge from the Challenger shuttle. He can decide to immediately detach the Space Challenger main tank and decide to continue with the voyage and still complete their mission in space. In this case, the failure will not be critical, and therefore there will be no consequences except the slight loss of fuel.

Risk Number 2 (Medium Risks): The Challenger Shuttle can lose fuel much faster than the crew expects it to last, but the shuttle can detach from the main fuel tank since it will be known by the crew through their fuel gage. The Space Challenger can return back to NASA and declare that the mission failed. They can reschedule their launch. The failure will be less critical, and NASA's consequences will be booed by the media. President Reagan will not be able to talk to Christa McAulliffe in space, which was their original plan.

Risk Number 1 (High Risks): The leak from the SRB, like a plume from a torch, can be directed in front of the main fuel tank, where the Space Shuttle was attached. The main fuel tank is composed of liquid oxygen and hydrogen. If this happens, the main tank will explode and totally collapse, creating a huge fireball. This massive explosion will instantly destroy the Challenger and kill all the Challenger crew. Therefore, if these will be the consequences of the risks, then the consequences are unacceptable. This will be a massive failure for NASA as their reputation will be at stake in this case. The failure is highly critical, and the consequences will be the destruction of the shuttle and the death of the crew. Unluckily, this is what the Challenger Shuttle experienced on January 28, 1986, as we all know. The explosion killed the crew, destroying the shuttle, and NASA being closed were the aftermath of the failure.

Although it was NASA's call to make the decision to launch, Lawrence Mulloy, the manager from NASA, was pressuring Morton Thiokol's management team to agree with him and approved the documents to launch even if the Thiokol's engineers warn of an impending danger to launch. To cut the story short, it was agreed by the Morton Thiokol management team, NASA's management, and Marshall Space Flight Center to launch the Space Challenger and on January 28, 1986, we all know the consequences of what happened. If you can see in figure 14.6, unluckily, the leak occurred directly in front of the main fuel tank where the shuttle was attached. Suppose the leak occurred on the opposite side, which was not directly facing the main fuel tank; definitely, the challenger will not explode. In this case, there was a 50-50 percent chance of success. NASA made a calculated risk, and they lost. The entire Challenger crew, including the teacher Christa McAullife, died, and there were no survivors. Their bodies were found beneath the sea after several days of search and rescue. The Challenger Shuttle was completely destroyed, and NASA was closed down, I believe, for a couple of years or more.

When we speak about risk, it is the possibility that something bad can happen. It means that the failure had not yet happened but is likely to happen. This means that if we do something, the failure can be averted. It is the probability in case it happens. If we have data, statistics, or previous experience, it can allow our judgment to indicate if the risk is imminent and there is a high probability that it will happen. If the risk is low, the failure will be less critical, and the consequences will be moderately low to none. If the risk is high, obviously, the failure will be critical, and the consequences of failure. Here are some cases to indicate if the failure will be critical:

• If the equipment is in an unmanned location
• If the failure can result in environmental consequences
• If the failure can result in safety consequences or can kill or injure someone
• If the failure can shut down the whole operations for an amount of time
• If the failure can permanently shut down the whole operations (loss of job)
• If the failure can result in industrial accidents
• If the failure can breach any known environmental regulations and laws
• If there is an absolute certainty that the failure cannot be detected
• If there are no redundancy or standby for the equipment
• If the cost of the failure will be extremely high and can result in a penalty
• If the company lawyers and insurance people will be involved in the failure
• The worst case will be if the media and those crooked old politicians will be involved

14.7: July 2020: An Advice to Storekeepers

Many industries are reluctant to implement changes, especially when there is some cost involved in the improvement process, especially when most industries have inquired about losses due to this **covid 19** pandemic, which is still ongoing as of this writing. The most common language used by management is, we have done this before, so why should we do it now? Besides, it cost something to do that. The requirement or request for having an additional workforce in the storeroom will originate from the storekeepers themselves. This is the case if you are operating 24 hours with only one shift from the storeroom. If we go on with the process, the storekeeper will request his supervisor or manager for an additional workforce to have a rotational shift. If the head of the storeroom is convinced, he will make his request to the plant's human resources. Simultaneously, the human resources will seek the CEO's approval or decision-makers for the additional manpower request. This is usually quite a long process that can take several months. Although we do not want to rush the hiring process as we want to hire the right person who can perform their job efficiently and effectively inside the storeroom. As long as your request is valid, things can go as we expect them to be as long as we have a solid justification for our request for additional manpower.

I was also an employee before reaching this stage in teaching industries. I always fight for what is right, even at the expense of losing my job; that is why my career shows that I have worked in several industries: shipping, woodworking, foundry, assembly plant, semiconductor, and mining was my last venture of employment. A thin line separates us, workers, from the decision-makers. We know the problem, but they are the ones to decide. I think differently since if I know I am right, then the boss should hear me out. Otherwise, they can have my resignation papers the first thing in the morning, but of course, I am not asking the reader to follow that road and do the same. Funny as you might think, some of these industries I used to work with previously hired me to spend my time once again in that plant and this time to improve not only their MRO storeroom but on how to do the correct maintenance on their equipment. In my thoughts, I was already once there in your plant before, telling you these things which you ignored then now you want me to come once again to teach you, what a complete waste of time, if you have just done this and that much sooner, then things will not turn ugly right now. If there is no doubt in your mind that you are right and that what you have in mind is for the benefit of the whole industry, then level with it and tell your boss. If the boss ignored you, bypass him and go on to the higher boss; if still, he ignores you, then it's time to tell El Chapo about this. If your recommendation falls on deaf ears, then at least you made your point precisely clear to them.

If you are a storekeeper reading this book, your job is not as easy as you think. If you operate for 24 hours and your storeroom only operates for one shift, no amount of improvement will help your storeroom. There will be a domino of problems, and expect more problems as the day passed by. My advice is to talk to your superiors and convince them to man the storeroom for 24 hours since your operation is 24 hours. You just cannot leave the key to the guard when you already sign-off for the day. Do not expect maintenance to be honest in filling up the withdrawal form every time they go to the storeroom as there is always pressure in repairing the equipment. The job of the storekeeper is just too overwhelming for just a single person to do. If we put this on the correct perspective from maintaining the spare parts, doing analysis on spares, regular cycle counts, identifying fast, slow, non-moving, obsolete parts, and daily activities dealing with withdrawals and accepting deliveries, checking the deliveries for defects and whatnot. The job of the storekeeper is simply not a routine function, as others might think it is. They have a big role and responsibility in managing the spare parts inventory and maintaining it, and this cannot be accomplished by providing only one shift for the storeroom. Industries might think that they are saving money by hiring only one person to manage everything in the storeroom, where the fact lies they are losing a massive amount of money on downtime, emergency buying, and squirreling of parts.

14.8: August 2020: Why the Term Usage is Different from the Parts Withdrawn

For 30 years, as of this writing, I have devoted my time teaching industries about reliability and maintenance. I have several training courses where MRO Spare Parts is just one of them, and I hear many problems in the storeroom from my students. If you plan to start an MRO Spare Parts Inventory Management improvement immediately, it would either be next to impossible or will not work as you expect it to be. We need to take care and clean up first those nitty-gritty problems, which have now turn up to become a cancerous stage before we can control or even manage the storeroom and spare parts.

While others might think that I'm nuts and might be thinking that the spares inventory can be controlled based upon its usage, the withdrawal, in this case, is monitored and recorded by the storekeeper. Let us say that during the previous year, a bearing whose parts code, for example, is BRG-1234J, had been withdrawn 10 times with a total quantity of 45 pieces last year. Hence, it is easy to conclude and purchase 45 bearings and perhaps place a buffer or safety stock of two bearings to ensure no stock out this year. Now that is the easy part, but this is not the actual situation and reality at hand. Let me ask these three questions for a start:

• Were there cases where maintenance withdraw an excess quantity of the item?
• Is someone squirreling this item for its own benefit?
• Does maintenance actually know the correct interval when to replace the bearing?

The quantities that had been withdrawn were 45 pieces, but the question is, what is the actual number that had been used by maintenance on their equipment? If only 10 pieces of the 45 had been used, then where are the rest? I am sure that they are hidden somewhere in the plant by someone, worst if some are not in the plant and kept for personal use. The term used in most MRO articles, books, blogs, newsletters, and other references is not the actual usage. The term usage actually refers to the parts withdrawn from the storeroom. They assume that the items withdrawn are equal to the actual usage, which in reality is not. I know the sentiments of maintenance since I am also a maintenance; their frame of mind is to better withdraw more since they assume that the failure will happen again soon. This means that if maintenance withdraws three pieces and only one item is used, the remaining two pieces will not be returned to the storeroom but will be kept by maintenance in their unofficial stores.

Another problem is that most industries do not understand the correct interval to replace parts, especially in their Preventive Maintenance activities. Sure, anyone can guess the interval monthly, every two months, or quarterly or follow OEM conservative recommendations. The question here will be, is it really the correct interval?

In most cases, replacements will be based on OEM recommendations or, if not on, guestimates as this is what our ancestors from WWII told us to do. Some will say that they have Vibration Monitoring in place to monitor the health of the bearing. The problem with Predictive Maintenance is that it will provide the symptoms that a bearing had reached its potential failure. Still, it is not a guarantee that the life of the bearing will be reached. The manufacturer has a life indicated on the bearing, but the environment in which the manufacturer performed the test might be the opposite of what your actual operating environment actually is. A maintenance planner can have a plan and BOM (Bill of Materials) on the activities to be done on Preventive

Maintenance such as inspection, cleaning, replacement, overhauling, whatnot, but does the maintenance planner really knows what parts need to be replaced since a part is nearing its rupture. There are actually two ways that can go wrong in this case. Replacing the part too early will definitely cost money while replacing the part too late will allow the failure to occur. If maintenance does not know the correct interval, then what more do we expect from the storekeeper. The storekeeper does not know this since they depend on the actual withdrawal of the parts, which assumes that the parts were actually consumed, but in several cases, the parts were withdrawn in excess and hidden away. This is because a maintenance can have some safety stock on hand since they know that they will be blamed for the failure even if they did not cause the equipment to fail.

One of the reasons why Preventive Maintenance is costly is because there are so many cases that the parts will be replaced even if it is still working just to comply with the checklist. Study shows that 33 cents out of every 1 dollar are put to waste by replacing perfectly good parts on the equipment just to comply with Preventive Maintenance schedule. Most of the decisions I see will be based on either the OEM or the equipment manual. This quote is extracted from Anthony Smith's book on RCM, Gateway to World Class Maintenance.

Historical evidence strongly suggests that some of our current PM activities are, in fact, not right. In some cases, PM tasks are unnecessary because they have little relationship to keeping the plant operational. It is not uncommon to examine a plant PM program and find 5 to 29 percent of the existing tasks discarded. The plant would never know the difference. The basis for many PM programs is the blind acceptance of OEM PM recommendations as the best course of action to pursue, even though the OEM recommendations are conservative and not necessarily applicable to the plant's operating profile.

Let me put it this way. If I have a mechanical seal with a life of 5 years, there is no OEM in the right frame of mind that will tell you to replace it every 5 years. They will tell you to replace it every 6 months or even less. I do not know how to stretch it further, but the way I see things in industries, what seems to be wrong somehow, is right since things have been done this way since the beginning of time. Only training and education can change this. Maintenance and storekeepers need knowledge so they can work things out together correctly. Doing the correct maintenance requires knowledge and not experience.

Here is another concern I have, especially for power plants, power distribution, and the like, where they need to comply with government regulatory policies. I have taught many power plants in the Philippines ever since I started teaching. Still, they find it difficult to use the learning's in their actual situation. Why? Because my teachings contradict government procedures and compliance, such as DOE (Department of Energy) that a plant needs to have a complete shutdown to avoid forced outages. What happens is that when an actual shutdown is due, maintenance will not only comply but will tend to replace parts even if the part is still in working condition. The tendency is to replace what they can replace. If a Vibration Monitoring indicates that it is still on its normal range of readings, it will be replaced to comply with the policy, thinking that the part will fail before the next shutdown is due. Most maintenance people from Power Plants perform intrusive maintenance. The truth of the matter is that Shutdown; Scheduled Outage, is not a guarantee of continuous operation of the equipment. As I stated many times in my other books, Preventive Maintenance alone cannot capture all failures, and performing too much PM or Outage will induce more Infant Mortality failures, and these PM outages, turn-around, or whatever you called it is only designed for age-related failures. For industries, it is not age-related but random failures that we regularly face in our day-to-day operations. For crying out loud, Preventive Maintenance shutdown will not and never address random failures since it is not designed for that. As long as the government procedures and policies on maintenance remain unchanged, nothing will change, and industrial accidents will continue due to wrong maintenance. Perhaps if I am in politics, then I will propose some bills to change this. As my good old friend, R. Keith Mobley, jokingly said that it takes decades to change policy from the government.

The actual quantity withdrawn does not reflect the actual usage that it had been actually used in the equipment, and further problems indicate that replacing parts, which are still working, which means that we are simply throwing away money for industries. Yes, it is easy to say that the seal had been replaced to comply with a Preventive Maintenance schedule. However, in this case, the seal that was replaced was still working and can still be used or can last for another year. I have repeatedly mentioned that improving the storeroom must be done both inside and outside the storeroom, and not only should the storeroom people be involved, but everyone in the entire organization, especially maintenance.

For a start, what is important is to do something about squirreling parts; remember that as long as squirrel stores exist, we cannot control and manage the storeroom. What the storekeeper can control is just what is inside and not outside the storeroom.

14.9: September 2020: What to Do with Non-Moving Parts

Many problems exist inside an industry's MRO storeroom. All I can say is that 99.99% of the problems in the storeroom are man-made. Although it is still premature to declare the top 3 or top 4 problems, as my survey of the top problems on MRO Spare parts is still ongoing, I am confident that one of them will be the case of non-moving parts. We shall update the survey and include that in my upcoming books. One sentiment with the maintenance people is that they do not want to throw away parts. They will hang to them until their remaining last breathe on earth since their thinking is that it might still be used someday. Non-moving parts are a real big problem inside the storeroom as almost all of these parts are big in size and big in cost. One of the main reasons they become non-moving is that the OEM recommended us to keep a spare part due to its criticality and lead-time. They told us ten years ago to keep this part, and we agree to their demands. Now the part is still in the storeroom and remained untouched. Although there are only a limited number of options on what we can do with non-moving parts. One of them is to sell them back to the OEM. I knew of one company here in the Philippines that did that. The cost of the item was worth one million pesos (19,600.00 USD), and the purchasing negotiated to return the part to their OEM after 5 years; the OEM agreed but at a much lower cost. It was sold for 100,000 pesos or 1,960.00 USD. Very nice business; it is much better than placing your money in the bank to earn interest.

Imagine that I am a salesperson inside an established department store in a mall. I am selling household appliances when suddenly a customer walked in and was looking forward to buying a refrigerator since their old refrigerator broke down. Using my sales training and sweet talk, I recommended the top-of-the-line brand, let us say, a two-door Kelvinator. The customer asked for the price and features of the refrigerator, which I gave to them. I even demonstrated the features to the customer. The customer was satisfied with my sweet talk and decided to purchase the refrigerator. I asked the customer that we have a spare door for some additional price; I gave him a good discount. I told the customer that if the door fails to close, it would probably take a minimum of 10 months to get a new door since the door will come from Somalia, Africa. I explain that if the door failed or fall down from the ref, the food inside can rot. If you were the customer, would you include the spare door with the purchase?

I guess definitely not. I also told the customer that we got a spare compressor, just in case the compressor failed. Again, the lead-time to get this compressor is long since it is just one of a kind, and the lead-time is 8 months since it will come from Antarctica or some faraway country. Well, I guess my sweet talk did not work this time, and you will really be annoyed with me or perhaps go away for good. Common sense tells us not to fall into the trap. If we go back to our equipment, why did we agree with the OEM to keep a stock of this and that which now lies sitting idle inside the storeroom for many years that have never been used? During my class, there is usually a door where the audio system is kept by the hotel staff. I always told the delegates not to open the door, and I asked them if they would like to know why. Everyone will agree and say yes. I told them that if you open the door, there is a ghost inside, which everyone will laugh at since they know it was all a bluff. If the OEM recommends that an expensive part be stock in the storeroom, we have no hesitation and agree immediately. Their scare tactics are that if you disagree, then the warranty can be void for good.

Non-moving parts are also termed as rarely used inventory. These items are difficult to obtain, or their lead-time usually takes several months, others even years, either because most of these parts are unavailable or imported from other countries. The difference between non-moving and obsolete parts is that the machine or asset is still used and operating in the plant. For obsolete parts, the equipment or asset has already been retired or decommissioned. An item or part becomes non-moving because the OEM recommended most of them to be stuck inside the storeroom. They claimed that the part is critical. Today, they sit idle inside the storeroom shelves for years and remain untouched. All parts inside the storeroom have a cost for storing them. They are called carrying costs or holding costs. Most maintenance thinks that there is no cost to hold or keep something in the storeroom. They do not want to get rid of the part by hanging on to them on the assumption that these parts might be needed some day; besides, their thinking is that it does not cost anything to keep them. This is where maintenance got it all wrong because all parts inside the storeroom have a cost on holding them.

Most non–moving parts are classified as insurance spares in which a spare part that you hold in the store that you would not expect to use in the normal operations of the equipment, but if they are not available when needed, it will result in a significant loss to the plant. Think of insurance spares like the insurance of your car or house. You do not want to claim them, but you do not want to be without them either. A classic example of this is the spare tire in your car. How many times have you used them, yet nobody wants to drive their car without one. The general thinking here is that industries stock them just in case we might need them someday. Unlike obsolete parts, the options we got for non-moving parts are very limited. An analysis of

100 MRO stores tells us that 60% or more of the total inventory cost has no usage for the past couple of years. Yet, most of these items should be on hand when needed because the impact and consequences of the failure are merely unacceptable.

It is recommended that the industry have a list of what items have not been used or withdrawn for one year, two years, three years, etc. It is best to do this with a team involving the maintenance, storekeepers, Purchasing, Finance, and Top Management with the Decision-Making process on what to do with these non–moving parts. Start with the oldest spare part. Know the consequences of the failure if it will occur. Will the equipment be running for the next 3 to 5 more years? I would also recommend that spare parts that have not moved for a couple of years must be inspected and thoroughly checked for any traces of deterioration, rust, or corrosion as they have remained idle for a long time. The most logical and practical way is to use them, starting with the oldest spare part. The original part removed is still functional, so we preserve it or even refurbish it. This now becomes our spare. This way, the part had been withdrawn from the storeroom and will definitely lower your inventory cost

My point is, why not start using these non-moving parts, starting with the old one. You might be having second thoughts regarding the consequences since the lead-time for these items usually take six months, and others may take years. We will be replacing a part that is still working, and we still have a spare, which was the original part that had been removed from the equipment. For example, I have an induction motor with specification Siemens 1LE1 Reversible Induction AC Motor, 5.5 kW, IE2, 3 Phase, 2 Pole, 400 V, 690 V, foot mounting horizontal motor in the storeroom. The cost of this motor is $ 3,500.00, and it had not moved for five consecutive years inside the storeroom. Your current cooling system setup is redundant or backup; hence, why not replace the backup motor with the one in the storeroom. The old motor will be inspected, cleaned, and refurbished when necessary. We now keep the old motor safely or return it back to the storeroom for safekeeping. Note that there will be no more holding or carrying cost on this motor since this is not considered a spare part since it had already been withdrawn a long time ago. The motor now becomes our spare. Although I contradict myself in my writings on my other books, we need to replace parts that are actually on the verge of failing and not on parts that are still working. This will be an exception to the case since our goal is to reduce the overall inventory of the storeroom. My point, in this case, is that if the part that will be replaced by the non-moving part can be preserved or refurbished by maintenance, let us considered that as a spare. This is the only option we got for non-moving parts, which is to use them.

14.10: October 2020: Six Myths about Lubrication

According to the Cambridge Dictionary, a myth is a traditional story, especially one which explains the early history, a cultural belief, or the practice of a group of people. A myth is also a common belief but a false idea. Here are some myths about lubrication that most maintenance people believe needs to be changed.

Myth 1: More grease, the better: This is a costly mistake. Over greasing can have many of the same negative side effects as under greasing, plus the added cost of high lubricant consumption. Whether for grease or oil, there is just one trade secret. If lubrication is excessive or inadequate, this will be detrimental and not good for the equipment. Maintenance might think that over-greasing is being kind to the equipment. Maintenance people must understand that both over greasing and under greasing can result in higher operating temperatures, premature bearing failure, and an increased risk of contaminant ingression.

Myth 2: New oil is clean oil: New oil is not clean oil and contains contaminants. Many industries do not know the cleanliness of the lubricants they receive from their suppliers, nor do they know the cleanliness of the fluids they deliver to the end-user. Many distributors do their own packaging both in drums, 55-gallon, or 5-gallon pails. Typically reconditioning these units can be a source of contamination. Consequently, they need to be carefully examined before filling. It has been demonstrated, however, that clean fluids extend the life of equipment several folds. In figure 14.7, I will consider 18/15/13 and lower to be pass; the rest are failed.

Average Cleanliness Of Fluids Offloaded To Distributor Tank Truck From Lube Supplier	
LUBRICANT TYPE	**ISO 4406 PARTICLE COUNT**
AW 32	16/14/11 to 19/17/14
AW 46	18/15/13 to 19/17/13
AW 68	18/16/13 to 21/19/15
15W-40 Diesel Engine Oil	21/18/13
85W-90 Gear Oil	20/18/13
5W-30 Automotive Engine Oil	20/17/14
TO-4 30	20/17/13
TO-4 10	18/16/12

Figure 14.7: Do Not Assume New Oil is Clean Oil

Myth 3: Grease is just grease, and oil is just oil: There is only one type of grease to use for all equipment, and that is the multi-purpose grease. Maintenance thinks it is fine to mix different grease inside the grease gun to serve various applications. This is how it has been done in this industry for 25 years of my stay. Grease has what we called incompatibility issues. We do not want to mix different additives, as it can lead to incompatibility issues that can shorten the life of either the lubricant or the part being lubricated.

Myth 4: I am saving money by bringing leaked oil back to the equipment: Our surroundings contain dust and dirt. Silica is environmental dust, and the smaller the contaminant, the more abrasive it can become. Also, open-type containers for topping or adding oil are also prone to contamination when exposed for several hours since it captures the leaked oil and the dust contaminants. These solid contaminants can be trapped and caught up in tight clearances inside the equipment, generating more contaminants in the process.

Myth 5: It is OK to store new lubricants outside in an open environment: With the limited storage space inside your plant, it is common to have an outside storage area for newly delivered oil drums exposed in the atmosphere. This practice creates a high risk for water and dust ingression into the new oil if the drums are not stored properly. This is especially true during the summer months when the temperature of the drums can reach 50 °C (122 °F) or beyond.

Myth 6: Just Follow the OEM when changing the oil and everything else: With little or no education on lubrication, most maintenance will rely heavily on OEM recommendations on changing oil, even if OEM recommendations are obviously conservative. While following OEM recommendation on changing oil may be correct, a better way of determining when to change the oil is actually not based on its running hours but based on the number of contaminants present in the oil and the actual condition of the oil itself through Oil Analysis.

14.11: November 2020: Effects of Excessive Moisture on Lubricating Oil

Moisture presence in lubricating oil can cause a wide array of problems in the oil and the component or part that it lubricates. It can cause premature breakdown of the lubricant itself through oxidation and deplete and render several additives useless. The effects of oxidation can increase the oil's viscosity and making the lubricating oil acidic. This can be traced when doing an oil analysis test on the Total Acid Number (TAN). These contaminants can lead to the formation of sludge and varnishes. Excessive moisture can also promote rust and corrosion on metal surfaces. Other problems from excessive moisture in the oil will accelerate surface fatigue, reduce the lubricating film thickness, and jamming components due to ice crystals forming. For transformer oil, this can cause a loss of dielectric strength in the oil.

Sources of Moisture Contamination in Lubricating Oil

• Leaks from within the system such as the cooling system
• Leaks from damaged coolers and heat exchangers
• Leaks through worn out or damaged seals
• Condensation due to humid air
• Reservoirs without desiccant breathers
• Ingression from condensation
• Precipitation due to fall in temperature
• Introduction of contaminated top-up fluid
• Rain in external reservoirs
• Rain leaking into incorrectly stowed barrels
• An oil barrel stored outside in a vertical position is likely to collect rainwater
• Through reservoir breather caps humid air. System fluid absorbs some of this moisture, while some are condensed on the inside surfaces of the reservoir.

Different Types of Moisture Present in Lubricating Oil

Three types of moisture can be present in the oil, which can be free, emulsified, and dissolved, which is explained here.

Free Water: This can be any water or gas which is not dissolved in the oil. Free water is when the water and the oil separates. Since the water is denser and heavier than the water, it will settle at the bottom. Once the water is settled at the bottom, we can easily untighten the drain plug and remove free water from the reservoir.

Emulsified Water: These can be microscopic droplets of water distributed in the form of an emulsion. Emulsified water is when the amount of dissolved water is greater than the saturation point. When the water is in an emulsified state in lubricating oil, its color becomes white or milky state in layman's terms. It is best to address the root cause of the problem for emulsified water and change the oil immediately.

Dissolved Water: While both free and emulsified water can be visible to the naked eye, dissolved water cannot be seen visibly, but it can be measured using a Karl Fisher Titration Test or Fourier Transform Infrared Thermography (FTIR). In this case, the water is dispersed in the oil in a homogeneous molecular solution. The gas or moisture cannot be removed from a conventional' nominal filtration; however, several absolute oil filtration on the market can remove a certain amount of moisture in the oil. Dissolved water is similar to humidity and cannot be visibly seen in the oil.

Effects of Moisture in Lubricating Oil: Dissolved water is similar to humidity and cannot be seen in the oil. Emulsified water is when the amount of dissolved water is greater than the saturation point. Free water is when the water content in oil separates as the water is heavier than oil. Free and dissolve water will cause the degradation of the oil. When the oil becomes emulsified, it changes its color and becomes milky-like in appearance. This means that the saturation limit at the oil temperature had been exceeded. When water is in an emulsified form, both free and dissolved water is present. Here are some effects of moisture on lubricating oil.

• Fluid breakdown due to additive depletion
• Water can cause cavitation in hydraulic pumps
• Promotes oxidation, hydrolysis, and aeration, causing the oil to form acid
• Increase in viscosity that can cause varnish and sludge.
• Moisture can accelerated metal surface fatigue
• Promotes rusts and corrosion on metal surface parts
• Jamming of components due to ice crystal formation at low temperatures
• Loss of dielectric strength in insulating fluids such as transformer oil

There are several ways to check the moisture content in oil. The simplest test is the crackle test, which can determine both free and emulsified water, but for dissolved moisture, a Karl Fischer titration test is much more accurate and can measure as little as 1 ppm of water the oil used correctly. This method uses two types of titration to identify trace amounts of water in a given sample: coulometric or volumetric, measured either in ppm or in percentage. The principles are the same for each, except volumetric uses a titrant solution. This testing method can be very beneficial because it allows you to analyze solids, liquids, or gases. The disadvantages of the Karl Fischer titration test include its cost and time required when water concentrations are high. There are several ways to remove moisture from the oil, such as coalescence, centrifuge, adsorption, and vacuum dehydration. There are also absolute filters in the market that can remove a certain amount of moisture from the oil. But the process of removing water in oil is not the solution to the problem, as the moisture will just return back to the oil. Interpreting the oil analysis reports can provide clues on where the moisture from the oil is coming from so we can address the root cause of the problem.

14.12: December 2020: The RCM and TPM Crossroads

RCM will identify all the possible failure modes on the equipment and create a sustainable process by identifying the most feasible and appropriate maintenance tasks. TPM wants that before we name all possible failure modes that can affect the asset, perhaps it is smarter to restore all the possible deteriorations in our equipment. This is accomplished by working parallel with Autonomous Maintenance for those implementing TPM as they have their activity on Initial Cleaning. In this case, the Autonomous Maintenance team will be focused on the external part of the equipment to identify abnormalities (Japanese called them fuguai), while the Planned Maintenance will address the interior part of the equipment by restoring deteriorations that are uncovered in the Restoration Phase.

If the equipment is restored first, then there will be lesser failure modes. What TPM is saying is that let's do the basics first. It's no rocket science program. This means that what good will a Vibration Monitoring proposed tasks be derived from RCM if the equipment lacks a couple of bolts, nuts, washers, and fasteners. Will it contribute to the overall vibration of the machine? I believe so. By restoring the equipment, we are also reducing the number of unplanned breakdowns on the equipment. This will provide a better result once RCM is implemented. RCM can be used to create a sustaining plan so that deteriorations restored by the Planned Maintenance team will no longer revert back to the equipment.

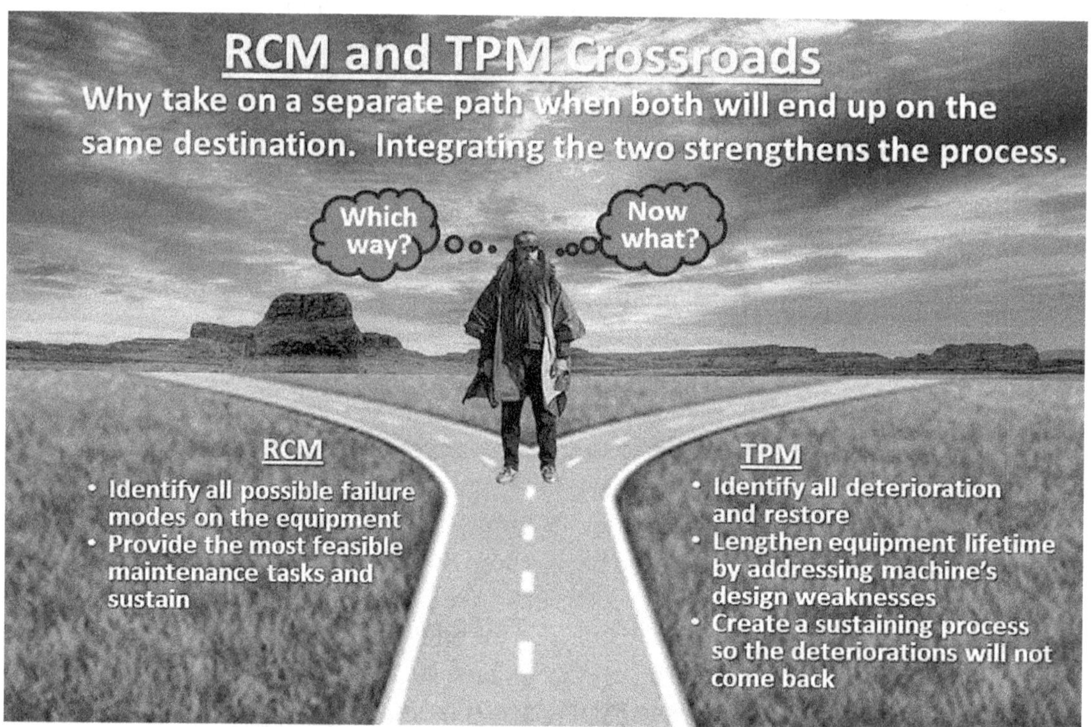

Figure 14.4: TPM and RCM Crossroads

Once the Planned Maintenance restoration phase is ongoing, a separate team consisting of the plant's Rolling Stones will head the RCM team. The team's goal is to create a sustainable process for RCM and TPM. Since the equipment is restored as close as possible to its original condition, TPM does not want the equipment to revert back to its old deteriorated state of condition. This is where RCM would be handy. Once the team completes its restoration activities, they will proceed with Planned Maintenance Phase 2, challenging the team to address its design weaknesses by lengthening its equipment lifetime. Now isn't it that RCM only focuses on the maintenance tasks, and those under Redesign or Modification will not be addressed in the RCM process since they may or may not be the exact team that will be doing this. In this case, those failure modes subject to redesign will just be passed on to the team performing Phase 2 of Planned Maintenance. This is what they will be doing, mostly redesign, modifications, and improvements. There is really no conflict between the two. The pioneers of the RCM process dramatically improved the cost of doing maintenance in their 747 aircraft and improved its safety by reducing the accidents per million take-offs, which is one of the indicators by the FAA (Federal Aviation Administration). The author of this book believes that if these two powerhouse methodologies can be integrated, they will yield a much better results than traditionally expected from implementing TPM. Doing the basics is just a matter of discipline, whether your plant is doing TPM or not. All people in the organization, not just maintenance, must understand that catastrophic failures started from small things. In fact, catastrophic failures are simply an accumulation of small problems neglected in the equipment. TPM and RCM are not contradicting methodologies, but rather, they complement one another. It is just the individual RCM and TPM consultants that do the contradiction and not the methodologies, and I think that is all I have to say about that.

Bibliography

- Anderson, T.L., **Fracture Mechanics, Fundamentals and Applications, Second Edition**, CRC Press Inc. Copyright 1995
- Angeles, Rolly S. **Feasibility Study on Adopting In-house Oil Analysis,** http://www.rsareliability.com/artciles.htm.
- Angeles, Rolly S. **RSA Newsletters Archive**, http://www.rsareliability.com/reliabilitynewsletter.htm.
- Angeles, Rolly, **Cutting Edge Maintenance Management Strategies,** Central books Printing, Philippines 2020
- Angeles, Rolly, **Decoding Reliability Centered Maintenance for Manufacturing Industries,** Central books Printing, Philippines 2021
- Angeles, Rolly, **Lubrication Tactics for Industries Made Simple,** Central books Printing, Philippines 2020
- Angeles, Rolly, **Maintenance Roadmap to Reliability**, Central books Printing, Philippines, 2016.
- Angeles, Rolly, **Problems, and Solutions on MRO Spare Parts and Storeroom,** Central books, Printing, Philippines 2020
- Angeles, Rolly, **Reliability, A Shared Responsibility for Operators and Maintenance,** Central books Printing, Philippines, 2018.
- Angeles, Rolly S. **World Class Maintenance Management – The 12 Disciplines**, 2009
- Bannister, Kenneth E., **Lubrication for Industries, Second Edition**, Industrial Press Inc., Copyright 2007
- Bazovsky, Igor, **Reliability Theory, and Practice**, Dover Publication Inc. Mineola, New York, published 2004
- Bloom, Neil, **Reliability Centered Maintenance Implementation Made Simple**, McGraw Hill Companies, Copyright 2006
- Brown, Michael V., **Managing Maintenance Storeroom**, Wiley Publishing Inc., 2004
- Covey, Stephen, **Principled Centered Leadership**, April 1989
- Drew Toyer and Jim Fitch, **Oil Analysis Basics** (Revised and Updated 2001), Noria Corporation, Tulsa Oklahoma, USA, 2001
- Gopalakrishnan, A.K. Banerji, **Maintenance and Spare Parts Management**, 2011
- Gotoh, Fumio, **Equipment Planning for TPM, Maintenance Prevention Design**, Portland, Oregon, Productivity Press, 1988.
- Japan Institute of Plant Maintenance, **Autonomous Maintenance for Operators**, CRC Press, Taylor and Francis Group, 1997
- Leading and Managing at Work. **Who is Dr. W. Edward Deming?** http://lii.net/deming.html
- Lim, Billi P.S. **Dare to Fail,** Selangor, Malaysia: Hard knocks Factory, 1996.
- Masaji Tajiri and Fumio Gotoh, **Autonomous Maintenance in Seven Steps – Implementing TPM on the Shop Floor**, Productivity Press, 1999.
- Molloy, Kevin. **Rick and Dick Hoyt, Team Hoyt**, http://www.teamhoyt.com.
- Mobley, Keith, **An Introductive to Predictive Maintenance, Second Edition**, Butterworth, Heinemann, 2002.
- Moubray John. **Reliability-Centred Maintenance Second Edition**, Great Britain: Butterworth, Heinemann, 1977.
- Nakajima, Seiichi. **TPM - Introduction to Total Productive Maintenance**, Productivity Press, Portland, Oregon, 1988.
- Nelms, C. Robert, **What You Can Learn From the Things That Go Wrong, A Guide Book to the Root Cause of Failure**, Virginia, USA: Failsafe, 2007.

Bibliography and References

- Noria Corporation, **Practicing Oil Analysis**, http://www.oilanalysis.com.
- Nowlan, Stanley and Heap, Howard. **Reliability-Centered Maintenance**. Springfield, US Department of Commerce National Technical Information Service, 1978.
- Online Source Wikipedia, **https://en.wikipedia.org/wiki/List_of_industrial_disasters**
- Pall Industries Hydraulics Co. **Contamination Control and Fundamentals,** New York, 1994.
- Reason, James and Hobbs Alan, **Managing Maintenance Error, A Practical Guide**, Ashgate Publishing Limited, 2003
- Robinson. Charles J. and Ginder, Andrew P. **Implementing TPM. The North American Experience**. Portland, Oregon: Productivity Press, 1995.
- Schein, Edgar. **Coming To a New Awareness of Organizational Culture**, Sloan Management Review, Winter, 1984.
- Shigeo Shingo**, A Revolution in Manufacturing, The SMED System,** Productivity Press, April 1985
- Shirose, Kunio, **TPM for Operators**, Productivity Press, Portland, Oregon, 1992
- SKF Condition Monitoring Marketing Communications, **How to Understand and Buy Condition Monitoring Equipment**, 1993.
- Shirose, Kimura, and Kaneda. **P-M Analysis, An Advance Step in TPM Implementation.** Portland Oregon: Productivity Press, 1995.
- Smith Anthony, **Reliability-Centered Maintenance, McGraw Hill, Inc.,** United States, 1993
- Smith, Anthony, Hinchcliffe, Glenn, **RCM Gateway to World Class Maintenance**, 2004
- Smith, Ricky, and Martin, David, CMRP, **Maintenance, and Reliability Metrics/KPIs101, Keeping it Simple**, Reliabilityweb.com, 2011
- Society of Tribologists and Lubrication Engineers, **Handbook on Lubrication and Tribology**, Volume. 1, Application & Maintenance, (CRC Press, Taylor, Francis Group LLC, 2006)
- Suzuki Tokutaro, **TPM in Process Industries**, Productivity Press, 1992
- Theo Mang and Wilfried Dresel, **Lubrication and Lubricants, Second Edition**, 2007 Wiley VCH, Wiley VCH, Federal Republic of Germany, 2007
- Wellins Richard, Byham, William and Wilson, Jeanne, **Empowered Teams, Creating Self-Directed Work Groups That Improve Quality, Productivity, and Participation,** Jossey-Bass Inc., Publishers, 1991.

<u>RSA Maintenance Books Collection in Series</u>

This is not only about a technical book on maintenance; it is a book that makes everyone in maintenance feel proud that they belong to the maintenance function." If you have been living through the day-to-day pressures of doing maintenance, then this is your story. *Rolly Angeles*

I have written these books inspired by the millions of maintenance mankind who want only the correct maintenance for their equipment and assets to function the way we want. If you have been living through the day-to-day pressures of doing maintenance, then you are included in these books. The books are in series, explaining each maintenance discipline's in detail based on its original concept on World Class Maintenance, The 12 Disciplines.

- Volume 1: World Class Maintenance Management – The 12 Disciplines
- Volume 2: Maintenance – Roadmap to Reliability
- Volume 3: Reliability – A Shared Responsibility for Both Operators and Maintenance
- Volume 4: Cutting – Edge Maintenance Management Strategies
- Volume 5: Problems and Solutions on MRO Spare Parts and Storeroom
- Volume 6: Lubrication Tactics for Industries Made Simple
- Volume 7: Decoding Reliability-Centered Maintenance Process for Manufacturing Industries
- Volume 8: RSA Reliability and Maintenance Newsletter Vault, Subscribers Edition

Figure F: Rolly Angeles Reliability and Maintenance Books

Thank you for purchasing this book. I hope you enjoy reading the newsletters. I would greatly appreciate it if you can provide an honest review of this.